Algebra and Applications 1

Contents

Foreword

Abdenacer MAKHLOUF
IRIMAS-Department of Mathematics, University of Haute Alsace, Mulhouse, France

We set out to compile several volumes pertaining to *Algebra and Applications* in order to present new research trends in algebra and related topics. The subject of algebra has grown spectacularly over the last several decades; algebra reasoning and combinatorial aspects turn out to be very efficient in solving various problems in different domains. Our objective is to provide an insight into the fast development of new concepts and theories. The chapters encompass surveys of basic theories on non-associative algebras, such as Jordan and Lie theories, using modern tools in addition to more recent algebraic structures, such as Hopf algebras, which are related to quantum groups and mathematical physics.

We provide self-contained chapters on various topics in algebra, each combining some of the features of both a graduate-level textbook and a research-level survey. They include an introduction with motivations and historical remarks, the basic concepts, main results and perspectives. Furthermore, the authors provide comments on the relevance of the results, as well as relations to other results and applications.

This first volume deals with non-associative and graded algebras (Jordan algebras, Lie theory, composition algebras, division algebras, pre-Lie algebras, Krichever–Novikov type algebras, C^*-algebras and H^*-algebras) and provides an introduction to derived categories.

I would like to express my deep gratitude to all the contributors of this volume and ISTE Ltd for their support.

1

Jordan Superalgebras

Consuelo MARTINEZ[1] **and Efim ZELMANOV**[2]
[1] *Department of Mathematics, University of Oviedo, Spain*
[2] *Department of Mathematics, University of California San Diego, USA*

1.1. Introduction

Superalgebras appeared in a physical context in order to study, in a unified way, supersymmetry of elementary particles. Jordan algebras grew out of quantum mechanics and gained prominence due to their connections to Lie theory. In this chapter, we survey Jordan superalgebras focusing on their connections to other subjects. In this section we introduce some basic definitions and in section 1.2 we give the Tits–Kantor–Koecher construction that shows the way in which Lie and Jordan structures are connected. In section 1.3, we show examples of some basic superalgebras (the so-called classical superalgebras). Section 1.4 is about the notion of brackets and explains how to construct superalgebras using different types of brackets. Section 1.5 explains Cheng–Kac superalgebras, an important class of superalgebras that appeared for the first time in the context of superconformal algebras. The classification of Jordan superalgebras is explained in section 1.6, and it includes the cases of an algebraically closed field of zero characteristics, the case of prime characteristic, both for Jordan superalgebras with semisimple even part and with non-semisimple even part, and the case of non-unital Jordan superalgebras. Finally, in section 1.7, we give some general ideas about Jordan superconformal algebras. Throughout the chapter, all algebras are considered over a field F, $\text{char}F \neq 2$.

Algebra and Applications 1,
coordinated by Abdenacer MAKHLOUF.

DEFINITION 1.1.– *A (linear) Jordan algebra is a vector space J with a linear binary operation $(x, y) \mapsto xy$ satisfying the following identities:*

(J1) $xy = yx$ *(commutativity);*

(J2) $(x^2y)x = x^2(yx)$ $\forall x, y \in J$ *(Jordan identity).*

Instead of (J2) we can consider the corresponding linearized identity:

(J'2) $(xy)(zu) + (xz)(yu) + (xu)(yz) = ((xy)z)u + ((xu)z)y + ((yu)z)x$ $\forall x, y, z, u \in J$.

REMARK 1.1.– A Lie algebra L is a vector space with a linear binary operation $(x, y) \mapsto [x, y]$ satisfying the following identities:

(L1) $[x, y] = -[y, x]$ (anticommutativity);

(L2) $[[x, y], z] + [[y, z], x] + [[z, x], y] = 0$ for arbitrary elements $x, y, z \in J$ (Jacobi identity).

EXAMPLE 1.1.– If A is an associative algebra, then $(A^{(+)}, \cdot)$, where $a \cdot b = ab + ba$ is a Jordan algebra, and $(A^{(-)}, [\ ,\ ])$, where $[a, b] = ab - ba$ is a Lie algebra. Both $A^{(+)}$ and $A^{(-)}$ have the same underlying vector space as A.

DEFINITION 1.2.– *A superalgebra A is an algebra with a $\mathbb{Z}/2\mathbb{Z}$-grading. So $A = A_{\bar{0}} + A_{\bar{1}}$ is a direct sum of two vector spaces and*

$$A_{\bar{0}}A_{\bar{0}} \subseteq A_{\bar{0}}, \quad A_{\bar{1}}A_{\bar{1}} \subseteq A_{\bar{0}}, \quad \textit{and} \quad A_{\bar{0}}A_{\bar{1}}, A_{\bar{1}}A_{\bar{0}} \subseteq A_{\bar{1}}.$$

Elements of $A_{\bar{0}} \cup A_{\bar{1}}$ are called homogeneous elements. The parity of a homogeneous element a, denoted $|a|$, is defined by $|a| = 0$ if $a \in A_{\bar{0}}$ and $|a| = 1$ if $a \in A_{\bar{1}}$.

Elements in $A_{\bar{0}}$ are called even and elements in $A_{\bar{1}}$ are called odd.

Note that $A_{\bar{0}}$ is a subalgebra of A, but $A_{\bar{1}}$ is not, instead it can be seen as a bimodule over $A_{\bar{0}}$.

EXAMPLE 1.2.– If V is a vector space of countable dimension, then $G = G(V)$ denotes the Grassmann (or exterior) algebra over V, that is, the quotient of the tensor algebra over the ideal generated by the symmetric tensors $v \otimes w + w \otimes v$, $v, w \in V$. This algebra $G(V)$ is $\mathbb{Z}/2\mathbb{Z}$-graded. Indeed, $G(V) = G(V)_{\bar{0}} + G(V)_{\bar{1}}$, where the "even part" is the linear span of all tensors of even length and the "odd part" $G(V)_{\bar{1}}$ is the linear span of all tensors of odd length.

$G(V)$ is an example of a superalgebra.

DEFINITION 1.3.– *Consider a variety of algebras $\mathcal{V}$ defined by homogeneous identities (see Jacobson (1968) or Zhevlakov* et al. *(1982)). We say that a superalgebra $A = A_{\bar{0}} + A_{\bar{1}}$ is a $\mathcal{V}$-superalgebra if the even part of $A \otimes_F G(V)$ lies in the variety, that is*

$$A_{\bar{0}} \otimes G(V)_{\bar{0}} + A_{\bar{1}} \otimes G(V)_{\bar{1}} \in \mathcal{V}.$$

DEFINITION 1.4.– *The algebra $A_{\bar{0}} \otimes G(V)_{\bar{0}} + A_{\bar{1}} \otimes G(V)_{\bar{1}}$ is called the Grassmann envelope of the superalgebra A and will be denoted as $G(A)$.*

Let us consider $\mathcal{V}$ the variety of associative, commutative, anticommutative, Jordan or Lie algebras, respectively. Then we get:

EXAMPLE 1.3.– A superalgebra $A = A_{\bar{0}} + A_{\bar{1}}$ is an associative superalgebra if and only if it is a $\mathbb{Z}/2\mathbb{Z}$-graded associative algebra.

EXAMPLE 1.4.– A superalgebra $A = A_{\bar{0}} + A_{\bar{1}}$ is a commutative superalgebra if it satisfies:

$$xy = (-1)^{|x||y|}yx$$

for any x, y homogeneous elements of A.

EXAMPLE 1.5.– A superalgebra A is an anticommutative superalgebra if

$$xy = -(-1)^{|x||y|}yx$$

for every x, y homogeneous elements of A.

EXAMPLE 1.6.– A Jordan superalgebra is a superalgebra that is commutative and satisfies the graded identity:

$$\begin{aligned}&(xy)(zu) + (-1)^{|y||z|}(xz)(yu) + (-1)^{|u||y|+|u||z|}(xu)(yz)\\ &= ((xy)z)u + (-1)^{|u||y|+|z||y|+|z||u|}((xu)z)y + (-1)^{|x|(|y|+|z|+|u|)+|z||u|}((yu)z)x\end{aligned}$$

for every homogeneous elements x, y, z, $u \in A_{\bar{0}} \cup A_{\bar{1}}$.

EXAMPLE 1.7.– An anticommutative superalgebra A is a Lie superalgebra if it satisfies:

$$[[x,y],z] + (-1)^{|x||y|+|x||z|}[[y,z],x] + (-1)^{|x||z|+|y||z|}[[z,x],y] = 0$$

for every x, y, $z \in A_{\bar{0}} + A_{\bar{1}}$.

DEFINITION 1.5.– *If* $J = J_{\bar{0}} + J_{\bar{1}}$ *is a Jordan superalgebra and* x, y, $z \in J_{\bar{0}} \cup J_{\bar{1}}$, *then their triple product is defined by:*

$$\{x, y, z\} = (xy)z + x(yz) - (-1)^{|x||y|}y(xz).$$

Note that every algebra is a superalgebra with the trivial grading, that is, $A = A_{\bar{0}}$.

1.2. Tits–Kantor–Koecher construction

Tits (1962, 1966) made an important observation that relates Lie and Jordan structures. Let L be a Lie superalgebra whose even part $L_{\bar{0}}$ contains an $\mathfrak{sl}_2$-triple $\{e, f, h\}$, that is,

$$[e, f] = h, \qquad [h, e] = 2e, \qquad [h, f] = -2f.$$

DEFINITION 1.6.– *An* $\mathfrak{sl}_2$*-triple* e, f, h *is said to be "good" if* $\mathrm{ad}(h) : L \to L$ *is diagonalizable and the eigenvalues are only* -2, 0, 2.

In such a case, $L = L_{-2} + L_0 + L_2$ decomposes as a direct sum of eigenspaces. We can define a new product in L_2 by:

$$a \circ b = [[a, f], b] \qquad \forall a, b \in L_2.$$

With this new product, $J = (L_2, \circ)$ becomes a Jordan superalgebra.

Moreover, (Tits 1962, 1966; Kantor 1972) and (Koecher 1967) showed that every Jordan superalgebra can be obtained in this way. The corresponding Lie superalgebra is not unique, but any two such Lie superalgebras are centrally isogenous, that is, they have the same central cover. Let us recall the construction of $L = \mathrm{TKK}(J)$, the universal Lie superalgebra in this class (see Martin and Piard (1992)).

CONSTRUCTION.– Consider J a unital Jordan superalgebra, and $\{e_i\}_{i \in I}$ a basis of J.

Let

$$\{e_i, e_j, e_k\} = \sum_t \gamma_{ijk}^t e_t, \qquad \text{where } \gamma_{ijk}^t \in F.$$

Define a Lie superalgebra K by generators $\{x_j^+, x_j^-\}_{j\in I}$ and relations

$$[x_i^+, x_j^+] = 0 = [x_i^-, x_j^-],$$
$$[[x_i^+, x_j^-], x_k^+] = \sum_t \gamma_{ijk}^t x_t^+,$$
$$[[x_i^-, x_j^+], x_k^-] = \sum_t \gamma_{ijk}^t x_t^-, \qquad \text{for every } i,j,k \in I.$$

This Lie superalgebra has a short grading $K = K_{-1} + K_0 + K_1$ where

$$K_{-1} = F(x_i^-)_{i\in I}, \quad K_1 = F(x_i^+)_{i\in I}, \quad K_0 = F([x_i^-, x_j^+])_{i,j\in I}.$$

K is the universal Tits–Kantor–Koecher Lie superalgebra of the unital Jordan superalgebra J:

$$K = \mathrm{TKK}(J).$$

1.3. Basic examples (classical superalgebras)

Let $A = A_{\bar 0} + A_{\bar 1}$ be an associative superalgebra. The new operation in the underlying vector space A given by:

$$a \circ b = \frac{1}{2}(ab + (-1)^{|a||b|}ba) \quad \forall a, b \in A_{\bar 0} \cup A_{\bar 1},$$

defines a structure of a Jordan superalgebra on A that is denoted $A^{(+)}$.

DEFINITION 1.7.– *Those Jordan superalgebras that can be obtained as subalgebras of a superalgebra $A^{(+)}$, with A an associative superalgebra, are called special. Superalgebras that are not special are called exceptional.*

REMARK 1.2.– If we consider in the original associative superalgebra the new product given by:

$$[a, b] = ab - (-1)^{|a||b|}ba \quad \forall a, b \in A_{\bar 0} \cup A_{\bar 1},$$

we get a Lie superalgebra that is denoted as $A^{(-)}$.

DEFINITION 1.8.– *A superalgebra A is simple if it does not have non-trivial graded ideals. A graded ideal is an ideal $I \trianglelefteq A$ such that for every $a = a_0 + a_1 \in I$, it follows that a_0, $a_1 \in I$. So every graded ideal I satisfies $I = (A_{\bar 0} \cap I) + (A_{\bar 1} \cap I)$.*

Wall (1963, 1964) proved that an arbitrary simple finite dimensional superalgebra over an algebraically closed field is isomorphic to one of the following two types:

$$\text{I)}\quad A = M_{m+n}(F) = A_{\bar{0}} + A_{\bar{1}}, \qquad A_{\bar{0}} = \left\{ \begin{pmatrix} * & 0 \\ 0 & * \end{pmatrix} \begin{matrix} m \\ n \end{matrix} \right\}, \qquad A_{\bar{1}} = \left\{ \begin{pmatrix} 0 & * \\ * & 0 \end{pmatrix} \right\}.$$

$$\text{II)}\quad A = Q(n) = \left\{ \begin{pmatrix} a & b \\ b & a \end{pmatrix} \;\middle|\; a, b \in M_n(F) \right\} \leq M_{n+n}(F).$$

Consequently, we can easily get the first examples of simple finite dimensional Jordan superalgebras as explained above.

EXAMPLE 1.8.– $J = M_{m+n}^{(+)}(F)$, $m \geq 1, n \geq 1$.

EXAMPLE 1.9.– $J = Q(n)^{(+)}$, $n \geq 2$.

DEFINITION 1.9.– *Let A be an associative superalgebra. A map $* : A \to A$ is a superinvolution if it satisfies:*

i) $(a^*)^* = a, \quad \forall a \in A$;

ii) $(ab)^* = (-1)^{|a||b|} b^* a^*, \quad \forall a, b \in A_{\bar{0}} \cup A_{\bar{1}}$.

If $* : A \to A$ is a superinvolution of the associative superalgebra A, then the set of symmetric elements $H(A, *)$ is a Jordan superalgebra of $A^{(+)}$. Similarly, the subspace of skew-symmetric elements $K(A, *) = \{a \in A \mid a^* = -a\}$ is a Lie subsuperalgebra of $A^{(-)}$.

The following two subsuperalgebras of $M_{m+n}^{(+)}$ are of this type.

EXAMPLE 1.10.– Let I_n, I_m be the identity matrices and let t be the transposition. Let us denote $U = \begin{pmatrix} 0 & -I_m \\ I_m & 0 \end{pmatrix}$.

Then $U^t = U^{-1} = -U$, and $* : M_{n+2m}(F) \to M_{n+2m}(F)$ given by

$$\begin{pmatrix} a & b \\ c & d \end{pmatrix}^* = \begin{pmatrix} I_n & 0 \\ 0 & U \end{pmatrix} \begin{pmatrix} a^t & -c^t \\ b^t & d^t \end{pmatrix} \begin{pmatrix} I_n & 0 \\ 0 & U^{-1} \end{pmatrix}$$

is a superinvolution.

The superalgebras

$$\mathrm{osp}_{n,2m} = K(M_{n+2m}(F), *) \quad \text{and} \quad \mathrm{Josp}_{n,2m}(F) = H(M_{n+2m}(F), *)$$

are the Lie and Jordan orthosymplectic superalgebras, respectively.

EXAMPLE 1.11.– The associative superalgebra $M_{n+n}(F)$ has another superinvolution given by:

$$\begin{pmatrix} a & b \\ c & d \end{pmatrix}^{\sigma} = \begin{pmatrix} d^t & -b^t \\ c^t & a^t \end{pmatrix}.$$

The Lie and Jordan superalgebras (respectively) that consist of skew-symmetric and symmetric elements, respectively, are denoted as $\mathrm{P}_n(F)$ and $\mathrm{JP}_n(F)$ (and are also called "strange series").

EXAMPLE 1.12.– The three-dimensional Kaplansky superalgebra $\mathrm{K}_3 = Fe + (Fx + Fy)$ with multiplication table:

$$e^2 = e, \quad ex = \frac{1}{2}x, \quad ey = \frac{1}{2}y, \quad x^2 = y^2 = 0, \quad x \cdot y = -y \cdot x = e$$

is a simple Jordan superalgebra. Note that K_3 is not unital.

EXAMPLE 1.13.– The one-parametric family of four-dimensional superalgebras $D(t)$ defined as $D(t) = (Fe_1 + Fe_2) + (Fx + Fy)$ with the product

$$e_i^2 = e_i, \; e_1e_2 = 0, \; e_ix = \frac{1}{2}x, \; e_iy = \frac{1}{2}y, \; xy = e_1 + te_2, \; t \in F, \; i = 1, 2.$$

The superalgebra $D(t)$ is simple if $t \neq 0$. In the case $t = -1$, the superalgebra $D(-1)$ is isomorphic to $M_{1+1}(F)^{(+)}$.

EXAMPLE 1.14.– Let V be a vector space that is $\mathbb{Z}/2\mathbb{Z}$-graded, $V = V_{\bar{0}} + V_{\bar{1}}$, and has a superform $(\;|\;) : V \times V \to F$, which is symmetric on $V_{\bar{0}}$, skew-symmetric in $V_{\bar{1}}$ and $(V_{\bar{0}}|V_{\bar{1}}) = 0 = (V_{\bar{1}}|V_{\bar{0}})$. Then $J = F1 + V = (F1 + V_{\bar{0}}) + V_{\bar{1}}$ is a Jordan superalgebra, where the product of two arbitrary elements $\alpha 1 + v$ and $\beta 1 + w$ in J is given by

$$(\alpha 1 + v) \cdot (\beta 1 + w) = \alpha\beta 1 + (v|w)1 + (\alpha w + \beta v)$$

for arbitrary $v, w \in V$.

We will refer to J as the superalgebra of a superform.

J is simple if and only if the form $(\;|\;)$ is non-degenerate.

EXAMPLE 1.15.– Kac (1977a) introduced a 10-dimensional Jordan superalgebra J whose even part has dimension 6 and splits as the direct sum of a superalgebra of a superform and a one-dimensional algebra. Thus, $J_{\bar{0}} = (Fe + \sum_{1\leq i\leq 4} Fv_i) \oplus F_f$ has the multiplication given by:

$$e^2 = e, \quad ev_i = v_i, \quad v_1v_2 = 2e = v_3v_4, \quad f^2 = f$$

and any other product of two basic elements is 0.

The odd part $J_{\bar{1}}$ has a basis $\{x_1, x_2, y_1, y_2\}$ and the following multiplication table:

$$x_iy_i = e - 3f, \quad x_i^2 = 0 = y_i^2, \quad x_1y_2 = v_3 = -y_2x_1, \quad x_2y_1 = v_4 = -y_1x_2,$$
$$x_1x_2 = 0 = x_2x_1, \quad y_1y_2 = v_2 = -y_2y_1, \qquad i = 1, 2.$$

Finally, the action of $J_{\bar{0}}$ over $J_{\bar{1}}$ is given by:

$$ex_j = \frac{1}{2}x_j, \quad ey_j = \frac{1}{2}y_j, \quad fx_j = \frac{1}{2}x_j, \quad fy_j = \frac{1}{2}y_j, \qquad j = 1, 2$$
$$v_1x_j = 0, \quad v_1y_1 = x_2, \quad v_1y_2 = -x_1, \quad v_2y_j = 0, \quad v_2x_1 = -y_2, \quad v_2x_2 = y_1$$
$$v_3x_1 = 0 = v_3y_2, \quad v_3x_2 = x_1, \quad v_3y_1 = y_2,$$
$$v_4x_2 = 0 = v_4y_1, \quad v_4x_1 = x_2, \quad v_4y_2 = y_1.$$

This superalgebra is simple if char $F \neq 3$. In case of char $F = 3$, it has an ideal of dimension 9 that is spanned by e, v_i, $1 \leq i \leq 4$, x_j, y_j, $1 \leq j \leq 2$. It is called degenerated Kac superalgebra and is denoted by K_9.

In Medvedev and Zelmanov (1992), it is proved that the Kac superalgebra K_{10} is not a homomorphic image of a special Jordan superalgebra.

Benkart and Elduque (2002) realized K_{10} as the space of $\mathrm{K}_3 \otimes \mathrm{K}_3 + F1$ (with a new product) where K_3 is the Kaplansky superalgebra.

Another (octonionic) construction of K_{10} was suggested in Racine and Zelmanov (2015).

1.4. Brackets

DEFINITION 1.10.– *Let A be an associative commutative superalgebra. A binary map $\{\ ,\ \} : A \times A \to A$ is called a Poisson bracket if*

1) $(A, \{\ ,\ \})$ is a Lie superalgebra;

2) $\{ab,c\} = a\{b,c\} + (-1)^{|b||c|}\{a,c\}b$ *for arbitrary* $a,b,c \in A_{\bar{0}} \cup A_{\bar{1}}$.

EXAMPLE 1.16.– Let $F[p_1,\ldots,p_n,q_1,\ldots,q_n]$ be a polynomial algebra in $2n$ variables. The classical Hamiltonian bracket:

$$\{f,g\} = \sum_{i=1}^{n}\left(\frac{\partial f}{\partial p_i}\frac{\partial g}{\partial q_i} - \frac{\partial g}{\partial p_i}\frac{\partial f}{\partial q_i}\right), \qquad f,g \in F[p_1,\ldots,p_n,q_1,\ldots,q_n]$$

is a Poisson bracket.

EXAMPLE 1.17.– Let

$$G(n) = \langle \xi_1,\ldots,\xi_n \rangle$$

be the Grassmann algebra over an n-dimensional vector space V. Then the bracket

$$\{f,g\} = \sum_{i=1}^{n}(-1)^{|f|}\frac{\partial f}{\partial \xi_i}\frac{\partial g}{\partial \xi_i}$$

for arbitrary $f,g \in G(V)_{\bar{0}} \cup G(V)_{\bar{1}}$ is a Poisson bracket.

DEFINITION 1.11.– *Given* $A = A_{\bar{0}} + A_{\bar{1}}$ *an associative commutative superalgebra, a bilinear map* $\{\ ,\ \} : A \times A \to A$ *is called a contact bracket if:*

i) $(A, \{\ ,\ \})$ *is a Lie superalgebra;*

ii) $\{ab,c\} = a\{b,c\} + (-1)^{|b||c|}\{a,c\}b + abD(c)$ *for arbitrary homogeneous elements a, b, c in A.*

Note that a Poisson bracket is a contact bracket with $D = 0$.

EXAMPLE 1.18.– Let $F[t]$ be the polynomial algebra. Then the bracket $\{f(t),g(t)\} = f'(t)g(t) - f(t)g'(t)$ is a contact bracket.

EXAMPLE 1.19.– Let $\Lambda(1:n)$ be the polynomial superalgebra in one (even) Laurent variable and n (odd) Grassmann variables $\xi_1,\ldots,\xi_n$, $\Lambda(1:n) = F[t,t^{-1},\xi_1,\ldots,\xi_n]$. Consider $D = \frac{\partial}{\partial t}$ or $D = t\frac{\partial}{\partial t}$ a derivation of $F[t,t^{-1}]$. The bilinear map defined on generators of $\Lambda(1:n)$ by:

$$[t,\xi_i] = \frac{1}{2}\xi_i D(t), \quad [\xi_i,\xi_j] = \delta_{ij}, \quad 1 \le i,j \le n$$

can be extended to a contact bracket on $\Lambda(1:n)$.

DEFINITION 1.12.– *[Kantor double] Let A be an associative commutative superalgebra with a bilinear map $\{\ ,\ \} : A \times A \to A$. Assume that $\{A_{\bar{i}}, A_{\bar{j}}\} \subseteq A_{\overline{i+j}}$. Consider a direct sum of vector spaces $\mathrm{KJ}(A, \{\ ,\ \}) = A + Av$ where $|v| = 1$. Define a new product in $J(A, \{\ ,\ \})$ that coincides with the original one in A and is given by:*

$$a(bv) = (ab)v, \quad (bv)a = (-1)^{|a|}(ba)v, \quad (av)(bv) = (-1)^{|b|}\{a, b\}.$$

The superalgebra $\mathrm{KJ}(A, \{\ ,\ \})$ is called the Kantor double of $(A, \{\ ,\ \})$.

Kantor (1990) proved that if the bracket $\{\ ,\ \}$ is a Poisson bracket, then $\mathrm{KJ}(A, \{\ ,\ \})$ is a Jordan superalgebra.

DEFINITION 1.13.– *The bilinear map $\{\ ,\ \}$ is called a Jordan bracket on the superalgebra A if $\mathrm{KJ}(A, \{\ ,\ \})$ is a Jordan superalgebra (see King and McCrimmon (1992)).*

Cantarini and Kac (2007) showed that there is a 1-1 correspondence between Jordan brackets and contact brackets. Indeed, if $[a, b]$ is a contact bracket with derivation D, $D(a) = [a, 1]$, then the new bracket

$$\{a, b\} = [a, b] + \frac{1}{2}(D(a)b - aD(b))$$

is a Jordan bracket.

Applying this to example 1.19, we get the following:

EXAMPLE 1.20.– The values $\{\xi_i, t\} = 0$, $\{\xi_i, \xi_j\} = \delta_{ij}$ for $1 \leq i, j \leq n$ extend to a Jordan bracket of $\Lambda(1 : n)$. Applying the Kantor double process to this bracket, we get Jordan superalgebras $J_n = \mathrm{KJ}(\Lambda(1 : n), \{\ ,\ \})$.

1.5. Cheng–Kac superalgebras

Given an arbitrary unital associative commutative (super)-algebra with an (even) derivation $d : Z \to Z$, Martínez and Zelmanov constructed (Martínez and Zelmanov 2010) a Jordan superalgebra $\mathrm{JCK}(Z, d)$ named the Cheng–Kac Jordan superalgebra.

The even part of $\mathrm{JCK}(Z, d)$ is a rank 4 free module over Z with basis $\{1, w_1, w_2, w_3\}$, $J_{\bar{0}} = Z + \sum\limits_{i=1}^{3} w_i Z$, and multiplication given by $w_i w_j = 0$ if $1 \leq i \neq j \leq 3$, $w_{\bar{1}}^2 = 1 = w_{\bar{2}}^2$, $w_{\bar{3}}^2 = -1$. The odd part of this superalgebra is also a rank 4 free module over Z with basis $\{x, x_1, x_2, x_3\}$, $J_{\bar{1}} = xZ + \sum_{i=1}^{3} x_i Z$.

The action of the even part on the odd part and the products of two elements, respectively, are given by the following multiplication tables:

	g	$w_j g$
xf	$x(fg)$	$x_j(fg^d)$
$x_i f$	$x_i(fg)$	$x_{i\times j}(fg)$

	xg	$x_j g$
xf	$f^d g - fg^d$	$-w_j(fg)$
$x_i f$	$w_i(fg)$	0

where $x_{i\times i} = 0$, $x_{1\times 2} = -x_{2\times 1} = x_3$, $x_{1\times 3} = -x_{3\times 1} = x_2$, $-x_{2\times 3} = x_1 = x_{3\times 2}$.

The superalgebra $\mathrm{JCK}(Z,d)$ is simple if and only if Z is d-simple, that is, Z does not contain proper d-invariant ideals (see Martínez and Zelmanov (2010)).

Let us remark that for $Z = \mathbb{C}[t,t^{-1}]$ the above construction leads to the Cheng–Kac superconformal algebra, that is, $\mathrm{CK}(6) = \mathrm{TKK}(\mathrm{JCK}(6))$, where

$$\mathrm{JCK}(6) = \mathrm{JCK}(\mathbb{C}[t,t^{-1}], D)$$

with $D = \frac{\partial}{\partial t}$ (see section 1.8).

1.6. Finite dimensional simple Jordan superalgebras

1.6.1. *Case F is algebraically closed and* $\mathrm{char}\, F = 0$

Let us assume now that F is algebraically closed and $\mathrm{char}\, F = 0$. Kac derived the classification of finite dimensional simple Jordan F-superalgebras from his classification of simple finite dimensional Lie superalgebras via the Tits–Kantor–Koecher construction.

THEOREM 1.1 (see Kac (1977a) and Kantor (1990)).– Let $J = J_{\bar{0}} + J_{\bar{1}}$ be a simple Jordan superalgebra over an algebraically closed field F, $\mathrm{char}\, F = 0$. Then J is isomorphic to one of the superalgebras in examples 1.8, 1.9 and 1.10–1.15 or it is the Kantor double of the Poisson bracket in example 1.17.

REMARK 1.3.– We will assume always in this section that $J_{\bar{1}} \neq (0)$.

1.6.2. *Case* $\mathrm{char}\, F = p > 2$*, the even part $J_{\bar{0}}$ is semisimple*

Let us assume next that $\mathrm{char}\, F = p > 2$ and the even part $J_{\bar{0}}$ is a semisimple Jordan algebra.

Recall that a semisimple Jordan algebra is a direct sum of finitely many simple ideals.

This case was addressed in Racine and Zelmanov (2003) and the classification essentially coincides with the one of zero characteristic, expect of some differences if $\mathrm{char}\, F = 3$.

EXAMPLE 1.21.– Let $H_3(F), K_3(F)$ denote the symmetric and skew-symmetric 3×3 matrices over F, $\mathrm{char}\, F = 3$. Consider $J_{\bar 0} = H_3(F)$ and $J_{\bar 1} = \overline{K_3(F)} \oplus \overline{\overline{K_3(F)}}$ the sum of two copies of $K_3(F)$. We have a Jordan superalgebra structure on $J = J_{\bar 0} + J_{\bar 1}$ via $a \cdot b = a \cdot b$ in $M_3(F)^+$ if $a, b \in H_3(F)$, that is,

$$a \cdot b = ab + ba,$$

$$\bar{c} \cdot \bar{\bar{d}} = cd + dc \in H_3(F) \quad \text{for } c, d \in K_3(F),$$

$$\overline{K_3(F)} \cdot \overline{K_3(F)} = (0) = \overline{\overline{K_3(F)}} \cdot \overline{\overline{K_3(F)}}, \quad \text{and}$$

$$a\bar{c} = \overline{\overline{ac + ca}}, \;\; a\bar{\bar{c}} = \overline{ac + ca} \quad \text{if } a \in H_3(F),\ c \in K_3(F).$$

This superalgebra is simple.

EXAMPLE 1.22.– Let $B = B_{\bar 0} + B_{\bar 1}$ with $B_{\bar 0} = M_2(F)$, $B_{\bar 1} = Fm_1 + Fm_2$, where F is a field, $\mathrm{char}\, F = 3$. The product in $B_{\bar 1}$ is given by:

$$m_1^2 = -e_{21}, \quad m_2^2 = e_{12}, \quad m_1 m_2 = e_{11}, \quad m_2 m_1 = -e_{22}.$$

The action of $B_{\bar 0}$ over $B_{\bar 1}$ is defined as follows:

$$e_{11}m_1 = m_1, \quad e_{11}m_2 = 0 = m_1 e_{11}, \quad m_2 e_{11} = m_2,$$

$$m_1 e_{22} = m_1, \quad e_{22}m_1 = 0 = m_2 e_{22}, \quad e_{22}m_2 = m_2,$$

$$e_{12}m_1 = m_2, \quad e_{12}m_2 = 0 = m_2 e_{12}, \quad m_1 e_{12} = -m_2,$$

$$e_{21}m_2 = m_1, \quad e_{21}m_1 = 0 = m_1 e_{21}, \quad m_2 e_{21} = -m_1.$$

Shestakov (1997) proved that B is an alternative superalgebra and has a natural involution $*$ given by $(a+m)^* = \bar{a} - m$, $a \in B_{\bar 0}$, where $a \mapsto \bar{a}$ is the symplectic involution, and $m \in B_{\bar 1}$.

If $H_3(B, *)$ denotes the symmetric matrices with respect to the involution $*$, then $H_3(B, *)$ is a simple Jordan superalgebra. It is i-exceptional, that is, it is not a homomorphic image of a special Jordan superalgebra.

THEOREM 1.2 (Racine and Zelmanov (2003)).– Let $J = J_{\bar 0} + J_{\bar 1}$ be a finite dimensional central simple Jordan superalgebra over an algebraically closed field F

of char $F = p > 2$. If $J_{\bar{1}} \neq (0)$ and $J_{\bar{0}}$ is semisimple, then J is isomorphic to one of the superalgebras in examples 1.8, 1.9, 1.10–1.14 or char $F = 3$ and J is the nine-dimensional degenerate Kac superalgebra (see example 1.15) or J is isomorphic to one of the superalgebras in examples 1.21 and 1.22.

1.6.3. *Case* char $F = p > 2$, *the even part* $J_{\bar{0}}$ *is not semisimple*

This case shows similarities with infinite dimensional superconformal Jordan algebras (see section 1.8) in characteristic 0.

Let us denote $B(m) = F[a_1, \ldots, a_m \mid a_i^p = 0]$ the algebra of truncated polynomials in m variables. Let $B(m,n) = B(m) \otimes G(n)$ be the tensor product of $B(m)$ with the Grassmann algebra $G(n) = \langle 1, \xi_1, \ldots, \xi_n \rangle$. Then $B(m,n)$ is an associative commutative superalgebra.

THEOREM 1.3 (Martínez and Zelmanov (2010)).– Let $J = J_{\bar{0}} + J_{\bar{1}}$ be a finite dimensional simple unital Jordan superalgebra over an algebraically closed field F of characteristic $p > 2$. If the even part $J_{\bar{0}}$ is not semisimple, then there exist integers m, n and a Jordan bracket $\{\ ,\ \}$ on $B(m,n)$ such that $J = B(m,n) + B(m,n)v = \mathrm{KJ}(B(m,n), \{\ ,\ \})$ is a Kantor double of $B(m,n)$ or J is isomorphic to a Cheng–Kac Jordan superalgebra $\mathrm{JCK}(B(m), d)$ for some derivation $d : B(m) \to B(m)$.

1.6.4. *Non-unital simple Jordan superalgebras*

Finally, let us consider non-unital simple Jordan superalgebras. As we have seen, K_3 the three-dimensional Kaplansky superalgebra and K_9 the nine-dimensional degenerate Kac superalgebra are examples of such superalgebras.

EXAMPLE 1.23.– Let Z be a unital associative commutative algebra, $D : Z \to Z$ a derivation. Assume that Z is D-simple and that the only constants are elements $\alpha 1$, $\alpha \in F$.

Let us consider in Z the bracket $\{\ ,\ \}$ given by:

$$\{a, b\} = (aD)b - a(bD) \quad \forall a, b \in Z.$$

The above bracket is a Jordan bracket, so the Kantor double $V(Z, D) = Z + Zv = \mathrm{KJ}(Z, \{\ ,\ \})$ is a simple unital Jordan superalgebra.

Now we will change the product in $V(Z, D)$, modifying only the action of the even part on the odd part and preserving the product of two even (respectively, two odd) elements. Denote with juxtaposition the product on $V(Z, D)$. Let $a, b \in Z$. We define the new product $\cdot$ by:

$$a \cdot b = ab, \quad a \cdot bv = \frac{1}{2}(ab)v, \quad av \cdot bv = \{a, b\} = (av)(bv).$$

In this way, we get another Jordan superalgebra $V_{1/2}(Z, D)$ that is simple but not unital.

It was proved in Zelmanov (2000) that:

THEOREM 1.4.– Let J be a finite dimensional simple central non-unital Jordan superalgebra over a field F. Then J is isomorphic to one of the superalgebras on the list:

i) the Kaplansky superalgebra K_3 (example 1.12);

ii) the field F has characteristic 3 and J is the degenerate Kac superalgebra (example 1.15);

iii) a superalgebra $V_{1/2}(Z, D)$ (example 1.23).

DEFINITION 1.14.– *Let A be a Jordan superalgebra and let N be its radical, that is, the largest solvable ideal of A. The superalgebra A is said to be semisimple if* $N = (0)$.

EXAMPLE 1.24.– Let B be a simple non-unital Jordan superalgebra and let $H(B) = B + F1$ be its unital hull. Then $H(B)$ is a semisimple Jordan superalgebra that is not simple.

THEOREM 1.5 (Zelmanov (2000)).– Let J be a finite dimensional Jordan superalgebra. Then J is semisimple if and only if

$$J \cong \bigoplus_{i=1}^{s} (J_{i1} \oplus \cdots \oplus J_{ir_i} + K_i 1) + J_{(1)} \oplus \cdots \oplus J_{(t)}$$

where $J_{(1)}, \ldots, J_{(t)}$ are simple Jordan superalgebras and for every $i = 1, \ldots, s$, the superalgebras J_{ij} are simple non-unital Jordan superalgebras over the field extension K_i of F.

1.7. Finite dimensional representations

Jacobson (1968) developed the theory of bimodules over semisimple finite dimensional Jordan algebras.

In this section, we discuss representations (bimodules) of finite dimensional Jordan superalgebras.

DEFINITION 1.15.– *The rank of a Jordan superalgebra J is the maximal number of pairwise orthogonal idempotents in the even part.*

Unless otherwise stated we will assume $\operatorname{char} F = 0$.

DEFINITION 1.16.– *Let V be a $\mathbb{Z}/2\mathbb{Z}$-graded vector space with bilinear mappings $V \times J \to V$, $J \times V \to V$. We call V a Jordan bimodule if the split null extension $V + J$ is a Jordan superalgebra.*

Recall that in the split null extension the multiplication extends the multiplication on J, products $V \cdot J$ and $J \cdot V$ are defined via the bilinear mappings above and $V \cdot V = (0)$.

Let $V = V_{\bar{0}} + V_{\bar{1}}$ be a Jordan bimodule over J. Consider the vector space $V^{\mathrm{op}} = V_{\bar{1}}^{\mathrm{op}} + V_{\bar{0}}^{\mathrm{op}}$, where $V_{\bar{i}}^{\mathrm{op}}$ is a copy of $V_{\bar{i}}$ with different parity. Define the action of J on V^{op} via

$$av^{\mathrm{op}} = (-1)^{|a|}(av)^{\mathrm{op}}, \quad v^{\mathrm{op}}a = (va)^{\mathrm{op}}.$$

Then V^{op} is also a Jordan bimodule over J. We call it the opposite module of V.

Let V be the free Jordan J-bimodule on one free generator.

DEFINITION 1.17.– *The associative subsuperalgebra $U(J)$ of $\operatorname{End}_F V$ generated by all linear transformations $R_V(a) : V \to V$, $v \mapsto va$, $a \in J$, is called the universal multiplicative enveloping superalgebra of J.*

Every Jordan bimodule over J is a right module over $U(J)$.

DEFINITION 1.18.– *A bimodule V over J is called a one-sided bimodule if $\{J, V, J\} = (0)$.*

Let $V(1/2)$ be the free one-sided Jordan J-bimodule on one free generator.

DEFINITION 1.19.– *The associative subsuperalgebra $S(J)$ of $\operatorname{End}_F V(1/2)$ generated by all linear transformations $R_{V(1/2)}(a) : V(1/2) \to V(1/2)$, $v \mapsto va$, $a \in J$, is called the universal associative enveloping algebra of J.*

Every one-sided Jordan J-bimodule is a right module over $S(J)$.

Finally, let J be a unital Jordan superalgebra with the identity element e. Let $V(1)$ denote the free unital J-bimodule on one free generator. The associative subsuperalgebra $U_1(J)$ of $\operatorname{End}_F V(1)$ generated by $\{R_{V(1/2)}(a)\}_{a \in J}$ is called the universal unital enveloping algebra of J.

For an arbitrary Jordan bimodule V, the Peirce decomposition

$$V = \{e, V, e\} \oplus \{e, V, 1 - e\} \oplus \{1 - e, V, 1 - e\}$$

is a decomposition of V into a direct sum of unital and one-sided bimodules. Hence $U(J) \cong U_1(J) \oplus S(J)$.

1.7.1. *Superalgebras of rank* ≥ 3

In this section, we consider Jordan bimodules over finite dimensional simple Jordan superalgebras of rank ≥ 3, that is, superalgebras $M_{m+n}^{(+)}$, $m + n \geq 3$; $\mathrm{Josp}(n, 2m)$, $n + m \geq 3$; $Q(n)^{(+)}$, $n \geq 3$; $\mathrm{JP}(n)$, $n \geq 3$.

In this case, the universal multiplicative enveloping superalgebra $U(J)$ is finite dimensional and semisimple (Martin and Piard 1992). Hence every Jordan bimodule is completely reducible, as in the case of Jordan algebras.

The superalgebras $\mathrm{Josp}(n, 2m)$ and $\mathrm{JP}(n)$ are of the type

$$H(A, *) = \{a \in A \mid a^* = a\}$$

where A is a simple finite dimensional associative superalgebra and $*$ is an involution.

EXAMPLE 1.25.– An arbitrary right module over A is a one-sided module over $H(A, *)$.

EXAMPLE 1.26.– The subspace $K(A, *) = \{k \in A \mid k^* = -k\}$ with the action $k \cdot a = ka + ak$; $k \in K(A, *)$, $a \in H(A, *)$ is a unital $H(A, *)$-bimodule.

THEOREM 1.6 (Martin and Piard (1992)).– An arbitrary irreducible Jordan bimodule over $\mathrm{Josp}(n, 2m)$, $n + m \geq 3$, or $\mathrm{JP}(n)$, $n \geq 3$, is a bimodule of examples 1.25 and 1.26 or the regular bimodule.

The superalgebras $M_{m+n}^{(+)}$, $Q(n)^{(+)}$ are of the type $A^{(+)}$, where A is a simple finite dimensional associative superalgebra.

EXAMPLE 1.27.– Every right module over A gives rise to a one-sided Jordan bimodule over $A^{(+)}$.

Suppose now that the superalgebra A is equipped with an involution $*$.

EXAMPLE 1.28.– The subspaces $H(A, *)$ and $K(A, *)$ become Jordan $A^{(+)}$-bimodules with the actions:

1) $h \circ a = ha + a^*h$;

2) $h \circ a = ha^* + ah$;

3) $k \circ a = ka + a^*k$;

4) $k \circ a = ka^* + ak$;

where $h \in H(A, *)$, $k \in K(A, *)$, $a \in A$.

The associative superalgebras $M_{m+n}(F)$, where both m, n are odd, and $Q(n)$, are not equipped with an involution, but they are equipped with a pseudoinvolution.

DEFINITION 1.20.– *A graded linear map* $* : A \to A$ *is called a pseudoinvolution if* $(ab)^* = (-1)^{|a|\cdot|b|}b^*a^*$, $(a^*)^* = (-1)^{|a|}a$ *for arbitrary elements* $a, b \in A_{\bar{0}} \cup A_{\bar{1}}$.

EXAMPLE 1.29.– The mapping $\begin{pmatrix} a & b \\ b & a \end{pmatrix} \mapsto \begin{pmatrix} a^t & \sqrt{-1}b^t \\ \sqrt{-1}b^t & a^t \end{pmatrix}$ is a pseudoinvolution on $Q(n)$.

EXAMPLE 1.30.– The mapping $\begin{pmatrix} a & b \\ c & d \end{pmatrix} \mapsto \begin{pmatrix} a^t & -c^t \\ b^t & d^t \end{pmatrix}$ is a pseudoinvolution on $M_{m+n}(F)$.

Replacing the involution $*$ in example 1.28 with the pseudoinvolutions of examples 1.29 and 1.30, we get unital Jordan bimodules over $M_{m+n}(F)$, where m, n are odd, and over $Q(n)^{(+)}$.

THEOREM 1.7 (see Martin and Piard (1992), Martínez *et al.* (2010)).– An arbitrary irreducible Jordan bimodule over $M_{m+n}^{(+)}$, $m + n \geq 3$ or $Q(n)^{(+)}$, $n \geq 3$, is one of examples 1.27 and 1.28 with an involution or a pseudoinvolution, or a regular bimodule.

The exceptional Jordan superalgebra K_{10} has rank 3. Jordan bimodules over K_{10} have been classified by Shtern (1987).

THEOREM 1.8 (Shtern (1987)).– All Jordan bimodules over K_{10} are completely reducible. The only irreducible Jordan bimodules over K_{10} are the regular bimodule and its opposite.

1.7.2. *Superalgebras of rank* ≤ 2

If J is a Jordan superalgebra of rank ≤ 2, then, generally speaking, it is no longer true that its universal multiplicative algebra is finite dimensional and that any Jordan bimodule is completely reducible.

1.7.2(a) In the case $J = Q(2)^{(+)}$, however, it is true (see Martínez *et al.* (2010)). The universal multiplicative enveloping algebra $U(Q(2)^{(+)})$ is finite dimensional and semisimple and the description of irreducible Jordan bimodules is similar to that of $Q(n)^{(+)}$, $n \geq 3$.

1.7.2(b) Let us discuss bimodules over Kantor superalgebras. Recall that the Kantor superalgebras $\mathrm{Kan}(n)$ are Kantor doubles of the Grassmann superalgebras $G(n)$, $n \geq 1$, with respect to the Poisson bracket

$$[f, g] = \sum_{i=1}^{n} (-1)^{|f|} \frac{\partial f}{\partial \xi_i} \frac{\partial g}{\partial \xi_i}.$$

Let $\mathrm{Kan}(n) = G(n) + G(n)v$. The Grassmann superalgebra $G(n)$ is embeddable in the associative commutative superalgebra $A = F[t, \xi_1, \ldots, \xi_n] = F[t] \otimes_F G(n)$.

For an arbitrary scalar $\alpha \in F$, the Poisson bracket $[\ ,\ ]$ extends to the Jordan bracket on A defined by $[t, \xi_i] = 0$, $[\xi_i, \xi_j] = -\delta_{ij}$, $[\xi_i, 1] = 0$, $[t, 1] = \alpha t$. The Kantor double $\mathrm{Kan}(n) = G(n) + G(n)v$ embeds in the Kantor double $\mathrm{Kan}(A, [\ ,\ ]) = A + Av$. The subspace $V(\alpha) = tG(n) + tG(n)v$ is an irreducible unital Jordan bimodule over $K(n)$. The square of the multiplication operator by the element v acts on $V(\alpha)$ as the scalar multiplication by α.

The simple superalgebras $\mathrm{Kan}(n)$, $n \geq 2$ are exceptional (see Martínez *et al.* (2001)). Therefore, they do not have non-zero one-sided Jordan bimodules.

THEOREM 1.9 (Stern (1995), Martínez and Zelmanov (2009), Solarte and Shestakov (2016)).– Every finite dimensional irreducible Jordan bimodule over $\mathrm{Kan}(n)$, $n \geq 2$ is isomorphic to $V(\alpha)$ or $V(\alpha)^{\mathrm{op}}$, $\alpha \in F$.

In Solarte and Shestakov (2016), the theorem above was proved for algebras over a field of characteristic $p > 2$.

1.7.2(c) Jordan superalgebras of a superform. Let $V = V_{\bar{0}} + V_{\bar{1}}$ be a $\mathbb{Z}/2\mathbb{Z}$-graded vector space with a non-degenerate supersymmetric bilinear form. Assume $\dim_F V_{\bar{0}} = m$, $\dim_F V_{\bar{1}} = 2n$ and choose a basis $e_1, \ldots, e_m$ in $V_{\bar{0}}$ with $\langle e_i, e_j \rangle = \delta_{ij}$ and a basis $v_1, w_1, \ldots, v_n, w_n$ in $V_{\bar{1}}$ such that

$$\langle v_i, w_j \rangle = \delta_{ij}, \quad \langle v_i, v_j \rangle = \langle w_i, w_j \rangle = 0, \quad 1 \leq i, j \leq n.$$

Let $\mathrm{Cl}(m)$ be the Clifford algebra of the restriction of the form $\langle\ ,\ \rangle$ to $V_{\bar{0}}$, and let

$$W_n = \langle 1, x_i, y_i, 1 \leq i \leq n \mid [x_i, y_j] = \delta_{ij}, \ [x_i, x_j] = [y_i, y_j] = 0 \rangle$$

be the simple Weyl algebra.

Then the tensor product $S = \mathrm{Cl}(m) \otimes_F W_n$ is the universal associative enveloping superalgebra of the Jordan superalgebra $J = V + F \cdot 1$.

Since the algebra W_n, $n \geq 1$ is infinite dimensional, it follows that the superalgebra J does not have non-zero finite dimensional one-sided Jordan bimodules unless $n = 0$.

Consider in the algebra S the chain of subspaces

$$\cdots \subseteq S_{-1} \subseteq S_0 = F \cdot 1 \subseteq S_1 \subseteq \ldots,$$

where $S_r = (0)$ for $r < 0$, $S_r = \sum_{i \leq r} \underbrace{V \cdots V}_{i}$ for $r \geq 1$. Clearly, $S = \bigcup_{r \geq 0} S_r$.

THEOREM 1.10 (Martin and Piard (1992)).–

1) For every $r \geq 1$, S_r/S_{r-2} is a unital irreducible Jordan bimodule over J.

2) Let $V' = Fu \oplus V$, where $|u| = 0$. Extend the bilinear form $\langle\ ,\ \rangle$ to V' via $\langle u, u \rangle = 1$, $\langle u, V \rangle = (0)$. Then for every even $r \geq 0$, the quotient uS_r/uS_{r-2} is a unital irreducible Jordan bimodule over J.

3) Every unital irreducible finite dimensional J-bimodule is isomorphic to S_r/S_{r-2} or to uS_r/uS_{r-2} for even r.

The classification of irreducible Jordan bimodules over $M_{1+1}(F)^{(+)}$, $D(t)$, K_3, $\mathrm{JP}(2)$ is too technical for an Encyclopedia survey. For a detailed description of finite dimensional irreducible Jordan bimodules, (see Martínez and Zelmanov (2003), Martin and Piard (1992), Martínez and Zelmanov (2006), Martínez and Shestakov (2020)). We will make only some general comments.

1.7.2(d) The universal associative enveloping superalgebra of $J = M_{1+1}(F)^{(+)}$ is infinite dimensional, and finite dimensional one-sided Jordan bimodules over J are not necessarily completely reducible.

There is a family of 4-dimensional unital Jordan J-bimodules $V(\alpha, \beta, \gamma)$, which are parameterized by scalars α, β, $\gamma \in F$. If $\gamma^2 - 1 - 4\alpha\beta \neq 0$, then the bimodule $V(\alpha, \beta, \gamma)$ is irreducible. If $\gamma^2 - 1 - 4\alpha\beta = 0$, then it has a composition series with 2-dimensional irreducible factors.

Every irreducible finite dimensional unital Jordan J-bimodule is isomorphic to $V(\alpha, \beta, \gamma)$, $\gamma^2 - 1 - 4\alpha\beta \neq 0$, or to a factor of a composition series of $V(\alpha, \beta, \gamma)$, $\gamma^2 - 1 - 4\alpha\beta = 0$ (see Martínez and Zelmanov (2009); Martínez and Shestakov (2020)).

1.7.2(e) Now let us discuss the superalgebras $D(t)$ and K_3. Recall that

$$D(t) = Fe_1 + Fe_2 + Fx + Fy;$$

$$e_1^2 = e_1, e_2^2 = e_2, e_1e_2 = 0, e_ix = \tfrac{1}{2}x, e_iy = \tfrac{1}{2}y, xy = e_1 + te_2.$$

Then $D(0) = F \cdot 1 + \mathrm{K}_3$, $D(-1) \cong M_{1+1}(F)^{(+)}$, $D(1)$ is a Jordan superalgebra of a superform.

We will assume therefore that $t \neq -1, 1$.

One-sided bimodules. The superalgebra K_3 does not have any non-zero one-sided Jordan bimodules (it has non-zero one-sided bimodules if char $F > 0$). All finite dimensional one-sided Jordan bimodules over $D(t)$, $t \neq -1, 1$, are completely reducible. The superalgebra $D(t)$ does not have non-zero one-sided bimodules unless $t = -\dfrac{m}{m+1}$ or $t = -\dfrac{m+1}{m}$, where $m \in \mathbb{Z}$, $m \geq 1$. For $t = -\dfrac{m}{m+1}$, there exists one (up to opposites) irreducible one-sided J-bimodule of dimension $2m+3$; for $t = -\dfrac{m+1}{m}$, there exists one (up to opposites) irreducible one-sided J-bimodule of dimension $2m+1$.

Unital bimodules. If $J = D(t)$ and t cannot be represented as $-\dfrac{m}{m+2}$, $m \in \mathbb{Z}$, $m \geq 0$ or $-\dfrac{m+2}{m}$, $m \in \mathbb{Z}$, $m \geq 1$, then there is one (up to opposites) series of irreducible finite dimensional unital J-bimodules parameterized by positive integers. All finite dimensional bimodules in this case are completely reducible.

If $t = -\dfrac{m+2}{m}$, $m \in \mathbb{Z}$, $m \geq 1$, or $t = -\dfrac{m}{m+2}$, $m \in \mathbb{Z}$, $m \geq 0$, then there is one (up to opposites) additional irreducible bimodule.

Remark. A finite dimensional irreducible bimodule over K_3 is a unital finite dimensional irreducible bimodule over $D(0)$. Hence the above description of unital finite dimensional irreducible bimodules over $D(0)$ applies to K_3.

The detailed description of irreducible and indecomposable $D(t)$-bimodules is contained in Martínez and Zelmanov (2003), Martínez and Zelmanov (2006). Trushina (2008) extended the description above to superalgebras over fields of positive characteristics.

1.7.2(f) Jordan bimodules over $\mathrm{JP}(2)$ Representation theory of $\mathrm{JP}(2)$ is essentially different from that of $\mathrm{JP}(n)$, $n \geq 3$.

The universal associative enveloping superalgebra $S(\mathrm{JP}(2))$ is isomorphic to $M_{2+2}(F[t])$, where $F[t]$ is the polynomial algebra in one variable (see Martínez and

Zelmanov (2003)). Hence, irreducible one-sided bimodules are parameterized by scalars $\alpha \in F$ and have dimension 4, whereas indecomposable bimodules are parameterized by Jordan blocks.

Let V be an irreducible finite dimensional bimodule over $\mathrm{JP}(2)$. Let $L = K(\mathrm{JP}(2))$ be the Tits–Kantor–Koecher Lie superalgebra of $\mathrm{JP}(2)$, $L = L_{-1} + L_0 + L_1$. The superalgebra L has one-dimensional center. Fix $0 \neq z \in L_0$, then $L/Fz \cong P(3)$ (see Martinez and Zelmanov (2001)). The Lie superalgebra L_0 acts on the module V (see Jacobson (1968); Martin and Piard (1992)), and the element z acts as a scalar multiplication.

DEFINITION 1.21.– *We say that V is a module of level $\alpha \in F$ if z acts on V as the scalar multiplication by α.*

For an arbitrary scalar $\alpha \in F$, there are exactly two (up to opposites) non-isomorphic unital irreducible finite dimensional Jordan bimodules over $\mathrm{JP}(2)$ of level α. For their explicit realization, see (Martínez and Zelmanov (2014)).

Kashuba and Serganova (2020) described indecomposable finite dimensional Jordan bimodules over $\mathrm{Kan}(n)$, $n \geq 1$ and $\mathrm{JP}(2)$.

1.8. Jordan superconformal algebras

In this section, we will discuss connections between infinite dimensional Jordan superalgebras and the so-called superconformal algebras that originated in mathematical physics.

In view of importance of the Virasoro algebra and (especially) its central extensions in physics, (Neveu and Schwarz 1971; Ramond 1971) and others considered superextensions of the algebra Vir. These superextensions became known as superconformal algebras. Kac and van de Leur (1989) put the theory on a more formal footing and recognized that all known superconformal algebras are in fact infinite dimensional superalgebras of Cartan type considered in Kac (1977b). Following (Kac and van de Leur 1989), we call a $\mathbb{Z}$-graded Lie superalgebra $L = \sum_{i \in \mathbb{Z}} L_i$ a *superconformal algebra* if

i) L is graded simple;

ii) $\mathrm{Vir} \subseteq L_{\bar{0}}$;

iii) the dimensions $\dim L_i$, $i \in \mathbb{Z}$ are uniformly bounded.

EXAMPLE 1.31.– The Lie superalgebra of superderivations

$$W(1:n) = \mathrm{Der}\,\mathbb{C}[t, t^{-1}, \xi_1, \ldots, \xi_n]$$

graded by degrees of t is a superconformal algebra.

EXAMPLE 1.32.– Let $\alpha \in \mathbb{C}$. Then $S(n, \alpha) = \{D \in W(1:n) \mid \operatorname{div}(t^\alpha D) = 0\} < W(1:n)$. Here if

$$D = f_0 \frac{\partial}{\partial t} + \sum_{i=1}^{n} f_i \frac{\partial}{\partial \xi_i}, \quad f_i \in \mathbb{C}[t, t^{-1}, \xi_1, \ldots, \xi_n]$$

then

$$\operatorname{div} D = \frac{\partial f_0}{\partial t} + \sum_{i=1}^{n} (-1)^{|f_i|} \frac{\partial f_i}{\partial \xi_i}.$$

If $\alpha \in \mathbb{Z}$, then $[S(n, \alpha), S(n, \alpha)]$ is a proper ideal in $S(n, \alpha)$ and the superalgebra $[S(n, \alpha), S(n, \alpha)]$ is simple. This family of superalgebras appeared in physics literature (Ademollo *et al.* 1976; Schwimmer and Seiberg 1987) under the name "SU_2-superconformal algebras".

EXAMPLE 1.33.– Consider the associative commutative superalgebra

$$\Lambda(1:n) = \mathbb{C}[t, t^{-1}, \xi_1, \ldots, \xi_n]$$

and the contact bracket $[\ ,\ ]$ of example 1.19. The superalgebra $K(n) = (\Lambda(1:n), [\ ,\ ])$ is simple unless $n = 4$. For $n = 4$, the commutator ideal $[K(4), K(4)]$ has codimension 1 and $[K(4), K(4)]$ is a simple superalgebra. This series is known in physics literature as "SO_n-superconformal algebras" (Ademollo *et al.* 1976; Schoutens 1987).

EXAMPLE 1.34.– In section 1.5, for an arbitrary associative commutative algebra Z with a derivation $d : Z \to Z$, we constructed the Jordan superalgebra $\mathrm{JCK}(Z, d)$. The Lie superalgebra $\mathrm{CK}(Z, d)$ is the Tits–Kantor–Koecher construction of $\mathrm{JCK}(Z, d)$. Taking $Z = \mathbb{C}[t, t^{-1}]$ and $d = \frac{\partial}{\partial t}$ or $d = t\frac{\partial}{\partial t}$, we get the Cheng–Kac superalgebras $\mathrm{CK}(6)$ discovered by Cheng and Kac (1997) and independently by Grozman *et al.* (2001).

Kac and van de Leur (1989) conjectured that examples 1.31–1.34 exhaust all superconformal algebras.

The superalgebras $K(n)$ and $\mathrm{CK}(6)$ appear as Tits–Kantor–Koecher constructions of Jordan superalgebras.

Let $[\,,\,]$ be the Jordan bracket on $\Lambda(1:n) = \mathbb{C}[t, t^{-1}, \xi_1, \ldots, \xi_n]$ of example 1.20, $J_n = K(\Lambda(1:n), [\,,\,])$ is the Kantor double of this bracket. Then

$$K(n+3) \cong \mathrm{TKK}(J_n), \quad n \geq 0.$$

In Kac *et al.* (2001), we classified "superconformal" Jordan superalgebras.

THEOREM 1.11 (Kac *et al.* (2001)).– Let $J = J_{\bar{0}} + J_{\bar{1}}$ be a $\mathbb{Z}$-graded Jordan superalgebra that is graded simple and the dimensions of all homogeneous components $\dim J_i$, $i \in \mathbb{Z}$ are uniformly bounded. Then either

1) J has finitely many negative (respectively, positive) non-zero homogeneous components;

2) or J is isomorphic to one of the superalgebras J_n, $\mathrm{JCK}(6)$, $n \geq 1$ or a twisted version of it.

This theorem agrees with the Kac–van de Leur conjecture on classification of superconformal algebras.

1.9. References

Ademollo, M., Brink, L., d'Adda, A., d'Auria, R., Napolitano, E., Scinto, S., Del Giudice, E., Di Vecchia, P., Ferrera, S., Gliozzi, F., Musto, R., Pettorino, R. (1976). Supersymmetric strings and color confinement. *Phys. Lett.*, 62B, 105–110.

Benkart, G., Elduque, A. (2002). A new construction of the Kac Jordan superalgebra. *Proc. Amer. Math. Soc.*, 130(11), 3209–3217.

Cantarini, N., Kac, V.G. (2007). Classification of linearly compact simple Jordan and generalized Poisson superalgebras. *J. Algebra*, 313(1), 100–124.

Cheng, S.-J., Kac, V.G. (1997). A new $N = 6$ superconformal algebra. *Comm. Math. Phys.*, 186(1), 219–231.

Grozman, P., Leites, D., Shchepochkina, I. (2001). Lie superalgebras of string theories. *Acta Math. Vietnam.*, 26(1), 27–63.

Jacobson, N. (1968). *Structure and Representations of Jordan Algebras*. American Mathematical Society Colloquium Publications, American Mathematical Society, Providence.

Kac, V.G. (1977a). Classification of simple Z-graded Lie superalgebras and simple Jordan superalgebras. *Comm. Algebra*, 5(13), 1375–1400.

Kac, V.G. (1977b). Lie superalgebras. *Adv. Math.*, 26(1), 8–96.

Kac, V.G., Martinez, C., Zelmanov, E. (2001). Graded simple Jordan superalgebras of growth one. *Mem. Amer. Math. Soc.*, 150(711).

Kac, V.G., van de Leur, J.W. (1989). On classification of superconformal algebras. In *Strings '88 (College Park, MD)*. World Science Publishing, Teaneck, 77–106.

Kantor, I.L. (1972). Certain generalizations of Jordan algebras. *Trudy Sem. Vektor. Tenzor. Anal.*, 16, 407–499.

Kantor, I.L. (1990). Connection between Poisson brackets and Jordan and Lie superalgebras. In *Lie Theory, Differential Equations and Representation Theory, Montreal, PQ, 1989*. University of Montréal, Montreal, 213–225.

Kashuba, I., Serganova, V. (2020). Representations of simple Jordan superalgebras. *Adv. Math.*, 370.

King, D., McCrimmon, K. (1992). The Kantor construction of Jordan superalgebras. *Comm. Algebra*, 20(1), 109–126.

Koecher, M. (1967). Imbedding of Jordan algebras into Lie algebras. I. *Amer. J. Math.*, 89, 787–816.

Martin, C., Piard, A. (1992). Classification of the indecomposable bounded admissible modules over the virasoro Lie algebra with weightspaces of dimension not exceeding two. *Comm. Math. Phys.*, 150(3), 465–493.

Martínez, C., Shestakov, I. (2020). Jordan bimodules over the superalgebra M_{1+1}. *Glasgow Math. J.*, 62(3). [Online].

Martínez, C., Shestakov, I., Zelmanov, E. (2001). Jordan superalgebras defined by brackets. *J. London Math. Soc.*, 64(2), 357–368.

Martínez, C., Shestakov, I., Zelmanov, E. (2010). Jordan bimodules over the superalgebras $P(n)$ and $Q(n)$. *Trans. Amer. Math. Soc.*, 362(4), 2037–2051.

Martinez, C., Zelmanov, E. (2001). Simple finite-dimensional Jordan superalgebras of prime characteristic. *J. Algebra*. 236(2), 575–629.

Martínez, C., Zelmanov, E. (2006). Unital bimodules over the simple Jordan superalgebra $D(t)$. *Trans. Amer. Math. Soc.*, 358(8), 3637–3649.

Martínez, C., Zelmanov, E. (2009). Jordan superalgebras and their representations. In *Algebras, Representations and Applications*. Futorney, V., Kac, V., Kashuba, I., Zelmanov, E. (eds). *American Mathematical Society*, Providence, 179–194.

Martínez, C., Zelmanov, E. (2010). Representation theory of Jordan superalgebras. I. *Trans. Amer. Math. Soc.*, 362(2), 815–846.

Martínez, C., Zelmanov, E. (2014). Irreducible representations of the exceptional Cheng-Kac superalgebra. *Trans. Amer. Math. Soc.*, 366(11), 5853–5876.

Martínez, C., Zelmanov, E.I. (2003). Lie superalgebras graded by $P(n)$ and $Q(n)$. *Proc. Natl. Acad. Sci. USA*, 100(14), 8130–8137.

Medvedev, Y.A., Zelmanov, E.I. (1992). Some counterexamples in the theory of Jordan algebras. In *Nonassociative Algebraic Models (Zaragoza, 1989)*. Nova Science Publishing, Commack, 1–16.

Neveu, A., Schwarz, J. (1971). Factorizable dual model of pions. *Nuclear Physics B*, 31(1), 86–112.

Racine, M.L., Zelmanov, E.I. (2003). Simple Jordan superalgebras with semisimple even part. *J. Algebra*, 270(2), 374–444.

Racine, M.L., Zelmanov, E.I. (2015). An octonionic construction of the Kac superalgebra K_{10}. *Proc. Amer. Math. Soc.*, 143(3), 1075–1083.

Ramond, P. (1971). Dual theory for free fermions. *Phys. Rev. D*, 3, 2415–2418.

Schoutens, K. (1987). A nonlinear representation of the $d = 2$ SO(4) extended superconformal algebra. *Phys. Lett.*, B194, 75–80.

Schwimmer, A., Seiberg, N. (1987). Comments on the n = 2,3,4 superconformal algebras in two dimensions. *Physics Letters B*, 184(2), 191–196.

Shestakov, I.P. (1997). Prime alternative superalgebras of arbitrary characteristic. *Algebra i Logika*. 36(6), 675–716, 722.

Shtern, A.S. (1987). Representations of an exceptional Jordan superalgebra. *Funktsional. Anal. i Prilozhen.*, 21(3), 93–94.

Solarte, O.F., Shestakov, I. (2016). Irreductible representations of the simple Jordan superalgebra of Grassmann Poisson bracket. *J. Algebra*, 455, 29–313.

Stern, A.S. (1995). Representations of finite-dimensional Jordan superalgebras of Poisson brackets. *Comm. Algebra*, 23(5), 1815–1823.

Tits, J. (1962). Une classe d'algèbres de Lie en relation avec les algèbres de Jordan. *Nederl. Akad. Wetensch. Proc.*, A 65, 24, 530–535.

Tits, J. (1966). Algèbres alternatives, algèbres de Jordan et algèbres de Lie exceptionnelles. I. Construction. *Nederl. Akad. Wetensch. Proc.*, A 69, 28, 223–237.

Trushina, M. (2008). Modular representations of the Jordan superalgebras $D(t)$ and K_3. *J. Algebra*, 320(4), 1327–1343.

Wall, C.T.C. (1963, 1964). Graded Brauer groups. *J. Reine Angew. Math.*, 213, 187–199.

Zelmanov, E. (2000). Semisimple finite dimensional Jordan superalgebras. In *Combinatorial and Computational Algebra*, Chan, K.Y., Mikhalev, A.A., Siu, M.-K., Yu, J.-T., Zelmanov, E. (eds). American Mathematical Society, Providence, 227–243.

Zhevlakov, K.A., Slin'ko, A.M., Shestakov, I.P., Shirshov, A.I. (1982). *Rings That are Nearly Associative* (Translated by H.F. Smith). Academic Press, Inc., New York.

2

Composition Algebras

Alberto ELDUQUE
Department of Mathematics, University of Zaragoza, Spain

2.1. Introduction

Unital composition algebras are the analogues of the classical algebras of the real and complex numbers, quaternions and octonions, but the class of composition algebras have been recently enriched with new algebras, mainly the so-called symmetric composition algebras, which play an effective role in understanding the triality phenomenon in dimension 8. The goal of this chapter is to examine the main definitions and results of these algebras.

Section 2.2 reviews the discovery of the real algebra of quaternions by Hamilton, surveys some of the applications of quaternions to deal with rotations in three- and four-dimensional euclidean spaces and then examines octonions, discovered shortly after Hamilton's breakthrough.

Section 2.3 is devoted to the classification of the unital composition algebras, also known as Hurwitz algebras. This is achieved by means of the Cayley–Dickson doubling process, which mimics the way in which Graves and Cayley constructed the octonions by doubling the quaternions. Hurwitz algebra itself, in 1898, proved that a positive definite quadratic form over the real numbers allows composition if and only if the dimension is restricted to $1, 2, 4$ or 8. These are the possible dimensions of Hurwitz algebras over arbitrary fields, and of finite-dimensional non-unital composition algebras. Some attention is paid to the case of Hurwitz algebras with

Algebra and Applications 1,
coordinated by Abdenacer MAKHLOUF.

isotropic norm, as this is instrumental in defining Okubo algebras later on, although the model of Zorn's vector matrices is not touched upon.

In section 2.4, symmetric composition algebras are defined. In dimension > 1, these are non-unital composition algebras, satisfying the extra condition of "associativity of the norm": $n(x * y, z) = n(x, y * z)$. The interest lies in dimension 8, where these algebras split into two disjoint families: para-Hurwitz algebras and Okubo algebras. The existence of the latter Okubo algebras justifies the introduction of the symmetric composition algebras.

Formulas for triality are simpler if one uses symmetric composition algebras instead of the classical Hurwitz algebras. This is examined in section 2.5. Triality is a broad subject. In projective geometry, there is duality relating points and hyperplanes. Given a vector space of dimension 8, endowed with a quadratic form q with maximal Witt index, the quadric of isotropic vectors $Q = \{\mathbb{F}v : v \in V, q(v) = 0\}$ in projective space $\mathbb{P}(V)$ contains points, lines, planes and "solids", and there are two kinds of "solids". Geometric triality relates points and the two kinds of solids in a cyclic way. This goes back to Study (1913) and Cartan (1925). Tits (1959) showed that there are two different types of geometric trialities; one is related to octonions (or para-octonions) and the exceptional group G_2, while the other is related to the classical groups of type A_2, unless the characteristic is 3. The algebras hidden behind this second type are the Okubo algebras. From the algebraic point of view, triality relates the natural and spin representations of the spin group on an eight-dimensional quadratic space, that is the three irreducible representations corresponding to the outer vertices of the Dynkin diagram D_4. Here, we will consider lightly the local version of triality, which gives a very symmetric construction of Freudenthal's magic square.

2.2. Quaternions and octonions

It is safe to say that the history of composition algebras starts with the discovery of the real quaternions by Hamilton.

2.2.1. *Quaternions*

Real and complex numbers are in the toolkit of any scientist. Complex numbers correspond to vectors in a Euclidean plane, so that addition and multiplication by real scalars are the natural ones for vectors. The length (or modulus, or norm) of the product of two complex numbers is the product of the lengths of the factors. In this way, multiplication by a norm 1 complex number is an isometry, actually a rotation, of the plane, and this allows us to identify the group of rotations of the Euclidean plane, that is, the special orthogonal group $SO_2(\mathbb{R})$, with the set of norm one complex numbers (the unit circle):

$$SO_2(\mathbb{R}) \simeq \{z \in \mathbb{C} : |z| = 1\} \simeq S^1.$$

In 1835, William Hamilton posed himself the problem of extending the domain of complex numbers to a system of numbers "of dimension 3". He tried to find a multiplication, analogous to the multiplication of complex numbers, but in dimension 3, that should respect the "law of moduli":

$$|z_1 z_2| = |z_1||z_2|. \qquad [2.1]$$

That is, he tried to get a formula for

$$(a + b\mathbf{i} + c\mathbf{j})(a' + b'\mathbf{i} + c'\mathbf{j}) = ??$$

and assumed $\mathbf{i}^2 = \mathbf{j}^2 = -1$. The problem is how to define $\mathbf{ij}$ and $\mathbf{ji}$.

With some "modern" insight, it is easy to see that this is impossible. A *non-associative* (i.e. not necessarily associative) *algebra* over a field $\mathbb{F}$ is just a vector space $\mathcal{A}$ over $\mathbb{F}$, endowed with a bilinear map (the multiplication) $m : \mathcal{A} \times \mathcal{A} \to \mathcal{A}$, $(x, y) \mapsto xy$. We will refer to the algebra $(\mathcal{A}, m)$ or simply $\mathcal{A}$ if no confusion arises. The algebra $\mathcal{A}$ is said to be a *division algebra* if the left and right multiplications, $L_x : y \mapsto xy$, $R_x : y \mapsto yx$, are bijective for any non-zero $x \in \mathcal{A}$. The "law of moduli" forces the algebra sought for by Hamilton to be a division algebra. But this is impossible.

PROPOSITION 2.1.– There are no real division algebras of odd dimension ≥ 3.

PROOF (See Petersson (2005)).– Given a real algebra $\mathcal{A}$ of odd dimension ≥ 3, and linearly independent elements $x, y \in \mathcal{A}$ with L_x and L_y bijective, $\det(L_x + tL_y)$ is a polynomial in t of odd degree, and hence it has a real root λ. Thus, the left multiplication $L_{x+\lambda y}$ is not bijective. □

Therefore, Hamilton could not succeed. But, after years of struggle, he discovered how to overcome the difficulties in dimension 3 by making a leap to dimension 4.

In a letter (Wilkins 2020) to his son Archibald, dated August 5, 1865, Hamilton explained how vividly he remembered the date: *October 16, 1843*, when he got the key idea. He wrote:

> Every morning in the early part of the above-cited month, on my coming down to breakfast, your (then) little brother William Edwin, and yourself, used to ask me, "Well, Papa, can you multiply triplets"? Whereto I was always obliged to reply, with a sad shake of the head: "No, I can only add and subtract them."
>
> But on the 16th day of the same month – which happened to be a Monday, and a Council day of the Royal Irish Academy – I was walking in to attend and preside, and your mother was walking with me, along the Royal Canal, to which she had perhaps driven; although she talked with me now and then, yet an undercurrent of thought was going on in my mind, which gave at last a result, whereof it is not too much to say that I felt at once the importance. An electric circuit seemed to close; and a spark flashed forth, the herald (as I foresaw, immediately) of many long years to come of definitely directed thought and work, by myself if spared, and at all events on the part of others, if I should even be allowed to live long enough distinctly to communicate the discovery.
>
> Nor could I resist the impulse – unphilosophical as it may have been – to cut with a knife on a stone of Brougham Bridge, as we passed it, the fundamental formula with the symbols, **i**, **j**, **k**, namely,
>
> $$\mathbf{i}^2 = \mathbf{j}^2 = \mathbf{k}^2 = \mathbf{ijk} = -1$$
>
> which contains the solution of the problem, but of course, as an inscription, has long since mouldered away. A more durable notice remains, however, on the Council Books of the Academy for that day (October 16, 1843), which records the fact, that I then asked for and obtained leave to read a Paper on 'Quaternions', at the first general meeting of the session: which reading took place accordingly, on Monday, November 13 following.

Hamilton realized that the product $\mathbf{ij}$ had to be linearly independent of $1, \mathbf{i}, \mathbf{j}$, and hence he defined the real algebra of *quaternions* as:

$$\mathbb{H} = \mathbb{R}1 \oplus \mathbb{R}\mathbf{i} \oplus \mathbb{R}\mathbf{j} \oplus \mathbb{R}\mathbf{k}$$

with multiplication determined by

$$\mathbf{i}^2 = \mathbf{j}^2 = \mathbf{k}^2 = -1,$$

$$\mathbf{ij} = -\mathbf{ji} = \mathbf{k}, \quad \mathbf{jk} = -\mathbf{kj} = \mathbf{i}, \quad \mathbf{ki} = -\mathbf{ik} = \mathbf{j}.$$

Some properties of this algebra are summarized as follows:

– For any $q_1, q_2 \in \mathbb{H}$, $|q_1 q_2| = |q_1||q_2|\ \forall q_1, q_2 \in \mathbb{H}$, where for $q = a+b\mathbf{i}+c\mathbf{j}+c\mathbf{k}$, $|a + b\mathbf{i} + c\mathbf{j} + d\mathbf{k}|^2 = a^2 + b^2 + c^2 + d^2$ is the standard Euclidean norm on $\mathbb{H} \simeq \mathbb{R}^4$.

– $\mathbb{H}$ is an associative division algebra, but we lose the commutativity valid for the real and complex numbers.

Therefore, $S^3 \simeq \{q \in \mathbb{H} : |q| = 1\}$ is a (Lie) group. This implies the parallelizability of the three-dimensional sphere S^3.

– $\mathbb{H}$ splits as $\mathbb{H} = \mathbb{R}1 \oplus \mathbb{H}_0$, where $\mathbb{H}_0 = \mathbb{R}\mathbf{i} \oplus \mathbb{R}\mathbf{j} \oplus \mathbb{R}\mathbf{k}$ is the subspace of *purely imaginary quaternions.*

Then, for any $u, v \in \mathbb{H}_0$:

$$uv = -u \bullet v + u \times v$$

where $u \bullet v$ and $u \times v$ denote the usual scalar and cross-products in $\mathbb{R}^3 \simeq \mathbb{H}_0$.

– For any $q = a1 + u \in \mathbb{H}$ ($u \in \mathbb{H}_0$), $q^2 = (a^2 - u \bullet u) + 2au$, so

$$q^2 - \mathrm{tr}(q)q + \mathrm{n}(q)1 = 0 \qquad [2.2]$$

with $\mathrm{tr}(q) = 2a$ and $\mathrm{n}(q) = |q|^2 = a^2 + u \bullet u$. That is, $\mathbb{H}$ is a *quadratic algebra.*

– The map $q = a + u \mapsto \overline{q} = a - u$ is an involution, with $q + \overline{q} = \mathrm{tr}(q) = 2a \in \mathbb{R}$ and $q\overline{q} = \overline{q}q = \mathrm{n}(q) = |q|^2 \in \mathbb{R}$.

– $\mathbb{H} = \mathbb{C} \oplus \mathbb{C}\mathbf{j} \simeq \mathbb{C}^2$ is a two-dimensional vector space over $\mathbb{C}$. The multiplication is then given by:

$$(p_1 + p_2\mathbf{j})(q_1 + q_2\mathbf{j}) = (p_1 q_1 - \overline{q_2} p_2) + (q_2 p_1 + p_2 \overline{q_1})\mathbf{j} \qquad [2.3]$$

for any $p_1, p_2, q_1, q_2 \in \mathbb{C}$.

2.2.2. *Rotations in three- (and four-) dimensional space*

Given a norm 1 quaternion q, there is an angle $\alpha \in [0, \pi]$ and a norm 1 imaginary quaternion u such that $q = (\cos\alpha)1 + (\sin\alpha)u$.

Consider the linear map:

$$\begin{aligned} \varphi_q : \mathbb{H}_0 &\longrightarrow \mathbb{H}_0, \\ x &\mapsto qxq^{-1} = qx\overline{q}. \end{aligned}$$

Complete u to an orthonormal basis $\{u, v, u \times v\}$. A simple computation gives:

$$\varphi_q(u) = quq^{-1} = u \qquad (\text{as } uq = qu),$$

$$\varphi_q(v) = \big((\cos\alpha)1 + (\sin\alpha)u\big)v\big((\cos\alpha)1 - (\sin\alpha)u\big)$$

$$
\begin{aligned}
&= ((\cos\alpha)v + (\sin\alpha)u \times v)((\cos\alpha)1 - (\sin\alpha)u) \\
&= (\cos^2\alpha)v + 2(\cos\alpha\sin\alpha)u \times v - (\sin^2\alpha)(u \times v) \times u \\
&= (\cos 2\alpha)v + (\sin 2\alpha)u \times v, \\
\varphi_q(u \times v) &= \ldots = -(\sin 2\alpha)v + (\cos 2\alpha)u \times v.
\end{aligned}
$$

Thus, the coordinate matrix of φ_q relative to the basis $\{u, v, u \times v\}$ is

$$
\begin{pmatrix} 1 & 0 & 0 \\ 0 & \cos 2\alpha & -\sin 2\alpha \\ 0 & \sin 2\alpha & \cos 2\alpha \end{pmatrix}
$$

In other words, φ_q is a rotation around the semi-axis $\mathbb{R}^+ u$ of angle 2α, and hence the map

$$
\begin{aligned}
\varphi : S^3 \simeq \{q \in \mathbb{H} : |q| = 1\} &\longrightarrow \mathrm{SO}_3(\mathbb{R}), \\
q &\mapsto \varphi_q
\end{aligned}
$$

is a surjective (Lie) group homomorphism with $\ker\varphi = \{\pm 1\}$. We thus obtain the isomorphism

$$
S^3/_{\{\pm 1\}} \simeq \mathrm{SO}_3(\mathbb{R})
$$

Actually, the group S^3 is the universal cover of $\mathrm{SO}_3(\mathbb{R})$.

Therefore, we get that rotations can be identified with conjugation by norm 1 quaternions modulo ± 1. The outcome is that it is quite easy now to compose rotations in three-dimensional Euclidean space, as it is enough to multiply norm 1 quaternions: $\varphi_p\varphi_q = \varphi_{pq}$. From this, one can very easily deduce the 1840 formulas by Olinde Rodrigues (Rodrigues 1840) for the composition of rotations.

But there is more about rotations and quaternions.

For any $p \in \mathbb{H}$ with $\mathrm{n}(p) = 1$, the left (respectively, right) multiplication L_p (respectively, R_p) by p is an isometry, due to the multiplicativity of the norm. Using [2.2], it follows that the characteristic polynomial of L_p and R_p is $\left(x^2 - \mathrm{tr}(p)x + 1\right)^2$ and, in particular, the determinant of the multiplication by p is 1, so both L_p and R_p are rotations.

Now, if ψ is a rotation in $\mathbb{R}^4 \simeq \mathbb{H}$, $a = \psi(1)$ is a norm 1 quaternion, and

$$
L_{\bar{a}}\psi(1) = \bar{a}a = \mathrm{n}(a) = 1,
$$

so the composition $L_{\overline{a}}\psi$ is actually a rotation in $\mathbb{R}^3 \simeq \mathbb{H}_0$. Hence, there is a norm 1 quaternion $q \in \mathbb{H}$ such that

$$\overline{a}\psi(x) = qxq^{-1}$$

for any $x \in \mathbb{H}$. That is, for any $x \in \mathbb{H}$,

$$\psi(x) = (aq)xq^{-1}$$

It follows that the map

$$\begin{aligned} \Psi : S^3 \times S^3 &\longrightarrow \mathrm{SO}_4(\mathbb{R}), \\ (p, q) &\mapsto \psi_{p,q} : x \mapsto pxq^{-1} \end{aligned}$$

is a surjective (Lie) group homomorphism with $\ker \Psi = \{\pm(1,1)\}$. We thus obtain the isomorphism

$$S^3 \times S^3/_{\{\pm(1,1)\}} \simeq \mathrm{SO}_4(\mathbb{R})$$

and from here we get the isomorphism $\mathrm{SO}_3(\mathbb{R}) \times \mathrm{SO}_3(\mathbb{R}) \simeq \mathrm{PSO}_4(\mathbb{R})$.

Again, this means that it is easy to compose rotations in four-dimensional space, as it reduces to multiplying pairs of norm 1 quaternions: $\psi_{p_1,q_1}\psi_{p_2,q_2} = \psi_{p_1p_2,q_1q_2}$.

2.2.3. *Octonions*

In a letter from Graves to Hamilton, dated October 26, 1843, only a few days after the "discovery" of quaternions, Graves writes:

> There is still something in the system which gravels me. I have not yet any clear views as to the extent to which we are at liberty arbitrarily to create imaginaries, and to endow them with supernatural properties.
>
> If with your alchemy you can make three pounds of gold, why should you stop there?

Actually, as we have seen in [2.3], the algebra of quaternions is obtained by doubling suitably the field of complex numbers: $\mathbb{H} = \mathbb{C} \oplus \mathbb{C}\mathbf{j}$.

Doubling again we get the *octonions* (Graves–Cayley):

$$\mathbb{O} = \mathbb{H} \oplus \mathbb{H}\mathbf{l} = \mathrm{span}_{\mathbb{R}} \{1, \mathbf{i}, \mathbf{j}, \mathbf{k}, \mathbf{l}, \mathbf{il}, \mathbf{jl}, \mathbf{kl}\}$$

with multiplication mimicked from [2.2]:

$$(p_1 + p_2\mathbf{l})(q_1 + q_2\mathbf{l}) = (p_1q_1 - \overline{q_2}p_2) + (q_2p_1 + p_2\overline{q_1})\mathbf{l}$$

and usual norm: $\mathrm{n}(p_1 + p_2\mathbf{l}) = \mathrm{n}(p_1) + \mathrm{n}(p_2)$, for $p_1, p_2, q_1, q_2 \in \mathbb{H}$.

This was already known to Graves, who wrote a letter to Hamilton on December 26, 1843 with his discovery of what he called *octaves*. Hamilton promised to announce Graves' discovery to the Irish Royal Academy, but did not do it in time. In 1845, independently, Cayley discovered the octonions and got the credit. Octonions are also called *Cayley numbers*.

Some properties of this new algebra of octonions are summarized here:

– The norm is multiplicative: $\mathrm{n}(xy) = \mathrm{n}(x)\mathrm{n}(y)$, for any $x, y \in \mathbb{O}$.

– $\mathbb{O}$ is a division algebra, and it is neither commutative nor associative!

But it is *alternative*, that is, any two elements generate an associative subalgebra.

A theorem by Zorn (1933) asserts that the only finite-dimensional real alternative division algebras are $\mathbb{R}$, $\mathbb{C}$, $\mathbb{H}$ and $\mathbb{O}$. And hence, as proved by Frobenius (1878), the only such associative algebras are $\mathbb{R}$, $\mathbb{C}$ and $\mathbb{H}$.

– The seven-dimensional Euclidean sphere $S^7 \simeq \{x \in \mathbb{O} : \mathrm{n}(x) = 1\}$ is not a group (associativity fails), but it constitutes the most important example of a *Moufang loop*.

– As for $\mathbb{H}$, for any two *imaginary octonions* $u, v \in \mathbb{O}_0 = \mathrm{span}_{\mathbb{R}}\{\mathbf{i}, \mathbf{j}, \mathbf{k}, \mathbf{l}, \mathbf{il}, \mathbf{jl}, \mathbf{kl}\}$ we have:

$$uv = -u \bullet v + u \times v.$$

for the usual scalar product $u \bullet v$ on $\mathbb{R}^7 \simeq \mathbb{O}_0$, and where $u \times v$ defines the usual cross-product in $\mathbb{R}^7$. This satisfies the identity $(u \times v) \times v = (u \bullet v)v - (v \bullet v)u$, for any $u, v \in \mathbb{R}^7$.

– $\mathbb{O}$ is again a *quadratic algebra*: $x^2 - \mathrm{tr}(x)x + \mathrm{n}(x)1 = 0$ for any $x \in \mathbb{O}$, where $\mathrm{tr}(x) = x + \overline{x}$ and $\mathrm{n}(x) = x\overline{x} = \overline{x}x$, where for $x = a1 + u$, $a \in \mathbb{R}$, $u \in \mathbb{O}_0$, $\overline{x} = a1 - u$.

And, as it happens for quaternions, octonions are also present in many interesting geometrical situations, here we mention a few:

– the groups Spin_7 and Spin_8 (universal covers of $\mathrm{SO}_7(\mathbb{R})$ and $\mathrm{SO}_8(\mathbb{R})$) can be described easily in terms of octonions;

– the fact that $\mathbb{O}$ is a division algebra implies the parallelizability of the seven-dimensional sphere S^7. Actually, S^1, S^3 and S^7 are the only parallelizable spheres (Adams 1958; Bott and Milnor 1958; Kervaire 1958);

– the six-dimensional sphere can be identified with the set of norm 1 imaginary units: $S^6 \simeq \{x \in \mathbb{O}_0 : \mathrm{n}(x) = 1\}$, and it is endowed with an *almost complex structure*, inherited from the multiplication of octonions.

S^2 and S^6 are the only spheres with such structures (Borel and Serre (1953));

– contrary to what happens in higher dimensions, projective planes do not need to be desarguesian. The simplest example of a non-desarguesian projective plane is the octonionic projective plane $\mathbb{O}P^2$.

David R. Wilkins has compiled a large amount of material on the work of Hamilton[1], and for complete expositions on quaternions and octonions, the interested reader may consult Ebbinghaus *et al.* (1991) and Conway and Smith (2003).

2.3. Unital composition algebras

Composition algebras constitute a generalization of the classical algebras of the real, complex, quaternion and octonion numbers.

A quadratic form $\mathrm{n} : V \to \mathbb{F}$ on a vector space V over a field $\mathbb{F}$ is said to be *non-degenerate* if its polar form

$$\mathrm{n}(x, y) := \mathrm{n}(x + y) - \mathrm{n}(x) - \mathrm{n}(y),$$

is so, that is, if its radical $V^\perp := \{v \in V : \mathrm{n}(v, V) = 0\}$ is trivial. Moreover, n is said to be *non-singular* either if it is non-degenerate or if it satisfies that the dimension of $V^\perp$ is 1 and $\mathrm{n}(V^\perp) \neq 0$. The last possibility only occurs over fields of characteristic 2.

DEFINITION 2.1.– *A* composition algebra *over a field* $\mathbb{F}$ *is a triple* $(\mathcal{C}, \cdot, \mathrm{n})$ *where*

– $(\mathcal{C}, \cdot)$ *is a non-associative algebra;*

– $\mathrm{n} : \mathcal{C} \to \mathbb{F}$ *is a non-singular quadratic form that is* multiplicative, *that is,*

$$\mathrm{n}(x \cdot y) = \mathrm{n}(x)\mathrm{n}(y) \qquad [2.4]$$

for any $x, y \in \mathcal{C}$.

The unital composition algebras are called *Hurwitz algebras*.

1 https://www.maths.tcd.ie/pub/HistMath/People/Hamilton/.

For simplicity, we will usually refer to the composition algebra $\mathcal{C}$.

Our goal in this section is to prove that Hurwitz algebras are quite close to $\mathbb{R}$, $\mathbb{C}$, $\mathbb{H}$ and $\mathbb{O}$.

By linearization of [2.4], we obtain:

$$\begin{aligned} &\mathrm{n}(x \cdot y, x \cdot z) = \mathrm{n}(x)\mathrm{n}(y, z), \\ &\mathrm{n}(x \cdot y, t \cdot z) + \mathrm{n}(t \cdot y, x \cdot z) = \mathrm{n}(x, t)\mathrm{n}(y, z), \end{aligned} \qquad [2.5]$$

for any $x, y, z, t \in \mathcal{C}$.

PROPOSITION 2.2.– Let $(\mathcal{C}, \cdot, \mathrm{n})$ be a Hurwitz algebra:

– either n is non-degenerate or $\mathrm{char}\,\mathbb{F} = 2$ and $\mathcal{C}$ is isomorphic to the ground field $\mathbb{F}$ (with norm $\alpha \mapsto \alpha^2$);

– the map $x \mapsto \bar{x} := \mathrm{n}(1, x)1 - x$ is an involution. That is, $\bar{\bar{x}} = x$ and $\overline{x \cdot y} = \bar{y} \cdot \bar{x}$ for any $x, y \in \mathcal{C}$. This involution is referred to as the *standard conjugation*;

– if $*$ denotes the conjugation of a linear endomorphism relative to n (i.e. $\mathrm{n}\big(f(x), y\big) = \mathrm{n}\big(x, f^*(y)\big)$ for any x, y), then for the left and right multiplications by elements $x \in \mathcal{C}$ we have $L_x^* = L_{\bar{x}}$ and $R_x^* = R_{\bar{x}}$;

– any $x \in \mathcal{C}$ satisfies the *Cayley–Hamilton equation*:

$$x^{\cdot 2} - \mathrm{n}(x, 1)x + \mathrm{n}(x)1 = 0$$

– $(\mathcal{C}, \cdot)$ is an alternative algebra: $x \cdot (x \cdot y) = x^{\cdot 2} \cdot y$ and $(y \cdot x) \cdot x = y \cdot x^{\cdot 2}$ for any $x, y \in \mathcal{C}$.

PROOF.– Plug $t = 1$ in [2.5] to get

$$\mathrm{n}(x \cdot y, z) = \mathrm{n}\big(y, (\mathrm{n}(x, 1)1 - x)z\big) = \mathrm{n}(y, \bar{x} \cdot z)$$

and symmetrically we get $\mathrm{n}(y \cdot x, z) = \mathrm{n}(y, z \cdot \bar{x})$.

Now, if $\mathrm{char}\,\mathbb{F} = 2$ and $\mathcal{C}^\perp = \mathbb{F}a$, with $\mathrm{n}(a) \neq 0$, then for any $x, y \in \mathcal{C}$, $\mathrm{n}(a \cdot x, y) = n(a, y \cdot \bar{x}) = 0$, so $a \cdot x \in \mathcal{C}^\perp$ and $a \cdot x = f(x)a$ for a linear map $f : \mathcal{C} \to \mathbb{F}$. But $\mathrm{n}(a)\mathrm{n}(x) = \mathrm{n}(a \cdot x) = f(x)^2\mathrm{n}(a)$. Hence $\mathrm{n}(x) = f(x)^2$ for any x and, by linearization, $\mathrm{n}(x, y) = 2f(x)f(y) = 0$ for any x, y. We conclude that $\mathcal{C} = \mathcal{C}^\perp = \mathbb{F}1$. In this case, all the assertions are trivial.

Assuming hence that $\mathcal{C}^\perp = 0$ (n is non-degenerate), since $x \mapsto \bar{x}$ is an isometry of order 2 (reflection relative to $\mathbb{F}1$) we get

$$\mathrm{n}(\overline{x \cdot y}, z) = \mathrm{n}(x \cdot y, \bar{z}) = \mathrm{n}(x, \bar{z} \cdot \bar{y}) = \mathrm{n}(z \cdot x, \bar{y}) = \mathrm{n}(z, \bar{y} \cdot \bar{x})$$

for any x, y, z, whence $\overline{x \cdot y} = \bar{y} \cdot \bar{x}$.

Finally, for any x, y, z, using again [2.5],

$$\mathrm{n}(x)\mathrm{n}(y, z) = \mathrm{n}(x \cdot y, x \cdot z) = \mathrm{n}(\overline{x} \cdot (x \cdot y), z)$$

so that, using left-right symmetry:

$$\overline{x} \cdot (x \cdot y) = \mathrm{n}(x)y = (y \cdot x) \cdot \overline{x}. \qquad [2.6]$$

With $y = 1$, this gives $\overline{x} \cdot x = \mathrm{n}(x)1$ and hence the Cayley–Hamilton equation. However, $\overline{x} \cdot (x \cdot y) = \mathrm{n}(x)y = (\overline{x} \cdot x) \cdot y$ and this shows $x \cdot (x \cdot y) = x^{\cdot 2} \cdot y$. Symmetrically we get $(y \cdot x) \cdot x = y \cdot x^{\cdot 2}$. □

2.3.1. *The Cayley–Dickson doubling process and the generalized Hurwitz theorem*

Let $(\mathcal{C}, \cdot, \mathrm{n})$ be a Hurwitz algebra, and assume that $\mathcal{Q}$ is a proper unital subalgebra of $\mathcal{C}$ such that the restriction of n to $\mathcal{Q}$ is non-degenerate. Our goal is to show that in this case $\mathcal{C}$ also contains a subalgebra obtained by "doubling" $\mathcal{Q}$, in a way similar to the construction of $\mathbb{H}$ from two copies of $\mathbb{C}$, or the construction of $\mathbb{O}$ from two copies of $\mathbb{H}$.

By non-degeneracy of n, $\mathcal{C} = \mathcal{Q} \oplus \mathcal{Q}^{\perp}$. Pick $u \in \mathcal{Q}^{\perp}$ with $\mathrm{n}(u) \neq 0$, and let $\alpha = -\mathrm{n}(u)$. As $1 \in \mathcal{Q}$, $\mathrm{n}(u, 1) = 0$ and hence $\overline{u} = -u$ and $u^{\cdot 2} = \alpha 1$ by the Cayley–Hamilton equation (proposition 2.2). This also implies that $R_u^2 = \alpha \mathrm{id}$, so the right multiplication R_u is bijective.

LEMMA 2.1.– Under the conditions above, the subspaces $\mathcal{Q}$ and $\mathcal{Q} \cdot u$ are orthogonal (i.e. $\mathcal{Q} \cdot u \subseteq \mathcal{Q}^{\perp}$) and the following properties hold for any $x, y \in \mathcal{Q}$:

1) $x \cdot u = u \cdot \overline{x}$;

2) $x \cdot (y \cdot u) = (y \cdot x) \cdot u$;

3) $(y \cdot u) \cdot x = (y \cdot \overline{x}) \cdot u$;

4) $(x \cdot u) \cdot (y \cdot u) = \alpha \overline{y} \cdot x$.

PROOF.– For any $x, y \in \mathcal{Q}$, $\mathrm{n}(x, y \cdot u) = \mathrm{n}(\overline{y} \cdot x, u) \in \mathrm{n}(\mathcal{Q}, u) = 0$, so $\mathcal{Q} \cdot u$ is a subspace orthogonal to $\mathcal{Q}$.

From $\mathrm{n}(x \cdot u, 1) = \mathrm{n}(u, \overline{x}) \in \mathrm{n}(u, \mathcal{Q}) = 0$, it follows that $\overline{x \cdot u} = -x \cdot u$. But $\overline{x \cdot u} = \overline{u} \cdot \overline{x} = -u \cdot \overline{x}$, whence $x \cdot u = u \cdot \overline{x}$.

Now $x \cdot (y \cdot u) = -x \cdot (\overline{y \cdot u}) = -x \cdot (\overline{u} \cdot \overline{y})$ and this is equal, because of [2.6], to $u \cdot (\overline{x} \cdot \overline{y}) = u \cdot (\overline{y \cdot x}) = (y \cdot x) \cdot u$.

In a similar vein, $(y \cdot u) \cdot x = -(y \cdot \overline{u}) \cdot x = (y \cdot \overline{x}) \cdot u$, and $(x \cdot u) \cdot (y \cdot u) = -(\overline{x \cdot u}) \cdot (y \cdot u) = \overline{y}\big((x \cdot u) \cdot u\big) = \overline{y} \cdot (x \cdot u^{\cdot 2}) = \alpha \overline{y} \cdot x$. □

Therefore, the subspace $\mathcal{Q} \oplus \mathcal{Q} \cdot u$ is also a subalgebra, and the restriction of n to it is non-degenerate. The multiplication and norm are given by (compared to [2.3]):

$$\begin{aligned} &(a + b \cdot u) \cdot (c + d \cdot u) = (a \cdot c + \alpha \overline{d} \cdot b) + (d \cdot a + b \cdot \overline{c}) \cdot u, \\ &\mathrm{n}(a + b \cdot u) = \mathrm{n}(a) - \alpha \mathrm{n}(b), \end{aligned} \qquad [2.7]$$

for any $a, b, c, d \in \mathcal{Q}$.

Moreover,

$$\mathrm{n}\big((a + b \cdot u) \cdot (c + d \cdot u)\big) = \mathrm{n}(a \cdot c + \alpha \overline{d} \cdot b) - \alpha \mathrm{n}(d \cdot a + b \cdot \overline{c}),$$

while on the other hand

$$\begin{aligned} \mathrm{n}(a + b \cdot u)\mathrm{n}(c + d \cdot u) &= \big(\mathrm{n}(a) - \alpha \mathrm{n}(b)\big)\big(\mathrm{n}(c) - \alpha \mathrm{n}(d)\big) \\ &= \mathrm{n}(a)\mathrm{n}(c) + \alpha^2 \mathrm{n}(b)\mathrm{n}(d) - \alpha\big(\mathrm{n}(d)\mathrm{n}(a) + \mathrm{n}(b)\mathrm{n}(c)\big) \\ &= \mathrm{n}(a \cdot c) + \alpha^2 \mathrm{n}(\overline{d} \cdot b) - \alpha\big(\mathrm{n}(d \cdot a) + \mathrm{n}(b \cdot \overline{c})\big). \end{aligned}$$

We conclude that $\mathrm{n}(a \cdot c, \alpha \overline{d} \cdot b) - \alpha \mathrm{n}(d \cdot a, b \cdot \overline{c}) = 0$, or $\mathrm{n}\big(d \cdot (a \cdot c), b\big) = \mathrm{n}\big((d \cdot a) \cdot c, b\big)$.

The non-degeneracy of the restriction of n to $\mathcal{Q}$ implies that $\mathcal{Q}$ is associative. In particular, any proper subalgebra of $\mathcal{C}$ with non-degenerate restricted norm is associative.

Conversely, given an *associative* Hurwitz algebra $\mathcal{Q}$ with non-degenerate n, and a non-zero scalar $\alpha \in \mathbb{F}$, consider the direct sum of two copies of $\mathcal{Q}$: $\mathcal{C} = \mathcal{Q} \oplus \mathcal{Q} \cdot u$, with multiplication and norm given by [2.7], extending those on $\mathcal{Q}$. The arguments above show that $(\mathcal{C}, \cdot, \mathrm{n})$ is again a Hurwitz algebra, which is said to be obtained by the *Cayley–Dickson doubling process* from $(\mathcal{Q}, \cdot, \mathrm{n})$ and α. This algebra is denoted by $\mathfrak{CD}(\mathcal{Q}, \alpha)$.

REMARK 2.1.– $\mathfrak{CD}(\mathcal{Q}, \alpha)$ is associative if and only if $\mathcal{Q}$ is commutative. This follows from $x \cdot (y \cdot u) = (y \cdot x) \cdot u$. If the algebra is associative, this equals $(x \cdot y) \cdot u$, and it forces $x \cdot y = y \cdot x$ for any $x, y \in \mathcal{Q}$. The converse is an easy exercise.

We arrive at the main result of this section.

THEOREM 2.1 (Generalized Hurwitz theorem).– Every Hurwitz algebra over a field $\mathbb{F}$ is isomorphic to one of the following:

1) the ground field $\mathbb{F}$;

2) a two-dimensional separable commutative and associative algebra: $\mathcal{K} = \mathbb{F}1 \oplus \mathbb{F}v$, with $v^{\cdot 2} = v + \mu 1$, $\mu \in \mathbb{F}$ with $4\mu + 1 \neq 0$, and $\mathrm{n}(\epsilon + \delta v) = \epsilon^2 - \mu\delta^2 + \epsilon\delta$, for $\epsilon, \delta \in \mathbb{F}$;

3) a *quaternion* algebra $\mathcal{Q} = \mathfrak{CD}(\mathcal{K}, \beta)$ for $\mathcal{K}$ as in (2) and $0 \neq \beta \in \mathbb{F}$;

4) a *Cayley* (or *octonion*) algebra $\mathcal{C} = \mathfrak{CD}(\mathcal{Q}, \gamma)$, for $\mathcal{Q}$ as in (3) and $0 \neq \gamma \in \mathbb{F}$.

In particular, the dimension of a Hurwitz algebra is restricted to $1, 2, 4$ or 8.

PROOF.– The only Hurwitz algebra of dimension 1 is, up to isomorphism, the ground field. If $(\mathcal{C}, \cdot, \mathrm{n})$ is a Hurwitz algebra and $\dim_{\mathbb{F}} \mathcal{C} > 1$, there is an element $v \in \mathcal{C} \setminus \mathbb{F}1$ such that $\mathrm{n}(v, 1) = 1$ and $\mathrm{n}|_{\mathbb{F}1+\mathbb{F}v}$ is non-degenerate. The Cayley–Hamilton equation shows that $v^{\cdot 2} - v + \mathrm{n}(v)1 = 0$, so $v^{\cdot 2} = v + \mu 1$, with $\mu = -\mathrm{n}(v)$. The non-degeneracy condition is equivalent to the condition $4\mu + 1 \neq 0$. Then $\mathcal{K} = \mathbb{F}1 + \mathbb{F}v$ is a Hurwitz subalgebra of $\mathcal{C}$ and, if $\dim_{\mathbb{F}} \mathcal{C} = 2$, we are done.

If $\dim_{\mathbb{F}} \mathcal{C} > 2$, we may take an element $u \in \mathcal{K}^{\perp}$ with $\mathrm{n}(u) = -\beta \neq 0$, and hence the subspace $\mathcal{Q} = \mathcal{K} \oplus \mathcal{K} \cdot u$ is a subalgebra of $\mathcal{C}$ isomorphic to $\mathfrak{CD}(\mathcal{K}, \beta)$. By the previous remark, $\mathcal{Q}$ is associative (as $\mathcal{K}$ is commutative), but it fails to be commutative, as $v \cdot u = u \cdot \bar{v} \neq u \cdot v$. If $\dim_{\mathbb{F}} \mathcal{C} = 4$, we are done.

Finally, if $\dim_{\mathbb{F}} \mathcal{C} > 4$, we may take an element $u' \in \mathcal{Q}^{\perp}$ with $\mathrm{n}(u') = -\gamma \neq 0$, and hence the subspace $\mathcal{Q} \oplus \mathcal{Q} \cdot u'$ is a subalgebra of $\mathcal{C}$ isomorphic to $\mathfrak{CD}(\mathcal{Q}, \gamma)$, which is not associative by remark 2.1, so it is necessarily the whole $\mathcal{C}$. □

Note that if $\mathrm{char}\, \mathbb{F} \neq 2$, the restriction of n to $\mathbb{F}1$ is non-degenerate, so we could have used the same argument for dimension > 1 in the proof above than the one used for $\dim_{\mathbb{F}} \mathcal{C} > 2$. Hence, we get:

COROLLARY 2.1.– Every Hurwitz algebra over a field $\mathbb{F}$ of characteristic not 2 is isomorphic to one of the following:

1) the ground field $\mathbb{F}$;

2) a two-dimensional algebra $\mathcal{K} = \mathfrak{CD}(\mathbb{F}, \alpha)$ for a non-zero scalar α;

3) a *quaternion* algebra $\mathcal{Q} = \mathfrak{CD}(\mathcal{K}, \beta)$ for $\mathcal{K}$ as in (2) and $0 \neq \beta \in \mathbb{F}$;

4) a *Cayley* (or *octonion*) algebra $\mathcal{C} = \mathfrak{CD}(\mathcal{Q}, \gamma)$, for $\mathcal{Q}$ as in (3) and $0 \neq \gamma \in \mathbb{F}$.

REMARK 2.2.– Over the real field $\mathbb{R}$, the scalars α, β and γ in corollary 2.1 can be taken to be ± 1. Note that [2.3] and the analogous equation for $\mathbb{C}$ and $\mathbb{O}$ give isomorphisms $\mathbb{C} \cong \mathfrak{CD}(\mathbb{R}, -1)$, $\mathbb{H} \cong \mathfrak{CD}(\mathbb{C}, -1)$ and $\mathbb{O} \cong \mathfrak{CD}(\mathbb{H}, -1)$.

REMARK 2.3.– Hurwitz (1898) only considered the real case with a positive definite norm. Over the years, this was extended in several ways. The actual version of the generalized Hurwitz theorem seems to appear for the first time in Jacobson (1958) (if $\operatorname{char}\mathbb{F} \neq 2$) and van der Blij and Springer (1959).

The problem of isomorphism between Hurwitz algebras of the same dimension relies on the norms:

PROPOSITION 2.3.– Two Hurwitz algebras over a field are isomorphic if and only if their norms are isometric.

PROOF.– Any isomorphism of Hurwitz algebras is, in particular, an isometry of the corresponding norms, due to the Cayley–Hamilton equation. The converse follows from Witt's cancellation theorem (see Elman *et al.* (2008, theorem 8.4)). □

A natural question is whether the restriction of the dimension of a Hurwitz algebra to be $1, 2, 4$ or 8 is still valid for arbitrary composition algebras. The answer is that this is the case for finite-dimensional composition algebras.

COROLLARY 2.2.– Let $(\mathcal{C}, \cdot, \mathrm{n})$ be a finite-dimensional composition algebra. Then its dimension is either 1, 2, 4 or 8.

PROOF.– Let $a \in \mathcal{C}$ be an element of non-zero norm. Then $u = \frac{1}{\mathrm{n}(a)} a^{\cdot 2}$ satisfies $\mathrm{n}(u) = 1$. Using the so-called *Kaplansky's trick* (Kaplansky 1953), consider the new multiplication

$$x \diamond y = R_u^{-1}(x) \cdot L_u^{-1}(y).$$

Note that since the left and right multiplications by a norm 1 element are isometries, we still have $\mathrm{n}(x \diamond y) = \mathrm{n}(x)\mathrm{n}(y)$, so $(\mathcal{C}, \diamond, \mathrm{n})$ is a composition algebra too. But $u^{\cdot 2} \diamond x = u \cdot L_u^{-1}(x) = x = x \diamond u^{\cdot 2}$ for any x, so the element $u^{\cdot 2}$ is the unity of $(\mathcal{C}, \diamond)$ and $(\mathcal{C}, \diamond, \mathrm{n})$ is a Hurwitz algebra, and hence $\dim_{\mathbb{F}} \mathcal{C}$ is restricted to $1, 2, 4$ or 8. □

However, contrary to the thoughts expressed in Kaplansky (1953), there are examples of infinite-dimensional composition algebras. For example (see Urbanik and Wright (1960)), let $\varphi : \mathbb{N} \times \mathbb{N} \to \mathbb{N}$ be a bijection (for instance, $\varphi(n, m) = 2^{n-1}(2m - 1)$), and let $\mathcal{A}$ be a vector space over a field $\mathbb{F}$ of characteristic not 2 with a countable basis $\{u_n : n \in \mathbb{N}\}$. Define a multiplication and a norm on $\mathcal{A}$ by

$$u_n \cdot u_m = u_{\varphi(n,m)}, \qquad \mathrm{n}(u_n, u_m) = 2\delta_{n,m}.$$

Then $(\mathcal{A}, \cdot, \mathrm{n})$ is a composition algebra.

In Elduque and Pérez (1997), one may find examples of infinite-dimensional composition algebras of arbitrary infinite dimension, which are even left unital.

2.3.2. *Isotropic Hurwitz algebras*

Assume now that the norm of a Hurwitz algebra $(\mathcal{C}, \cdot, \mathrm{n})$ represents 0. That is, there is a non-zero element $a \in \mathcal{C}$ such that $\mathrm{n}(a) = 0$. This is always the case if $\dim_{\mathbb{F}} \mathcal{C} \geq 2$ and $\mathbb{F}$ is algebraically closed.

With a as above, take $b \in \mathcal{C}$ such that $\mathrm{n}(a, \bar{b}) = 1$, so that $\mathrm{n}(a \cdot b, 1) = 1$. Also $\mathrm{n}(a \cdot b) = \mathrm{n}(a)\mathrm{n}(b) = 0$. By the Cayley–Hamilton equation, the non-zero element $e_1 := a \cdot b$ satisfies $e_1^{\cdot 2} = e_1$, that is, e_1 is an idempotent. Consider too the idempotent $e_2 := 1 - e_1 = \overline{e_1}$, and the subalgebra $\mathcal{K} = \mathbb{F}e_1 \oplus \mathbb{F}e_2$ ($\cong \mathbb{F} \times \mathbb{F}$) generated by e_1. $(1 = e_1 + e_2)$.

For any $x \in \mathcal{K}^\perp$, $x \cdot e_1 + \overline{x \cdot e_1} = \mathrm{n}(x \cdot e_1, 1) = \mathrm{n}(x, \overline{e_1}) = \mathrm{n}(x, e_2) = 0$ and, as $\bar{x} = -x$ and $\overline{e_1} = e_2$, we conclude that $x \cdot e_1 = e_2 \cdot x$, and in the same way, $x \cdot e_2 = e_1 \cdot x$, for any $x \in \mathcal{K}^\perp$.

But $x = 1 \cdot x = e_1 \cdot x + e_2 \cdot x$, and $e_2 \cdot (e_1 \cdot x) = (1 - e_1) \cdot (e_1 \cdot x) = 0 = e_1 \cdot (e_2 \cdot x)$. It follows that $\mathcal{K}^\perp$ splits as $\mathcal{K}^\perp = \mathcal{U} \oplus \mathcal{V}$ with

$$\mathcal{U} = \{x \in \mathcal{C} : e_1 \cdot x = x = x \cdot e_2,\ e_2 \cdot x = 0 = x \cdot e_1\},$$
$$\mathcal{V} = \{x \in \mathcal{C} : e_2 \cdot x = x = x \cdot e_1,\ e_1 \cdot x = 0 = x \cdot e_2\}.$$

For any $u \in \mathcal{U}$, $\mathrm{n}(u) = \mathrm{n}(e_1 \cdot u) = \mathrm{n}(e_1)\mathrm{n}(u) = 0$, so $\mathcal{U}$, and $\mathcal{V}$ too, are totally isotropic subspaces of $\mathcal{K}^\perp$ paired by the norm. In particular, $\dim_{\mathbb{F}} \mathcal{U} = \dim_{\mathbb{F}} \mathcal{V}$, and this common value is either 0, 1 or 3, depending on $\dim_{\mathbb{F}} \mathcal{C}$ being 2, 4 or 8. The case of $\dim_{\mathbb{F}} \mathcal{U} = 0$ is trivial, and the case of $\dim_{\mathbb{F}} \mathcal{U} = 1$ is quite easy (and subsumed in the arguments below). Hence, let us assume that $\mathcal{C}$ is a Cayley algebra (dimension 8), so $\dim_{\mathbb{F}} \mathcal{U} = \dim_{\mathbb{F}} \mathcal{V} = 3$.

For any $u_1, u_2 \in \mathcal{U}$ and $v \in \mathcal{V}$, using [2.5] we get

$$\mathrm{n}(u_1 \cdot u_2, \mathcal{K}) \subseteq \mathrm{n}(u_1, \mathcal{K} \cdot u_2) \subseteq \mathrm{n}(\mathcal{U}, \mathcal{U}) = 0,$$
$$\mathrm{n}(u_1 \cdot u_2, v) = \mathrm{n}(u_1 \cdot u_2, e_2 \cdot v) = -\mathrm{n}(e_2 \cdot u_2, u_1 \cdot v) + \mathrm{n}(u_1, e_2)\mathrm{n}(u_2, v) = 0.$$

Hence, $\mathcal{U}^{\cdot 2}$ is orthogonal to both $\mathcal{K}$ and $\mathcal{V}$, so it must be contained in $\mathcal{V}$. Also $\mathcal{V}^{\cdot 2} \subseteq \mathcal{U}$.

Besides,

$$\mathrm{n}(\mathcal{U}, \mathcal{U} \cdot \mathcal{V}) \subseteq \mathrm{n}(\mathcal{U}^{\cdot 2}, \mathcal{V}) \subseteq \mathrm{n}(\mathcal{V}, \mathcal{V}) = 0 = \mathrm{n}(\mathcal{V}, \mathcal{U} \cdot \mathcal{V}),$$

so that $\mathcal{U} \cdot \mathcal{V} \subseteq (\mathcal{U} + \mathcal{V})^{\perp} = \mathcal{K}$, and also $\mathcal{V} \cdot \mathcal{U} \subseteq \mathcal{K}$. But for any $u \in \mathcal{U}$ and $v \in \mathcal{V}$, we have

$$\mathrm{n}(u \cdot v, e_2) = -\mathrm{n}(u, e_2 \cdot v) = -\mathrm{n}(u, v), \quad \mathrm{n}(u \cdot v, e_1) = -\mathrm{n}(u, e_1 \cdot v) = 0,$$

and the analogues for $v \cdot u$. We conclude that

$$u \cdot v = -\mathrm{n}(u, v)e_1, \quad v \cdot u = -\mathrm{n}(v, u)e_2.$$

Now, for linearly independent elements $u_1, u_2 \in \mathcal{U}$, let $v \in \mathcal{V}$ with $\mathrm{n}(u_1, v) \neq 0 = \mathrm{n}(u_2, v)$. Then the alternative law gives $(u_1 \cdot u_2) \cdot v = -(u_1 \cdot v) \cdot u_2 + u_1 \cdot (u_2 \cdot v + v \cdot u_2) = -\mathrm{n}(u_1, v)u_2 \neq 0$, so that $u_1 \cdot u_2 \neq 0$. In particular, $\mathcal{U}^{\cdot 2} \neq 0$, and the same happens with $\mathcal{V}$.

Consider the trilinear map:

$$\begin{aligned} \mathcal{U} \times \mathcal{U} \times \mathcal{U} &\longrightarrow \mathbb{F} \\ (x, y, z) &\mapsto \mathrm{n}(x \cdot y, z). \end{aligned}$$

This is alternating because $x^{\cdot 2} = 0$ for any $x \in \mathcal{U}$ by the Cayley–Hamilton equation, and $\mathrm{n}(x \cdot y, y) = -\mathrm{n}(x, y \cdot y) = 0$. It is also non-zero, because $\mathcal{U}^{\cdot 2} \subseteq \mathcal{V}$, so that $\mathrm{n}(\mathcal{U}^{\cdot 2}, \mathcal{U}) = \mathrm{n}(\mathcal{U}^{\cdot 2}, \mathcal{C}) \neq 0$.

Fix a basis $\{u_1, u_2, u_3\}$ of $\mathcal{U}$ with $\mathrm{n}(u_1 \cdot u_2, u_3) = 1$ and take $v_1 := u_2 \cdot u_3$, $v_2 := u_3 \cdot u_1$, $v_3 := u_1 \cdot u_2$. Then $\{v_1, v_2, v_3\}$ is the dual basis in $\mathcal{V}$ relative to the norm, and the multiplication of the basis $\{e_1, e_2, u_1, u_2, u_3, v_1, v_2, v_3\}$ is completely determined. For instance, $v_1 \cdot v_2 = v_1 \cdot (u_3 \cdot u_1) = -u_3 \cdot (v_1 \cdot u_1) = -u_3 \cdot \big(-\mathrm{n}(v_1, u_1)e_2\big) = u_3, \ldots$

The multiplication table is given in Figure 2.1.

The Cayley algebra with this multiplication table is called the *split Cayley algebra* and denoted by $\mathcal{C}_s(\mathbb{F})$. The subalgebra spanned by e_1, e_2, u_1, v_1 is isomorphic to the algebra $\mathrm{Mat}_2(\mathbb{F})$ of 2×2 matrices.

We summarize the above arguments in the next result.

THEOREM 2.2.– There are, up to isomorphism, only three Hurwitz algebras with isotropic norm: $\mathbb{F} \times \mathbb{F}$, $\mathrm{Mat}_2(\mathbb{F})$ and $\mathcal{C}_s(\mathbb{F})$.

COROLLARY 2.3.– The real Hurwitz algebras are, up to isomorphism, the following algebras:

– the classical division algebras $\mathbb{R}$, $\mathbb{C}$, $\mathbb{H}$, and $\mathbb{O}$;

– the algebras $\mathbb{R} \times \mathbb{R}$, $\mathrm{Mat}_2(\mathbb{R})$ and $\mathcal{C}_s(\mathbb{R})$.

	e_1	e_2	u_1	u_2	u_3	v_1	v_2	v_3
e_1	e_1	0	u_1	u_2	u_3	0	0	0
e_2	0	e_2	0	0	0	v_1	v_2	v_3
u_1	0	u_1	0	v_3	$-v_2$	$-e_1$	0	0
u_2	0	u_2	$-v_3$	0	v_1	0	$-e_1$	0
u_3	0	u_3	v_2	$-v_1$	0	0	0	$-e_1$
v_1	v_1	0	$-e_2$	0	0	0	u_3	$-u_2$
v_2	v_2	0	0	$-e_2$	0	$-u_3$	0	u_1
v_3	v_3	0	0	0	$-e_2$	u_2	$-u_1$	0

Figure 2.1. *Multiplication table of the split Cayley algebra*

PROOF.– It is enough to take into account that a non-degenerate quadratic form over $\mathbb{R}$ is either isotropic or definite. Hence, the norm of a real Hurwitz algebra is either isotropic or positive definite (as $\mathrm{n}(1) = 1$). □

2.4. Symmetric composition algebras

In this section, a new important family of composition algebras will be described.

DEFINITION 2.2.– *A composition algebra* $(\mathcal{S}, *, \mathrm{n})$ *is said to be a* symmetric composition algebra *if* $L_x^* = R_x$ *for any* $x \in \mathcal{S}$ *(that is,* $\mathrm{n}(x * y, z) = \mathrm{n}(x, y * z)$ *for any* $x, y, z \in \mathcal{S}$*).*

THEOREM 2.3.– Let $(\mathcal{S}, *, \mathrm{n})$ be a composition algebra. The following conditions are equivalent:

a) $(\mathcal{S}, *, \mathrm{n})$ is symmetric;

b) for any $x, y \in \mathcal{S}$, $(x * y) * x = x * (y * x) = \mathrm{n}(x)y$.

The dimension of any symmetric composition algebra is finite, and hence restricted to $1, 2, 4$ or 8.

PROOF.– If $(\mathcal{S}, *, \mathrm{n})$ is symmetric, then for any $x, y, z \in \mathcal{S}$,

$$\mathrm{n}\big((x * y) * x, z\big) = \mathrm{n}(x * y, x * z) = \mathrm{n}(x)\mathrm{n}(y, z) = \mathrm{n}\big(\mathrm{n}(x)y, z\big)$$

so that $(x * y) * x - \mathrm{n}(x)y \in \mathcal{S}^{\perp}$. Also

$$\begin{aligned}\mathrm{n}\big((x * y) * x - \mathrm{n}(x)y\big) &= \mathrm{n}\big((x * y) * x\big) + \mathrm{n}(x)^2\mathrm{n}(y) - \mathrm{n}(x)\mathrm{n}\big((x * y) * x, y\big)\\ &= 2\mathrm{n}(x)^2\mathrm{n}(y) - \mathrm{n}(x)\mathrm{n}(x * y, x * y) = 0,\end{aligned}$$

whence (b), since n is non-singular.

Conversely, take $x, y, z \in \mathcal{S}$ with $\mathrm{n}(y) \neq 0$, so that L_y and R_y are bijective, and hence there is an element $z' \in \mathcal{S}$ with $z = z' * y$. Then:

$$\mathrm{n}(x * y, z) = \mathrm{n}(x * y, z' * y) = \mathrm{n}(x, z')\mathrm{n}(y) = \mathrm{n}\big(x, y * (z' * y)\big) = \mathrm{n}(x, y * z).$$

This proves (a) assuming $\mathrm{n}(y) \neq 0$, but any isotropic element is the sum of two non-isotropic elements, so (a) follows.

Finally, we can use a modified version of Kaplansky's trick (see corollary 2.2) as follows. Let $a \in \mathcal{S}$ be a norm 1 element and define a new product on $\mathcal{S}$ by:

$$x \diamond y := (a * x) * (y * a),$$

for any $x, y \in \mathcal{S}$. Then $(\mathcal{S}, \diamond, \mathrm{n})$ is also a composition algebra. Let $e = a^{*2}$. Then, using (b) we have $e \diamond x = \big(a * (a * a)\big) * (x * a) = a * (x * a) = x$, and similarly $x \diamond e = x$ for any x. Thus, $(\mathcal{S}, \diamond, \mathrm{n})$ is a Hurwitz algebra with unity e, and hence it is finite-dimensional. □

REMARK 2.4.– Condition (b) above implies that $((x * y) * x) * (x * y) = \mathrm{n}(x * y)x$, but also $((x * y) * x) * (x * y) = \mathrm{n}(x)y * (x * y) = \mathrm{n}(x)\mathrm{n}(y)x$, so that condition (b) already forces the quadratic form n to be multiplicative.

EXAMPLES 2.1 (Okubo (1978)).–

– *Para-Hurwitz algebras*: let $(\mathcal{C}, \cdot, \mathrm{n})$ be a Hurwitz algebra and consider the composition algebra $(\mathcal{C}, \bullet, \mathrm{n})$ with the new product given by

$$x \bullet y = \overline{x} \cdot \overline{y}.$$

Then $\mathrm{n}(x \bullet y, z) = \mathrm{n}(\overline{x} \cdot \overline{y}, z) = \mathrm{n}(\overline{x}, z \cdot y) = \mathrm{n}(x, \overline{z \cdot y}) = \mathrm{n}(x, y \bullet z)$, for any x, y, z, so that $(\mathcal{C}, \bullet, \mathrm{n})$ is a symmetric composition algebra (note that $1 \bullet x = x \bullet 1 = \overline{x} = \mathrm{n}(x, 1)1 - x$ for any x: 1 is a *para-unit* of $(\mathcal{C}, \bullet, \mathrm{n})$).

– *Okubo algebras*: assume $\mathrm{char}\,\mathbb{F} \neq 3$ (the case of $\mathrm{char}\,\mathbb{F} = 3$ requires a different definition), and let $\omega \in \mathbb{F}$ be a primitive cubic root of 1. Let $\mathcal{A}$ be a central simple associative algebra of degree 3 with trace tr, and let $\mathcal{S} = \mathcal{A}_0 = \{x \in \mathcal{A} :$

$\operatorname{tr}(x) = 0\}$. For any $x \in \mathcal{S}$, the quadratic form $\frac{1}{2}\operatorname{tr}(x^2)$ make sense even if $\operatorname{char}\mathbb{F} = 2$ (check this!). Now define a multiplication and norm on $\mathcal{S}$ by:

$$\begin{aligned} x * y &= \omega xy - \omega^2 yx - \tfrac{\omega-\omega^2}{3}\operatorname{tr}(xy)1, \\ \operatorname{n}(x) &= -\tfrac{1}{2}\operatorname{tr}(x^2), \end{aligned}$$

Then, for any $x, y \in \mathcal{S}$:

$$\begin{aligned} (x*y)*x &= \omega(x*y)x - \omega^2 x(x*y) - \tfrac{\omega-\omega^2}{3}\operatorname{tr}\big((x*y)x\big)1 \\ &= \omega^2 xyx - yx^2 - \tfrac{\omega^2-1}{3}\operatorname{tr}(xy)x - x^2y + \omega xyx + \tfrac{1-\omega}{3}\operatorname{tr}(xy)x \\ &\qquad - \tfrac{\omega-\omega^2}{3}\operatorname{tr}\Big((\omega-\omega^2)x^2y\Big)1 \quad (\operatorname{tr}(x) = 0) \\ &= -\big(x^2y + yx^2 + xyx\big) + \operatorname{tr}(xy)x + \operatorname{tr}(x^2y)1 \quad ((\omega-\omega^2)^2 = -3). \end{aligned}$$

But if $\operatorname{tr}(x) = 0$, then $x^3 - \frac{1}{2}\operatorname{tr}(x^2)x - \det(x)1 = 0$, so

$$x^2y + yx^2 + xyx - \big(\operatorname{tr}(xy)x + \frac{1}{2}\operatorname{tr}(x^2)y\big) \in \mathbb{F}1.$$

Since $(x*y)*x \in \mathcal{A}_0$, we have $(x*y)*x = -\frac{1}{2}\operatorname{tr}(x^2)y = x*(y*x)$.

Therefore, $(\mathcal{S}, *, \operatorname{n})$ is a symmetric composition algebra.

In case $\omega \notin \mathbb{F}$, take $\mathbb{K} = \mathbb{F}[\omega]$ and a central simple associative algebra $\mathcal{A}$ of degree 3 over $\mathbb{K}$ endowed with a $\mathbb{K}/\mathbb{F}$-involution of second kind J. Then take $\mathcal{S} = K(\mathcal{A}, J)_0 = \{x \in \mathcal{A}_0 : J(x) = -x\}$ (this is an $\mathbb{F}$-subspace) and use the same formulas above to define the multiplication and the norm.

REMARK 2.5.– For $\mathbb{F} = \mathbb{R}$, take $\mathcal{A} = \operatorname{Mat}_3(\mathbb{C})$, and then there appears the Okubo algebra $(\mathcal{S}, *, \operatorname{n})$ with $\mathcal{S} = \mathfrak{su}_3 = \{x \in \operatorname{Mat}_3(\mathbb{C}) : \operatorname{tr}(x) = 0,\ x^* = -x\}$ (x^* denotes the conjugate transpose of x). This algebra was termed the algebra of *pseudo-octonions* by Okubo (1978), who studied these algebras and classified them, under some restrictions, in joint work with Osborn Okubo and Osborn (1981a,b).

The name *Okubo algebras* was given in Elduque and Myung (1990). Faulkner (1988) discovered Okubo's construction independently, in a more general setting, related to separable alternative algebras of degree 3, and gave the key idea for the classification of the symmetric composition algebras in Elduque and Myung (1993) ($\operatorname{char}\mathbb{F} \neq 2, 3$). A different, less elegant, classification was given in Elduque and Myung (1991), based on the fact that Okubo algebras are *Lie-admissible*.

The term *symmetric composition algebra* was given in Knus *et al.* (1998, Chapter VIII).

REMARK 2.6.– Given an Okubo algebra, note that for any $x, y \in \mathcal{S}$,

$$\begin{aligned} x * y &= \omega xy - \omega^2 yx - \tfrac{\omega-\omega^2}{3}\operatorname{tr}(xy)1, \\ y * x &= \omega yx - \omega^2 xy - \tfrac{\omega-\omega^2}{3}\operatorname{tr}(xy)1, \end{aligned}$$

so that

$$\omega x * y + \omega^2 y * x = (\omega^2 - \omega)xy - (\omega + \omega^2)\frac{\omega - \omega^2}{3}\operatorname{tr}(xy)1,$$

and

$$xy = \frac{\omega}{\omega^2 - \omega} x * y + \frac{\omega^2}{\omega^2 - \omega} y * x + \frac{1}{3}\operatorname{n}(x, y)1, \qquad [2.8]$$

so the product in $\mathcal{A}$ is determined by the product in the Okubo algebra.

Also, as noted by Faulkner, the construction above is valid for separable alternative algebras of degree 3.

THEOREM 2.4 (Elduque and Myung (1991, 1993)).– Let $\mathbb{F}$ be a field of characteristic not 3.

– If $\mathbb{F}$ contains a primitive cubic root ω of 1, then the symmetric composition algebras of dimension ≥ 2 are, up to isomorphism, the algebras $(\mathcal{A}_0, *, \operatorname{n})$ for $\mathcal{A}$ a separable alternative algebra of degree 3.

Two such symmetric composition algebras are isomorphic if and only if the corresponding alternative algebras are too.

– If $\mathbb{F}$ does not contain primitive cubic roots of 1, then the symmetric composition algebras of dimension ≥ 2 are, up to isomorphism, the algebras $\big(K(\mathcal{A}, J)_0, *, \operatorname{n}\big)$ for $\mathcal{A}$ a separable alternative algebra of degree 3 over $\mathbb{K} = \mathbb{F}[\omega]$, and J a $\mathbb{K}/\mathbb{F}$-involution of the second kind.

Two such symmetric composition algebras are isomorphic if and only if the corresponding alternative algebras, as algebras with involution, are too.

Sketch of proof: we can go in the reverse direction of Okubo's construction. Given a symmetric composition algebra $(\mathcal{S}, *, \operatorname{n})$ over a field containing ω, define the algebra $\mathcal{A} = \mathbb{F}1 \oplus \mathcal{S}$ with multiplication determined by formula [2.8]. Then $\mathcal{A}$ turns out to be a separable alternative algebra of degree 3.

In case $\omega \notin \mathbb{F}$, then we must consider $\mathcal{A} = \mathbb{F}[\omega]1 \oplus \big(\mathbb{F}[\omega] \otimes \mathcal{S}\big)$, with the same formula for the product. In $\mathbb{F}[\omega]$, we have the Galois automorphism $\omega^\tau = \omega^2$. Then the conditions $J(1) = 1$ and $J(s) = -s$ for any $s \in \mathcal{S}$ induce a $\mathbb{F}[\omega]/\mathbb{F}$-involution of the second kind in $\mathcal{A}$. □

COROLLARY 2.4.– The algebras in examples 2.1 essentially exhaust, up to isomorphism, the symmetric composition algebras over a field $\mathbb{F}$ of characteristic not 3.

Sketch of proof: let ω be a primitive cubic root of 1 in an algebraic closure of $\mathbb{F}$, and let $\mathbb{K} = \mathbb{F}[\omega]$, so that $\mathbb{K} = \mathbb{F}$ if $\omega \in \mathbb{F}$. A separable alternative algebra over $\mathbb{K}$ is, up to isomorphism, one of the following:

– a central simple associative algebra, and hence we obtain the Okubo algebras in examples 2.1;

– $\mathcal{A} = \mathbb{K} \times \mathcal{C}$ for a Hurwitz algebra $\mathcal{C}$, in which case $(\mathcal{A}_0, *, \mathrm{n})$ is shown to be isomorphic to the para-Hurwitz algebra attached to $\mathcal{C}$ if $\mathbb{K} = \mathbb{F}$, and $\big(K(\mathcal{A}, J)_0, *, \mathrm{n}\big)$ to the para-Hurwitz algebra attached to $\widehat{\mathcal{C}} = \{x \in \mathcal{C} : J(x) = \overline{x}\}$ if $\mathbb{K} \neq \mathbb{F}$;

– $\mathcal{A} = \mathbb{K} \otimes_{\mathbb{F}} \mathbb{L}$, for a cubic field extension $\mathbb{L}$ of $\mathbb{F}$ (if $\omega \notin \mathbb{F}$, $\mathbb{L} = \{x \in \mathcal{A} : J(x) = x\}$), in which case the symmetric composition algebra is shown to be a twisted form of a two-dimensional para-Hurwitz algebra. □

One of the clues to understand symmetric composition algebras over fields of characteristic 3 is the following result of Petersson (1969) (dealing with $\operatorname{char}\mathbb{F} \neq 2, 3$!).

THEOREM 2.5.– Let $\mathbb{F}$ be an algebraically closed field of characteristic $\neq 2, 3$. Then any simple finite-dimensional algebra satisfying

$$(xy)x = x(yx), \quad ((xz)y)(xz) = \big(x((zy)z)\big)x \qquad [2.9]$$

for any x, y, z is, up to isomorphism, one of the following:

– the algebra $(\mathcal{B}, \bullet)$, where $(\mathcal{B}, \cdot, \mathrm{n})$ is a Hurwitz algebra and $x \bullet y = \overline{x} \cdot \overline{y}$ (that is, a para-Hurwitz algebra);

– the algebra $(\mathcal{C}_s(\mathbb{F}), *)$, where $\mathcal{C}_s(\mathbb{F})$ is the split Cayley algebra, and $x * y = \varphi(\overline{x})\varphi^2(\overline{y})$, where φ is a precise order 3 automorphism of $\mathcal{C}_s(\mathbb{F})$ given, in the basis in Figure 2.1 by

$$e_i \mapsto e_i,\ i = 1, 2, \qquad u_j \mapsto \omega^{j-1}u_j,\ v_j \mapsto \omega^{1-j}v_j,\ j = 1, 2, 3,$$

where ω is a primitive cubic root of 1.

Note that any symmetric composition algebra $(\mathcal{S}, *, \mathrm{n})$ satisfies [2.9] so the unique, up to isomorphism, Okubo algebra over an algebraically closed field of characteristic $\neq 2, 3$ must be isomorphic to the last algebra in the theorem above, and this seems to be the first appearance of these algebras in the literature.

This results in the next definition:

DEFINITION 2.3 (Knus *et al.* (1998, §34.b)).– *Let* $(\mathcal{C}, \cdot, \mathrm{n})$ *be a Hurwitz algebra, and let* $\varphi \in \mathrm{Aut}(\mathcal{C}, \cdot, \mathrm{n})$ *be an automorphism with* $\varphi^3 = \mathrm{id}$. *The composition algebra* $(\mathcal{C}, *, \mathrm{n})$, *with*

$$x * y = \varphi(\bar{x}) \cdot \varphi^2(\bar{y})$$

is called a Petersson algebra, *and denoted by* $\mathcal{C}_\varphi$.

In case $\varphi = \mathrm{id}$, the Petersson algebra is the para-Hurwitz algebra associated with $(\mathcal{C}, \cdot, \mathrm{n})$.

Modifying the automorphism in theorem 2.5, consider the order 3 automorphism φ of the split Cayley algebra given by:

$$\varphi(e_i) = e_i,\ i = 1, 2, \qquad \varphi(u_j) = u_{j+1},\ \varphi(v_j) = v_{j+1} \text{ (indices } j \text{ modulo 3).}$$

With this automorphism, we may define Okubo algebras over arbitrary fields (see Elduque and Pérez (1996)).

DEFINITION 2.4.– *Let* $(\mathcal{C}_s(\mathbb{F}), \cdot, \mathrm{n})$ *be the split Cayley algebra over an arbitrary field* $\mathbb{F}$. *The Petersson algebra* $\mathcal{C}_s(\mathbb{F})_\varphi$ *is called the* split Okubo algebra *over* $\mathbb{F}$.

Its twisted forms (i.e. those composition algebras $(\mathcal{S}, *, \mathrm{n})$ *that become isomorphic to the split Okubo algebra after extending scalars to an algebraic closure) are called* Okubo algebras.

In the basis in Figure 2.1, the multiplication table of the split Okubo algebra is given in Figure 2.2.

Over fields of characteristic $\neq 3$, our new definition of Okubo algebras coincide with the definition in examples 2.1, due to corollary 2.4. Okubo and Osborn (1981b) had given an *ad hoc* definition of the Okubo algebra over an algebraically closed field of characteristic 3.

Note that the split Okubo algebra does not contain any non-zero element that commutes with every other element, that is, its commutative center is trivial. This is not so for the para-Hurwitz algebra, where the para-unit lies in the commutative center.

Let $\mathbb{F}$ be a field of characteristic 3 and let $0 \neq \alpha, \beta \in \mathbb{F}$. Consider the elements

$$e_1 \otimes \alpha^{1/3}, \quad u_1 \otimes \beta^{1/3}$$

in $\mathcal{C}_s(\mathbb{F}) \otimes_{\mathbb{F}} \overline{\mathbb{F}}$ ($\overline{\mathbb{F}}$ being an algebraic closure of $\mathbb{F}$). These elements generate, by multiplication and linear combinations over $\mathbb{F}$, a twisted form of the split Okubo algebra $(\mathcal{C}_s(\mathbb{F}), *, \mathrm{n})$. Denote by $\mathcal{O}_{\alpha,\beta}$ this twisted form.

	e_1	e_2	u_1	v_1	u_2	v_2	u_3	v_3
e_1	e_2	0	0	$-v_3$	0	$-v_1$	0	$-v_2$
e_2	0	e_1	$-u_3$	0	$-u_1$	0	$-u_2$	0
u_1	$-u_2$	0	v_1	0	$-v_3$	0	0	$-e_1$
v_1	0	$-v_2$	0	u_1	0	$-u_3$	$-e_2$	0
u_2	$-u_3$	0	0	$-e_1$	v_2	0	$-v_1$	0
v_2	0	$-v_3$	$-e_2$	0	0	u_2	0	$-u_1$
u_3	$-u_1$	0	$-v_2$	0	0	$-e_1$	v_3	0
v_3	0	$-v_1$	0	$-u_2$	$-e_2$	0	0	u_3

Figure 2.2. *Multiplication table of the split Okubo algebra*

The classification of the symmetric composition algebras in characteristic 3, which completes the classification of symmetric composition algebras over fields, is as follows (Elduque (1997), see also Chernousov *et al.* (2013)):

THEOREM 2.6.– Any symmetric composition algebra $(\mathcal{S}, *, \mathrm{n})$ over a field $\mathbb{F}$ of characteristic 3 is either:

– a para-Hurwitz algebra. Two such algebras are isomorphic if and only if the associated Hurwitz algebras are too;

– a two-dimensional algebra with a basis $\{u, v\}$ and multiplication given by

$$u * u = v, \quad u * v = v * u = u, \quad v * v = \lambda u - v,$$

for a non-zero scalar $\lambda \in \mathbb{F} \setminus \mathbb{F}^3$. These algebras do not contain idempotents and are twisted forms of the para-Hurwitz algebras.

Algebras corresponding to the scalars λ and λ' are isomorphic if and only if $\mathbb{F}^3\lambda + \mathbb{F}^3(\lambda^2+1) = \mathbb{F}^3\lambda' + \mathbb{F}^3((\lambda')^2+1)$.

– Isomorphic to $\mathcal{O}_{\alpha,\beta}$ for some $0 \neq \alpha, \beta \in \mathbb{F}$. Moreover, $\mathcal{O}_{\alpha,\beta}$ is isomorphic or anti-isomorphic to $\mathcal{O}_{\gamma,\delta}$ if and only if $\mathrm{span}_{\mathbb{F}^3}\{\alpha^{\pm1}, \beta^{\pm1}, \alpha^{\pm1}\beta^{\pm1}\} = \mathrm{span}_{\mathbb{F}^3}\{\gamma^{\pm1}, \delta^{\pm1}, \gamma^{\pm1}\delta^{\pm1}\}$.

A more precise statement for the isomorphism condition in the last item is given in (Elduque 1997). A key point in the proof of this theorem is the study of idempotents on Okubo algebras. If there are non-zero idempotents, then these algebras are Petersson algebras. The most difficult case appears in the absence of idempotents. This is only possible if the ground field $\mathbb{F}$ is not perfect.

2.5. Triality

The importance of symmetric composition algebras lies in their connections with the phenomenon of triality in dimension 8, related to the fact that the Dynkin diagram D_4 is the most symmetric one. The details of much of what follows can be found in (Knus *et al.* (1998), Chapter VIII).

Let $(\mathcal{S}, *, \mathrm{n})$ be an eight-dimensional symmetric composition algebra over a field $\mathbb{F}$, that is, $\mathcal{S}$ is either a para-Hurwitz algebra or an Okubo algebra. Write $L_x(y) = x * y = R_y(x)$ as usual. Then, due to theorem 2.3, $L_xR_x = R_xL_x = \mathrm{n}(x)\mathrm{id}$ for any $x \in \mathcal{S}$ so that, inside $\mathrm{End}_{\mathbb{F}}(\mathcal{S} \oplus \mathcal{S}) \simeq \mathrm{Mat}_2\big(\mathrm{End}_{\mathbb{F}}(\mathcal{S})\big)$, we have

$$\begin{pmatrix} 0 & L_x \\ R_x & 0 \end{pmatrix}^2 = \mathrm{n}(x)\mathrm{id}.$$

Therefore, the map

$$x \mapsto \begin{pmatrix} 0 & L_x \\ R_x & 0 \end{pmatrix}$$

extends to an isomorphism of associative algebras with involution:

$$\Phi : (\mathfrak{Cl}(\mathcal{S}, \mathrm{n}), \tau) \longrightarrow (\mathrm{End}_{\mathbb{F}}(\mathcal{S} \oplus \mathcal{S}), \sigma_{n\perp n})$$

where $\mathfrak{Cl}(\mathcal{S}, \mathrm{n})$ is the Clifford algebra on the quadratic space $(\mathcal{S}, \mathrm{n})$, τ is its canonical involution ($\tau(x) = x$ for any $x \in \mathcal{S}$) and $\sigma_{n\perp n}$ is the orthogonal involution on $\mathrm{End}_{\mathbb{F}}(\mathcal{S} \oplus \mathcal{S})$ induced by the quadratic form where the two copies of $\mathcal{S}$ are orthogonal and the restriction on each copy coincides with the norm. The multiplication in the Clifford algebra will be denoted by juxtaposition.

Consider the *spin group*:

$$\mathrm{Spin}(\mathcal{S}, n) = \{u \in \mathfrak{Cl}(\mathcal{S}, \mathrm{n})_{\bar{0}}^{\times} : uxu^{-1} \in \mathcal{S},\ u\tau(u) = 1,\ \forall x \in \mathcal{C}\}.$$

For any $u \in \mathrm{Spin}(\mathcal{S}, n)$,

$$\Phi(u) = \begin{pmatrix} \rho_u^- & 0 \\ 0 & \rho_u^+ \end{pmatrix}$$

for some $\rho_u^\pm \in \mathrm{O}(\mathcal{S}, n)$ such that

$$\chi_u(x * y) = \rho_u^+(x) * \rho_u^-(y)$$

for any $x, y \in \mathcal{S}$, where $\chi_u(x) = uxu^{-1}$ gives the natural representation of $\mathrm{Spin}(\mathcal{S}, \mathrm{n})$, while $\rho_u^{\pm 1}$ give the two half-spin representations, and the formula above links the three of them.

The last condition is equivalent to:

$$\langle \chi_u(x), \rho_u^+(y), \rho_u^-(z) \rangle = \langle x, y, z \rangle$$

for any $x, y, z \in \mathcal{S}$, where

$$\langle x, y, z \rangle = \mathrm{n}(x, y * z),$$

and this has cyclic symmetry:

$$\langle x, y, z \rangle = \langle y, z, x \rangle.$$

THEOREM 2.7.– Let $(\mathcal{S}, *, \mathrm{n})$ be an eight-dimensional symmetric composition algebra. Then:

$$\mathrm{Spin}(\mathcal{S}, \mathrm{n}) \simeq \{(f_0, f_1, f_2) \in \mathrm{O}^+(\mathcal{S}, n)^3 : f_0(x * y) = f_1(x) * f_2(y)\ \forall x, y \in \mathcal{S}\}.$$

Moreover, the set of *related triples* (the set on the right hand side) has cyclic symmetry.

The cyclic symmetry on the right-hand side induces an outer automorphism of order 3 (*trialitarian automorphism*) of $\mathrm{Spin}(\mathcal{S}, \mathrm{n})$. Its fixed subgroup is the group of automorphisms of the symmetric composition algebra $(\mathcal{S}, *, \mathrm{n})$, which is a simple algebraic group of type G_2 in the para-Hurwitz case, and of type A_2 in the Okubo case if $\mathrm{char}\,\mathbb{F} \neq 3$.

The group(-scheme) of automorphisms of an Okubo algebra over a field of characteristic 3 is not smooth (Chernousov *et al.* 2013).

At the Lie algebra level, assume char $\mathbb{F} \neq 2$, and consider the associated orthogonal Lie algebra

$$\mathfrak{so}(\mathcal{S}, \mathrm{n}) = \{d \in \mathrm{End}_{\mathbb{F}}(\mathcal{S}) : \mathrm{n}\big(d(x), y\big) + \mathrm{n}\big(x, d(y)\big) = 0 \; \forall x, y \in \mathcal{S}\}.$$

The *triality Lie algebra* of $(\mathcal{S}, *, \mathrm{n})$ is defined as the following Lie subalgebra of $\mathfrak{so}(\mathcal{S}, \mathrm{n})^3$ (with componentwise bracket):

$$\begin{aligned}&\mathfrak{tri}(\mathcal{S}, *, \mathrm{n})\\&= \{(d_0, d_1, d_2) \in \mathfrak{so}(\mathcal{S}, \mathrm{n})^3 : d_0(x*y) = d_1(x)*y+x*d_2(y) \; \forall x, y, z \in \mathcal{S}\}.\end{aligned}$$

Note that the condition $d_0(x * y) = d_1(x) * y + x * d_2(y)$ for any $x, y \in \mathcal{S}$ is equivalent to the condition

$$\mathrm{n}\big(x * y, d_0(z)\big) + \mathrm{n}\big(d_1(x) * y, z\big) + \mathrm{n}\big(x * d_2(y), z\big) = 0,$$

for any $x, y, z \in \mathcal{S}$. But $\mathrm{n}(x * y, z) = \mathrm{n}(y * z, x) = \mathrm{n}(z * x, y)$. Therefore, the linear map:

$$\begin{aligned}\theta : \mathfrak{tri}(\mathcal{S}, *, \mathrm{n}) &\longrightarrow \mathfrak{tri}(\mathcal{S}, *, \mathrm{n})\\(d_0, d_1, d_2) &\mapsto (d_2, d_0, d_1),\end{aligned}$$

is an automorphism of the Lie algebra $\mathfrak{tri}(\mathcal{S}, *, \mathrm{n})$.

THEOREM 2.8.– Let $(\mathcal{S}, *, \mathrm{n})$ be an eight-dimensional symmetric composition algebra over a field of characteristic $\neq 2$. Then:

– *Principle of local triality*: the projection map:

$$\begin{aligned}\pi_0 : \mathfrak{tri}(\mathcal{S}, *, \mathrm{n}) &\longrightarrow \mathfrak{so}(\mathcal{S}, \mathrm{n})\\(d_0, d_1, d_2) &\mapsto d_0\end{aligned}$$

is an isomorphism of Lie algebras.

– For any $x, y \in \mathcal{S}$, the triple

$$t_{x,y} = \Big(\sigma_{x,y} = \mathrm{n}(x, .)y - \mathrm{n}(y, .)x, \frac{1}{2}\mathrm{n}(x, y)id - R_x L_y, \frac{1}{2}\mathrm{n}(x, y)id - L_x R_y\Big)$$

belongs to $\mathfrak{tri}(\mathcal{S}, *, \mathrm{n})$, and $\mathfrak{tri}(\mathcal{S}, *, \mathrm{n})$ is spanned by these elements. Moreover, for any $a, b, x, y \in \mathcal{S}$:

$$[t_{a,b}, t_{x,y}] = t_{\sigma_{a,b}(x),y} + t_{x,\sigma_{a,b}(y)}.$$

PROOF.– Let us first check that $t_{x,y} \in \mathfrak{tri}(\mathcal{S}, *, \mathrm{n})$:

$$\begin{aligned} \sigma_{x,y}(u * v) &= \mathrm{n}(x, u * v)y - \mathrm{n}(y, u * v)x \\ R_x L_y(u) * v &= \big((y * u) * x\big) * v = -(v * x) * (y * u) + \mathrm{n}(y * u, v)x, \\ u * L_x R_y(v) &= u * \big(x * (v * y)\big) = -u * \big(y * (v * x)\big) + \mathrm{n}(x, y)u * v \\ &= (v * x) * (y * u) + \mathrm{n}(u, v * x)y + \mathrm{n}(x, y)u * v, \end{aligned}$$

and hence

$$\sigma_{x,y}(u * v) - \Big(\frac{1}{2}\mathrm{n}(x, y)id - R_x L_y\Big)(u) * v - u * \Big(\frac{1}{2}\mathrm{n}(x, y)id - L_x R_y\Big)(v) = 0.$$

Also $\sigma_{x,y} \in \mathfrak{so}(\mathcal{S}, \mathrm{n})$ and $\left(\frac{1}{2}\mathrm{n}(x, y)id - R_x L_y\right)^* = \frac{1}{2}\mathrm{n}(x, y)id - R_y L_x$ (adjoint relative to the norm n), but $R_x L_x = \mathrm{n}(x)id$, so $R_x L_y + R_y L_x = \mathrm{n}(x, y)id$ and hence $\left(\frac{1}{2}\mathrm{n}(x, y)id - R_x L_y\right)^* = -\left(\frac{1}{2}\mathrm{n}(x, y)id - R_x L_y\right)$, so that $\frac{1}{2}\mathrm{n}(x, y)id - R_x L_y \in \mathfrak{so}(\mathcal{S}, \mathrm{n})$, and $\frac{1}{2}\mathrm{n}(x, y)id - L_x R_y \in \mathfrak{so}(\mathcal{S}, \mathrm{n})$ too. Therefore, $t_{x,y} \in \mathfrak{tri}(\mathcal{S}, *, \mathrm{n})$.

Since the Lie algebra $\mathfrak{so}(\mathcal{S}, \mathrm{n})$ is spanned by the $\sigma_{x,y}$'s, it is clear that the projection π_0 is surjective (and hence so are π_1 and π_2). It is not difficult to check that $\ker \pi_0 = 0$, and therefore, π_0 is an isomorphism.

Finally, the formula $[t_{a,b}, t_{x,y}] = t_{\sigma_{a,b}(x),y} + t_{x,\sigma_{a,b}(y)}$ follows from the "same" formula for the σ's and the fact that π_0 is an isomorphism. □

Given two symmetric composition algebras $(\mathcal{S}, *, \mathrm{n})$ and $(\mathcal{S}', \star, \mathrm{n}')$, consider the vector space:

$$\mathfrak{g} = \mathfrak{g}(\mathcal{S}, \mathcal{S}') = \big(\mathfrak{tri}(\mathcal{S}) \oplus \mathfrak{tri}(\mathcal{S}')\big) \oplus \Big(\oplus_{i=0}^{2} \iota_i(\mathcal{S} \otimes \mathcal{S}')\Big),$$

where $\iota_i(\mathcal{S} \otimes \mathcal{S}')$ is just a copy of $\mathcal{S} \otimes \mathcal{S}'$ ($i = 0, 1, 2$) and we write $\mathfrak{tri}(\mathcal{S})$, $\mathfrak{tri}(\mathcal{S}')$ instead of $\mathfrak{tri}(\mathcal{S}, *, \mathrm{n})$ and $\mathfrak{tri}(\mathcal{S}', \star, \mathrm{n}')$ for short. Define now an anticommutative bracket on $\mathfrak{g}$ by means of:

- the Lie bracket in $\mathfrak{tri}(\mathcal{S}) \oplus \mathfrak{tri}(\mathcal{S}')$, which thus becomes a Lie subalgebra of $\mathfrak{g}$;
- $[(d_0, d_1, d_2), \iota_i(x \otimes x')] = \iota_i\big(d_i(x) \otimes x'\big)$;
- $[(d'_0, d'_1, d'_2), \iota_i(x \otimes x')] = \iota_i\big(x \otimes d'_i(x')\big)$;
- $[\iota_i(x \otimes x'), \iota_{i+1}(y \otimes y')] = \iota_{i+2}\big((x * y) \otimes (x' \star y')\big)$ (indices modulo 3);
- $[\iota_i(x \otimes x'), \iota_i(y \otimes y')] = \mathrm{n}'(x', y')\theta^i(t_{x,y}) + \mathrm{n}(x, y)\theta'^i(t'_{x',y'}) \in \mathfrak{tri}(\mathcal{S}) \oplus \mathfrak{tri}(\mathcal{S}')$.

THEOREM 2.9 (Elduque (2004)).– Assume char $\mathbb{F} \neq 2,3$. With the bracket above, $\mathfrak{g}(\mathcal{S},\mathcal{S}')$ is a Lie algebra and, if $\mathcal{S}_r$ and $\mathcal{S}'_s$ denote symmetric composition algebras of dimension r and s, then the Lie algebra $\mathfrak{g}(\mathcal{S}_r,\mathcal{S}'_s)$ is a (semi) simple Lie algebra whose type is given by Freudenthal's magic square:

	$\mathcal{S}_1$	$\mathcal{S}_2$	$\mathcal{S}_4$	$\mathcal{S}_8$
$\mathcal{S}'_1$	A_1	A_2	C_3	F_4
$\mathcal{S}'_2$	A_2	$A_2 \oplus A_2$	A_5	E_6
$\mathcal{S}'_4$	C_3	A_5	D_6	E_7
$\mathcal{S}'_8$	F_4	E_6	E_7	E_8

Different versions of this result using Hurwitz algebras instead of symmetric composition algebras have appeared over the years (see Elduque (2004) and the references therein). The advantage of using symmetric composition algebras is that new constructions of the exceptional simple Lie algebras are obtained, and these constructions highlight interesting symmetries due to the different triality automorphisms.

A few changes are needed for characteristic 3. Also, quite interestingly, over fields of characteristic 3 there are non-trivial symmetric composition *superalgebras*, and these can be plugged into the previous construction to obtain an extended Freudenthal's magic square that includes some new simple finite dimensional Lie superalgebras (see Cunha and Elduque (2007)).

2.6. Concluding remarks

It is impossible to give a thorough account of composition algebras in a few pages, so many things have had to be left out: Pfister forms and the problem of composition of quadratic forms (see Shapiro (2000)), composition algebras over rings (or even over schemes), where Hurwitz algebras are no longer determined by their norms (see Gille (2014)), the closely related subject of absolute valued algebras (see Rodríguez-Palacios (2004)), etc.

The interested reader may consult the following studies: (Conway and Smith 2003; Springer and Veldkamp 2000; Ebbinghaus *et al.* 1991; Knus *et al.* 1998; Okubo 1995). Baez (2002) is a beautiful introduction to octonions and some of their many applications.

Let us conclude with the first words of Okubo in his introduction to the monograph (Okubo 1995):

The saying that God is the mathematician, so that, even with meager experimental support, a mathematically beautiful theory will ultimately have a greater chance of being correct, has been attributed to Dirac. Octonions algebra may surely be called a beautiful mathematical entity. Nevertheless, it has never been systematically utilized in physics in any fundamental fashion, although some attempts have been made toward this goal. However, it is still possible that non-associative algebras (other than Lie algebras) may play some essential future role in the ultimate theory, yet to be discovered.

2.7. Acknowledgments

This work has been supported by grants MTM2017-83506-C2-1-P (AEI/FEDER, UE) and E22 17R (Gobierno de Aragón, Grupo de referencia "Álgebra y Geometría", co-funded by Feder 2014–2020 "Construyendo Europa desde Aragón").

2.8. References

Adams, J.F. (1958). On the nonexistence of elements of Hopf invariant one. *Bull. Amer. Math. Soc.*, 64, 279–282.

Baez, J.C. (2002). The octonions. *Bull. Amer. Math. Soc. (N.S.)*, 39(2), 145–205.

Borel, A., Serre, J.-P. (1953). Groupes de Lie et puissances réduites de Steenrod. *Amer. J. Math.*, 75, 409–448.

Bott, R., Milnor, J. (1958). On the parallelizability of the spheres. *Bull. Amer. Math. Soc.*, 64, 87–89.

Cartan, E. (1925). Le principe de dualité et la théorie des groupes simples et semi-simples. *Bull. Sci. Math.*, 49, 361–374.

Chernousov, V., Elduque, A., Knus, M.-A., Tignol, J.-P. (2013). Algebraic groups of type D_4, triality, and composition algebras. *Doc. Math.*, 18, 413–468.

Conway, J.H., Smith, D.A. (2003). *On Quaternions and Octonions: Their Geometry, Arithmetic, and Symmetry*. A K Peters, Ltd., Natick.

Cunha, I., Elduque, A. (2007). An extended Freudenthal magic square in characteristic 3. *J. Algebra*, 317(2), 471–509.

Ebbinghaus, H.-D., Hermes, H., Hirzebruch, F., Koecher, M., Mainzer, K., Neukirch, J., Prestel, A., Remmert, R. (1991). *Numbers*, Ewing, J.H. (ed.). With an introduction by Lamotke, K. Translated by Orde, H.L.S. Springer-Verlag, New York.

Elduque, A. (1997). Symmetric composition algebras. *J. Algebra*, 196(1), 282–300.

Elduque, A. (2004). The magic square and symmetric compositions. *Rev. Mat. Iberoamericana*, 20(2), 475–491.

Elduque, A., Myung, H.C. (1990). On Okubo algebras. In *From Symmetries to Strings: Forty Years of Rochester Conferences*, Das, E. (ed.). World Science Publishing, River Edge, 299–310.

Elduque, A., Myung, H.C. (1991). Flexible composition algebras and Okubo algebras. *Comm. Algebra*, 19(4), 1197–1227.

Elduque, A., Myung, H.C. (1993). On flexible composition algebras. *Comm. Algebra*, 21(7), 2481–2505.

Elduque, A., Pérez, J.M. (1996). Composition algebras with associative bilinear form. *Comm. Algebra*, 24(3), 1091–1116.

Elduque, A., Pérez, J.M. (1997). Infinite-dimensional quadratic forms admitting composition. *Proc. Amer. Math. Soc.*, 125(8), 2207–2216.

Elman, R., Karpenko, N., Merkurjev, A. (2008). *The Algebraic and Geometric Theory of Quadratic Forms*. American Mathematical Society, Providence.

Faulkner, J.R. (1988). Finding octonion algebras in associative algebras. *Proc. Amer. Math. Soc.*, 104(4), 1027–1030.

Frobenius, F.G. (1878). Über lineare substitutionen und bilineare formen. *J. Reine Angew. Math.*, 84, 1–63.

Gille, P. (2014). Octonion algebras over rings are not determined by their norms. *Canad. Math. Bull.*, 57(2), 303–309.

Hurwitz, A. (1898). Über die komposition der quadratischen formen von beliebig vielen variablen. *Nachr. Ges. Wiss. Göttingen*, 309–316.

Jacobson, N. (1958). Composition algebras and their automorphisms. *Rend. Circ. Mat. Palermo*, 7(2), 55–80.

Kaplansky, I. (1953). Infinite-dimensional quadratic forms admitting composition. *Proc. Amer. Math. Soc.*, 4, 956–960.

Kervaire, M.A. (1958). Non-parallelizability of the n-sphere for $n > 7$. *Proc. Natl. Acad. Sci.* 44(3), 280–283.

Knus, M.-A., Merkurjev, A., Rost, M., Tignol, J.-P. (1998). *The Book of Involutions*. With a preface in French by J. Tits. American Mathematical Society, Providence.

Okubo, S. (1978). Pseudo-quaternion and pseudo-octonion algebras. *Hadronic J.*, 1(4), 1250–1278.

Okubo, S. (1995). *Introduction to Octonion and Other Non-associative Algebras in Physics*. Cambridge University Press, Cambridge.

Okubo, S., Osborn, J.M. (1981a). Algebras with nondegenerate associative symmetric bilinear forms permitting composition. *Comm. Algebra*, 9(12), 1233–1261.

Okubo, S., Osborn, J.M. (1981b). Algebras with nondegenerate associative symmetric bilinear forms permitting composition II. *Comm. Algebra*, 9(20), 2015–2073.

Petersson, H.P. (1969). Eine Identität fünften Grades, der gewisse Isotope von Kompositions-Algebren genügen. *Math. Z.*, 109, 217–238.

Petersson, H.P. (2005). Letter to the editor: An observation on real division algebras. *European Math. Soc. Newsletter*, 57, 20.

Rodrigues, O. (1840). Des lois géométriques qui régissent les déplacements d'un système solide dans l'espace, et de la variation des coordonnées provenant de ces déplacements considérés indépendamment des causes qui peuvent les produire. *J. de Mathématiques Pures et Appliquées*, 5, 380–440.

Rodríguez-Palacios, A. (2004). *Absolute-valued Algebras, and Absolute-valuable Banach Spaces*, Advanced Courses of Mathematical Analysis. World Science Publishing, Hackensack.

Shapiro, D.B. (2000). *Compositions of Quadratic Forms*. Walter de Gruyter & Co., Berlin.

Springer, T.A., Veldkamp, F.D. (2000). *Octonions, Jordan Algebras and Exceptional Groups*. Springer Monographs in Mathematics, Springer-Verlag, Berlin.

Study, E. (1913). Grundlagen und ziele der analytischer kinematik. *Sitz. Ber. Berliner Math. Gesellschaft*, 12, 36–60.

Tits, J. (1959). Sur la trialité et certains groupes qui s'en déduisent. *Inst. Hautes Études Sci. Publ. Math.*, 2, 13–60.

Urbanik, K., Wright, F.B. (1960). Absolute-valued algebras. *Proc. Amer. Math. Soc.*, 11, 861–866.

van der Blij, F., Springer, T.A. (1959). The arithmetics of octaves and of the group G_2. *Nederl. Akad. Wetensch. Proc. Ser. A 62 = Indag. Math.*, 21, 406–418.

Zorn, M. (1933). Alternativkörper und quadratische systeme. *Abh. Math. Sem. Univ. Hamburg*, 9(1), 395–402.

3

Graded-Division Algebras

Yuri BAHTURIN,[1] Mikhail KOCHETOV[1] and Mikhail ZAICEV[2]

[1] *Department of Mathematics and Statistics, Memorial University of Newfoundland, St. John's, Canada*

[2] *Faculty of Mechanics and Mathematics, Lomonosov Moscow State University, Russia*

3.1. Introduction

Let $\mathcal{A}$ be an algebra over a field $\mathbb{F}$, and let G be a group. A *G-grading* on $\mathcal{A}$ is a vector space decomposition $\Gamma : \mathcal{A} = \bigoplus_{g\in G} \mathcal{A}_g$ such that $\mathcal{A}_g\mathcal{A}_h \subset \mathcal{A}_{gh}$ for all $g, h \in G$. The subspaces $\mathcal{A}_g$ are called the *homogeneous components*. A *G-graded algebra* is an algebra with a fixed G-grading. The non-zero elements $x \in \mathcal{A}_g$ are said to be *homogeneous of degree g*, which can be written as $\deg x = g$, and the *support* of Γ (or of $\mathcal{A}$) is the set $\operatorname{Supp}\Gamma := \{g \in G \mid \mathcal{A}_g \neq 0\}$. A subspace $\mathcal{B}$ in $\mathcal{A}$ (in particular, a subalgebra or ideal) is called *graded* if $\mathcal{B} = \bigoplus_{g\in G} \mathcal{B}_g$, where $\mathcal{B}_g := \mathcal{B} \cap \mathcal{A}_g$.

In the context of G-graded algebras, it is natural to consider graded analogues of the standard concepts. For example, a *homomorphism of G-graded algebras* $\psi : \mathcal{A} \to \mathcal{B}$ is a homomorphism of algebras that preserves degrees: $\psi(\mathcal{A}_g) \subset \mathcal{B}_g$ for all $g \in G$. In particular, $\mathcal{A}$ and $\mathcal{B}$ are *graded-isomorphic* if there exists an isomorphism of graded algebras $\mathcal{A} \to \mathcal{B}$. A G-graded algebra $\mathcal{A}$ is called *graded-simple*, or *simple as a graded algebra*, if $\mathcal{A}^2 \neq 0$ and $\mathcal{A}$ has no graded ideals different from 0 and $\mathcal{A}$.

Among these, we are especially interested in studying the *graded-division* algebras, which are defined by the property that every non-zero homogeneous element is invertible (in an appropriate sense). Their importance in the case of

Algebra and Applications 1,
coordinated by Abdenacer MAKHLOUF.

Artinian algebras is underscored by the following analogue of the classical Wedderburn Theorem (see Nastasescu and Van Oystaeyen (2004) and Elduque and Kochetov (2013)); in Bahturin *et al.* (2011), the importance of graded-division algebras is also shown in the case of infinite-dimensional *finitary* algebras).

THEOREM 3.1 (Graded Wedderburn Theorem).– Any G-graded-Artinian G-graded-simple associative algebra $\mathcal{R}$ is graded isomorphic to $\mathrm{End}_{\mathcal{D}}(V)$, where V is a G-graded finite-dimensional right vector space (more precisely, a free right $\mathcal{D}$-module) over a G-graded-division algebra $\mathcal{D}$.

The grading on $\mathcal{E} = \mathrm{End}_{\mathcal{D}}(V)$ in theorem 3.1 is defined as follows:

$$\mathcal{E}_g = \{\varphi \mid \varphi(V_h) \subset V_{gh} \text{ for any } h \in G\}.$$

In the case $\mathcal{D} = \mathbb{F}$, the above grading on $\mathrm{End}_{\mathbb{F}}(V)$ is called *elementary*. In general, the grading on $\mathrm{End}_{\mathcal{D}}(V)$ is determined by the grading on $\mathcal{D}$, called a *division grading*, and the grading on V. The support of the grading on $\mathcal{D}$ is a subgroup T of G. If $v_1, \ldots, v_k$ is a graded basis of V over $\mathcal{D}$, with $\deg(v_i) = g_i$, then the grading on V is completely defined by the function $\kappa : G/T \to \mathbb{Z}$ such that $\kappa(gT) = \#\{i \mid g_iT = gT\}$.

Thus, the most important component is the determination of graded-division algebras. In the case where $\mathcal{R}$ is simple as an ungraded algebra, the graded-division algebra $\mathcal{D}$ is also simple.

A typical example of a graded-division algebra is the group algebra $\mathcal{R} = \mathbb{F}G$, which is naturally G-graded: $\mathcal{R}_g = \mathbb{F}g$. More generally, the twisted group algebra $\mathbb{F}^\gamma G$ (where $\gamma : G \times G \to \mathbb{F}^\times$ is a 2-cocycle) is a graded-division associative algebra. If the field is algebraically closed, the converse is also true in the finite-dimensional case (for more details, see section 3.3). A classification of associative graded-division algebras over an arbitrary field was given in Karrer (1973) in terms of group actions on ordinary division algebras and group cohomology, but even when these actions and cohomology can be computed, it is not an easy matter to obtain the corresponding graded-division algebras explicitly. Over an algebraically closed field, an explicit (cohomology-free) description of finite-dimensional associative graded-division algebras that have abelian support, and are simple as ungraded algebras, was given in Bahturin *et al.* (2001) and Bahturin and Zaicev (2003).

For non-simple graded-simple algebras, the first result relating graded-simple and simple graded was obtained in Bahturin *et al.* (2001), in the case of finite-dimensional algebras graded by finite abelian groups over algebraically closed fields of characteristic zero. It also works in the case where the field contains primitive nth roots of 1, n being the order of the grading group. A considerably more general result was obtained in a later paper: Allison *et al.* (2008). This paper provided the so-called *loop algebra* construction for graded-simple, not neccesarily associative algebras. The base field does not need to be algebraically closed but the centroid of the algebra must be isomorphic to a group algebra, in which case the centroid is said to be *split*. It is always the case if the algebra is graded-central over an algebraically closed field. This construction generalizes the so-called multiloop algebras in Lie theory and, if $\mathbb{F}$ is algebraically closed, yields a classification of graded- central-simple (i.e. graded-central and graded-simple) algebras in a given class, provided that a classification of gradings is known for central simple algebras in this class (see section 3.2.6 for details).

Finite-dimensional graded-division associative algebras over the field of real numbers $\mathbb{R}$ have been classified in Bahturin and Zaicev (2018) up to equivalence (see section 3.2). That work was preceded by Bahturin and Zaicev (2016), where the *division gradings* on finite-dimensional simple real associative algebras (i.e. the gradings that turn them into graded-division algebras) have been classified up to equivalence, and Rodrigo-Escudero (2016), where these gradings have been classified up to isomorphism, as well as up to equivalence.

In Bahturin and Kochetov (2019), the authors have tackled the classification problem in the case of real graded-division alternative algebras and Jordan algebras. The loop algebra construction does not always work in this case, so they suggested a modified real loop algebra construction. Using this construction, they also provided the classification of real graded-simple and graded-division alternative algebras and so-called Jordan algebras of degree 2. Note that the modified loop construction also allows one to classify associative graded-division algebras over $\mathbb{R}$, up to isomorphism. It should be mentioned that an essential part in the classification of real graded-division alternative algebras is the classification of division gradings on the octonion algebras. Group gradings on these algebras over any field have been described in Elduque (1998) (see also Elduque and Kochetov (2013)). A classification of gradings up to isomorphism is also given in Elduque and Kochetov (2013).

Abelian group gradings on finite-dimensional simple special Jordan algebras (for general references about Jordan algebras, see Jacobson (1968)) over an algebraically closed field of characteristic 0 were described in Bahturin *et al.* (2005) and classified up to isomorphism and up to equivalence in Elduque and Kochetov (2013) for any characteristic different from 2. The finite-dimensional central simple special Jordan algebras of degree ≥ 3 over $\mathbb{R}$ can be dealt with following the approach of Bahturin *et al.* (2018), where a classification of gradings for classical central simple Lie algebras over $\mathbb{R}$ was obtained by transferring the problem to associative algebras with involution. Gradings on simple Jordan algebras of degree 2, i.e. Jordan algebras of bilinear forms, were described in Bahturin and Shestakov (2001) and classified in Elduque and Kochetov (2013).

If we deal with graded-simple Jordan algebras that are not special, we need to know gradings on Albert algebra. It should be noted that abelian group gradings on the Albert algebra over $\mathbb{C}$ were classified in Draper and Martín González (2009) and over any algebraically closed field of characteristic different from 2 in Elduque and Kochetov (2012) (see also Elduque and Kochetov (2013)). In particular, it has a division grading by the group $\mathbb{Z}_3^3$ (which appeared in Griess (1990)) if the characteristic is not 3. However, none of the three real forms of the complex Albert algebra admits a division grading, as follows from the classification of fine gradings in Calderón Martín *et al.* (2010) (since the identity component always contains non-trivial idempotents).

In conclusion, let us note that one might ask: can one expand the classification of graded-division algebras to real algebras that are not alternative or Jordan? The classical topological theorem given by Bott and Milnor (1958) stating that the sphere S^n is parallelizable if and only if $n = 1$, 3 or 7 implies that the dimensions of any real division algebra are equal to 1, 2, 4 or 8. At the same time, the classification of eight-dimensional real division algebras is nowhere near completion.

Another question: what can we say about the classification of graded-division algebras over the fields that are not algebraically or real closed? Here, we come to a much more specific question: what are the gradings of field extensions? At the end of this survey, we give the answer in the case of finite fields.

3.2. Background on gradings

In this section, we give some definitions relevant to the classification of graded-division algebras and describe some tools that proved useful.

3.2.1. *Gradings induced by a group homomorphism*

Given a G-grading $\Gamma : \mathcal{A} = \bigoplus_{g\in G} \mathcal{A}_g$ and a group homomorphism $\alpha : G \to G'$, we can define a G'-grading on $\mathcal{A}$ as follows: for any $g' \in G'$, set $\mathcal{A}'_{g'} := \bigoplus_{g\in\alpha^{-1}(g')} \mathcal{A}_g$. Then $\mathcal{A} = \bigoplus_{g'\in G'} \mathcal{A}'_{g'}$ is a G'-grading, which will be denoted by ${}^\alpha\Gamma$. Thus, if α is fixed, every G-graded algebra $\mathcal{A}$ can be considered as a G'-graded algebra, which will be denoted by ${}^\alpha\mathcal{A}$ or just $\mathcal{A}$ if there is no risk of confusion. In particular, if G is a subgroup of G', then every G-graded algebra is also G'-graded, with the same support and homogeneous components. More generally, if α is injective on the support, then ${}^\alpha\Gamma$ has the same components as Γ, but relabeled through α. Otherwise, ${}^\alpha\Gamma$ is a so-called *proper coarsening* of Γ.

If $\psi : \mathcal{A} \to \mathcal{B}$ is a homomorphism of G-graded algebras and we consider $\mathcal{A}$ and $\mathcal{B}$ as G'-graded through α, as above, then ψ is also a homomorphism of G'-graded algebras. Thus, α determines a functor F_α from the category of G-graded algebras to the category of G'-graded algebras: $F_\alpha(\mathcal{A}) = {}^\alpha\mathcal{A}$ and $F_\alpha(\psi) = \psi$.

3.2.2. *Weak isomorphism and equivalence*

We call a G-graded algebra $\mathcal{A}$ and a G'-graded algebra $\mathcal{B}$ *graded-weakly isomorphic* if ${}^{\alpha}\mathcal{A}$ is graded-isomorphic to $\mathcal{B}$ for some group isomorphism $\alpha : G \to G'$, i.e. there exists an isomorphism of algebras $\psi : \mathcal{A} \to \mathcal{B}$ such that $\psi(\mathcal{A}_g) = \mathcal{B}_{\alpha(g)}$ for all $g \in G$.

There is a more general kind of relabeling: we say that a G-graded algebra $\mathcal{A}$ and a G'-graded algebra $\mathcal{B}$ are *graded-equivalent* if there exists an isomorphism of algebras $\psi : \mathcal{A} \to \mathcal{B}$ such that, for each $g \in G$, there is $g' \in G'$ such that $\psi(\mathcal{A}_g) = \mathcal{B}_{g'}$.

The support of a division grading on an alternative (in particular, associative) algebra is a subgroup. If, for two such graded-division algebras, we replace the grading groups by the supports, then their equivalence is the same as weak isomorphism.

3.2.3. *Basic properties of division gradings*

In this section, we fix a few well-known useful properties of graded-division algebras (see Elduque and Kochetov (2013, Chapter 2)).

LEMMA 3.1.– Let $\Gamma : \mathcal{R} = \bigoplus_{g\in G} \mathcal{R}_g$ be a grading by a group G on an (associative) algebra $\mathcal{R}$ over a field $\mathbb{F}$. If Γ is a division grading then the following holds:

1) the identity component $\mathcal{R}_e$ of Γ is a division algebra over $\mathbb{F}$;

2) given $g \in G$ and a non-zero $a \in \mathcal{R}_g$, we have $\mathcal{R}_g = aR_e$;

3) for any $g \in G$, $\dim \mathcal{R}_g = \dim \mathcal{R}_e$ and $\dim \mathcal{R} = |\operatorname{Supp}\Gamma| \dim \mathcal{R}_e$;

4) $\operatorname{Supp}(\Gamma)$ is a subgroup of G isomorphic to the universal group $U(\Gamma)$. □

In the case where the base field is $\mathbb{R}$, it follows that $\mathcal{R}_e$ is one of $\mathbb{R}$, $\mathbb{C}$ or $\mathbb{H}$, the division algebra of quaternions.

As mentioned above, the support of the division grading is a subgroup in the grading group. This makes it natural to always assume that the support of $\mathcal{R}$ equals the whole of G. Thus, when we speak about gradings on finite-dimensional graded-division algebras, we may assume that the grading group G is finite.

One notational remark: given an element g of order n in a group G, we denote by $(g)_n$ the cyclic subgroup generated by g. Given vectors $v_1, \ldots, v_m$ in a vector space V over a field $\mathbb{F}$, we denote by $\langle v_1, \ldots, v_m\rangle_{\mathbb{F}}$, or simply $\langle v_1, \ldots, v_m\rangle$ the linear span of $v_1, \ldots, v_m$, with coefficients in $\mathbb{F}$. To avoid confusion with $1 \in \mathbb{F}$, we often denote the identity element of a graded-division algebra $\mathcal{R}$ by I.

3.2.4. *Graded presentations of associative algebras*

Let $\mathcal{F}(X)$ be the free unital associative algebra with the set X of free generators over a field $\mathbb{F}$. One often calls the elements of $\mathcal{F}(X)$ non-commutative polynomials in the set of (non-commutative) variables X. One can evaluate them in any unital associative algebra $\mathcal{A}$. If $f(x_1,\dots,x_n) \in \mathcal{F}(X)$ is a non-commutative polynomial, depending on $x_1,\dots,x_n \in X$, and $a_1,\dots,a_n \in \mathcal{A}$, then $f(a_1,\dots,a_n)$ is the image of $f(x_1,\dots,x_n)$ under the homomorphism of $\mathcal{F}$ to $\mathcal{A}$ extending the map $x_1 \mapsto a_1,\dots,x_n \mapsto a_n$.

Now let R be a subset of $\mathcal{F}(X)$ and $I(R)$ the two-sided ideal of $\mathcal{F}(X)$ generated by R. Then we say that the algebra $\mathcal{A} = \mathcal{F}(X)/I(R)$ is presented in terms of generators X and defining relations $\{r = 0 \mid r \in R\}$. The set of elements $\overline{x} = x + I(R)$, where $x \in X$, generates $\mathcal{A}$, and if $r = r(x_1,\dots,x_n) \in R$, then $r(\overline{x}_1,\dots,\overline{x}_n) = 0$ in $\mathcal{A}$. We write $\mathcal{A} = \langle X \mid R\rangle$. If $X = \{x_1,\dots,x_n\}$ is finite, then $\mathcal{A}$ is *finitely generated.* If $R = \{r_1,\dots,r_m\}$ is finite, then $\mathcal{A}$ is called *finitely related.* A finitely generated and finitely related algebra is called *finitely presented.* Its presentation is often written as

$$\mathcal{A} = \langle x_1,\dots,x_n \mid r_1 = 0,\dots,r_m = 0\rangle.$$

An important observation is that if $\mathcal{B}$ is another algebra generated by some elements $b_1,\dots,b_n$ so that $r_1(b_1,\dots,b_n) = 0,\dots,r_m(b_1,\dots,b_n) = 0$, then the mapping $\overline{x}_1 \mapsto b_1,\dots,\overline{x}_n \mapsto b_n$ extends to a surjective homomorphism φ of $\mathcal{A}$ onto $\mathcal{B}$. If $\mathcal{A}$ and $\mathcal{B}$ are finite-dimensional and $\dim\mathcal{B} \ge \dim\mathcal{A}$, then φ is an isomorphism.

Given a group G and a mapping $\delta : X \to G$, one can give $\mathcal{F} = \mathcal{F}(X)$ a G-grading

$${}^{\delta}\Gamma : \mathcal{F} = \bigoplus_{g\in G} \mathcal{F}_g$$

if we set

$$\mathcal{F}_g = \operatorname{span}\{x_1x_2\cdots x_n \mid x_1,x_2,\dots x_n \in X \text{ and } g = \delta(x_1)\delta(x_2)\cdots\delta(x_n)\}.$$

This grading is induced from the natural F-grading on $\mathcal{F}$, where F is the free group with the set X of free generators, by the group homomorphism $F \to G$ that extends the mapping δ.

If $\mathcal{A} = \langle X \mid R\rangle$ and the elements of R are all homogeneous with respect to ${}^{\delta}\Gamma$, then $I(R)$ is a graded subspace of $\mathcal{F}$, so $\mathcal{A}$ naturally acquires a G-grading ${}^{\delta}\Gamma_R : \mathcal{A} = \bigoplus_{g\in G}\mathcal{A}_g$ with $\mathcal{A}_g = (\mathcal{F}_g + I(R))/I(R)$. In this case, we say that $\langle X \mid R\rangle$ is a *graded presentation* for $\mathcal{A}$ in terms of graded generating set X and graded defining relations R.

It is easy to see that any graded algebra $\mathcal{A} = \bigoplus_{g\in G} \mathcal{A}_g$ can be given a graded presentation $\langle X \mid R\rangle$. Indeed, choose a graded basis E of $\mathcal{A}$ as the set X of free generators and define $\delta(x) = g$ if $x \in \mathcal{A}_g$. Now, in $\mathcal{A}$, for $x \in E\cap\mathcal{A}_g$ and $y \in E\cap\mathcal{A}_h$, we have $xy = \sum_{z\in E} c_{xy}^z z \in \mathcal{A}_{gh}$, where $c_{xy}^z \in \mathbb{F}$ are the structure constants of $\mathcal{A}$. As a result, all the defining relations $xy - \sum_{z\in E} c_{xy}^z z \in \mathfrak{F}$ are homogeneous elements with respect to ${}^{\delta}\Gamma$. It is clear that ${}^{\delta}\Gamma_R$ is the original grading on $\mathcal{A}$.

3.2.4.1. *Power norm residue symbols*

An example that follows is borrowed from Milnor (1971, Chapter 15). By providing this example, we would like to show the ubiquity of the approach via generators and defining relations.

Let $\mathbb{F}$ be a field containing a primitive nth root of 1, denoted by ω. The *Steinberg symbol* a_ω on $\mathbb{F}$ with values in the Brauer group of $\mathbb{F}$ is defined as follows. Given elements $\alpha, \beta \in \mathbb{F}^\times$, let

$$A_\omega(\alpha, \beta)$$

be the associative algebra with unit, of dimension n^2 over $\mathbb{F}$, which is generated by two elements x and y, subject to the relations

$$x^n = \alpha 1,\ y^n = \beta 1,\ yx = \omega xy;$$

where 1 denotes the identity element of A. Thus the monomials $x^i y^j$ with $0 \le i, j < n$ form a basis for A over $\mathbb{F}$.

It is clear that $A_\omega(\alpha, \beta)$ can be canonically given a graded presentation

$$\langle x, y \mid xy = \omega yx, x^n = \alpha, y^n = \beta\rangle$$

with the grading group $\mathbb{Z}_n \times \mathbb{Z}_n$ and the degree function $\delta(x) = (1, 0), \delta(y) = (0, 1)$.

A unital algebra over $\mathbb{F}$ is called *central* if its center is $\mathbb{F}\cdot 1$. A finite-dimensional central simple algebra $\mathcal{A}$ is isomorphic to the matrix algebra $M_n(\mathcal{D})$ where the (central) division algebra $\mathcal{D}$ is uniquely determined by $\mathcal{A}$, up to isomorphism. The tensor product of two central simple algebras, $M_n(\mathcal{D}_1)$ and $M_m(\mathcal{D}_2)$, is again a central simple algebra, $M_k(\mathcal{D}_3)$. Defining multiplication $\mathcal{D}_1 \cdot \mathcal{D}_2 = \mathcal{D}_3$ and taking into account that $\mathcal{D} \cdot \mathcal{D}^{\mathrm{op}} \cong \mathrm{End}_{\mathbb{F}}(\mathcal{D})$, one makes the set $\mathrm{Br}(\mathbb{F})$ of (isomorphism classes of) finite-dimensional central division algebras over $\mathbb{F}$ into an abelian group, called the *Brauer group* of $\mathbb{F}$.

Now since $A_\omega(\alpha, \beta)$ is central simple, it is isomorphic to a matrix algebra over a division algebra, which is denoted by $a_\omega(\alpha, \beta)$. A curious fact, when compared with the construction of complex graded-division algebras in section 3.3, is that the mapping $a_\omega : \mathbb{F}^\times \times \mathbb{F}^\times \to \mathrm{Br}(\mathbb{F})$ is as alternating bilinear function. In addition, it satisfies the so-called *Steinberg identity* $a_\omega(1-\beta, \beta) = 1$.

Note that $A_\omega(\alpha, \beta)$ itself is a graded-division algebra with respect to the above grading by $G = \mathbb{Z}_n \times \mathbb{Z}_n$. It belongs to the so-called *graded Brauer group* $\mathrm{Br}_G(\mathbb{F})$, which consists of (isomorphism classes of) finite-dimensional central simple algebras over $\mathbb{F}$ equipped with a division G-grading. The multiplication in this group is induced by the tensor product of G-graded algebras. For an abelian group G, the tensor product $\mathcal{A} \otimes \mathcal{B}$ of G-graded algebras $\mathcal{A}$ and $\mathcal{B}$ is again a G-graded algebra if we define the homogeneous component of degree g to be the span of all elements of the form $a \otimes b$ where $a \in \mathcal{A}_{g_1}$, $b \in \mathcal{B}_{g_2}$, and $g_1 g_2 = g$. (A different concept of the tensor product of graded algebras, which is graded by the direct product of groups, will be presented in section 3.2.5.) For details on and far-reaching generalizations of the graded Brauer group, the reader may refer to Caenepeel (1998).

Examples of the above construction are $\mathbb{Z}_2 \times \mathbb{Z}_2$-graded real algebras given by presentations $\mathbb{H}^{(4)} = \langle x, y \mid xy = -yx, x^2 = y^2 = -1\rangle$ and $M_2^{(4)} = \langle x, y \mid xy = -yx, x^2 = y^2 = 1\rangle$. Their quaternionic matrix representations are given in [3.6] and [3.7], respectively.

3.2.4.2. *Clifford gradings*

Another source of gradings, which is essential for the case of algebras over the field $\mathbb{R}$ of real numbers is as follows. Let (V, Q) be a real finite-dimensional vector space endowed with a non-singular quadratic form $Q : V \to \mathbb{R}$. Let $T(V)$ be the tensor algebra of V and J_Q the ideal generated by all elements $x \otimes x + Q(x)1$, where $x \in V$. Then $C(V, Q) = T(V)/J_Q$ is called the *Clifford algebra* of (V, Q). It is well known that if Q is non-singular, with positive inertia index p and negative inertia index q, and V has even dimension $n = 2k$, then $C(V, Q)$ is central simple, that is, either $M_{2^k}(\mathbb{R})$ or $M_{2^{k-1}}(\mathbb{H})$. More precisely, if $p - q \equiv 0$ or $6 \pmod 8$, then $C(V, Q) \cong M_{2^k}(\mathbb{R})$, and if $p - q \equiv 2$ or $4 \pmod 8$, then $C(V, Q) \cong M_{2^{k-1}}(\mathbb{H})$. Also, if $n = 2k + 1$ and $p - q \equiv 1 \pmod 4$, then $C(V, Q) \cong M_{2^k}(\mathbb{C})$.

Now let G be an elementary abelian 2-group and $g_1, \ldots, g_n \in G$. Note that the tensor algebra $T(V)$ is a free associative algebra. Any basis of V can serve as the free generating set for $T(V)$. Choose an orthogonal basis $e_1, \ldots, e_n$ in V with respect to the bilinear form associated with Q. We may assume that $Q(e_1) = \ldots = Q(e_p) = 1$ and $Q(e_{p+1}) = \ldots = Q(e_n) = -1$. Assign $\delta(e_i) = g_i$. The defining relations of $C(V, Q)$ are $e_i e_j + e_j e_i = 0$, for all $1 \leq i < j \leq n$, $e_1^2 = \ldots = e_p^2 = -1$ and $e_{p+1}^2 = \ldots = e_n^2 = 1$. As a result, $C(V, Q)$ becomes a graded algebra. It is well known that the ordered monomials $e_{i_1} \cdots e_{i_k} = e_{i_1} \otimes \cdots \otimes e_{i_k} + J_q$, with $i_1 < \ldots < i_k$, form a basis of $C(V, Q)$. The degree of $e_{i_1} \cdots e_{i_k}$ is the product $g_{i_1} \cdots g_{i_k}$. This shows that if $g_1, \ldots, g_n$ are linearly independent in G, as a vector space over $\mathbb{Z}_2$, then all homogeneous components are one-dimensional and spanned by invertible elements. As a result, $C(V, Q)$ is a graded-division algebra. We will sometimes write $C(p, q)$ instead of $C(V, Q)$.

PROPOSITION 3.1.– Each simple algebra $R = M_{2^k}(D)$, $D = \mathbb{R}, \mathbb{C}, \mathbb{H}$ acquires a fine division grading by $\mathbb{Z}_2^{2k}$ if one identifies R with an appropriate $C(p, q)$.

Clifford algebras can also serve as a source of division gradings on $M_n(\mathbb{C})$, viewed as an algebra over $\mathbb{R}$. For example, $C(p, q)$ with $p - q \equiv 1 \pmod 4$, can be endowed with a division grading by $\mathbb{Z}_4 \times \mathbb{Z}_2^m$. Let us illustrate with $M_2(\mathbb{C})$ and $M_4(\mathbb{C})$. In the case of $R = C(2, 1)$, we choose $G = (\alpha)_4 \times (\beta)_2$, set $R_\alpha = \langle e_1 + e_2e_3 \rangle$ and $R_\beta = \langle e_2 \rangle$. Then, since $R_gR_h = R_{gh}$, we obtain

$$R_{\alpha^2} = \langle e_1e_2e_3 \rangle, \; R_{\alpha^3} = \langle e_1 - e_2e_3 \rangle.$$

Also

$$R_{\alpha\beta} = \langle e_3 + e_1e_2 \rangle, \; R_{\alpha^2\beta} = \langle e_1e_3 \rangle, \; R_{\alpha^3\beta} = \langle e_3 - e_1e_2 \rangle.$$

If we use one of the standard identifications (called *Clifford maps*, see (Garding 2011, p.121)) of $C(2, 1)$ with $M_2(\mathbb{C})$, given by $e_1 \mapsto iB$, $e_2 \mapsto C$, $e_3 \mapsto A$, then we will find the homogeneous components as

$$R_e = \langle I \rangle, \; R_\alpha = \langle \omega A \rangle, \; R_{\alpha^2} = \langle iI \rangle, \; R_{\alpha^3} = \langle i\omega A \rangle.$$

$$R_\beta = \langle C \rangle, \; R_{\alpha\beta} = \langle \omega B \rangle, \; R_{\alpha^2\beta} = \langle iC \rangle, \; R_{\alpha^3\beta} = \langle i\omega B \rangle.$$

Here, for brevity, instead of $1 + i$, we used $\omega = \frac{1}{\sqrt{2}}(1 + i)$, the eighth root of 1 such that $\omega^2 = i$. We denote this grading by $M_2^{(8)}$. A $\mathbb{Z}_4$-division grading of $M_2(\mathbb{C})$, which is not fine, appears as the coarsening of $M_2^{(8)}$ by means of identifying β and e.

$$R_e = \langle I, C \rangle, \; R_\alpha = \langle \omega A, \omega B \rangle, \; R_{\alpha^2} = \langle iI, iC \rangle, \; R_{\alpha^3} = \langle i\omega A, i\omega B \rangle.$$

In this case, the identity component of the grading is isomorphic to $\mathbb{C}$. We denote this algebra as $M_2(\mathbb{C}, \mathbb{Z}_4)$. The algebras $M_2^{(8)}$ and $M_2(\mathbb{C}, \mathbb{Z}_4)$ play an important role in the classification of division gradings on simple real algebras.

The graded presentations of algebras, in terms of graded generators and defining relations, are given as

$$\begin{aligned} M_2^{(8)} &= \langle x_1, x_2 \mid x_1^4 = x_2^2 = -1, \; x_1x_2 = -x_2x_1 \rangle, \delta(x_1) = \beta, \delta(x_2) = \alpha, \\ M_2(\mathbb{C}, \mathbb{Z}_4) &= \langle x_1, x_2 \mid x_1^4 = x_2^2 = -1, \; x_1x_2 = -x_2x_1 \rangle, \delta(x_1) = e, \delta(x_2) = \alpha. \end{aligned} \qquad [3.1]$$

3.2.5. *Tensor products of division gradings*

Given groups $G_1, G_2, \ldots, G_m$ and G_k-graded algebras $\mathcal{R}_1, \mathcal{R}_2, \ldots, \mathcal{R}_m$, $k = 1, \ldots, m$, one can endow the tensor product of algebras $\mathcal{R} = \mathcal{R}_1 \otimes \mathcal{R}_2 \otimes \cdots \otimes \mathcal{R}_m$ with a $G_1 \times G_2 \times \cdots \times G_m$-grading, called the *(outer) tensor product of gradings* if one sets

$$\mathcal{R}_{(g_1, g_2, \ldots, g_m)} = (\mathcal{R}_1)_{g_1} \otimes (\mathcal{R}_2)_{g_2} \otimes \cdots \otimes (\mathcal{R}_m)_{g_m},$$

where $g_k \in G_k$, for all $k = 1, \ldots, m$.

This should not be confused with the tensor product of G-graded algebras for an abelian group G. If $\mathcal{R}_1$ and $\mathcal{R}_2$ are G-graded algebras, then $\mathcal{R}_1 \otimes \mathcal{R}_2$ is a G-graded algebra, with respect to the G-grading induced from the tensor product of gradings in the above sense by the group homomorphism $G \times G \to G$ sending $(g_1, g_2) \mapsto g_1 g_2$. Note that if G is the (inner) direct product of its subgroups $G_1, G_2, \ldots, G_m$, then the two concepts coincide: each G_k-graded algebra $\mathcal{R}_k$, $k = 1, \ldots, m$, can be considered as a G-graded algebra (with support contained in G_k) and their tensor product $\mathcal{R}_1 \otimes \mathcal{R}_2 \otimes \cdots \otimes \mathcal{R}_m$ as G-graded algebras has the tensor product of gradings in the above sense.

When we work with finite-dimensional associative algebras over an algebraically closed field $\mathbb{F}$, the tensor product of division gradings is a division grading. This is no longer true if $\mathbb{F}$ is not algebraically closed. For example, if $\mathbb{F} = \mathbb{R}$, then $D_1 \otimes D_2$, where each $D_i = \mathbb{R}, \mathbb{C}, \mathbb{H}$ is a division algebra with trivial grading, becomes a division algebra with trivial grading if and only if at least one of D_i is $\mathbb{R}$. If $\mathcal{R}$ is a G-graded-division algebra and $\mathcal{S}$ is an H-graded-division algebra, then all non-zero elements in a homogeneous component $(\mathcal{R} \otimes \mathcal{S})_{(g,h)} = \mathcal{R}_g \otimes \mathcal{S}_h$ are invertible if this is true for the identity component. As a result, a tensor product of two division gradings is a division grading if and only if the identity component of one of them is one-dimensional. This condition is not sufficient if we want to obtain a *simple* graded-division algebra. Clearly, it is necessary that both tensor factors be simple algebras. And even this is not sufficient, as shown by the example of $\mathbb{C}^{(2)} \otimes \mathbb{C}^{(2)}$ (see the definition of $\mathbb{C}^{(2)}$ and similar notations in section 3.4).

Another question is the equivalence of different tensor product gradings. In the ungraded case, it is well known (see (Garding 2011, Section 3.6)) that $\mathbb{H} \otimes \mathbb{C} \cong M_2(\mathbb{R}) \otimes \mathbb{C} \cong M_2(\mathbb{C})$ and $\mathbb{H} \otimes \mathbb{H} \cong M_2(\mathbb{R}) \otimes M_2(\mathbb{R}) \cong M_4(\mathbb{R})$.

These isomorphisms combine well with the gradings in the case where the grading of each factor is fine. In the first case, the isomorphism is given by $\varphi : i \otimes 1 \mapsto B \otimes i$, $j \otimes 1 \mapsto C \otimes 1$, $k \otimes 1 \mapsto A \otimes i$. Now $\mathcal{R} = \mathbb{H}^{(4)} \otimes \mathbb{C}^{(2)}$ is graded by $G = (\alpha)_2 \times (\beta)_2) \times (\gamma)_2$ so that $\deg i \otimes 1 = \alpha$, $\deg j \otimes 1 = \beta$, $\deg 1 \otimes i = \gamma$.

Let us choose the same group to grade $\mathcal{S} = M_2^{(4)} \otimes \mathbb{C}^{(2)}$ by $\deg A \otimes 1 = \alpha\beta\gamma$, $\deg B \otimes 1 = \alpha\gamma$, $\deg I \otimes i = \gamma$. Then the supports of $M_2^{(4)}$, equal to $(\alpha\beta\gamma)$, $(\alpha\gamma))$, and $\mathbb{C}^{(2)}$, equal to (γ), have trivial intersection, so this grading is a tensor product grading. Also, $\alpha \mapsto \alpha\beta\gamma$, $\beta \mapsto \alpha\gamma$, $\gamma \mapsto \gamma$ is an automorphism of G. This proves that $\varphi : \mathcal{R} \to \mathcal{S}$ is a weak isomorphism of $\mathbb{H}^{(4)} \otimes \mathbb{C}^{(2)}$ and $M_2^{(4)} \otimes \mathbb{C}^{(2)}$.

In a similar fashion, one establishes the weak isomorphism of $\mathcal{R} = \mathbb{H}^{(4)} \otimes \mathbb{H}^{(4)}$ and $\mathcal{S} = M_2^{(4)} \otimes M_2^{(4)}$. In this case, the graded equivalence $\psi : \mathcal{R} \to \mathcal{S}$ is given by $\psi(i \otimes 1) = C \otimes I$, $\psi(j \otimes 1) = A \otimes C$, $\psi(1 \otimes i) = I \otimes C$, $\psi(1 \otimes j) = C \otimes A$.

At the same time, there is no graded equivalence between $\mathcal{R} = \mathbb{H}^{(4)} \otimes \mathbb{H}$, graded by $\mathbb{Z}_2 \times \mathbb{Z}_2$ and $\mathcal{S} = M_2(\mathbb{R}) \otimes M_2(\mathbb{R})$, as graded tensor products. Indeed, if we give trivial grading to the second factor in $\mathcal{S}$, then it is not a graded-division algebra. If we grade both factors by $\mathbb{Z}_2$, then the identity component of the tensor product will be isomorphic to $\mathbb{C} \otimes \mathbb{C}$, which is not a division algebra. Still, the map $\psi : \mathbb{H}^{(4)} \otimes \mathbb{H} \to M_4(\mathbb{R})$ transfers the structure of a graded-division algebra, which is not a tensor product of graded algebras of smaller dimension. We will denote this grading by $M_4^{(4)}$. Its identity component is four-dimensional, hence isomorphic to the quaternion algebra $\mathbb{H}$.

Note one more equivalence of graded tensor products: $\mathcal{R} = M_2(\mathbb{C}, \mathbb{Z}_4) \otimes \mathbb{H}^{(4)}$ and $\mathcal{S} = M_2(\mathbb{C}, \mathbb{Z}_4) \otimes M_2^{(4)}$. The presentation of $\mathcal{R}$ in terms of generators and defining relations is

$$\langle x_1, x_2, y_1, y_2 \mid x_1^4 = x_2^2 = y_1^2 = y_2^2 = -1,$$
$$x_1x_2 = -x_2x_1, y_1y_2 = -y_2y_1, x_iy_j = y_jx_i\rangle.$$

Here $\delta(x_1) = \alpha$, $\delta(x_2) = e$, $\delta(y_1) = \beta$, $\delta(y_2) = \gamma$, where $o(\alpha) = 4$, $o(\beta) = o(\gamma) = 2$.

The presentation of $\mathcal{S}$ in terms of generators and defining relations is

$$\langle x_1, x_2, z_1, z_2 \mid x_1^4 = x_2^2 = -z_1^2 = -z_2^2 = -1,$$
$$x_1x_2 = -x_2x_1, z_1z_2 = -z_2z_1, x_iz_j = z_jx_i\rangle.$$

Here $\delta(x_1) = \alpha$, $\delta(x_2) = e$, $\delta(z_1) = \beta$, $\delta(z_2) = \gamma$, where $o(\alpha) = 4$, $o(\beta) = o(\gamma) = 2$. Clearly, in the first presentation, we can perform a graded change of generators $z_1 = y_1x_1^2$ and $z_2 = y_2x_1^2$, which will lead to the equivalence of $\mathcal{R}$ and $\mathcal{S}$.

A much more general approach to the equivalence of graded tensor products will be discussed in section 3.4.

3.2.6. *Loop construction*

3.2.6.1. *Centroid of a graded algebra*

Given a G-graded algebra $\mathcal{B}$ over a field $\mathbb{F}$, the centroid $\mathcal{C}(\mathcal{B})$ is the subalgebra in $\mathrm{End}(\mathcal{B})$ consisting of the maps $\varphi : \mathcal{B} \to \mathcal{B}$ such that, for all $a, b, x \in \mathcal{B}$, one has $a\varphi(x) = \varphi(ax)$ and $\varphi(xb) = \varphi(x)b$. If $\mathcal{B}$ is unital with identity element 1, then the map $\varphi \mapsto \varphi(1)$ identifies $\mathcal{C}(\mathcal{B})$ with the center of $\mathcal{B}$, that is, the set of all elements of $\mathcal{B}$ that commute and associate with all other elements.

A linear map $f : \mathcal{B} \to \mathcal{B}$ is said to be homogeneous of degree $g \in G$ if $f(\mathcal{B}_h) \subset \mathcal{B}_{gh}$ for all $h \in G$. In particular, the set of all elements of the centroid $C(\mathcal{B})$ that satisfy this condition will be denoted by $C(\mathcal{B})_g$.

If $\psi : \mathcal{B} \to \mathcal{B}'$ is an isomorphism of G-graded algebras, then we have an isomorphism of the centroids $C(\psi) : C(\mathcal{B}) \to C(\mathcal{B}')$ defined by $C(\psi)(c) = \psi \circ c \circ \psi^{-1}$ for all $c \in C(\mathcal{B})$. (If the algebras are unital and we identify the centroids with the centers, then $C(\psi)$ is just a restriction of ψ.) It is clear that $C(\psi)$ maps $C(\mathcal{B})_g$ onto $C(\mathcal{B}')_g$ for any $g \in G$.

Suppose $\mathcal{B}$ is graded-simple. Then, by (Benkart and Neher 2006, Proposition 2.16), we have $C(\mathcal{B}) = \bigoplus_{g\in G} C(\mathcal{B})_g$, $C(\mathcal{B})$ is a commutative associative graded-division algebra (a *graded-field*), and $\mathcal{B}$ is a graded $C(\mathcal{B})$-module (hence free). Moreover, the graded-simple algebra $\mathcal{B}$ is simple if and only if $C(\mathcal{B})$ is a field (Allison *et al.* 2008, Lemma 4.2.2). In any case, the identity component $C(\mathcal{B})_e$ is a field containing $\mathbb{F}$, and $\mathcal{B}$ is a graded-central algebra over $C(\mathcal{B})_e$, with exactly the same homogeneous components, but now viewed as $C(\mathcal{B})_e$-subspaces. For this reason, it is natural to restrict ourselves to the graded-central case.

So, suppose $\mathcal{B}$ is a graded-central-simple algebra over $\mathbb{F}$. Then the homogeneous components of $C(\mathcal{B})$ are one-dimensional: $C(\mathcal{B})_h = \mathbb{F}u_h$ for every $h \in H$ where H is the support of $C(\mathcal{B})$, which is a subgroup of G, and u_h is an invertible element. Thus, $C(\mathcal{B})$ is a twisted group algebra of H, with its natural H-grading considered as a G-grading. Following Allison *et al.* (2008), we will say that the centroid is *split* if $C(\mathcal{B})$ is graded-isomorphic to the group algebra $\mathbb{F}H$. This is always the case if $\mathbb{F}$ is algebraically closed (Allison *et al.* 2008, Lemma 4.3.8), but not if $\mathbb{F}$ is the field of real numbers. For example, $C(\mathcal{B})$ can be the field $\mathbb{C}$ with a non-trivial grading: $\mathbb{C} = \mathbb{C}_e \oplus \mathbb{C}_h$ where $\mathbb{C}_e = \mathbb{R}$, $\mathbb{C}_h = \mathbb{R}i$, and h is an element in G of order 2. The loop construction in Allison *et al.* (2008), which we are now going to review, produces graded-central-simple algebras with split centroid. In section 3.5, we will extend this construction to cover all possible centroids in the case $\mathbb{F} = \mathbb{R}$ (or any real closed field). Another (more complicated) extension is given in Elduque (2019) for an arbitrary field $\mathbb{F}$.

3.2.6.2. *Loop algebras with a split centroid*

Let $\pi : G \to \overline{G}$ be an epimorphism of abelian groups and let H be the kernel of π. As discussed in section 3.2.1, any G-graded algebra $\mathcal{B}$ can be regarded as a $\overline{G}$-graded algebra, by setting $\mathcal{B}_{\overline{g}} = \bigoplus_{g\in\pi^{-1}(\overline{g})} \mathcal{B}_g$ for any $\overline{g} \in \overline{G}$, and this gives us a functor from the category of G-graded algebras to the category of $\overline{G}$-graded algebras. The loop construction is the right adjoint of this functor, defined as follows. For a given $\overline{G}$-graded algebra $\mathcal{A}$, consider the tensor product $\mathcal{A} \otimes \mathbb{F}G$. The loop algebra $L_\pi(\mathcal{A})$ is the following subalgebra of $\mathcal{A} \otimes \mathbb{F}G$:

$$L_\pi(\mathcal{A}) = \bigoplus_{g\in G} \mathcal{A}_{\pi(g)} \otimes g,$$

which is naturally G-graded: $L_\pi(\mathcal{A})_g = \mathcal{A}_{\pi(g)} \otimes g$. If $\psi : \mathcal{A} \to \mathcal{A}'$ is a homomorphism of $\overline{G}$-graded algebras, then the linear map $L_\pi(\psi) : L_\pi(\mathcal{A}) \to L_\pi(\mathcal{A}')$ that sends $a \otimes g \mapsto \psi(a) \otimes g$, for all $a \in \mathcal{A}_{\pi(g)}$ and $g \in G$, is a homomorphism of G-graded algebras.

Note that $L_\pi(\mathcal{A})$ is unital if and only if so is $\mathcal{A}$. Also, if $\mathfrak{V}$ is a variety of algebras, i.e. a class defined by polynomial identities, and $\mathbb{F}$ is infinite, then $L_\pi(\mathcal{A})$ belongs to $\mathfrak{V}$ if and only if so does $\mathcal{A}$. Indeed, if $\mathcal{A} \in \mathfrak{V}$, then $\mathcal{A} \otimes \mathbb{F}G \in \mathfrak{V}$ (since $\mathbb{F}G$ is commutative, see, for example, (Giambruno and Zaicev 2005, p. 10)) and hence $L_\pi(\mathcal{A}) \in \mathfrak{V}$; the converse follows from the fact that $\mathcal{A}$ is a quotient of $L_\pi(\mathcal{A})$ (by means of $a \otimes g \mapsto a$).

One of the assertions of the "Correspondence Theorem" (see (Allison *et al.* 2008, Theorem 7.1.1)) is the following.

THEOREM 3.2.– Any G-graded algebra $\mathcal{B}$ that is graded-central-simple and has a split centroid is graded-isomorphic to $L_\pi(\mathcal{A})$ for some central simple algebra $\mathcal{A}$ equipped with a $\overline{G}$-grading where $\overline{G} = G/H$, H is the support of $C(\mathcal{B})$, and $\pi : G \to \overline{G}$ is the natural homomorphism.

Note that the condition on the centroid of $\mathcal{B}$ is necessary because, for any central simple algebra $\mathcal{A}$ with a $\overline{G}$-grading, $L_\pi(\mathcal{A})$ is a graded-central-simple algebra whose centroid can be identified with $L_\pi(\mathbb{F}) = \mathbb{F}H$, where $h \in H$ acts on $L_\pi(\mathcal{A})$ as follows:

$$h(a \otimes g) = a \otimes hg \text{ for all } a \in \mathcal{A}_{\pi(g)},\ g \in G.$$

Another assertion of the "Correspondence Theorem" tells us that the *G-graded algebra $\mathcal{B}$ determines the $\overline{G}$-graded algebra $\mathcal{A}$ up to isomorphism and twist* in the following sense. Fix a transversal for the subgroup H in G, so we have a section ξ :

$\overline{G} \to G$ (which is not necessarily a group homomorphism). This defines a 2-cocycle on $\overline{G}$ with values in H, namely,

$$\sigma(\overline{g}_1, \overline{g}_2) = \xi(\overline{g}_1)\xi(\overline{g}_2)\xi(\overline{g}_1\overline{g}_2)^{-1}.$$

Now, given a character $\lambda : H \to \mathbb{F}^\times$ and a $\overline{G}$-graded algebra $\mathcal{A}$, we define the *twist* $\mathcal{A}^\lambda$ to be $\mathcal{A}$ as a $\overline{G}$-graded space, but with a twisted multiplication:

$$a_1 * a_2 = \lambda(\sigma(\overline{g}_1, \overline{g}_2))a_1 a_2 \text{ for all } a_1 \in \mathcal{A}_{\overline{g}_1},\ a_2 \in \mathcal{A}_{\overline{g}_2}. \qquad [3.2]$$

In the above formula, $\lambda \circ \sigma$ is a 2-cocycle on $\overline{G}$ with values in $\mathbb{F}^\times$, so this is actually a standard cocycle twist of a graded algebra. Note that the isomorphism class of $\mathcal{A}^\lambda$ does not depend on the choice of the transversal. Also, if λ extends to a character $G \to \mathbb{F}^\times$ (e.g. if $\mathbb{F}$ is algebraically closed or if H is a direct summand of G) then $\mathcal{A}^\lambda$ is graded-isomorphic to $\mathcal{A}$.

REMARK 3.1.– As a graded $L_\pi(\mathbb{F})$-module, $L_\pi(\mathcal{A}) \cong L_\pi(\mathbb{F}) \otimes \underline{\mathcal{A}}$, where $\underline{\mathcal{A}}$ is $\mathcal{A}$ regarded as a G-graded vector space with $\deg a = \xi(\overline{g})$ for all non-zero $a \in \mathcal{A}_{\overline{g}}$, $\overline{g} \in \overline{G}$. An isomorphism $L_\pi(\mathbb{F}) \otimes \underline{\mathcal{A}} \to L_\pi(\mathcal{A})$ is given by $h \otimes a \mapsto a \otimes h\xi(\overline{g})$, where we have identified $L_\pi(\mathbb{F})$ with $\mathbb{F}H$. In particular, if ξ is a group homomorphism (which can happen if and only if H is a direct summand of G), then $\underline{\mathcal{A}} = {}^\xi\mathcal{A}$ is a G-graded algebra, and $L_\pi(\mathcal{A}) \cong \mathbb{F}H \otimes {}^\xi\mathcal{A}$ as graded algebras.

3.2.7. *Another construction of graded-simple algebras*

In this section, we give a construction originating in Bahturin *et al.* (2001), which gives another description of G-graded-simple and graded-division algebras over an algebraically closed field $\mathbb{F}$ of characteristic 0, or, more generally, over any field containing nth primitive roots of 1, where n is the order of G.

First, we recall the important duality between the gradings and the automorphism actions on algebras (see, for example, Montgomery (1993)). Let $\mathcal{B}$ be an algebra graded by a finite abelian group G. Consider the dual group (group of irreducible characters) for G defined as

$$\widehat{G} = \mathrm{Hom}(G, \mathbb{F}^\times).$$

One can define an action of $\widehat{G}$ on $\mathcal{B}$ by setting $\chi * a = \chi(g)$ for each $a \in \mathcal{B}_g$. It is clear that in this way we obtain an automorphism of $\mathcal{B}$. A subspace $V \subset \mathcal{B}$ is graded if and only if $\widehat{G} * V = V$. Since the radical of $\mathcal{B}$ is stable under all automorphisms of $\mathcal{B}$, it follows that the graded-simple algebra $\mathcal{B}$ must be semisimple. Moreover, if we choose a non-zero minimal two-sided ideal $\mathcal{A}$ of $\mathcal{R}$, then $\sum_{\chi \in \widehat{G}} \chi * \mathcal{A}$ is a non-zero graded ideal of $\mathcal{B}$. Hence any graded-simple associative algebra is the direct sum of isomorphic simple ideals.

If $\mathcal{A}$ is one of these ideals, then it is not G-graded (unless $\mathcal{B}$ is simple). Consider

$$\Lambda = \{\chi \in \widehat{G} \mid \chi * \mathcal{A} = \mathcal{A}\}.$$

This is a subgroup of $\widehat{G}$ that can be canonically identified with $\widehat{G/H}$ where

$$H = \Lambda^{\perp} := \{g \mid \lambda(g) = 1 \text{ for all } \lambda \in \Lambda\}.$$

Thus, the action of Λ on $\mathcal{A}$ makes $\mathcal{A}$ a G/H-graded algebra:

$$\mathcal{A}_{gH} = \{a \in \mathcal{A} \mid \lambda * a = \lambda(g)a \text{ for all } \lambda \in \Lambda\}.$$

Now we describe all G-graded-simple algebras for the finite abelian group G. Note that in this case a G-graded algebra $\mathcal{B} = \bigoplus_{g \in G} \mathcal{B}_g$ is a left module over the group ring $\mathbb{F}\widehat{G}$ of the group $\widehat{G}$ of all irreducible characters on G and any $\chi \in \widehat{G}$ acts on $\mathcal{B}$ by an automorphism.

THEOREM 3.3.– Let G be a finite abelian group and $\mathcal{B} = \bigoplus_{g \in G} \mathcal{B}_g$ a finite-dimensional G-graded algebra over an algebraically closed field of characteristic zero. If $\mathcal{B}$ is a G-graded-simple, then the following are true.

1) $\mathcal{B}$ is semisimple with isomorphic simple components;

2) there exists a subgroup $H \subseteq G$ and a simple ideal $\mathcal{A} \subseteq R$ such that $\mathcal{B}$ is G/H-homogeneous;

3) as a G-graded algebra, $\mathcal{B}$ is isomorphic to the left $\mathbb{F}\widehat{G}$-module

$$\mathcal{B} = \mathbb{F}\widehat{G} \otimes_{\mathbb{F}\Lambda} \mathcal{A}$$

with the multiplication

$$(\chi \otimes b)(\psi \otimes c) = \begin{cases} 0 & \text{if } \psi\chi^{-1} \notin \Lambda \\ \chi \otimes b(\lambda * c) & \text{if } \psi = \chi\lambda \text{ with } \lambda \in \Lambda \end{cases}$$

where Λ is the subgroup of $\widehat{G}$ that consists of all χ such that $\chi * \mathcal{A} = \mathcal{A}$. Moreover, $H = \{g \in G \mid \lambda(g) = 1 \text{ for all } \lambda \in \Lambda\}$.

As it turns out, the approach of the above theorem is equivalent to the loop construction of the previous section, with less stringent restrictions than those imposed in Bahturin *et al.* (2001). We will only assume that H is finite and $\mathbb{F}$ is algebraically closed and its characteristic does not divide n. Let $\overline{G} = G/H$. A $\overline{G}$-graded vector space V becomes a module over the group algebra $\mathbb{F}(H^{\perp})$ where the subgroup

$$H^{\perp} := \{\chi \in \widehat{G} : \chi(h) = 1\, \forall h \in H\}$$

is naturally isomorphic to the group of characters of $\overline{G}$. Assume for now that $|H| = n$. Then we have $|\widehat{H}| = n$ and, moreover, any character of H extends to a character of G. Fix such extensions, $\chi_1, \ldots, \chi_n$, for all characters of H, so $\widehat{H} = \{\chi_1|_H, \ldots, \chi_n|_H\}$. Then $\{\chi_1, \ldots, \chi_n\}$ is a transversal of $H^{\perp}$ in $\widehat{G}$ (i.e. a set of coset representatives of $H^{\perp}$ in $\widehat{G}$), hence $\mathbb{F}\widehat{G} = \chi_1\mathbb{F}(H^{\perp}) \oplus \cdots \oplus \chi_n\mathbb{F}(H^{\perp})$.

If V is a $\overline{G}$-graded vector space, then we can consider the induced $\mathbb{F}\widehat{G}$-module,

$$I_\pi(V) := \mathrm{Ind}_{H^{\perp}}^{\widehat{G}}(V) = \mathbb{F}\widehat{G} \otimes_{\mathbb{F}(H^{\perp})} V = \chi_1 \otimes V \oplus \cdots \oplus \chi_n \otimes V,$$

which is clearly $\overline{G}$-graded, with the homogeneous component of degree $\bar{g}$ being $\chi_1 \otimes V_{\bar{g}} \oplus \cdots \oplus \chi_n \otimes V_{\bar{g}}$. In fact, this $\overline{G}$-grading on $I_\pi(V)$ can be refined to a G-grading:

$$I_\pi(V)_g := \{x \in \chi_1 \otimes V_{\pi(g)} \oplus \cdots \oplus \chi_n \otimes V_{\pi(g)} : \chi \cdot x = \chi(g)x\, \forall \chi \in \widehat{G}\}.$$

Now, if $\mathcal{A}$ is a $\overline{G}$-graded algebra, then $I_\pi(\mathcal{A})$ is a G-graded algebra with multiplication defined by $(\chi_i \otimes a')(\chi_j \otimes a'') := \delta_{ij}\chi_i \otimes a'a''$ for $1 \le i, j \le n$ and $a', a'' \in A$, so each of the direct summands $\chi_j \otimes \mathcal{A}$ is a $\overline{G}$-graded ideal isomorphic to $\mathcal{A}$ as a $\overline{G}$-graded algebra.

It turns out that, under the above assumptions on H and $\mathbb{F}$, $I_\pi(\mathcal{A})$ is isomorphic to $\mathcal{L}_\pi(\mathcal{A})$ as a G-graded algebra.

An isomorphism $L_\pi(\mathcal{A}) \to I_\pi(\mathcal{A})$ is given by

$$v \otimes g \mapsto \sum_{j=1}^{n} \chi_j(g)^{-1}\chi_j \otimes v \quad \forall v \in \mathcal{A}_{\pi(g)},\, g \in G,$$

it does not depend on the choice of the transversal $\{\chi_1, \ldots, \chi_n\}$, and its inverse $I_\pi(V) \to L_\pi(V)$ is given by

$$\chi_j \otimes v \mapsto \frac{1}{n} \sum_{h \in H} \chi_j(gh) v \otimes gh \quad \forall v \in V_{\pi(g)},\, g \in G,\, 1 \le j \le n.$$

3.3. Graded-division algebras over algebraically closed fields

We have seen in section 3.1 that the classification of G-graded algebras that are graded-simple and satisfy the descending chain condition on graded left ideals can be reduced to the classification of graded-division algebras whose support is a subgroup of G. We are going to give a classification of finite-dimensional graded-division algebras assuming that the ground field $\mathbb{F}$ is algebraically closed.

Let G be a finite abelian group, $\mathcal{A}$ a G-graded-division algebra with coefficients from an algebraically closed field. We know that, in this case, each homogeneous component of $\mathcal{A}$ is one-dimensional. Given $g, h \in G$, there is $\beta(g,h) \in \mathbb{F}$ such that for any $x \in \mathcal{A}_g$ and $y \in \mathcal{A}_h$, $xy = \beta(g,h)yx$. One can easily check the following relations, for any $g, h, k \in G$.

1) $\beta(gh,k) = \beta(g,k)\beta(h,k)$;

2) $\beta(g,hk) = \beta(g,h)\beta(g,k)$;

3) $\beta(g,g) = 1$.

Such a function is called an *alternating bicharacter*.

Now let us write G as the direct product of cyclic subgroups $G = (g_1)_{n_1} \times \cdots \times (g_m)_{n_m}$ of orders $n_1, \ldots, n_m$. Then consider an algebra $\mathcal{R}$ given by G-graded generators $x_1, \ldots, x_m$ of G-degrees $g_1, \ldots, g_m$ and G-graded relations of two kinds: $x_1^{n_1} = 1, \ldots, x_m^{n_m} = 1$ and $x_i x_j = \beta(g_i, g_j)x_j x_i$. Clearly, $\dim \mathcal{A} \le n_1 \cdots n_m$. On the other hand, if $o(g) = n$, then in $\mathcal{A}_g$ one can choose an element u such that $u^n = z \in \mathbb{C}$; replacing u by $v = \frac{1}{\sqrt[n]{z}}u$, we obtain an element v in $\mathcal{A}_g$ with $v^n = 1$. Thus, the generators $X_{g_1}, \ldots, X_{g_m}$ of degrees $g_1, \ldots, g_m$ for A can be chosen so that $X_{g_i}^{n_i} = I$. In this case, A maps G-graded homomorphically onto $\mathcal{R}$, and $\dim \mathcal{R} = n_1 \cdots n_m$. As a result, $\mathcal{R} \cong \mathcal{A}$.

Let us denote $\mathcal{R}$ as above, by $\mathcal{P}(\beta)$. Since in $\mathcal{P}(\beta)$, we always have

$$X_g X_h = \beta(g,h) X_h X_g,$$

for any $g, h \in G$, it follows that $\mathcal{P}(\beta')$ is isomorphic to $\mathcal{P}(\beta)$ if and only if $\beta' = \beta$. In the case of equivalence (=weak isomorphism) $\mathcal{P}(\beta') \sim \mathcal{P}(\beta)$, accompanied by a group automorphism $\alpha : G \to G$, we must have $\beta'(\alpha(g), \alpha(h)) = \beta(g,h)$, for all $g, h \in G$. In other words, $\mathcal{P}(\beta')$ is equivalent to $\mathcal{P}(\beta)$ if and only if β' belongs to the same orbit as β under the natural action of $\operatorname{Aut} G$ on the set of alternating bicharacters $G \times G \to \mathbb{C}^\times$. For instance, $\mathcal{P}(\beta)$ is a simple algebra if and only if β is non-degenerate. As indicated in Elduque and Kochetov (2013, Chapter 2), all non-degenerate alternating bicharacters on G form one orbit, hence, given a finite abelian group G, all division

G-gradings on $M_n(\mathbb{C})$ are equivalent. In fact, such gradings exist if and only if $G \cong H \times H$, where $|H| = n$.

In the case of commutative algebras, $\mathcal{R}$ is just the group algebra $\mathbb{C}G$. So any commutative G-graded-division algebra $\mathcal{R}$ over $\mathbb{C}$ is isomorphic to the group algebra $\mathbb{C}G$. In other words, $\mathcal{R}$ is isomorphic to the graded tensor product $\mathbb{C}(g_1)_{n_1} \otimes \cdots \otimes \mathbb{C}(g_m)_{n_m}$ of group algebras of cyclic groups and is completely determined by the same invariants as the abelian group G.

It is instructive to apply the loop algebra construction to complex division algebras. Suppose our algebra is $\mathcal{B} = \mathcal{P}(\beta)$. Let us consider the centroid (=center) of $\mathcal{B}$. Clearly, this is the complex graded commutative algebra whose support is the kernel of β, $H = \text{Ker}\ \beta$. The central factor is the matrix algebra $\mathcal{A} = M_k(\mathbb{C})$ with the division grading by the group G/H. By the correspondence theorem from section 3.2.6, $\mathcal{B}$ can be viewed as the subalgebra in $M_k(\mathbb{C}) \otimes \mathbb{C}G$, spanned by the elements of degree g of the form $X_{\overline{g}} \otimes g$, for all $g \in G$.

Finally – but historically this was one of the first examples of graded-division algebras – let us explicitly exhibit a division grading on the matrix algebra $\mathcal{A} = M_n(\mathbb{C})$. Again, let $G = (g_1)_{n_1} \times \cdots \times (g_m)_{n_m}$ and let $X_{g_1}, \ldots, X_{g_m}$ be the generating elements of $\mathcal{A}$. Since $\mathcal{A}$ is central simple, the associated bicharacter β is non-degenerate. A procedure, similar to reducing an alternating bilinear form on a vector space to the canonical form (for details, see Elduque and Kochetov (2013, Section 2.2)); but also compare with the argument in section 3.4.4) one can assume that the set of generators $g_1, \ldots, g_m$ of the group G and their corresponding generators $X_{g_1}, \ldots, X_{g_m}$ of the algebra $\mathcal{A}$ can be chosen to have the form $g_1, g_1', g_2, g_2', \ldots, g_s, g_s'$ so that $o(g_i) = o(g_i') = n_i$, $X_{g_i}^{n_i} = X_{g_i'}^{n_i} = I$ and $X_{g_i}X_{g_j} = \omega_i X_{g_j}X_{g_i}$, where $\omega_i = \beta(g_i, g_i')$ is an n_ith primitive root of 1. This means that $\mathcal{A} = \mathcal{A}_1 \otimes \mathcal{A}_2 \otimes \cdots \otimes \mathcal{A}_m$, where each tensor factor $\mathcal{A}_i$ is isomorphic to one of the algebras of the form

$$\mathcal{A} = \langle X, Y \mid X^n = Y^n = 1,\ XY = \omega YX \rangle, \qquad [3.3]$$

where ω is an nth primitive root of 1.

Clearly, this algebra has dimension at most n^2. Since there is an algebra, given just below, that satisfies the same relations and having dimension n^2, it follows that $\dim \mathcal{A} = n^2$.

To use [3.3], we require that the base field $\mathbb{F}$ has characteristic 0 or p such that p is coprime to n. A matrix model for the algebra in equation [3.3] is given with the help of classical Sylvester *"clock"* and *"shift"* matrices. Sometimes, they are called (generalized) *Pauli matrices* but this is more appropriate for the case $n = 2$. So we set

$$X = \begin{bmatrix} \omega^{n-1} & 0 & 0 \dots 0 \; 0 \\ 0 & \omega^{n-2} & 0 \dots 0 \; 0 \\ \dots & & \\ 0 & 0 & 0 \dots \omega \; 0 \\ 0 & 0 & 0 \dots 0 \; 1 \end{bmatrix} \text{ and } Y = \begin{bmatrix} 0 \; 1 \; 0 \dots 0 \; 0 \\ 0 \; 0 \; 1 \dots 0 \; 0 \\ \dots \\ 0 \; 0 \; 0 \dots 0 \; 1 \\ 1 \; 0 \; 0 \dots 0 \; 0 \end{bmatrix} \quad [3.4]$$

One easily checks that X, Y satisfy the defining relations of [3.3] and the products $X^i Y^j$ span $M_n(\mathbb{C})$. Thus, $\mathcal{A} \cong M_n(\mathbb{C})$ as graded algebras.

3.4. Real graded-division associative algebras

In this section, we survey the classification of real graded-division algebras.

3.4.1. *Simple graded-division algebras*

We recall some low dimensional simple graded-division algebras and the classification of simple graded-division algebras, up to equivalence (see Bahturin and Zaicev (2016) and Rodrigo-Escudero (2016)).

The simplest examples of graded-division algebras are $\mathbb{R}$, $\mathbb{C}$ and the quaternions $\mathbb{H}$. These are graded by the trivial group. Given any group G, the group algebras $\mathbb{R}G$, $\mathbb{C}G$ and $\mathbb{H}G = \mathbb{H} \otimes \mathbb{R}G$ are further examples of real graded-division algebras of dimensions $|G|$, $2|G|$ and $4|G|$, respectively. Also, we can refine the trivial gradings on $\mathbb{C}$ and $\mathbb{H}$ to obtain the $\mathbb{Z}_2$-gradings $\mathbb{C}^{(2)}$ and as follows. If we set $\mathbb{Z}_2 \cong (\alpha)_2$, then

$$\mathbb{C}^{(2)} = \langle 1 \rangle_e \oplus \langle i \rangle_\alpha, \; \mathbb{H}^{(2)} = \langle 1, i \rangle_e \oplus \langle j, k \rangle_\alpha. \quad [3.5]$$

Also, if we set $\mathbb{Z}_2 \times \mathbb{Z}_2 \cong (\alpha)_2 \times (\beta)_2$, then a $\mathbb{Z}_2 \times \mathbb{Z}_2$-refinement, $\mathbb{H}^{(4)}$, on $\mathbb{H}$ will look like the following:

$$\mathbb{H}^{(4)} = \langle 1 \rangle_e \oplus \langle i \rangle_\alpha \oplus \langle j \rangle_\beta \oplus \langle k \rangle_{\alpha\beta}. \quad [3.6]$$

The matrix algebra $M_2 = M_2(\mathbb{R})$, with trivial grading, is not a graded-division algebra, but it has $\mathbb{Z}_2$- and $\mathbb{Z}_2^2$-refinements $M_2^{(2)}$ and $M_2^{(4)}$, which are graded-division algebras. Recall the Pauli (Sylvester) matrices

$$A = \begin{pmatrix} 1 & 0 \\ 0 & -1 \end{pmatrix}, \; B = \begin{pmatrix} 0 & 1 \\ 1 & 0 \end{pmatrix}, \; C = \begin{pmatrix} 0 & 1 \\ -1 & 0 \end{pmatrix}. \quad [3.7]$$

Then $A^2 = B^2 = I$, $AB = -BA = C$. It follows then that $C^2 = -I$, $AC = -CA = B$, $BC = -CB = -A$. So we have the following

$$M_2^{(2)} = \langle I, C\rangle_e \oplus \langle A, B\rangle_\alpha,\ M_2^{(4)} = \langle I\rangle_e \oplus \langle C\rangle_\alpha \oplus \langle A\rangle_\beta \oplus \langle B\rangle_{\alpha\beta}. \quad [3.8]$$

The only eight-dimensional real simple algebra is $R = M_2(\mathbb{C})$. Since $M_2(\mathbb{R}) \otimes \mathbb{C} \cong M_2(\mathbb{C}) \cong \mathbb{H} \otimes \mathbb{C}$, we can obtain division gradings on R in various ways. A division $\mathbb{Z}_2$-grading can be obtained as $\mathbb{H} \otimes \mathbb{C}^{(2)}$. A division $\mathbb{Z}_2^2$-grading on R can be obtained as $\mathbb{H}^{(4)} \otimes \mathbb{C}$, $M_2^{(4)} \otimes \mathbb{C}$, $\mathbb{H}^{(2)} \otimes \mathbb{C}^{(2)}$ and $M_2^{(2)} \otimes \mathbb{C}^{(2)}$. It is known from Bahturin and Zaicev (2016) that the first two gradings are equivalent, and also the last two are equivalent. A division $\mathbb{Z}_2^3$-grading on R can be obtained in two equivalent ways: $\mathbb{H}^{(4)} \otimes \mathbb{C}^{(2)}$ or $M_2^{(4)} \otimes \mathbb{C}^{(2)}$. Note that the natural isomorphism $\varphi : \mathbb{H} \otimes \mathbb{C} \to M_2(\mathbb{R}) \otimes \mathbb{C}$ defined by $\varphi(i \otimes z) = A \otimes iz$, $\varphi(j \otimes z) = B \otimes iz$, for any $z \in \mathbb{C}$, induces a graded isomorphism for the refinements $\varphi : H^{(2)} \otimes \mathbb{C}^{(2)} \to M_2^{(2)} \otimes \mathbb{C}^{(2)}$ and $\varphi : H^{(4)} \otimes \mathbb{C}^{(2)} \to M_2^{(4)} \otimes \mathbb{C}^{(2)}$.

So far, all division gradings on $M_2(\mathbb{C})$ have appeared as tensor product gradings. However, there are gradings on this algebra which are not tensor products. Let us fix a complex number ω such that $\omega^2 = i$. Then a $\mathbb{Z}_4$-grading on $M_2(\mathbb{C})$, denoted by $M_2(\mathbb{C}, \mathbb{Z}_4)$ can be obtained as follows. We set $\mathbb{Z}_4 \cong (\gamma)_4$. Then

$$M_2(\mathbb{C}, \mathbb{Z}_4) = \langle I, C\rangle_e \oplus \langle \omega A, \omega B\rangle_\gamma \oplus \langle iI, iC\rangle_{\gamma^2} \oplus \langle \omega^3 A, \omega^3 B\rangle_{\gamma^3}.$$

A $\mathbb{Z}_2 \times \mathbb{Z}_4$-refinement of $M_2(\mathbb{C}, \mathbb{Z}_4)$ is denoted by $M_2^{(8)}$.We set $\mathbb{Z}_2 \times \mathbb{Z}_4 \cong (\alpha)_2 \times (\gamma)_4$. Then

$$\begin{aligned} M_2^{(8)} &= \langle I\rangle_e \oplus \langle C\rangle_\alpha \oplus \langle \omega A\rangle_\gamma \oplus \langle \omega B\rangle_{\alpha\gamma} \oplus \langle iI\rangle_{\gamma^2} \oplus \langle iC\rangle_{\alpha\gamma^2} \\ &\oplus \langle \omega^3 A\rangle_{\gamma^3} \oplus \langle \omega^3 B\rangle_{\alpha\gamma^3}. \end{aligned}$$

We know that $\mathbb{H} \otimes \mathbb{H} \cong M_2(\mathbb{R}) \otimes M_2(\mathbb{R})$. This isomorphism transfers the structure of graded-division algebra from $\mathbb{H}^{(4)} \otimes \mathbb{H}$ to $M_4(\mathbb{R}) \cong M_2(\mathbb{R}) \otimes M_2(\mathbb{R})$. This grading of $M_4(\mathbb{R})$ is not a tensor product of gradings on the tensor factors $M_2(\mathbb{R})$. Thus $R = M_4(\mathbb{R})$ acquires a division $\mathbb{Z}_2 \times \mathbb{Z}_2 \cong (\alpha)_2 \times (\beta)_2$-grading $M_4^{(4)}$ whose components are as follows:

$$\begin{aligned} &R_e = \langle I \otimes I, C \otimes I, A \otimes C, B \otimes C\rangle, \quad [3.9] \\ &R_\alpha = (I \otimes C)R_e,\ R_\beta = (C \otimes A)R_e,\ R_{\alpha\beta} = (C \otimes B)R_e, \end{aligned}$$

Finally, given a non-degenerate alternating bicharacter $\beta : G \times G \to \mathbb{C}$, $|G| = n^2$, we denote by $\mathcal{P}(\beta)$ the unique, up to equivalence, fine grading on $M_n(\mathbb{C})$ defined by

β (see details below). Now let us denote by $\mathcal{P}(\beta)_{\mathbb{R}}$ the same grading, viewed as a grading of an algebra over $\mathbb{R}$. We call $\mathcal{P}(\beta)_{\mathbb{R}}$ a *Pauli grading*.

THEOREM 3.4.– Any division grading on a real simple algebra $M_n(\Delta)$, Δ a real division algebra, is equivalent to one of the following types:

1) $\Delta = \mathbb{R}$:

i) $(M_2^{(4)})^{\otimes k}$;

ii) $M_2^{(2)} \otimes (M_2^{(4)})^{\otimes (k-1)}$, a coarsening of (i);

iii) $M_4^{(4)} \otimes (M_2^{(4)})^{\otimes (k-2)}$, a coarsening of (i);

2) $\Delta = \mathbb{H}$:

iv) $\mathbb{H}^{(4)} \otimes (M_2^{(4)})^{\otimes k}$;

v) $\mathbb{H}^{(2)} \otimes (M_2^{(4)})^{\otimes k}$ a coarsening of (iv);

vi) $\mathbb{H} \otimes (M_2^{(4)})^{\otimes k}$, a coarsening of (v);

3) $\Delta = \mathbb{C}$:

vii) $\mathbb{C}^{(2)} \otimes (M_2^{(4)})^{\otimes k}$;

viii) $\mathbb{C}^{(2)} \otimes M_2^{(2)} \otimes (M_2^{(4)})^{\otimes (k-1)}$, a coarsening of (vii);

ix) $\mathbb{C}^{(2)} \otimes \mathbb{H} \otimes (M_2^{(4)})^{\otimes (k-1)}$, a coarsening of (vii);

x) $M_2^{(8)} \otimes (M_2^{(4)})^{\otimes (k-1)}$;

xi) $M_2(\mathbb{C}, \mathbb{Z}_4) \otimes (M_2^{(4)})^{\otimes (k-1)}$, a coarsening of (x);

xii) $M_2^{(8)} \otimes M_2^{(2)} \otimes (M_2^{(4)})^{\otimes (k-2)}$, a coarsening of (x);

xiii) $M_2^{(8)} \otimes \mathbb{H} \otimes (M_2^{(4)})^{\otimes (k-2)}$, a coarsening of (x);

xiv) Pauli gradings.

None of the gradings of different types or of the same type but with different values of k is equivalent to the other.

Note that (xii) is missing on the list in Bahturin and Zaicev (2016) but appears in Rodrigo-Escudero (2016). It is useful to mention that the components of the gradings in theorem 3.4 are one-dimensional in the cases (i), (iv), (vii) and (x). They are two-dimensional in (ii), (v), (viii), (xi), (xii) and (xiv) and four-dimensional in the remaining cases (iii), (vi), (ix) and (xiii).

3.4.2. *Pauli gradings*

Let $\Gamma : \mathcal{R} = \bigoplus_{g\in G} \mathcal{R}_g$ be a division grading of a real algebra $\mathcal{R}$ such that $\dim \mathcal{R}_e = 2$ and $\mathcal{R}_e$ is a central subalgebra of $\mathcal{R}$. Then there is an isomorphism of algebras $\mu : \mathbb{C} \to \mathcal{R}_e$. If $I = \mu(1)$ and $J = \mu(i)$, then setting $(a + bi)X = (aI + bJ)X$ where $a, b \in \mathbb{R}$, $X \in \mathcal{R}$ turns $\mathcal{R}$ to a complex algebra $\mathcal{R}^{\mathbb{C}}$ endowed with a complex grading $\Gamma^{\mathbb{C}}$, which is a division grading. In this grading, $(\mathcal{R}^{\mathbb{C}})_g = \mathcal{R}_g$. By the previous section, there is an alternating bicharacter $\beta : G \times G \to \mathbb{C}^\times$ such that $\Gamma^{\mathbb{C}} \cong \mathcal{P}(\beta)$. Since, obviously, $(\Gamma^{\mathbb{C}})_{\mathbb{R}} \cong \Gamma$, we have that any real grading with two-dimensional central identity component is isomorphic to $\mathcal{P}(\beta)_{\mathbb{R}}$, for an appropriate alternating complex bicharacter β on G. We call the gradings of the type $\mathcal{P}(\beta)$ *Pauli gradings*.

Thus, $\mathcal{R} = \mathcal{P}(\beta)_{\mathbb{R}}$, as a unital algebra with identity element I, in terms of graded generators and defining relations can be given as follows. We choose the canonical decomposition $G = (g_1)_{n_1} \times \ldots \times (g_m)_{n_m}$, $n_1 | \cdots | n_m$, and choose the generators J of trivial degree and X_i of degree g_i, for all $i = 1, \ldots, m$. Then we impose the relations $J^2 = -I$, $X_i^{n_i} = I$, $JX_i = X_iJ$, for all $i = 1, \ldots, m$, and $X_iX_j = \mu(\beta(g_i, g_j))X_jX_i$. Here, $\mu : \mathbb{C} \to \langle I, J\rangle$ is given, as above.

THEOREM 3.5.– If $\Gamma : \mathcal{R} = \bigoplus_{g\in G} \mathcal{R}_g$ is a real G-graded-division algebra such that $\dim \mathcal{R}_e = 2$ and $\mathcal{R}_e$ is central, then there exists an alternating bicharacter $\beta : G\times G \to \mathbb{C}^\times$ such that $\Gamma = \mathcal{P}(\beta)_{\mathbb{R}}$. Furthermore, $\mathcal{P}(\beta)_{\mathbb{R}}$ is isomorphic to $\mathcal{P}(\beta')_{\mathbb{R}}$ if and only if either $\beta' = \beta$ or $\beta' = \overline{\beta}$. Finally, $\mathcal{P}(\beta)_{\mathbb{R}}$ is equivalent to $\mathcal{P}(\beta')_{\mathbb{R}}$ if and only if the orbit of β' under the action of $\operatorname{Aut} G$ contains β or $\overline{\beta}$.

The study of non-degenerate alternating bicharacters over an algebraically closed field (often called the *commutation factors*) is performed in various works: Zmud (1971) and Zolotykh (1997). At the same time, there are much earlier papers where the authors came across alternating bicharacters with values in multiplicative abelian groups in other areas of mathematics (see de Rham (1931) or Tignol (1982)). Bicharacters are also important in Hopf's theory (see Montgomery (1993)).

3.4.3. *Commutative case*

In the case of real commutative graded-division algebras, the situation is different from the case of complex commutative graded-division algebras. Given a natural number $n > 1$ and a number $\varepsilon = \pm 1$, we denote by $\mathcal{C}(m; \varepsilon)$ the real graded subalgebra in the complex group algebra $\mathbb{C}(g)_m$ of the cyclic group of order m generated by μg, where $\mu^m = \varepsilon$. Alternatively, one can view $\mathcal{C}(m; \varepsilon)$ as a graded algebra generated by one element x of degree g, with defining relation $x^m = \varepsilon I$. We will call this algebra a *basic algebra of the first kind*. Of course, $\mathcal{C}(2; -1) \cong \mathbb{C}^{(2)}$.

PROPOSITION 3.2.– Let $\mathcal{R}$ be a G-graded commutative division algebra over the field $\mathbb{R}$ of real numbers. Suppose $\dim \mathcal{R}_e = 1$. Let $G = (g_1)_{m_1} \times \cdots \times (g_k)_{m_k}$ be the direct product of primary cyclic subgroups of orders $m_1, \ldots, m_k$. Then

$$\mathcal{R} \cong \mathcal{C}(m_1; \varepsilon_1) \otimes \cdots \otimes \mathcal{C}(m_k; \varepsilon_k),$$

for a sequence of numbers $\varepsilon_1, \ldots, \varepsilon_k = \pm 1$. Here, additionally, we can assume that $\varepsilon_i = 1$ in the case where m_i is an odd number.

It follows from the above proposition that if we use a primary factorization of G as the product of cyclic subgroups, then we could write $G = H \times K$ where H is a 2-subgroup of G and K is a subgroup of odd order. As a result, any commutative graded-division algebra with one-dimensional homogeneous components and support G can be written as the graded tensor product $\mathcal{R} \cong \mathcal{S} \otimes \mathbb{R}K$, where $\mathcal{S}$ is the commutative graded-division algebra whose support is an abelian 2-group H and $\mathbb{R}K$ is the real group algebra of K.

The ungraded structure of real graded commutative division algebras is given by the following.

PROPOSITION 3.3.–

1) $\mathcal{C}(2^q; 1) \cong \mathbb{R}\mathbb{Z}_{2^q} \cong \mathbb{C} \oplus \cdots \oplus \mathbb{C} \oplus \mathbb{R} \oplus \mathbb{R}$, for $q \geq 1$;

2) $\mathcal{C}(2^q; -1) \cong \mathbb{C} \oplus \cdots \oplus \mathbb{C}$, because $x^2 = -1$ is not solvable in $\mathbb{R}$;

3) $\mathcal{C}(2^{m_1}; 1) \otimes \cdots \otimes \mathcal{C}(2^{m_t}; 1) \cong \mathbb{C} \oplus \cdots \oplus \mathbb{C} \oplus \underbrace{\mathbb{R} \oplus \cdots \oplus \mathbb{R}}_{2^t}$;

4) $\mathcal{C}(2^{m_1}; \eta_1) \otimes \cdots \otimes \mathcal{C}(2^{m_t}, \eta_t) \cong \mathbb{C} \oplus \cdots \oplus \mathbb{C}$ if at least one $\eta_i = -1$.

To determine the equivalence classes of real graded commutative division algebras, one needs the following result. This is one of the typical tricks used for the determination of equivalence classes of graded-simple algebras. We give the proof as an example.

LEMMA 3.2.– Let m, n be any natural numbers, $m \leq n$. Then

$$\mathcal{C}(2^m; -1) \otimes \mathcal{C}(2^n; -1) \sim \mathcal{C}(2^m; 1) \otimes \mathcal{C}(2^n; -1).$$

PROOF.– We write

$$\mathcal{C}(2^m; -1) \otimes \mathcal{C}(2^n; -1) = \text{alg}\{u, v \,|\, u^{2^m} = -1,\ v^{2^n} = -1; uv = vu\}$$
$$\mathcal{C}(2^m; 1) \otimes \mathcal{C}(2^n; -1) = \text{alg}\{u_1, v_1 \,|\, u_1^{2^m} = 1,\ v_1^{2^n} = -1; uv = vu\}$$

Here $\mathcal{C}(2^m;-1)$ is graded by the cyclic group $G=(a)_{2^m}$ and $\mathcal{C}(2^n;-1)$ by the cyclic group $H=(b)_{2^n}$, so that $\mathcal{C}(2^m;-1)\otimes\mathcal{C}(2^n;-1)$ is graded by $G\times H=(a)_{2^m}\times(b)_{2^n}$. The same group grades $\mathcal{C}(2^m;1)\otimes\mathcal{C}(2^n;-1)$. Since $m\leq n$, then there is $k\leq n$ such that $(v^{2^k})^{2^m}=-1$ and $(uv^{2^k})^{2^m}=1$. The mapping $u_1\to uv^{2^k}$, $v_1\to v$ extends to an ungraded algebra isomorphism of $\mathcal{C}(2^m;1)\otimes\mathcal{C}(2^n;-1)$ to $\mathcal{C}(2^m;-1)\otimes\mathcal{C}(2^n;-1)$. Actually, this is an isomorphism, accompanied by an automorphism of $G\times H$, mapping $a\to ab^{2^k}$, $b\to b$. The proof is complete. □

Now we are ready to state the main result about the structure of real commutative graded-division algebras.

THEOREM 3.6.– Any real finite-dimensional commutative graded-division algebra is equivalent to exactly one of the following:

1) a naturally graded real group algebra $\mathbb{R}G$, for a finite abelian group G;

2) a naturally graded complex group algebra $\mathbb{C}G$, viewed as a real algebra, for a finite abelian group G;

3) a tensor product of graded algebras $\mathcal{C}(2^m;-1)\otimes\mathbb{R}G$.

For the proof of the "non-equivalence" part, one could use proposition 3.3. One can also say that in the first and second cases, the homogeneous components are one-dimensional, whereas in the second they are two-dimensional. At the same time, in the first case there are no homogeneous solutions of the equation $x^2=-1$, whereas in the second they exist. The idea of comparing the supports in the grading group for the sets of homogeneous equations of the form $x^{2^m}=\pm 1$ turned out to be useful for the determination of the equivalence classes of real graded-division algebras.

3.4.4. *Non-commutative graded-division algebras with one-dimensional homogeneous components*

Let $\Gamma:\mathcal{R}=\bigoplus_{g\in G}\mathcal{R}_g$ be a division grading of an algebra $\mathcal{R}$ over $\mathbb{R}$ by an abelian group G. For a subgroup H of G, we denote by $\mathcal{R}_H$ the sum of all graded components $\mathcal{R}_g$, where $g\in H$. Then $\mathcal{R}_H$ is a graded subalgebra of $\mathcal{R}$. If $G=H\times K$, then, as a vector space, $\mathcal{R}=\mathcal{R}_H\otimes\mathcal{R}_K$. If, additionally, $\mathcal{R}_H$ and $\mathcal{R}_K$ commute, then $\mathcal{R}\cong\mathcal{R}_H\otimes\mathcal{R}_K$ is the tensor product of algebras, endowed with the tensor product of gradings.

Now let us assume $\dim\mathcal{R}_e=1$. Let $G=(g_1)_{n_1}\times\cdots\times(g_q)_{n_q}$ be a primary cyclic factorization of G, where each g_i is an element of order n_i, $i=1,\ldots,q$. Let $g,h\in\{g_1,\ldots,g_q\}$. Then for any non-zero $a\in\mathcal{R}_g$ and $b\in\mathcal{R}_h$ we should have $aba^{-1}b^{-1}=\lambda\in\mathbb{R}$. Since $a^n\in\mathcal{R}_e\cong\mathbb{R}$, we must have $a^nba^{-n}=b=\lambda^n b$. Hence $\lambda^n=1$ and then also $\lambda=\pm 1$. If n is odd, then we must have $\lambda=1$ so that $\mathcal{R}_g$ is in the center of $\mathcal{R}$. It is also possible that $\mathcal{R}_g$ is in the center of $\mathcal{R}$ for some more $g \in \{g_1, \dots, g_q\}$. Let H be the subgroup of G generated by all these elements, K the subgroup generated by the remaining elements in $\{g_1, \dots, g_q\}$. Then $G = H \times K$, where $\mathcal{R}_H$ is a commutative H-graded-division algebra and $\mathcal{R}_K$ is a K-graded-division algebra, where K is an abelian 2-group. None of the graded components $\mathcal{R}_{g_i}$, $g_i \in K$, is in the center of $\mathcal{R}_K$.

The structure of $\mathcal{R}_H$ has been described in theorem 3.6. So from now on we assume that $G = (g_1)_{n_1} \times \cdots \times (g_q)_{n_q}$ is an abelian 2-group, $n_1|n_2|\dots|n_q$ and none of $\mathcal{R}_{g_i}$ is central. Suppose that k be the least number such that $\mathcal{R}_{g_1}$ does not commute with $\mathcal{R}_{g_k}$. If $\mathcal{R}_{g_1}$ also does not commute with $\mathcal{R}_{g_{t_1}}, \dots, \mathcal{R}_{g_{t_p}}$, for some $k < t_1 < \cdots < t_p$, we replace g_{t_j} by $g_k g_{t_j}$ and $\mathcal{R}_{g_{t_j}}$ by $\mathcal{R}_{g_k g_{t_j}}$. Since $o(g_{t_j}) = o(g_k g_{t_j})$ and $\mathcal{R}_{g_k g_{t_j}} = \mathcal{R}_{g_k}\mathcal{R}_{g_{t_j}}$, we will now have that $\mathcal{R}_{g_1}$ commutes with all new $\mathcal{R}_{g_j}$ but one, which is $\mathcal{R}_{g_k}$. Now let $1 < s_1 < s_2 < \dots < s_p$ be such that $\mathcal{R}_{g_k}$ does not commute with $\mathcal{R}_{g_{s_j}}$. We replace g_{s_j} by $g_1 g_{s_j}$ and $\mathcal{R}_{g_{s_j}}$ by $\mathcal{R}_{g_1 g_{s_j}}$. Then again $o(g_{s_j}) = o(g_1 g_{s_j})$ and now $\mathcal{R}_{g_k}$ commutes with $\mathcal{R}_{g_1 g_{s_j}} = \mathcal{R}_{g_1}\mathcal{R}_{g_{s_j}}$. As a result, we find that $\mathcal{R}$ is a graded tensor product of graded subalgebras $\mathcal{R}_{(g_1)\times(g_k)}$ and $\mathcal{R}_K$, where $K = (g_2)_{n_2} \times \cdots \times (g_{k-1})_{n_{k-1}} \times (g_{k+1})_{n_{k+1}} \times \cdots \times (g_q)_{n_q}$. This allows us to proceed by induction to finally write a graded tensor product $\mathcal{R} = \mathcal{R}_1 \otimes \mathcal{R}_2 \otimes \cdots \otimes \mathcal{R}_n$, where each $\mathcal{R}_i$ is a graded-division algebra whose support is the product of at most two cyclic 2-group.

Those algebras with cyclic support have been described in proposition 3.2. A graded-division algebra $\mathcal{R}$ with $\dim \mathcal{R}_e = 1$, whose support is the product of two cyclic 2-group $G = (g)_k \times (h)_\ell$ can be described as follows. We choose $a \in \mathcal{R}_g$ and $b \in \mathcal{R}_h$ so that $a^k = \mu I$ and $b^\ell = \nu I$, where $\mu, \nu = \pm 1$. These elements are graded generators of the whole of $\mathcal{R}$, and they anticommute: $ab = -ba$. Since $\mathcal{R}$ is graded, we have $\dim \mathcal{R} = k\ell$. It remains to produce a $(k\ell)$-dimensional G-graded algebra with generators u, v of the same degrees as a, b, respectively, satisfying the same relations. To do so, we consider $S = \mathcal{R}G \otimes M_2(\mathbb{C})$ and inside a subalgebra $\mathcal{D}(k, \ell; \mu, \nu)$ generated by $u = g \otimes (\varepsilon A), v = h \otimes (\eta B)$. Here $\varepsilon^k = \mu, \eta^\ell = \nu$, $A = \begin{pmatrix} 1 & 0 \\ 0 & -1 \end{pmatrix}$ and $B = \begin{pmatrix} 0 & 1 \\ 1 & 0 \end{pmatrix}$. The reader will easily check that the required relations are satisfied and that the set of elements $u^i v^j = g^i h^j \otimes \varepsilon^i \eta^j A^i B^j$, where $0 \le i < k$, $0 \le j < \ell$ is linearly independent because this is true for the set of elements $g^i h^j$, $0 \le i < k$, $0 \le j < \ell$, forming a basis of $\mathcal{R}G$. So $\mathcal{R}$ admits a graded homomorphism onto S. Comparing dimensions, we can see that $\mathcal{R} \cong \mathcal{D}(k, \ell; \mu, \nu)$. As we see, if $G = (g)_k \times (h)_\ell$, then, in terms of G-graded generators and defining relations, this algebra can be given by

$$\mathcal{D}(k, \ell; \mu, \nu) = \langle x, y \mid x^k = \mu I,\ y^\ell = \nu I, xy = -yx,\ \deg x = g,\ \deg y = h \rangle.$$

Recall that $\mu, \nu = \pm 1$. In what follows, we will always be assuming that $k \le \ell$.

We will call $\mathcal{D}(k, \ell; \mu, \nu)$ a *basic algebra of the second kind*. A generator x is called *even* if $\mu = 1$. Otherwise, x is called *odd*. We write $so(x) = k$ and call k, which is the order of the degree of x in the group grading, the *special order* of u. The same terminology will be used for y and for the generator of the basic (commutative) algebra $\mathcal{C}(m; \eta)$. The relations of the form $x^k = \mu I$ will be called the *power relations* while those of the form $xy = \pm yx$ the *commutation relations*.

Note that the algebras $\mathcal{R} \cong \mathcal{D}(k, \ell; \mu, \nu)$ generalize both $\mathbb{H}^{(4)}$ and $M_2^{(4)}$. We have

$$\mathbb{H}^{(4)} = \mathcal{D}(2, 2; -1, -1), \; M_2^{(4)} = \mathcal{D}(2, 2; 1, 1) \sim \mathcal{D}(2, 2; -1, 1) \sim \mathcal{D}(2, 2; 1, -1). \quad [3.10]$$

They can also be viewed as *generalized quaternion* algebras. One can trace them back to algebras described in section 3.2.4.1 or even further, to what is called *cyclic division algebras*.

Taking into account theorem 3.6, we obtain the following.

THEOREM 3.7.– Let G be a finite abelian group. Then any real non-commutative finite-dimensional G-graded-division algebra with one-dimensional graded components is equivalent to the graded tensor product of several copies of $\mathcal{D}(2^k, 2^\ell; \mu, \nu)$, at most one copy of $\mathcal{C}(2^m; -1)$ and a group algebra $\mathcal{R}H$, where k, ℓ, m are natural numbers, $k \leq \ell$, $\mu, \nu = \pm 1$ and H a subgroup, which is a direct factor of G.

3.4.5. *Equivalence classes of graded-division algebras with one-dimensional homogeneous components*

3.4.5.1. *Basic algebras*

The case of basic algebras of the type $\mathcal{C}(2^m; \eta)$ was discussed earlier in theorem 3.6; we now discuss the question of when $\mathcal{R}_1 = \mathcal{D}(k_1, \ell_1; \mu_1, \nu_1)$ is equivalent to $\mathcal{R}_2 = \mathcal{D}(k_2, \ell_2; \mu_2, \nu_2)$. Let us set $k_i = 2^{r_i}, \ell_i = 2^{s_i}, i = 1, 2$.

Note a useful technical remark.

LEMMA 3.3.– Let u, v be two elements of an algebra $\mathcal{R}$ such that $uv = -vu$. Then, for any natural m, n, p, we have

$$(u^m v^n)^p = u^{mp} v^{np} (-1)^{mn\frac{p(p-1)}{2}}.$$

In particular, if one of m or n is even or p is divisible by 4, then

$$(u^m v^n)^p = u^{mp} v^{np}.$$

The complete classification of the basic algebras, up to equivalence, is given in the following.

PROPOSITION 3.4.– The following are the equivalence classes of algebras of the form $\mathcal{D}(2^r, 2^s; \mu, \nu)$:

1) $\{\mathbb{H}^{(4)} \cong \mathcal{D}(2,2;-1,-1)\}$;

2) $\{M_2^{(4)} \cong \mathcal{D}(2,2;1,1), \mathcal{D}(2,2;1,-1), \mathcal{D}(2,2;-1,1)\}$;

3) $\{\mathcal{D}(2^r, 2^r; 1, 1)\}$, where $r > 1$;

4) $\{\mathcal{D}(2^r, 2^r; -1, -1), \mathcal{D}(2^r, 2^r; 1, -1), \mathcal{D}(2^r, 2^r, -1, 1)\}$, where $r > 1$;

5) $\{\mathcal{D}(2^r, 2^s; 1, 1)\}$, where $1 \leq r < s$;

6) $\{\mathcal{D}(2^r, 2^s; -1, 1)\}$, where $1 \leq r < s$;

7) $\{\mathcal{D}(2^r, 2^s; -1, -1), \mathcal{D}(2^r, 2^s; 1, -1)\}$, where $1 \leq r < s$.

3.4.5.2. *Tensor products of basic algebras: commutation relations*

A much more complicated case is the equivalence of the *tensor products* of algebras of the type $\mathcal{D}(k, \ell; \mu, \nu)$ or $\mathcal{C}(m; \eta)$. It was mentioned in Bahturin and Zaicev (2016) that $\mathbb{H}^{(4)} \otimes \mathbb{H}^{(4)}$ is equivalent to $M_2^{(4)} \otimes M_2^{(4)}$. In other words, $\mathcal{D}(2,2;-1,-1) \otimes \mathcal{D}(2,2;-1,-1)$ is equivalent to $\mathcal{D}(2,2;1,1) \otimes \mathcal{D}(2,2;1,1)$. Also, $\mathbb{H}^{(4)} \otimes \mathbb{C}^{(2)}$ is equivalent to $M_2^{(4)} \otimes \mathbb{C}^{(2)}$. In other words, $\mathcal{D}(2,2;-1,-1) \otimes \mathcal{C}(2;-1)$ is equivalent to $\mathcal{D}(2,2;1,1) \otimes \mathcal{C}(2;-1)$. Thus, an additional work is necessary to determine when two tensor products of gradings mentioned in theorem 3.7 are equivalent to each other.

Before we proceed with our classification, let us remark that if

$$G = G_1 \times \cdots \times G_k$$

is the factorization of the group G through its Sylow subgroups, then

$$\mathcal{R} \cong \mathcal{R}_{G_1} \otimes \cdots \otimes \mathcal{R}_{G_k}.$$

Moreover, if $\mathcal{R} \sim \mathcal{R}'$, where

$$\mathcal{R}' \cong \mathcal{R}'_{G_1} \otimes \cdots \otimes \mathcal{R}'_{G_k},$$

then $\mathcal{R}_{G_1} \sim \mathcal{R}'_{G_1}, \ldots, \mathcal{R}_{G_k} \sim \mathcal{R}'_{G_k}$. At the same time, if the order of G_i is odd then $\mathcal{R}_{G_i} \sim \mathbb{R}G_i$. *This enables us to restrict ourselves to the case, where the support of the grading is an abelian 2-group*. Now each tensor product

$$\mathcal{R} = \mathcal{D}(2^{k_1}, 2^{\ell_1}; \mu_1, \nu_1) \otimes \cdots \otimes \mathcal{D}(2^{k_s}, 2^{\ell_s}; \mu_s, \nu_s) \otimes \mathcal{C}(2^{m_1}; \eta_1) \otimes \cdots \otimes \mathcal{C}(2^{m_t}; \eta_t) \qquad [3.11]$$

gives rise to a sequence

$$\theta = \{(k_1, \ell_1; \mu_1, \nu_1), \ldots, (k_s, \ell_s; \mu_s, \nu_s), (m_1; \eta_1), \ldots, (m_t; \eta_t)\}. \qquad [3.12]$$

Such a sequence will be called the *characteristic* of $\mathcal{R}$ and denoted by $\theta(\mathcal{R})$. We will also define the *truncated characteristic* of $\mathcal{R}$ by setting

$$\overline{\theta}(\mathcal{R}) = \{(k_1, \ell_1), \ldots, (k_s, \ell_s), m_1, \ldots, m_t\}. \qquad [3.13]$$

Since permuting the components in the characteristic does not change the equivalence class of a related graded-division algebra, the equality of two characteristics will be always understood up to a permutation of its tuples.

An important classification result is the following.

PROPOSITION 3.5.– If $\mathcal{R}_1$ and $\mathcal{R}_2$ are equivalent graded-division algebras of the form [3.11], then $\overline{\theta}(\mathcal{R}_1) = \overline{\theta}(\mathcal{R}_2)$.

3.4.5.3. *Tensor products of basic algebras: equivalence*

We keep assuming that the grading group of all graded-division algebras is a finite abelian 2-group. Note that the statement of theorem 3.7 can be made more precise as follows. If, among the generators of an algebra $\mathcal{R}$ in [3.11], there is a central odd generator whose special order is greater than or equal to the special orders of all other odd generators, then this algebra is equivalent to an algebra where there is only one odd generator, and this is central. This can be done using lemma 3.2 when we deal with the tensor products of basic algebras of the first kind.

At the same time, the argument of that lemma easily extends to the case where one of the algebras is of the first kind and the second is of the second kind. To simplify this and many further calculations, we use the following notation:

$$[\mu k, \nu\ell] = \mathcal{D}(2^k, 2^\ell; \mu, \nu),\ [\eta k] = \mathcal{C}(2^k; \eta).$$

For instance, we write $[1, -2]$ instead $\mathcal{D}(2, 4; 1, -1)$ or $[-5]$ instead of $\mathcal{C}(32; -1)$. We may also omit the sign of the tensor product when using the above notation. Given the tensor product of two basic algebras $\mathcal{R} \otimes S$, we will denote by $\{u, v\}\{w, z\}$ the generating set for $\mathcal{R} \otimes S$, where $\{u, v\}$ is the canonical generating set of $\mathcal{R} = \mathcal{D}(2^k, 2^\ell; \mu, \nu)$ and $\{w, z\}$ is the same for S. So we have $u^{2^k} = \mu I$, $v^{2^\ell} = \nu I$, $uv = -vu$. As usual, $\mathcal{R}$ is naturally graded by $\mathbb{Z}_{2^k} \times \mathbb{Z}_{2^\ell}$. We have $uv = -vu$, $wz = -zw$ and the elements in $\{u, v\}$ commute with those in $\{w, z\}$. Likewise, if we deal with the tensor product $\mathcal{D}(2^k, 2^\ell; \mu, \nu) \otimes \mathcal{C}(2^m; \eta) = [\mu k, \nu \ell][\eta m]$, the canonical generating set will be $\{u, v\}\{w\}$, with the relations $u^{2^k} = \mu I$, $v^{2^\ell} = \nu I$, $w^{2^m} = \eta I$, $uv = -vu$, $uw = wu$, $vw = wv$.

Let us check, for instance, that $[-k, \ell][-m] \sim [k, \ell][-m]$, provided that $k \leq m$. Indeed, if the canonical set of generators for $\mathcal{R} = [-k, \ell][-m]$ is $\{x, y\}\{z\}$, $x^{2^k} = -I$, $y^{2^\ell} = I$, $z^{2^m} = -I$, $\mathcal{R}_a = \mathbb{R}x$, $\mathcal{R}_b = \mathbb{R}y$, $\mathcal{R}_c = \mathbb{R}z$, $G = (a)_{2^k} \times (b)_{2^k} \times (c)_{2^m}$, then the new generators $x_1 = xz^{2^{m-k}}$, $y_1 = y$, $z_1 = z$ whose gradings $a_1 = ac^{2^{m-k}}$, $b_1 = b$ and $c_1 = c$, are obtained from a, b, c by an automorphism of the group G, satisfy all the defining relations of $\mathcal{R}_1 = [k, \ell][-m]$ and thus provide us with a weak isomorphism of $\mathcal{R}_1$ and $\mathcal{R}$.

Thus, one of the cases to be considered in the classification of non-commutative graded-division algebras with one-dimensional components is where the algebras have the form of

$$\mathcal{D}(2^{k_1}, 2^{\ell_1}; 1, 1) \otimes \cdots \otimes \mathcal{D}(2^{k_s}, 2^{\ell_s}; 1, 1) \otimes \mathcal{C}(2^m; -1) \otimes \mathcal{R}H, \qquad [3.14]$$

where H is a direct factor of G and $s, m \geq 1$. If two algebras of this kind are equivalent, then the parameter m must be the same because in an algebra with this parameter there is a central homogeneous solution of the equation $x^{2^m} = -1$ but there is no solution of $x^{2^n} = -1$, if $n > m$. Once m is fixed, the isomorphism class of the group H is also fixed. Since the truncated characteristic is an invariant by proposition 3.5, it follows that $\{(k_1, \ell_1), \ldots, (k_s, \ell_s)\}$ is also uniquely defined, up to permutation.

Now if the highest special order of an odd generator in [3.11] does not appear among the central generators, then there is no need to consider central odd generators. The argument is essentially the same as above. For example, let us prove that $\mathcal{R} = [k, -\ell][-m] \sim \mathcal{R}_1 = [k, -\ell][m]$ if $\ell > m$. If $\{x, y\}\{z\}$ is the canonical generating set for $\mathcal{R}$. Then the element $y^{2^{\ell-m}}$ is a central element of degree $b^{2^{\ell-m}}$, so switching to $\{x, y\}\{y^{2^{\ell-m}}z\}$ provides us with the desired equivalence. As a result, in this case, we need to deal only with the algebras of the form

$$\mathcal{D}(2^{k_1}, 2^{\ell_1}; \mu_1, \nu_1) \otimes \cdots \otimes \mathcal{D}(2^{k_s}, 2^{\ell_s}; \mu_s, \nu_s) \otimes \mathbb{R}H, \qquad [3.15]$$

where $\mu_i, \nu_i = \pm 1$ and H is a direct factor of G.

In fact, these algebras admit a further significant reduction.

PROPOSITION 3.6.– Let G be a finite abelian 2-group. Then any graded-division algebra $\mathcal{R}$ with one-dimensional homogeneous components is equivalent to one of the following types:

Type 1 Algebras as in [3.15] where none of the parameters μ_i, ν_i equals -1;

Type 2 Algebras as in [3.14];

Type 3 Algebras as in [3.15] where exactly one of the parameters μ_i, ν_i equals -1 and where there are no factors of the form $\mathcal{D}(2,2;-1,1)$ or $\mathcal{D}(2,2;1,-1)$;

Type 4 Algebras of the form $\mathcal{D}(2,2;-1-1) \otimes S$, where S is of the Type 1.

Note the following:

$$\mathcal{D}(2,2;-1-1) \sim \mathbb{H}^{(4)},$$

$$\mathcal{D}(2,2;-1,1) \sim \mathcal{D}(2,2;1,-1) \sim \mathcal{D}(2,2;1,1) \sim M_2^{(4)}.$$

We illustrate the proof of proposition 3.6 by an important technical argument as follows. This argument allows one to eliminate odd generators, even in the case where their special orders are the same.

Consider $\mathcal{R} = [-k,\ell][m,-k]$, where $m \leq k \leq \ell$. We will prove that $\mathcal{R} \sim \mathcal{R}_1 = [-k,\ell][m,k]$. Let $\{u,v\}\{w,z\}$ be the canonical set of generators for $\mathcal{R}$. Let $u \in \mathcal{R}_a$, $v \in \mathcal{R}_b$, $w \in \mathcal{R}_c$ and $z \in \mathcal{R}_d$. Here $o(a) = 2^k$, $o(b) = 2^\ell$, $o(c) = 2^m$ and $o(d) = 2^k$. We choose the new generating set $\{u, vw\}\{w, uz\}$ for $\mathcal{R}$. Let us check the power relations: $u^{2^k} = -I$, $(vw)^{2^\ell} = (v^{2^\ell})(w^{2^\ell}) = I$, because $d(w) \leq d(v)$, $w^{2^m} = I$ and $(uz)^{2^k} = (u)^{2^k}(z)^{2^k} = (-I)(-I) = I$. Now we check the commutation relations: $u(vw) = -v(uw) = -v(wu) = -(vw)u$, $w(uz) = w(zu) = -z(wu) = -(zu)w = -(uz)w$. Finally, $uw = wu$, $u(uz) = (uz)u$, $(vw)w = w(vw)$ and $(vw)(uz) = (vu)(wz) = (-(uv))(-(zw)) = (uz)(vw)$. The gradings of the elements of the new generating set are $\{a, bc, c, ad\}$ such that $o(a) = 2^k$, $o(bc) = 2^\ell$, $o(c) = 2^m$ and $o(ad) = 2^k$. As a result, we see that $\mathcal{R} \sim \mathcal{R}_1$. Similar arguments work in the other cases. Thus, in the case $k > 1$, the algebra $\mathcal{R}$ is equivalent to an algebra of Type 2 or 3.

3.4.5.4. *Tensor products of basic algebras: non-equivalence*

We know from proposition 3.5 that the truncated characteristics of two algebras with one-dimensional components are the same.

Our first goal in this section is to prove that the algebras of different types in proposition 3.6 are not equivalent.

In our proofs, we will consider (systems) of equations of the form $x^2 = \pm 1$ and compare the number of elements in the supports of the sets of homogeneous solutions.

If these numbers for two algebras are different, then the algebras are not equivalent. For instance, the following remark is very useful in the forthcoming arguments.

REMARK 3.2.– If $k \geq 2$, then the equation $x^{2^k} = -1$ has no homogeneous solutions in the algebras of Type 1 in proposition 3.6.

The following is an easy exercise from the domain of Clifford algebras.

LEMMA 3.4.– Let us denote by $d_\pm^m$ the number of elements in the support of the set of homogeneous solutions for $x^2 = \pm 1$ in $[1,1]^{\otimes m}$, $m \geq 1$. We also set $d_+^0 = 1$ and $d_-^0 = 0$. Then, for all $m \geq 0$, we have

$$d_+^{m+1} = 3d_+^m + d_-^m, \quad d_-^{m+1} = d_+^m + 3d_-^m \qquad [3.16]$$

$$d_+^m = \frac{4^m + 2^m}{2}, \quad d_-^m = \frac{4^m - 2^m}{2} \qquad [3.17]$$

In particular, for each $m > 0$, one of d_+^m, d_-^m is congruent to $0 \bmod 3$ while the other is congruent to $1 \bmod 3$. Also, $d_+^m > d_-^m$, for all $m \geq 0$. □

A thorough analysis of homogeneous solutions of various equations allows us to conclude the following.

PROPOSITION 3.7.– Two graded-division algebras P and Q, of the form [3.11], and asuming the restrictions of proposition 3.6, are equivalent if and only if $\theta(P) = \theta(Q)$.

Now we can state our final result about the equivalence classes of real graded-division algebras with one-dimensional homogeneous components. We denote by $\mathcal{D}(\theta)$ the tensor product of basic algebras of the second kind with characteristic θ. If none of μ_i, ν_j equals -1, we call θ even. If only one of μ_i, ν_j equal -1 and θ does not have terms $(1,1;-1,1)$ or $(1,1;1,-1)$, we call it odd.

THEOREM 3.8.– Any finite-dimensional non-commutative real graded-division algebra with one-dimensional components is equivalent to exactly one algebra on the following list:

1) $\mathcal{D}(\theta) \otimes \mathbb{R}G$, where θ can be even or odd and G a finite abelian group;

2) $\mathcal{C}(2^m; -1) \otimes \mathcal{D}(\theta) \otimes \mathbb{R}G$, where θ is even and G a finite abelian group;

3) $\mathbb{H}^{(4)} \otimes \mathcal{D}(\theta) \otimes \mathbb{R}G$, where θ is even and G a finite abelian group.

3.4.6. *Graded-division algebras with non-central two-dimensional identity components*

Again, we have $\mathcal{R} = \bigoplus_{g\in G} \mathcal{R}_g$ but now $\dim \mathcal{R}_e = 2$ and $\mathcal{R}_e \not\subset Z(\mathcal{R})$. As before, let $G = (g_1)_{n_1} \times \ldots \times (g_q)_{n_q}$ be a primary cyclic factorization of G, where each g_i is an element of order n_i, $i = 1, \ldots, q$.

Since $\mathcal{R}_e \cong \mathbb{C}$, there is an element $J \in \mathcal{R}_e$ such that $J^2 = -1$ and since $\mathcal{R}_e$ is non-central, $J \notin Z(\mathcal{R})$. In this case, for any $a \in \mathcal{R}_g$, the map $x \to axa^{-1}$, where $x \in \mathcal{R}_e$ is an automorphism of $\mathcal{R}_e \cong \mathbb{C}$, hence $aJa^{-1} = \pm J$. Also, $a^2 J = Ja^2$.

If $g \in G$ is an element of odd order $o(g) = 2s - 1$, then $(g^2)^s = g$. Given $a \in \mathcal{R}_g$, we then have $(a^2)^s \in \mathcal{R}_g$. Then $\mathcal{R}_g = \mathcal{R}_e(a^2)^s = \langle (a^2)^s, J(a^2)^s \rangle$. Since a^2 commutes with J, it follows that the whole of $\mathcal{R}_g$ commutes with J. Thus, $\bigoplus_{o(g)\text{ odd}} \mathcal{R}_g$ commutes with J.

Therefore, only $a \in \mathcal{R}_{g_i}$, with $o(g_i)$ a 2-power, can anticommute with J. Using the same method as in the proof of theorem 3.7, we can modify our primary cyclic factorization of G so that

$$G = (g_1)_{n_1} \times (g_2)_{n_2} \times \cdots \times (g_p)_{n_p} \times (g_{p+1})_{n_{p+1}} \times \cdots \times (g_q)_{n_q}$$

is such that n_1 and $n_2| \ldots |n_p$ are 2-powers, there is $a \in \mathcal{R}_{g_1}$ such that $aJ = -Ja$ while the elements in $\mathcal{R}_{g_2}, \ldots, \mathcal{R}_{g_q}$ commute with J. We can assume that n_1 is the minimal number with this property. Let us set $H = (g_2)_{n_2} \times \ldots \times (g_q)_{n_q}$, $T = (g_1)_{n_1}$. We have that $Q = \mathcal{R}_H$ commutes with J while $P = \mathcal{R}_T$ does not.

Clearly, P, as an algebra, is generated by J and a. Let S be the span of the set of homogeneous elements x of Q such that $xax^{-1} = \lambda a$, where λ is a positive real number. Note that for such an element $x \in \mathcal{R}_h$, if we take another $x' \in \mathcal{R}_h$ with the same property, we must have $x' \in \mathbb{R}x$. Indeed, since $\mathcal{R}_h$ commutes with J, the subspace $\mathcal{R}_h$ is a one-dimensional *complex* space, $\mathcal{R}_e \cong \mathbb{C}$ being the complex coefficients. So there is $z \in \mathbb{C}$ such that $x' = xz$. Also, the conjugation of $\mathcal{R}_e$ by a is a complex conjugation in $\mathcal{R}_e$. So if $x'a(x')^{-1} = \mu a$, where μ is a positive real number, then we must have $xzaz^{-1}x^{-1} = \mu a$, or $xz\overline{z}^{-1}ax^{-1} = \mu a$. Hence $\frac{1}{|z|^2} z^2(xax^{-1}) = \mu a$. Finally, we find that z^2 is a positive real number. It follows then that z is a real number and so $x' \in \mathbb{R}x$. Also, if $x_1 \in \mathcal{R}_{h_1}$, $x_2 \in \mathcal{R}_{h_2}$ are such that $x_i a(x_i)^{-1} = \lambda_i a$, where $\lambda_i > 0$, for $i = 1, 2$, then $(x_1x_2)a(x_1x_2)^{-1} = (\lambda_1\lambda_2)a$, proving that the space S spanned by such homogeneous elements is a graded subalgebra with one-dimensional components.

Finally, note that such x exists in every homogeneous component of Q. Indeed, if x is an arbitrary element of $\mathcal{R}_h$, then $xax^{-1} = ua$, for some $u \in \mathcal{R}_e$. If we replace x

by $x' = xz$, then, as previously, $x'a(x')^{-1} = \frac{1}{|z|^2}z^2(xax^{-1}) = \frac{1}{|z|^2}z^2ua$. Taking z such that $z^2 = u^{-1}$, we will have $x'a(x')^{-1} = \frac{1}{|z|^2}a$ so that x' is the desired element of $\mathcal{R}_h$.

As a result, S is a graded-division subalgebra of Q and $\dim S = |H|$. We have described such subalgebras in section 3.4.4. Now we have two graded subalgebras in $\mathcal{R}$: P of dimension $2n_1$ and S of dimension $|H|$. The bilinear map $f : P \times S \to \mathcal{R}$ given by $f(a^k u, x) = a^k ux$, where $u \in \mathcal{R}_e$, $k = 0, 1, 2, \dots n_1 - 1$, $x \in S$, extends to the linear map $\bar{f} : P \otimes S \to \mathcal{R}$. Clearly, this is an onto map. Since the dimensions of $\mathcal{R}$ and $P \otimes S$ are the same, we have that $\bar{f}$ is a *vector space* isomorphism. If we prove that S and P commute, then $\bar{f}^{-1}$ is a G-graded algebra isomorphism $\mathcal{R} \cong P \otimes S$.

To prove that, indeed, P and S commute, we consider the action of a on the complex space Q. Pick a homogeneous $x \in Q_h$, then $axa^{-1} = zx$ where $z \in \mathcal{R}_e$. Since a acts by conjugation on $\mathcal{R}_e \cong \mathbb{C}$, we have that $a^2xa^{-2} = z\bar{z}x = |z|^2x$. Now if $n_1 = 2m$, then $g_1^{2m} = e$, and so $(a^2)^m \in \mathcal{R}_e$. Since x commutes with $\mathcal{R}_e$, we have that $(|z|^2)^m x = (a^2)^m x(a^2)^{-m} = x$. It then follows that $|z|^{2m} = 1$ and hence $|z| = 1$. Thus, a^2 commutes with the whole of Q. Since a^2 commutes with a and J, it follows that a^2 is in the center of $\mathcal{R}$.

Now let us take $x \in S_h$. There is a positive real λ such that $xax^{-1} = \lambda a$. Then $a^2 = (xax^{-1})(xax^{-1}) = \lambda^2 a^2$ so that $\lambda = 1$, as needed.

To determine a *realization* of P, let us set $g = g_1$, $n = n_1$. We remember that n is a 2-power. Then, for any homogeneous $b \in P$, we must have $b^n = \alpha I + \beta J \in \mathcal{R}_e$. If $b \in P_g$ is such that also $bJ = -Jb$, then $b^n = \alpha I + \beta J$ so that, if we conjugate by b, we obtain $\alpha I - \beta J = \alpha I + \beta J$. As a result, $\beta = 0$ and $b^n = \alpha I$, where α is a real number. If we replace b by $\frac{1}{\sqrt[n]{|\alpha|}}b$, we will obtain $b^n = \pm I$.

Now let us endow $M = \mathbb{R}(g)_n \otimes M_2(\mathbb{C})$, with natural grading by $(g)_n$. Then choose $u = 1 \otimes C$ and $v = g \otimes \omega A$, where ω is a complex number such that $\omega^n = \varepsilon$, where $a^n = \varepsilon I$. Also, A and C are standard Pauli matrices, as defined by [3.7]. Then consider the subalgebra $\widehat{M} = \operatorname{alg}\{u, v\}$. Comparing defining relations and dimensions we easily obtain $P \cong \widehat{M}$ as G-graded algebras. Let us denote the algebra thus described by $\mathcal{E}(n; \varepsilon)$. This is a *basic algebra of the third kind*. One can give this algebra in terms of generators and define relations as follows:

$$\mathcal{E}(n; \varepsilon) = \langle x, y \mid x^2 = -1, y^{2^n} = \varepsilon, xy = -yx \rangle,$$

where $\delta(x) = e$, $\delta(y) = g$, $G = (g)_{2^n} \cong \mathbb{Z}_{2^n}$.

THEOREM 3.9.– Any division grading on a finite-dimensional real algebra with two-dimensional non-central identity components is equivalent to the tensor product of one grading of the form $\mathcal{E}(n;\varepsilon)$, several gradings of the form $\mathcal{D}(k,\ell;\mu,\nu)$ and several gradings $\mathcal{C}(m;\kappa)$ where m,n,k,ℓ are natural numbers > 1, n,k,ℓ are 2-powers, and $\varepsilon,\mu,\nu,\kappa = \pm 1$. Moreover, n is the minimal positive integer such that the equation $x^n = \pm 1$ has a homogeneous solution not lying in the centralizer of $\mathcal{R}_e$. □

3.4.6.1. *Equivalence of graded-division algebras with non-central two-dimensional identity components*

Clearly, $\mathcal{E}(2^k;\mu) \sim \mathcal{E}(2^\ell;\nu)$ if and only if $k = \ell$ and $\mu = \nu$. Indeed, the first equation is guaranteed because the grading group of $\mathcal{E}(2^k;\mu)$ is $\mathbb{Z}_{2^k}$. As for the second equation, if $k > 1$ then an equation $x^{2^k} = -1$ has a homogeneous solution in $\mathcal{E}(2^k;-1)$ but not in $\mathcal{E}(2^k;1)$ (see lemma 3.6). If $k = 1$, then $\mathcal{E}(2;-1) \sim \mathbb{H}^{(2)}$, while $\mathcal{E}(2;1) \sim M_2^{(2)}$.

In what follows, we write $(\rho k]$ for $\mathcal{E}(2^k;\rho)$, where $\rho = \pm 1$. In particular, $(\rho]$ stands for $(1] \sim M_2^{(2)}$ if $\rho = 1$ and $(-1] \sim \mathbb{H}^{(2)}$ if $\rho = -1$. By $\{J,u\}$, we denote the standard generating set for $\mathcal{E}(2^k;\mu)$, that is $u^2 = -I$, $v^{2^k} = \mu I$ and $uv = -vu$. Now $\deg u = e, \deg v = g$, where $\mathbb{Z}_{2^k} = (g)_{2^k}$. We have the following equivalences for the tensor products of basic algebras $\mathcal{E}(2^k;\rho) = (\rho k]$ and $\mathcal{D}(2^\ell, 2^m;\mu,\nu) = [\mu\ell,\nu m]$ or $\mathcal{C}(2^\ell;\varepsilon) = [\varepsilon\ell]$. We write $\{v,w\}$ for the standard generating set of $[\varepsilon]$ and $\{w\}$ for $[\varepsilon\ell]$.

LEMMA 3.5.–

1) $(-k][-\ell] \sim (k][-\ell]$ if $k \le \ell$, new generating set $\{J, uv^{2^{\ell-k}}\}\{v\}$;

2) $(-k][-\ell] \sim (-k][\ell]$ if $k > \ell$, new generating set $\{J,u\}\{u^{2^{k-\ell}}v\}$;

3) $(-k][\ell,-m] \sim (k][\ell,-m]$ if $k < m$, new generating set $\{J, uw^{2^{m-k}}\}\{v,w\}$;

4) $(-k][\ell,-m] \sim (-k][\ell,m]$ if $k > m$, new generating set $\{J,u\}\{v, u^{2^{k-m}}w\}$;

5) $(-k][\ell,-k] \sim (k][\ell,-k]$, if $\ell \ge 2$, new generating set $\{J,uw\}\{Jv,w\}$;

6) $(-k][1,-k] \sim (k][1,-k]$ if $k \ge 2$, new generating set $\{J,uw\}\{Jvw^{2^{k-1}},w\}$;

7) $(-k][-\ell,m] \sim (-k][\ell,m]$ if $k > \ell$, new generating set $\{J,u\}\{u^{2^{k-\ell}}v,w\}$;

8) $(-k][-\ell,m] \sim (k][-\ell,m]$ if $k < \ell$, new generating set $\{J,uv^{2^{\ell-k}}\}\{v,w\}$;

9) $(-k][-k,m] \sim (k][-k,m]$, if $m \ge 2$, new generating set $\{J,uv\}\{v,Jw\}$;

10) $(\rho k][-1,\ell] \sim (\rho k][1,\ell]$, if $\ell \le k$, new generating set $\{J,uw\}\{Jv,w\}$;

11) $(\rho k][-1,-1] \sim (\rho k][1,-1]]$, if $k \ge 2$, new generating set $\{J,uw\}\{Jv,w\}$;

12) $(\rho][-1,-1] \sim (-\rho][1,-1]$, new generating set $\{J,uw\}\{Jv,w\}$.

Because of this lemma, tensor factors of Type 4 in proposition 3.6 do not appear in the following.

PROPOSITION 3.8.– Any graded-division algebra with non-central two-dimensional identity component is equivalent to one of the following:

1) $\mathcal{E}(2^k; 1) \otimes S$, where S is of Type 1 in proposition 3.6;

2) $\mathcal{E}(2^k; -1) \otimes S$, where S is of Type 1 in proposition 3.6;

3) $\mathcal{E}(2^k; 1) \otimes S$, where S is of Type 2 or 3 in proposition 3.6, without factors of the form $\mathcal{D}(2, 2^\ell; -1, 1)$ with $\ell \leq k$.

Let $\{J, u\}$ be the canonical generating set of $(\mu k] = \mathcal{E}(2^k; \mu)$, where $\mu = \pm 1$. So $Ju = -uJ$, $J^2 = -I$, $u^{2^k} = \mu I$. Let a be the generating element of the grading group $G \cong \mathbb{Z}_{2^k}$. Any non-zero homogeneous element of degree a^s can be uniquely written in the form zu^s, where $z = \alpha I + \beta J$ and $\alpha, \beta \in \mathbb{R}$, with $\alpha^2 + \beta^2 \neq 0$. To facilitate further arguments, we do some calculations in $(\mu k]$. We set $\overline{z} = \alpha I - \beta J$.

$$u^s z = \begin{cases} zu^s & \text{if } s \text{ is even} \\ \overline{z}u^s & \text{if } s \text{ is odd} \end{cases}$$

$$(z_1 u^r)(z_2 u^s) = \begin{cases} (z_1 z_2) u^{r+s} & \text{if } r \text{ is even} \\ z_1 \overline{z}_2 u^{r+s} & \text{if } r \text{ is odd} \end{cases}$$

If $1 \leq m \leq k$,

$$(zu^s)^{2^m} = \begin{cases} z^{2^m} u^{2^{m+1}t} & \text{if } s = 2t \\ (z\overline{z})^{2^{m-1}} u^{2^{m+1}t} u^{2^m} & \text{if } s = 2t+1 \end{cases}$$

In particular,

$$(zu^s)^{2^k} = \begin{cases} z^{2^k} & \text{if } s = 2t \\ (z\overline{z})^{2^{k-1}} \mu & \text{if } s = 2t+1 \end{cases}$$

Let us assume $k \geq 2$ and determine the homogeneous solutions of the equations $x^{2^k} = \pm 1$ in $(\pm k]$.

LEMMA 3.6.– The homogeneous solutions of the equation $x^{2^k} = 1$ are as follows.

In $(k]$, we have finitely many solutions of the form zu^{2t}, with $z^{2^k} = I$, and infinitely many solutions of the form zu^{2t+1}, where $|z| = 1$.

In $(-k]$, we have finitely many solutions of the form zu^{2t}, with $z^{2^t} = -I$.

The homogeneous solutions of the equation $x^{2^k} = -1$ are as follows.

In $(k]$, we have finitely many solutions of the form zu^{2t}, with $z^{2^k} = -I$.

In $(-k]$, we have finitely many solutions of the form zu^{2t}, with $z^{2^t} = -I$ and infinitely many solutions of the form zu^{2t+1}, where $|z| = 1$.

A thorough analysis along the same lines as in the case of graded-division algebras with one-dimensional components leads to the following.

THEOREM 3.10.– Every graded-division algebra with non-central two-dimensional identity component is equivalent to exactly one of the following:

1) $\mathcal{E}(2^k; 1) \otimes \mathcal{D}(\theta) \otimes \mathbb{R}G$, θ even and G a finite abelian group;

2) $\mathcal{E}(2^k; 1) \otimes \mathcal{D}(\theta) \otimes \mathcal{C}(2^m; -1) \otimes \mathbb{R}G$, θ even and G a finite abelian group;

3) $\mathcal{E}(2^k; -1) \otimes \mathcal{D}(\theta) \otimes \mathbb{R}G$, θ even and G a finite abelian group;

4) $\mathcal{E}(2^k; 1) \otimes \mathcal{D}(\theta) \otimes \mathbb{R}G$, θ is odd, without quadruples of the form $(1, \ell; -1, 1)$ with $\ell \leq k$ and G a finite abelian group;

3.4.7. *Graded-division algebras with four-dimensional identity components*

Let $\mathcal{R} = \bigoplus_{g \in G} \mathcal{R}_g$ be a graded-division algebra such that $\mathcal{R}_e \cong \mathbb{H}$. It follows from the double centralizer theorem (Jacobson 1989, Theorem 4.7) that $\mathcal{R} \cong \mathcal{R}_e \otimes C$ where C is the centralizer of $\mathcal{R}_e$ in $\mathcal{R}$.

If $\mathcal{R}_1$ and $\mathcal{R}_2$ are two graded-division algebras with four-dimensional components, then $(\mathcal{R}_i)_e \cong \mathbb{H}$ and $\mathcal{R}_i \cong \mathcal{R}_e \otimes C_{\mathcal{R}_i}((\mathcal{R}_1)_e)$, for $i = 1, 2$. If we set $A_i = C_{\mathcal{R}_i}(\mathcal{R}_e)$, then A_i is a graded-division algebra with one-dimensional components, for $i = 1, 2$. Let $f : \mathcal{R}_1 \to \mathcal{R}_2$ be a weak isomorphism of graded division algebras. Then $f((\mathcal{R}_1)_e) = (\mathcal{R}_2)_e$ and

$$f(A_1) = f(C_{\mathcal{R}_1}((\mathcal{R}_1)_e)) = C_{f(\mathcal{R}_1)}(f((\mathcal{R}_1)_e)) = C_{\mathcal{R}_2}((\mathcal{R}_2)_e) = A_2.$$

Hence $A_1 \sim A_2$.

THEOREM 3.11.– If $\mathcal{R} = \bigoplus_{g \in G} \mathcal{R}_g$ is a finite-dimensional real graded division algebra with $\dim \mathcal{R}_e = 4$, then $\mathbb{R}$ is equivalent to exactly one of the following:

1) $\mathbb{H} \otimes \mathcal{D}(\theta) \otimes \mathbb{R}G$, where θ can be even or odd and G a finite abelian group;

2) $\mathbb{H} \otimes \mathcal{C}(2^m; -1) \otimes \mathcal{D}(\theta) \otimes \mathbb{R}G$, θ even and G a finite abelian group;

3) $\mathbb{H} \otimes \mathbb{H}^{(4)} \otimes \mathcal{D}(\theta) \otimes \mathbb{R}G$, θ even and G a finite abelian group.

3.4.8. *Classification of real graded-division algebras, up to isomorphism*

In principle, to obtain the classification of real graded-division algebras up to isomorphism, one could use the classification up to equivalence, given above, and apply all possible automorphisms of the support G. However, this approach is impractical, because it is not clear how to determine which of the resulting gradings are isomorphic to each other.

The difficulty lies with the 2-primary component (=2-Sylow subgroup) of G, which we denote by G_0, since the product of the odd primary components, which we denote by G_1, just contributes its group algebra $\mathbb{R}G_1$. So, assume for a moment that G is a 2-group. Denote by $G_{[2]}$ the subgroup consisting of e and the elements of order 2 (=the socle of G). We will also need the subgroup $G^{[2]} = \{g^2 \mid g \in G\}$.

For graded-division algebras with homogeneous components of dimension 1, we introduce two parameters: the alternating bicharacter $\beta : G \times G \to \{\pm 1\}$ determined, as in section 3.3, by the commutation relations, and a function $\mu : G_{[2]} \to \{\pm 1\}$ sending e to 1 and any element g of order 2 to the sign of X_g^2. Here, as before, X_g spans the homogeneous component of degree g, for any $g \in G$, with the convention that $X_e = 1$. It is easy to check that $\mu(gh) = \mu(g)\mu(h)\beta(g,h)$ for all $g, h \in G_{[2]}$. In other words, regarded as a map of vector spaces over the field $\mathbb{Z}_2$, μ is a quadratic form whose polarization is the restriction of β to $G_{[2]}$.

Up to isomorphism, the graded-division algebra can be recovered from β and μ in terms of generators and defining relations. More precisely, we introduce a G-graded-division algebra $\mathcal{D}(T, \beta, \mu)$ as follows. Let us write G as the direct product of cyclic subgroups $G = (g_1)_{n_1} \times \cdots \times (g_m)_{n_m}$ of orders $n_1, \ldots, n_m$ (by our assumption, powers of 2). Then define $\mathcal{D}(T, \beta, \mu)$ by G-graded generators $x_1, \ldots, x_m$ of degrees $g_1, \ldots, g_m$ and G-graded relations of two kinds: $x_i^{n_i} = \mu_i$ $(1 \le i \le m)$ and $x_i x_j = \beta_{ij} x_j x_i$ $(1 \le i < j \le m)$, where $\mu_i = \mu(g_i^{n_i/2})$ and $\beta_{ij} = \beta(g_i, g_j)$. One checks that, for an arbitrary choice of the scalars $\mu_i \in \{\pm 1\}$ and $\beta_{ij} \in \{\pm 1\}$, this is indeed a graded-division algebra with one-dimensional components and support G. Hence, any alternating bicharacter β on G and any quadratic form μ on $G_{[2]}$ (whose polarization is the restriction of β) are allowed.

As we have seen in the previous section, the case of homogeneous components of dimension 4 easily reduces to the case of dimension 1. The case of dimension 2 requires considerably more work. We will state the classification result from Bahturin *et al.* (2019). The parameterization of isomorphism classes with homogeneous components of dimension 2 depends on a fixed representative $t_0 = t_0(K)$ of the non-trivial coset for every subgroup K of index 2. The choice of t_0 is arbitrary, but it is convenient to make $t_0^2 = e$ if K is a direct summand.

THEOREM 3.12.– Any finite-dimensional real graded-division algebra with abelian support $G = G_0 \times G_1$, with $|G_0|$ being a power of 2 and $|G_1|$ odd, is isomorphic to one of the following:

1) $\mathcal{D}(G_0, \beta, \mu) \otimes \mathbb{R}G_1$, where $\beta : G_0 \times G_0 \to \{\pm 1\}$ is an alternating bicharacter and $\mu : G_{[2]} \to \{\pm 1\}$ is a quadratic form whose polarization is the restriction of β;

2) $\mathcal{D}(G_0, \beta, \mu) \otimes \mathbb{R}G_1 \otimes \mathbb{H}$, where $\beta : G_0 \times G_0 \to \{\pm 1\}$ and $\mu : G_{[2]} \to \{\pm 1\}$ are as above and $\mathbb{H}$ is the quaternion algebra with trivial grading;

3) The real form $\mathcal{D} = \bigoplus_{g \in G} \mathcal{D}_g$ of $M_2(\mathcal{C})$ with two-dimensional components $\mathcal{D}_g = \langle Y_g, JY_g \rangle$, where $\mathcal{C}$ is the complexification of $\mathcal{D}(K, \beta, \mu) \otimes \mathbb{R}G_1 = \bigoplus_{g \in K \times G_1} \mathbb{R}X_g$, K is a subgroup of index 2 in G_0, $\beta : K \times K \to \{\pm 1\}$ and $\mu : K_{[2]} \to \{\pm 1\}$ are as above, with $G_0^{[2]} \subset \operatorname{rad} \beta$, $J = \begin{bmatrix} i & 0 \\ 0 & -i \end{bmatrix}$, and the elements Y_g are defined as follows:

$$Y_g = \begin{bmatrix} X_g & 0 \\ 0 & X_g \end{bmatrix} \text{ and } Y_{t_0 g} = \begin{bmatrix} 0 & X_{t_0^2} X_g \\ \delta X_g & 0 \end{bmatrix} \quad \text{for all } g \in K \times G_1 = \operatorname{Supp} \mathcal{C},$$

where $t_0 = t_0(K)$, $\delta \in \{\pm 1\}$ if K is a direct summand of G_0, and $\delta = 1$ if K is not a direct summand of G_0;

4) $\mathcal{P}(\beta)$ (viewed as a real algebra), where $\beta : G \times G \to \mathbb{C}^\times$ is an alternating bicharacter.

Two algebras in different items are not graded-isomorphic to one another, nor two algebras in the same item if they have different parameters (β, μ, K, δ, as applicable), except that, in item 4), β and β^{-1} give graded-isomorphic algebras.

The simple real graded-division algebras were classified up to isomorphism in Rodrigo-Escudero (2016). They correspond to non-degenerate bicharacters β (the cases $M_n(\mathbb{R})$ and $M_n(\mathbb{H})$) and β with radical of order at most 2 (the case $M_n(\mathbb{C})$). On the other hand, commutative graded-division algebras (=graded-fields) occur in items 1) and 4) when β is trivial, so we get just $\mathbb{C}G$ in item 4) and the real forms of $\mathbb{C}G$ parameterized by characters $\mu : G_{[2]} \to \{\pm 1\}$ in item 1).

3.5. Real loop algebras with a non-split centroid

Let $\pi : G \to \overline{G}$ be an epimorphism of abelian groups, $H = \ker \pi$. We will apply the loop construction from section 3.2.6 over the algebraic closure $\mathbb{C}$ and then define a suitable Galois descent to $\mathbb{R}$. Denote the generator of $\mathrm{Gal}(\mathbb{C}/\mathbb{R})$ by ι, i.e. $\iota(z) = \bar{z}$ for all $z \in \mathbb{C}$.

Let $\chi : G \to \mathbb{C}^\times$ be a character with values in the unit circle $U \subset \mathbb{C}$ (which is automatic for finite groups). Then the $\mathbb{R}$-linear operator on $\mathbb{C}G = \mathbb{C} \otimes \mathbb{R}G$ defined by $z \otimes g \mapsto \bar{z}\chi(g) \otimes g$ is involutive, so we obtain a semilinear action of $\mathrm{Gal}(\mathbb{C}/\mathbb{R})$ on $\mathbb{C}G$ and hence a real form of $\mathbb{C}G$ as a G-graded algebra. (In fact, any such real form is obtained in this way.) The homogeneous component of degree g is $\mathbb{R}u_g$, where $u_g = z_g \otimes g$ and $z_g \in \mathbb{C}^\times$ is any element satisfying $z_g/\bar{z}_g = \chi(g)$. We have

$$u_{g_1}u_{g_2} = \gamma(g_1, g_2)u_{g_1g_2} \text{ for all } g_1, g_2 \in G,$$

so our real form is the twisted group algebra $\mathcal{D} = \mathbb{R}^\gamma G$ where the 2-cocycle $\gamma : G \times G \to \mathbb{R}^\times$ is given by $\gamma(g_1, g_2) = z_{g_1}z_{g_2}/z_{g_1g_2}$. We will choose z_g to be in the unit circle, i.e. a square root of $\chi(g)$. Then γ takes values in $\{\pm 1\}$.

Now let $\mathcal{A}$ be a $\overline{G}$-graded algebra over $\mathbb{R}$. Similarly to the above, $\mathcal{A} \otimes \mathbb{C} \otimes \mathbb{R}G$ acquires a semilinear action of $\mathrm{Gal}(\mathbb{C}/\mathbb{R})$ defined by $\iota \cdot (a \otimes z \otimes g) = a \otimes \bar{z}\chi(g) \otimes g$, which restricts to a degree-preserving action on $L_\pi(\mathcal{A}\otimes\mathbb{C}) = \bigoplus_{g\in G} \mathcal{A}_{\pi(g)} \otimes \mathbb{C} \otimes \mathbb{R}G$. As a result, we obtain the following G-graded $\mathbb{R}$-form of $L_\pi(\mathcal{A} \otimes \mathbb{C})$:

$$L_\pi^\chi(\mathcal{A}) := \{x \in L_\pi(\mathcal{A} \otimes \mathbb{C}) \mid \iota \cdot x = x\} = \bigoplus_{g\in G} \mathcal{A}_{\pi(g)} \otimes u_g \subset \mathcal{A} \otimes \mathbb{R}^\gamma G.$$

By construction, $L_\pi^\chi(\mathcal{A}) \otimes \mathbb{C} \cong L_\pi(\mathcal{A} \otimes \mathbb{C})$. It follows that $L_\pi^\chi(\mathcal{A})$ is unital if and only if so is $\mathcal{A}$, and $L_\pi^\chi(\mathcal{A})$ belongs to a variety $\mathfrak{V}$ if and only if so does $\mathcal{A}$. Indeed, we already know that L_π preserves these properties, and they are not affected by field extensions. By the same argument, if $\mathcal{A}$ is central simple, then $L_\pi^\chi(\mathcal{A})$ is graded-central-simple.

The action of $\mathrm{Gal}(\mathbb{C}/\mathbb{R})$ passes on to the centroid of $L_\pi(\mathcal{A} \otimes \mathbb{C})$: $(\iota \cdot c)(x) = \iota \cdot c(\iota^{-1} \cdot x)$, for any $x \in L_\pi(\mathcal{A} \otimes \mathbb{C})$ and c in the centroid. Assuming that $\mathcal{A}$ is central simple, we have identified the centroid with $\mathbb{C}H = \mathbb{C} \otimes \mathbb{R}H$. With this identification, we have $\iota \cdot (z \otimes h) = \bar{z}\chi(h) \otimes h$ for all $z \in \mathbb{C}$ and $h \in H$. Therefore, the centroid of $L_\pi^\chi(\mathcal{A})$ can be identified with $\mathcal{D}_H := \bigoplus_{h\in H} \mathcal{D}_h$, which is a graded subalgebra of the real form $\mathcal{D}$ of $\mathbb{C}G$ defined above. Note that $\mathcal{D}_H = L_\pi^\chi(\mathbb{R})$.

We are now ready to extend a part of the "Correspondence Theorem" to the case of the non-split centroid.

THEOREM 3.13.– Let $\mathcal{B}$ be a G-graded algebra that is graded-central-simple over $\mathbb{R}$. Let H be the support of the centroid $C(\mathcal{B})$ and $\pi : G \to \overline{G} = G/H$ be the natural homomorphism. Then, for some character $\chi : G \to U \subset \mathbb{C}^\times$, there exists a central simple algebra $\mathcal{A}$ over $\mathbb{R}$ and a $\overline{G}$-grading on $\mathcal{A}$ such that $\mathcal{B} \cong L_\pi^\chi(\mathcal{A})$ as G-graded algebras. Moreover, we can take any character satisfying $L_\pi^\chi(\mathbb{R}) \cong C(\mathcal{B})$. If $\mathcal{A}$ and $\mathcal{A}'$ are two central simple algebras over $\mathbb{R}$ equipped with $\overline{G}$-gradings, then $L_\pi^\chi(\mathcal{A})$ and $L_\pi^\chi(\mathcal{A}')$ are isomorphic as G-graded algebras if and only if the $\overline{G}$-graded algebra $\mathcal{A}'$ is isomorphic to a twist of $\mathcal{A}$.

Recall that the twist of a simple graded algebra was defined in section 3.2.6.2.

The non-zero homogeneous elements of degree g in $L_\pi^\chi(\mathcal{A})$ have the form $a \otimes u_g$ where $0 \neq a \in \mathcal{A}_{\pi(g)}$, so the support of $L_\pi^\chi(\mathcal{A})$ is the inverse image under π of the support of $\mathcal{A}$. Remark 3.1 has the following analogue:

REMARK 3.3.– As a graded $L_\pi^\chi(\mathbb{F})$-module, $L_\pi^\chi(\mathcal{A}) \cong L_\pi^\chi(\mathbb{F}) \otimes \underline{\mathcal{A}}$ where $\underline{\mathcal{A}}$ is $\mathcal{A}$ regarded as a G-graded vector space with $\deg a = \xi(\overline{g})$ for all non-zero $a \in \mathcal{A}_{\overline{g}}$, $\overline{g} \in \overline{G}$. An isomorphism $L_\pi^\chi(\mathbb{F}) \otimes \underline{\mathcal{A}} \to L_\pi^\chi(\mathcal{A})$ is given by $u_h \otimes a \mapsto a \otimes u_h u_{\xi(\overline{g})}$, where we have identified $L_\pi^\chi(\mathbb{F})$ with $\mathcal{D}_H$. In particular, if ξ is a group homomorphism then we can choose χ to be trivial on the complement $\xi(\overline{G})$ of H in G, which entails $L_\pi^\chi(\mathcal{A}) \cong L_\pi^\chi(\mathbb{F}) \otimes {}^\xi\!\mathcal{A}$ as graded algebras.

Since the elements u_g are invertible, we also observe the following:

REMARK 3.4.– If $\mathcal{A}$ is alternative (in particular, associative) or Jordan, then $\mathcal{A}$ is a graded-division algebra with respect to $\overline{G}$ if and only if $L_\pi^\chi(\mathcal{A})$ is a graded-division algebra with respect to G.

EXAMPLE 3.1.– If $\mathcal{B} = \bigoplus_{g \in G} \mathcal{B}_g$ is a graded-field with $\mathcal{B}_e = \mathbb{R}1$, then $\mathcal{B} \cong L_\pi^\chi(\mathbb{R})$ where $H = G$ and $\pi : G \to \{1\}$. If G is finite, we can write it as a direct product of cyclic groups: $G = \langle g_1 \rangle \times \cdots \times \langle g_s \rangle$, and choose roots of unity z_j, $j = 1, \ldots, s$, satisfying $z_j^2 = \chi(g_j)$. Then

$$L_\pi^\chi(\mathbb{R}) = \mathbb{R}[u_1] \otimes \cdots \otimes \mathbb{R}[u_s] \text{ where } \deg u_j = g_j \text{ and } u_j^{o(g_j)} = z_j^{o(z_j)} \in \{\pm 1\}.$$

This is the description of such algebras given in proposition 3.2.

3.6. Alternative algebras

Recall that an algebra is said to be *alternative* if the associator $(x, y, z) := (xy)z - x(yz)$ is an alternating function of the variables x, y, z or, equivalently, the left and right alternative identities hold: $x(xy) = x^2y$ and $(yx)x = yx^2$. By Artin's theorem, this implies that any subalgebra generated by two elements is associative. The definition of inverses in a unital alternative algebra is the same as in the associative case, and the set of invertible elements is closed under multiplication (it is a so-called *Moufang loop*). Kleinfeld (1953, 1957) (see also

Zhevlakov *et al.* (1982)) proved that any simple alternative algebra is either associative or an octonion algebra over a field (we give more background about the simple alternative in section 3.6.1.

Group gradings on octonion algebras over any field were described in Elduque (1998) (see also Elduque and Kochetov (2013)). (Here, as in the case of simple Lie algebras, there is no loss of generality in assuming the grading group abelian because the elements of the support always commute.) A classification of gradings up to isomorphism is given in Elduque and Kochetov (2013).

Now let $\mathcal{C}$ be a real octonion algebra, i.e. either the octonion division algebra $\mathbb{O}$ or the split octonion algebra $\mathbb{O}_s$. Suppose $\pi : G \to \overline{G}$ is an epimorphism of abelian groups and $\mathcal{C}$ is given a $\overline{G}$-grading. Then, for any character $\chi : G \to U$, the G-graded algebra $L_\pi^\chi(\mathcal{C})$ is a real alternative non-associative algebra that is graded-central-simple. Conversely, every such algebra is graded-isomorphic to an algebra of the form $L_\pi^\chi(\mathcal{C})$, where $\mathcal{C}$ is a simple alternative non-associative algebra with centroid $\mathbb{R}$, which is a real octonion algebra by Kleinfeld's theorem.

Thus, our extension of the loop algebra construction can be used to derive a classification up to isomorphism of graded-simple alternative algebras that are graded-central over $\mathbb{R}$ and not associative (theorem 3.15). Among these, we will also classify the graded-division algebras up to equivalence (corollary 3.1). For the classification up to isomorphism, the only remaining question is to calculate the twists, but first we review the relevant facts about octonions.

3.6.1. *Cayley–Dickson doubling process*

Recall that a *Hurwitz algebra* is a unital composition algebra, i.e. a unital algebra equipped with a multiplicative non-singular quadratic form, which is called the *norm* and will be denoted by n. Except in the case $\operatorname{char}\mathbb{F} = 2$,"non-singular" means that the polar form of n, defined by $n(x, y) := n(x+y) - n(x) - n(y)$, is non-degenerate. The *standard involution* of a Hurwitz algebra is given by $\bar{x} := n(x, 1)1 - x$.

It is well known that the dimension of a Hurwitz algebra can only be 1, 2, 4 or 8. Hurwitz algebras of dimension 4 are referred to as *quaternion algebras* and those of dimension 8 as *octonion* or *Cayley algebras.*

Given an *associative* Hurwitz algebra $\mathcal{A}$ such that the polar form of n is non-degenerate and any scalar $\alpha \in \mathbb{F}^\times$, let $\mathfrak{CD}(\mathcal{A}, \alpha)$ be the direct sum of two copies of $\mathcal{A}$, where we may formally write the element (x, y) as $x + yw$, so $\mathfrak{CD}(\mathcal{A}, \alpha) = \mathcal{A} \oplus \mathcal{A}w$. This is a Hurwitz algebra with multiplication and norm given by:

$$\begin{aligned}(a + bw)(c + dw) &= (ac + \alpha\bar{d}b) + (da + b\bar{c})w,\\ n\big(a + bw\big) &= n(a) - \alpha n(b).\end{aligned}$$

For example, the real division algebras of complex numbers, quaternions, and octonions are obtained as $\mathbb{C} = \mathfrak{CD}(\mathbb{R}, -1)$, $\mathbb{H} = \mathfrak{CD}(\mathbb{C}, -1)$, and $\mathbb{O} = \mathfrak{CD}(\mathbb{H}, -1)$. Since $\mathbb{O}$ is not associative, we cannot obtain a Hurwitz algebra by doubling it. We will abbreviate $\mathfrak{CD}(\mathcal{A}, \alpha, \beta) := \mathfrak{CD}\big(\mathfrak{CD}(\mathcal{A}, \alpha), \beta\big)$, and similarly for $\mathfrak{CD}(\mathbb{F}, \alpha, \beta, \gamma)$. For any $\delta \in \mathbb{F}^\times$, the mapping $(x, y) \mapsto (x, \delta^{-1}y)$ is an isomorphism $\mathfrak{CD}(\mathcal{A}, \alpha) \to \mathfrak{CD}(\mathcal{A}, \alpha\delta^2)$. Hence, over $\mathbb{R}$, the isomorphism class of these algebras depends only on the sign of the parameters α, β and γ. It turns out that if any of the parameters is positive, we get the "split complex numbers" $\mathbb{C}_s = \mathbb{R} \times \mathbb{R}$, split quaternions $\mathbb{H}_s \cong M_2(\mathbb{R})$, and split octonions $\mathbb{O}_s$.

Conversely, given any Hurwitz algebra $\mathcal{C}$ with norm n and a subalgebra $\mathcal{A}$ such that the restriction to $\mathcal{A}$ of the polar form of n is non-degenerate, and given any non-isotropic element $w \in \mathcal{A}^\perp$, it follows that $n(\mathcal{A}, \mathcal{A}w) = 0$ and that $\mathcal{A} \oplus \mathcal{A}w$ is a subalgebra of $\mathcal{C}$ isomorphic to $\mathfrak{CD}(\mathcal{A}, \alpha)$ with $\alpha = -n(w) = w^2$.

3.6.2. *Gradings on octonion algebras*

Let $\mathcal{Q}$ be a quaternion subalgebra of a Cayley algebra $\mathcal{C}$ over a field $\mathbb{F}$, and let $w \in \mathcal{Q}^\perp$ with $n(w) = -\gamma \neq 0$. Then $\mathcal{C} = \mathcal{Q} \oplus \mathcal{Q}w$ is isomorphic to $\mathfrak{CD}(\mathcal{Q}, \gamma)$, which gives a $\mathbb{Z}_2$-grading on $\mathcal{C}$ with $\mathcal{C}_{\bar{0}} = \mathcal{Q}$ and $\mathcal{C}_{\bar{1}} = \mathcal{Q}^\perp = \mathcal{Q}w$.

In its turn, the quaternion subalgebra $\mathcal{Q}$ can be obtained from a two-dimensional subalgebra $\mathcal{K}$ (either $\mathbb{F} \times \mathbb{F}$ or a separable quadratic field extension of $\mathbb{F}$) as $\mathcal{Q} = \mathcal{K} \oplus \mathcal{K}v \cong \mathfrak{CD}(\mathcal{K}, \beta)$, with $v \in \mathcal{Q} \cap \mathcal{K}^\perp$ and $n(v) = -\beta \neq 0$. Then $\mathcal{C}$ is isomorphic to $\mathfrak{CD}(\mathcal{K}, \beta, \gamma)$, which gives a $\mathbb{Z}_2^2$-grading on $\mathcal{C}$ with $\mathcal{C}_{(\bar{0},\bar{0})} = \mathcal{K}$, $\mathcal{C}_{(\bar{1},\bar{0})} = \mathcal{K}v$, $\mathcal{C}_{(\bar{0},\bar{1})} = \mathcal{K}w$, and $\mathcal{C}_{(\bar{1},\bar{1})} = (\mathcal{K}v)w$.

If $\operatorname{char} \mathbb{F} \neq 2$, then $\mathcal{K}$ can be obtained by doubling $\mathbb{F}$: $\mathcal{K} = \mathbb{F} \oplus \mathbb{F}u \cong \mathfrak{CD}(\mathbb{F}, \alpha)$, with $u \in \mathcal{K} \cap \mathbb{F}^\perp$ and $n(u) = -\alpha \neq 0$, so $\mathcal{C} \cong \mathfrak{CD}(\mathbb{F}, \alpha, \beta, \gamma)$, which gives a $\mathbb{Z}_2^3$-grading on $\mathcal{C}$, with $\deg u = (\bar{1}, \bar{0}, \bar{0})$, $\deg v = (\bar{0}, \bar{1}, \bar{0})$, and $\deg w = (\bar{0}, \bar{0}, \bar{1})$.

The above gradings by $\mathbb{Z}_2^r$, $r = 1, 2, 3$, are called the *gradings induced by the Cayley–Dickson doubling process* (Elduque and Kochetov 2013, p. 131) ($r \neq 3$ if $\operatorname{char} \mathbb{F} = 2$).

Finally, over any field, there is a unique (up to isomorphism) split Cayley algebra $\mathcal{C}_s$, and it admits a $\mathbb{Z}^2$-grading, called the *Cartan grading*, whose homogeneous components are the eigenspaces for the action of a maximal torus of $\mathbf{Aut}_{\mathbb{F}}(\mathcal{C}_s)$.

The following is a stronger version of the main result of Elduque (1998), where it was used to classify gradings on Cayley algebras over algebraically closed fields (see also Elduque and Kochetov (2013), Theorem 4.12).

THEOREM 3.14 (Elduque and Kochetov (2013)).– Any non-trivial grading on a Cayley algebra is, up to equivalence, either a grading induced by the Cayley–Dickson doubling process, starting with a Hurwitz division subalgebra, or a coarsening of the Cartan grading on the split Cayley algebra.

Starting from this point, a classification of gradings up to isomorphism is obtained in Elduque and Kochetov (2013) for any Cayley algebra. Here, we will state the result only for $\mathbb{F} = \mathbb{R}$ and $\mathbb{F} = \mathbb{C}$. To be consistent with our previous notation, we will denote the grading group by $\overline{G}$.

The coarsenings of the Cartan grading on $\mathbb{O}_s$ are induced by arbitrary homomorphisms $\mathbb{Z}^2 \to \overline{G}$ (see section 3.2.1). Let $\overline{g}_1$, $\overline{g}_2$ and $\overline{g}_3$ be the images of the elements $(1,0)$, $(0,1)$, and $(-1,-1)$, respectively, so $\overline{g}_1\overline{g}_2\overline{g}_3 = \bar{e}$ (the identity element of $\overline{G}$). Then the grading is determined by the triple $(\overline{g}_1, \overline{g}_2, \overline{g}_3)$, and two triples yield isomorphic gradings if and only if they are in the same orbit under the action of the group $S_3 \times \mathbb{Z}_2$ (the Weyl group of type G_2), where S_3 permutes the entries of the triple and the generator of $\mathbb{Z}_2$ inverts them simultaneously. We will denote the resulting graded algebras by $\mathcal{C}(\overline{g}_1, \overline{g}_2, \overline{g}_3)$. None of them is a graded-division algebra (e.g. because the identity component contains a non-trivial idempotent).

The remaining non-trivial gradings on $\mathbb{O}$ and $\mathbb{O}_s$ are obtained by arbitrary monomorphisms $\mathbb{Z}_2^r \to \overline{G}$, $1 \le r \le 3$, from the gradings induced by the Cayley–Dickson doubling process starting from $\mathbb{H}$ if $r = 1$, from $\mathbb{C}$ if $r = 2$, and from $\mathbb{R}$ if $r = 3$. All of them are division gradings. To include the trivial grading on $\mathbb{O}$, we will allow $r = 0$. The support of the grading is the image of $\mathbb{Z}_2^r$, which we denote by $\overline{T}$. The parameters used in the doubling process determine a character $\mu : \overline{T} \to \{\pm 1\}$ as follows: for any non-zero homogeneous element x of degree $\bar{t}$, we have $x^2 \in -\mu(\bar{t})\mathbb{R}_{>0}$. The trivial characters give gradings on $\mathbb{O}$ and the non-trivial ones on $\mathbb{O}_s$. We will denote the resulting graded-division algebras by $\mathcal{C}(\overline{T}, \mu)$. The algebras corresponding to different pairs $(\overline{T}, \mu)$ are not graded-isomorphic.

A similar classification of division gradings is valid for $\mathbb{H}$ and $\mathbb{H}_s$, on the one hand, and $\mathbb{C}$ and $\mathbb{C}_s$, on the other hand. We will denote the resulting graded-division algebras by $\mathcal{Q}(\overline{T}, \mu)$ and $\mathcal{K}(\overline{T}, \mu)$, respectively.

Over $\mathbb{C}$, there is only one Cayley algebra (up to isomorphism), and the classification of gradings is the same as above, except in the Cayley–Dickson case there is no parameter μ, and r must be equal to 3. We will denote these graded algebras by $\mathcal{C}_{\mathbb{C}}(\overline{g}_1, \overline{g}_2, \overline{g}_3)$ and $\mathcal{C}_{\mathbb{C}}(\overline{T})$. Similarly, we also have $\mathcal{Q}_{\mathbb{C}}(\overline{T})$ with $r = 2$ for $M_2(\mathbb{C})$ (the complex quaternion algebra) and $\mathcal{K}_{\mathbb{C}}(\overline{T})$ with $r = 1$ for $\mathbb{C} \times \mathbb{C}$.

3.6.3. *Graded-simple real alternative algebras*

Let $\mathcal{B}$ be a G-graded algebra over $\mathbb{R}$ that is graded-simple and alternative but not associative. Assume also that the identity component of the centroid $\mathbb{L} := C(\mathcal{B})$

is finite dimensional, so it is either $\mathbb{R}$ or $\mathbb{C}$. As before, let H be the support of $\mathbb{L}$, $\overline{G} = G/H$, and π be the natural homomorphism $G \to \overline{G}$.

If $\mathbb{L}_e \cong \mathbb{C}$, then $\mathbb{L} \cong \mathbb{C}H$ and, by theorem 3.2 ("Correspondence Theorem"), we have $\mathcal{B} \cong L_\pi(\mathcal{C})$, where $\mathcal{C}$ is the unique complex Cayley algebra, equipped with a $\overline{G}$-grading, and this grading is determined uniquely up to isomorphism. Thus, in this case, the algebras $\mathcal{B}$ can be of two kinds: those with $\mathcal{C} \cong \mathcal{C}_{\mathbb{C}}(\overline{g}_1, \overline{g}_2, \overline{g}_3)$ are classified up to isomorphism by the $(S_3 \times \mathbb{Z}_2)$-orbits of the triples $(\overline{g}_1, \overline{g}_2, \overline{g}_3)$ and those with $\mathcal{C} \cong \mathcal{C}_{\mathbb{C}}(\overline{T})$ are classified by the subgroups $\overline{T} \subset \overline{G}$ isomorphic to $\mathbb{Z}_2^3$. The graded-division algebras are the latter.

Now let us assume $\mathbb{L}_e = \mathbb{R}$ and fix a character $\chi : G \to U \subset \mathbb{C}^\times$ such that $\mathbb{L} \cong L_\pi^\chi(\mathbb{R})$. By theorem 3.13, we have $\mathcal{B} \cong L_\pi^\chi(\mathcal{C})$ where $\mathcal{C}$ is either $\mathbb{O}$ or $\mathbb{O}_s$, equipped with a $\overline{G}$-grading, so $\mathcal{C} \cong \mathcal{C}(\overline{g}_1, \overline{g}_2, \overline{g}_3)$ or $\mathcal{C} \cong \mathcal{C}(\overline{T}, \mu)$. To apply the isomorphism part of theorem 3.13, we have to classify these $\overline{G}$-graded algebras up to twist. As a result, we have the following.

THEOREM 3.15.– Let $\mathcal{B}$ be a finite-dimensional G-graded algebra over $\mathbb{R}$ that is graded-simple and alternative but not associative. Let H be the support of $\mathbb{L} = C(\mathcal{B})$, $\overline{G} = G/H$, and π be the natural homomorphism $G \to \overline{G}$. If $\mathbb{L}_e = \mathbb{R}$, pick a character $\chi : G \to U \subset \mathbb{C}^\times$ such that $\mathbb{L} \cong L_\pi^\chi(\mathbb{R})$.

1) If $\mathbb{L}_e = \mathbb{R}$, then $\mathcal{B}$ is graded-isomorphic to one of the following:

a) $L_\pi^\chi(\mathcal{C}(\overline{g}_1, \overline{g}_2, \overline{g}_3))$ for a triple $(\overline{g}_1, \overline{g}_2, \overline{g}_3) \in \overline{G}^3$ with $\overline{g}_1\overline{g}_2\overline{g}_3 = \bar{e}$;

b) $L_\pi^\chi(\mathcal{C}(\overline{T}, \mu))$ for a subgroup $\overline{T} \subset \overline{G}$ isomorphic to $\mathbb{Z}_2^r$, $0 \leq r \leq 3$, and a character $\mu : \overline{T} \to \{\pm 1\}$.

2) If $\mathbb{L}_e \cong \mathbb{C}$, then $\mathcal{B}$ is graded-isomorphic to one of the following:

a) $L_\pi(\mathcal{C}_{\mathbb{C}}(\overline{g}_1, \overline{g}_2, \overline{g}_3))$ for a triple $(\overline{g}_1, \overline{g}_2, \overline{g}_3) \in \overline{G}^3$ with $\overline{g}_1\overline{g}_2\overline{g}_3 = \bar{e}$;

b) $L_\pi(\mathcal{C}_{\mathbb{C}}(\overline{T}))$ for a subgroup $\overline{T} \subset \overline{G}$ isomorphic to $\mathbb{Z}_2^3$.

Two algebras in different items are not graded-isomorphic to one another. Two algebras in item 1a) or 2a) are graded-isomorphic if and only if their triples belong to the same $(S_3 \times \mathbb{Z}_2)$-orbit. Two algebras in item 2b) are graded-isomorphic if and only if their subgroups are equal. Finally, two algebras in item 1b) are graded-isomorphic if and only if their subgroups are equal and their characters have the same restriction to $\overline{T}_0 := \ker \tau$ where $\tau : \overline{T} \to H/H^{[2]}$ is given by $\tau(tH) = t^2 H^{[2]}$.

3.6.4. *Graded-division real alternative algebras*

To summarize our classification, it is convenient to introduce the following notation. Let T be an abelian group with a subgroup H such that $\overline{T} := T/H$ is an elementary 2-group of rank $r \in \{0, 1, 2, 3\}$. Let $\mathcal{O}$ be an orbit of non-trivial characters $H_{[2]} \to \{\pm 1\}$ under the action of the stabilizer of H in $\mathrm{Aut}(T)$. For each triple $(T, H, \mathcal{O})$, we fix a representative $\chi_0 \in \mathcal{O}$ and extend it to a character $\chi : T \to U \subset \mathbb{C}^\times$. Let

$$\mathcal{DA}(T, H, \mathcal{O}) := L_\pi^\chi(\mathcal{C}(\overline{T}, 1)),$$

where $\pi : T \to \overline{T}$ is the natural homomorphism. For each pair (T, H) as above, fix a character $\mu : \overline{T} \to \{\pm 1\}$ whose restriction to $\overline{T}_0 := \ker \tau$ (see theorem 3.15) is non-trivial, and set

$$\mathcal{DA}(T, H) := L_\pi(\mathcal{C}(\overline{T}, 1)) \text{ and } \mathcal{DA}(T, H)' := L_\pi(\mathcal{C}(\overline{T}, \mu)).$$

(If $\overline{T}_0$ is trivial, then only $\mathcal{DA}(T, H)$ is defined.) Finally, for each pair (T, H) with $r = 3$, let

$$\mathcal{DA}_\mathbb{C}(T, H) := L_\pi(\mathcal{C}_\mathbb{C}(\overline{T})).$$

Then we have the following:

THEOREM 3.16.– Let $\mathcal{B}$ be a real graded-division algebra with support T that is alternative but not associative. Let $\mathbb{L} = C(\mathcal{B})$ and assume that $\mathbb{L}_e$ is finite-dimensional, so it is either $\mathbb{R}$ or $\mathbb{C}$, and let H be the support of $\mathbb{L}$.

1) If $\mathbb{L}_e = \mathbb{R}$ and $\mathbb{L} \ncong \mathbb{R}H$, then $\mathcal{B}$ is equivalent to $\mathcal{DA}(T, H, \mathcal{O})$ for a unique orbit $\mathcal{O}$ of non-trivial characters $H_{[2]} \to \{\pm 1\}$ under the action of the stabilizer of H in $\mathrm{Aut}(T)$.

2) If $\mathbb{L}_e = \mathbb{R}$ and $\mathbb{L} \cong \mathbb{R}H$, then $\mathcal{B}$ is equivalent to $\mathcal{DA}(T, H)$ or $\mathcal{DA}(T, H)'$, but not both.

3) If $\mathbb{L}_e \cong \mathbb{C}$, then $\mathcal{B}$ is equivalent to $\mathcal{DA}_\mathbb{C}(T, H)$.

We can make the classification more explicit in the finite-dimensional case by classifying the pairs (T, H) and the triples (T, H, χ_0) up to isomorphism, where T is a finite abelian group and, as before, H is a subgroup such that $\overline{T} := T/H$ is an elementary 2-group of rank $r \in \{0, 1, 2, 3\}$ and χ_0 is a non-trivial character $H_{[2]} \to \{\pm 1\}$. First of all, we have a canonical decomposition $T = T' \times T''$ where T' is a 2-group and T'' has odd order. Then $H = H' \times T''$ and $H_{[2]} = H'_{[2]}$, so the problem reduces to 2-groups.

A subset $\{g_1, \ldots, g_s\}$ of a finite abelian group G will be called a *basis* if the orders of g_j are non-trivial prime powers and $G = \langle g_1 \rangle \times \cdots \times \langle g_s \rangle$. If G is an elementary p-group, this is indeed a basis of G as a vector space over the field of p elements.

LEMMA 3.7.– Let $\pi : G \to \overline{G}$ be an epimorphism of finite abelian p-groups where $\overline{G}$ is elementary. Then there exists a basis $\{g_1, \dots, g_s\}$ of G such that the elements $\pi(g_j)$ that are different from the identity form a basis of $\overline{G}$.

In particular, our pairs (T, H) are described as follows: there is a basis of T in which r elements are marked, the orders of the marked elements are powers of 2, and H is generated by the unmarked elements and the squares of the marked elements. Up to isomorphism, a pair (T, H) is represented by a tuple of non-trivial prime powers (the orders of the basis elements) in which r powers of 2 are marked: for example, the pair $(\mathbb{Z}_2 \times \mathbb{Z}_2 \times \mathbb{Z}_4, 0 \times \mathbb{Z}_2 \times 2\mathbb{Z}_4)$ is represented by $(\underline{2}, 2, \underline{4})$. Two pairs are isomorphic if and only if the tuples are the same up to permutation. (The "only if" part can be seen by looking at the quotients $T^{[2^k]}/H^{[2^k]}$ for $k = 1, 2, \dots$)

Now we want to bring χ_0 into consideration. The classification of pairs (H, χ_0) up to isomorphism is equivalent to the classifications of graded-fields in example 3.1 up to equivalence, which was studied in Bahturin and Zaicev (2018) (see section 3.4.3). Such a pair is described by the tuple consisting of the orders of the basis elements where each power of 2 is given a sign according to the value of χ_0 on the unique element of order 2 in the cyclic group, generated by the basis element. Since we consider non-trivial χ_0, at least one entry of the tuple must have a negative sign. Similar to what was done earlier in section 3.4, by changing the basis, one can always achieve exactly one negative sign. This is done using the following moves: if two distinct basis elements h_i and h_j with $o(h_i) = 2^m \leq 2^n = o(h_j)$ give a negative sign, we replace h_i with $h_i h_j^{2^{n-m}}$, which gives a positive sign. Hence, a pair (H, χ_0) is represented, up to isomorphism, by a tuple of non-trivial prime powers in which exactly one power of 2 is given a negative sign: for example, the tuple $(2, -8)$ represents the pair $(\mathbb{Z}_2 \times \mathbb{Z}_8, \chi_0)$ where $\chi_0((\bar{1}, \bar{0})) = 1$ and $\chi_0((\bar{0}, \bar{4})) = -1$. Two pairs are isomorphic if and only if the tuples are the same up to permutation. (The "only if" part can be seen by looking at the restrictions of χ_0 to the subgroups $H^{[2^k]} \cap H_{[2]}$ for $k = 1, 2, \dots\dots$). Hence, every graded-field in example 3.1 is equivalent to either $\mathbb{R}H$ (corresponding to the trivial χ_0) or $\mathbb{R}K \otimes [-n])$ where $H \cong K \times \mathbb{Z}_{2^n}$ and the $\mathbb{Z}_{2^{|m|}}$-graded algebra $[m]$ was defined in section 3.4.5.3, for any $m \in \mathbb{Z}$ as follows:

$$[n] := \mathbb{R}[x]/(x^{2^n} - 1) \text{ and } [-n] := \mathbb{R}[x]/(x^{2^n} + 1) \text{ where } \deg x = \bar{1} \in \mathbb{Z}_{2^n}.$$

Note that $[0] = \mathbb{R}$ (trivially graded) and, if H is a 2-group, it is clear from the structure of H that $\mathbb{R}H$ is equivalent to the tensor product of graded algebras of the form $[n]$, $n \in \mathbb{N}$.

Finally, for the triples (T, H, χ_0), we combine the previous arguments: we have a basis of H in which $r_0 \leq r$ elements are marked (being the non-trivial squares of the marked basis elements of T), and each basis element gives a sign. We can reduce the number of negative signs to one by applying the same moves as above to the basis elements of T, except that we are not allowed to replace t_i with $t_i t_j^{2^{n-m}}$ if t_i is in H, t_j is not in H, and $m = n$. This is not an obstacle, because in this situation we can replace t_j with $t_i t_j$. Therefore, the triples (T, H, χ_0) are classified by tuples of non-trivial prime powers in which r powers of 2 are marked and one of the powers of 2 (marked or unmarked) is given a negative sign. This labeling is arbitrary except that a marked power of 2 can be given a negative sign only if it is at least 4.

Now we can describe more explicitly the finite-dimensional graded algebras. For integers n_1, n_2, n_3, consider $T = \mathbb{Z}_{2^{|n_1|+1}} \times \mathbb{Z}_{2^{|n_2|+1}} \times \mathbb{Z}_{2^{|n_3|+1}}$ and its standard basis $t_1 = (\bar{1}, \bar{0}, \bar{0})$, $t_2 = (\bar{0}, \bar{1}, \bar{0})$, $t_3 = (\bar{0}, \bar{0}, \bar{1})$. Define the following T-graded algebra:

$$\mathbb{R}\{n_1, n_2, n_3\} := \mathfrak{CD}\big(\mathbb{L}, (x_1, t_1), (x_2, t_2), (x_3, t_3)\big) \text{ where } \mathbb{L} = [n_1] \otimes [n_2] \otimes [n_3],$$

with x_j being the generator of the jth factor of $\mathbb{L}$, with $\deg x_j = 2t_j$. Thus, the Cayley–Dickson generators satisfy $u^2 = x_1$, $v^2 = x_2$, $w^2 = x_3$ and are assigned degrees $\deg u = t_1$, $\deg v = t_2$, $\deg w = t_3$. Here, we make the convention that $x_j = -1$ if $n_j = 0$. We define in a similar manner the graded algebras

$$\mathbb{C}\{n_2, n_3\} := \mathfrak{CD}\big(\mathbb{C} \otimes \mathbb{L}, (x_2, t_2), (x_3, t_3)\big) \text{ where } \mathbb{L} = [n_2] \otimes [n_3];$$

$$\mathbb{H}\{n_3\} := \mathfrak{CD}\big(\mathbb{H} \otimes \mathbb{L}, (x_3, t_3)\big) \text{ where } \mathbb{L} = [n_3].$$

For non-negative integers n_1, n_2, n_3, we also define the graded algebras $\mathbb{R}\{n_1, n_2, n_3\}'$, $\mathbb{C}\{n_2, n_3\}'$ and $\mathbb{H}\{n_3\}'$ as above, but with the convention $x_j = 1$ if $n_j = 0$. Note that in our notation the letters $\mathbb{R}$, $\mathbb{C}$ and $\mathbb{H}$ indicate the identity component. We can also have $\mathbb{O}$ as the identity component. In summary:

COROLLARY 3.1.– Let $\mathcal{B}$ be a finite-dimensional real graded-division algebra that is alternative but not associative.

1) If $\mathcal{B}$ is graded-central with a non-split centroid, then $\mathcal{B}$ is equivalent to one of the following:

a) $\mathbb{R}K \otimes \mathbb{R}\{n_1, n_2, n_3\}$, $\mathbb{R}K \otimes \mathbb{C}\{n_2, n_3\}$, or $\mathbb{R}K \otimes \mathbb{H}\{n_3\}$, where K is a finite abelian group and exactly one of the integers n_j is negative;

b) $\mathbb{R}K \otimes (-n) \otimes \mathbb{R}\{n_1, n_2, n_3\}$, $\mathbb{R}K \otimes (-n) \otimes \mathbb{C}\{n_2, n_3\}$, $\mathbb{R}K \otimes (-n) \otimes \mathbb{H}\{n_3\}$ or $\mathbb{R}K \otimes (-n) \otimes \mathbb{O}$, where K is a finite abelian group, $n > 0$ and $n_j \geq 0$.

2) If $\mathcal{B}$ is graded-central with a split centroid, then $\mathcal{B}$ is equivalent to one of the following:

a) $\mathbb{R}K \otimes \mathbb{R}\{n_1, n_2, n_3\}$, $\mathbb{R}K \otimes \mathbb{C}\{n_2, n_3\}$, $\mathbb{R}K \otimes \mathbb{H}\{n_3\}$ or $\mathbb{R}K \otimes \mathbb{O}$, where K is a finite abelian group and $n_j \geq 0$;

b) $\mathbb{R}K \otimes \mathbb{R}\{n_1, n_2, n_3\}'$, $\mathbb{R}K \otimes \mathbb{C}\{n_2, n_3\}'$ or $\mathbb{R}K \otimes \mathbb{H}\{n_3\}'$, where K is a finite abelian group and $n_j \geq 0$ with at least one being 0.

3) If $\mathcal{B}$ is not graded-central, then $\mathcal{B}$ is equivalent to $\mathbb{C}K \otimes \mathbb{R}\{n_1, n_2, n_3\}$, where K is a finite abelian group and $n_j \geq 0$.

Two algebras in different items are not equivalent to one another. Two algebras in the same item are equivalent if and only if they have isomorphic abelian groups K and, if applicable, the same value of n and the same n_j up to permutation.

3.7. Gradings of fields

A natural question that arises when we study gradings on algebras over arbitrary fields is what gradings are possible on fields. First of all, it is clear that the support of the grading must be an abelian group. Indeed, if $g, h \in \operatorname{Supp}\mathbb{K}$, $\mathbb{K}$ a field, $0 \neq a \in \mathbb{K}_g$, $0 \neq b \in \mathbb{K}_h$, then $0 \neq ab = ba \in \mathbb{K}_{gh} \cap \mathbb{K}_{hg}$. It follows that $gh = hg$, as claimed. More precise information about the support is given by the next result from Bahturin *et al.* (2019). It is convenient to use the term *G-graded extension* for a field extension $\mathbb{K}/\mathbb{F}$ where $\mathbb{K}$ is a G-graded algebra over $\mathbb{F}$ such that the support is G and $\mathbb{K}_e = \mathbb{F}$.

PROPOSITION 3.9.– Let $\mathbb{K}/\mathbb{F}$ be a G-graded field extension. Then G is a torsion abelian group and there is a subgroup M in the multiplicative group $\mathbb{K}^\times$ such that $\mathbb{F}^\times \subset M$ and G is isomorphic to the quotient $M/\mathbb{F}^\times$. In particular,

i) if $\mathbb{K}$ is a finite field, then G is a cyclic group;

ii) $\mathbb{K}/\mathbb{F}$ is an algebraic extension.

It should be noted that a G-graded field extension need not be normal or separable.

If $|\mathbb{F}| = p^\ell$ and $|\mathbb{K}| = p^{k\ell}$, then for $\mathbb{K}/\mathbb{F}$ to be a G-graded extension, it is necessary that G be cyclic of order k and that k be a divisor of $\dfrac{p^{k\ell}-1}{p^\ell - 1}$. The following result gives necessary and sufficient conditions.

THEOREM 3.17.– Let $\mathbb{K} = GF(p^n)$ and $\mathbb{F} = GF(p^\ell)$ be Galois fields where ℓ divides n. Let $k = n/\ell$. Then there exists a G-grading on $\mathbb{K}$ with support G and the identity component $\mathbb{F}$ if and only if the following two conditions hold:

i) if q is a prime divisor of k, then q divides $p^\ell - 1$;

ii) if 4 divides k, then 4 divides $p^\ell - 1$.

If such a grading exists, the group G must be cyclic of order k.

The idea of the proof is the following. Similar to our argument in section 3.4.3, a field extension of a field $\mathbb{F}$, graded by a cyclic group of order k, must have the form $\mathbb{F}[x]/(x^k - a)$. So we need to determine if there exists $0 \neq a \in \mathbb{F}$ such that the binomial $x^k - a$ is irreducible. According to Lang (1965, Theorem 16 in Section VIII.9), $x^k - a$ is irreducible if and only if $a \notin \mathbb{F}^q$, for any prime divisor q of k, and also $a \notin -4\mathbb{F}^4$ if 4 divides k. This leads to conditions (i) and (ii); see Bahturin *et al.* (2019) for details.

In general, a field extension $\mathbb{K}/\mathbb{F}$ graded by a finite abelian group $G = (g_1)_{n_1} \times (g_2)_{n_2} \times \cdots \times (g_m)_{n_m}$ must look like

$$\mathbb{K} = \mathbb{F}[x_1]/(x_1^{n_1} - a_1) \otimes \mathbb{F}[x_2]/(x_1^{n_2} - a_2) \otimes \cdots \otimes \mathbb{F}[x_m]/(x_m^{n_m} - a_m). \quad [3.18]$$

For any $0 \neq a_i \in \mathbb{F}$, this tensor product is a commutative G-graded-division algebra. If we can find $a_1, a_2, \ldots, a_m$ such that it is a field, we have constructed a G-graded extension $\mathbb{K}$ of $\mathbb{F}$. In a particular case, where G is an elementary abelian 2-group, we have the following criterion:

PROPOSITION 3.10.– Let $\mathbb{F}$ be a field of characteristic not 2, $G = \prod_{i=1}^m (g_i)_2$, $\{a_i \mid i = 1, 2, \ldots, m\}$ a set of non-zero elements of $\mathbb{F}$, and $\mathbb{K} = \bigotimes_{i=1}^m \mathbb{F}[x_i]/(x_i^2 - a_i)$ a G-graded algebra in which $\deg x_i = g_i$. Then $\mathbb{K}$ is the field if and only if the set $\{a_i \mid i = 1, 2, \ldots, m\}$ is linearly independent (in the multiplicative sense) in $\mathbb{F}^\times/(\mathbb{F}^\times)^2$.

Since there is a lot of information about the multiplicative group of fields in particular cases, we have some instructive consequences such as

i) over $\mathbb{Q}$, there are $\mathbb{Z}_2^m$ -graded field extensions for any $m = 1, 2, \ldots$;

ii) over the field $\mathbb{Q}_p$ of p-adic numbers, $p \neq 2$, $\mathbb{Z}_2^m$ -graded field extensions exist only for $m = 1, 2$;

iii) over the field $\mathbb{Q}_2$ of 2-adic numbers, $\mathbb{Z}_2^m$ -graded field extensions exist only for $m = 1, 2, 3$.

It should be noted than any Kummer extension can be made a graded field extension in a canonical way. We conclude with a couple of examples illustrating theorem 3.17.

EXAMPLE 3.2.– Let $\mathbb{K}$ be the Galois field $GF(2^{q^n})$, where q is a prime number. Then any grading on $\mathbb{K}$ is trivial.

EXAMPLE 3.3.– Let $\mathbb{K} = GF(p^{q\ell})$ where p and q are prime numbers and q divides $p^\ell - 1$. Then $\mathbb{K}$ admits a $\mathbb{Z}_q$-grading with identity component $GF(p^\ell)$.

3.8. References

Allison, B., Berman, S., Faulkner, J., Pianzola, A. (2008). Realization of graded-simple algebras as loop algebras. *Forum Math.*, 20, 395–432.

Bahturin, Y., Brear, M., Kochetov, M. (2011). Group gradings on finitary simple Lie algebras. *Int. J. Algebra Appl.*, 321, 1250046.

Bahturin, Y., Elduque, A., Kochetov, M. (2019). Graded-division algebras over arbitrary fields. *arXiv math.*, (1912.11911).

Bahturin, Y., Kochetov, M. (2019). On nonassociative graded-simple algebras over the field of real numbers. *Contemporary Mathematics*, 728, 25–48.

Bahturin, Y., Kochetov, M., Rodrigo-Escudero, A. (2018). Gradings on classical central simple real Lie algebras. *J. Algebra*, 506, 1–42.

Bahturin, Y., Shestakov, I. (2001). Gradings of simple Jordan algebras and their relation to the gradings of simple associative algebras. *Comm. Algebra*, 29(9), 4095–4102. (Special issue dedicated to Alexei Ivanovich Kostrikin.)

Bahturin, Y., Sehgal, S., Zaicev, M. (2001). Group gradings on associative algebras. *J. Algebra*, 241, 677–698.

Bahturin, Y., Zaicev, M. (2003). Graded algebras and graded identities. Polynomial identities and combinatorial methods. *Pure and Appl. Math.*, 235, 101–139.

Bahturin, Y., Zaicev, M. (2016). Simple graded-division algebras over the field of real numbers. *Linear Algebra and its Applications*, 490, 102–123.

Bahturin, Y., Zaicev, M. (2018). Graded-division algebras over the field of real numbers. *J. Algebra*, 514, 273–309.

Bahturin, Y., Shestakov, I., Zaicev, M. (2005). Gradings on simple Jordan and Lie algebras. *J. Algebra*, 283(2), 849–868.

Benkart, G., Neher, E. (2006). The centroid of extended affine and root graded Lie algebras. *J. Pure Appl. Algebra*, 205, 117–145.

Bott, R., Milnor, J. (1958). On the parallelizability of the spheres. *Bull. Amer. Math. Soc.*, 64, 87–89.

Caenepeel, S. (1998). *Brauer Groups, Hopf Algebras and Galois Theory*. K-Monographs in Mathematics Volume 4. Kluwer Academic Publishers, Dordrecht.

Calderón Martín, A., Draper, C., Martín Gonzalez, C. (2010). Gradings on the real forms of the Albert algebra, of $\mathfrak{g}_2$ and of $\mathfrak{f}_4$. *J. Math. Phys.*, 51(5), 053516.

Draper, C., Martín González, C. (2009). Gradings on the Albert algebra and on $\mathfrak{f}_4$. *Rev. Mat. Iberoam.*, 25(3), 841–908.

Elduque, A. (1998). Gradings on octonions. *J. Algebra*, 207(1), 342–354.

Elduque, A. (2019). Graded-simple algebras and cocycle twisted loop algebras. *Proc. Amer. Math. Soc.*, 147(7), 2821–2833.

Elduque, A., Kochetov, M. (2012). Gradings on the exceptional Lie algebras f_4 and g_2 revisited. *Rev. Mat. Iberoam.*, 28(3), 773–813.

Elduque, A., Kochetov, M. (2013). Gradings on the simple real Lie algebras of types G2 and D4. J. Algebra, 512(2018), 382–426.

Elduque, A., Kochetov, M. (2018). Gradings on the simple real Lie algebras of types g_2 and d_4, *J. Algebra*, 512, 382–426.

Garling, D. (2011). Clifford algebras: An introduction. *London Mathematical Society, Student Texts*. Cambridge University Press, Cambridge.

Giambruno, A., Zaicev, M. (2005). *Polynomial Identities and Asymptotics Methods*. Mathematical Surveys and Monographs, American Mathematical Society, Providence.

Griess, R.L., Jr. (1990). A Moufang loop, the exceptional Jordan algebra, and a cubic form in 27 variables. *J. Algebra*, 131(1), 281–293.

Jacobson, N. (1968). *Structure and Representations of Jordan Algebras*. AMS Coll. Publ., American Mathematical Society, Providence, 39, x+453.

Jacobson, N. (1989). *Basic Algebra. II. 2nd edition* W. H. Freeman and Co., New York.

Karrer, G. (1973). Graded division algebras. *Math. Z.*, 133, 67–73.

Kleinfeld, E. (1953). Simple alternative rings. *Ann. of Math*, 58(2), 544–547.

Kleinfeld, E. (1957). Alternative nil rings. *Ann. of Math*. 66(2), 395–399.

Lang, S. (1965). *Algebra*. Addison-Wesley, Reading.

Milnor, J. (1971). Introduction to algebraic K-theory. *Annals of Mathematics Studies*. Princeton University Press, Princeton, University of Tokyo Press, Tokyo, 72, xiii+184.

Montgomery, S. (1993). Hopf algebras and their actions on rings. *CBMS Regional Conference Series*. In *Mathematics*. American Mathematical Society, Providence, 82, xiv+238.

Nastasescu, C., Van Oystaeyen, F. (2004). *Methods of Graded Rings*. Lecture Notes in Mathematics. Springer, Berlin.

de Rham, G. (1931). Sur l'analysis situs des variétés à n-dimensions. *J. Math. Pures Appl.*, 10, 115–200.

Rodrigo-Escudero, A. (2016). Classification of division gradings on finite-dimensional simple real algebras. *Linear Algebra and its Applications*, 493, 164–182.

Tignol, J.-P. (1982). Sur les décompositions des algèbres à division en produit tensoriel d'algèbres cycliques. In *Brauer Groups in Ring Theory and Algebraic Geometry*, van Oystaeyen F.M.J., Verschoren A.H.M.J. (eds). Lecture Notes in Mathematics. Springer, Berlin.

Zhevlakov, K., Slin'ko, A., Shestakov, I., Shirshov, A. (1982). *Rings that are Nearly Associative. Pure and Applied Mathematics*. Translated by Smith, H.F. Academic Press Inc., New York, London.

Zmud, E. (1971). Symplectic geometries on finite abelian groups. *Mat. Sb. (N.S)*, 86(128), 9–33.

Zolotykh, A. (1997). Commutation factors and varieties of associative algebras. *Fundam. Prikl. Mat.*, 3, 453–468.

4

Non-associative C^*-algebras

Ángel RODRÍGUEZ PALACIOS and Miguel CABRERA GARCÍA
Department of Mathematical Analysis, Granada University, Spain

4.1. Introduction

In this chapter, we deal with the different approaches to non-associative models trying to generalize (associative) C^*-algebras (namely JB-algebras, JB^*-triples and non-commutative JB^*-algebras), and discuss how these notions are related. As the reader can realize, by looking at proposition 4.7, theorems 4.8 and 4.24, proposition 4.23 and corollary 4.17, in the end all these approaches give rise to essentially the same mathematical creature.

4.2. *JB*-algebras

Throughout this chapter, we consider (possibly non-associative and non-unital) algebras A over a field $\mathbb{K}$ of characteristic 0, and we understand that $\mathbb{K} = \mathbb{R}$ or $\mathbb{C}$ whenever A is a normed algebra. *Jordan algebras* over $\mathbb{K}$ are defined as those commutative algebras over $\mathbb{K}$ satisfying the so-called "Jordan identity", namely $(ab)a^2 = a(ba^2)$. The basic references for Jordan algebras are as follows: (Braun and Koecher 1966; Jacobson 1968; McCrimmon 2004; Zhevlakov *et al.* 1982). *JB-algebras* are defined as those complete normed Jordan real algebras A satisfying

$$\|x\|^2 \leq \|x^2 + y^2\| \tag{4.1}$$

This chapter is dedicated to the memory of Wilhelm Kaup.

Algebra and Applications 1,
coordinated by Abdenacer MAKHLOUF.

for all x, y in A. We note that if x is any element in a *JB*-algebra, then we have $\|x^2\| = \|x\|^2$ (indeed, take $y = 0$ in (4.1)). The motivating example for *JB*-algebras is as follows.

EXAMPLE 4.1.– Let A be a C^*-algebra. Then, the self-adjoint part, $H(A, *)$, of A becomes a *JB*-algebra under the Jordan product

$$x \bullet y := \frac{1}{2}(xy + yx). \quad [4.2]$$

More examples of *JB*-algebras can be obtained by considering the so-called *JC-algebras*, namely closed subalgebras of the *JB*-algebra $H(A, *)$ for some C^*-algebra A. A fundamental example of a *JB*-algebra, which is not a *JC*-algebra, is as follows.

EXAMPLE 4.2.– Consider the algebra $M_3(\mathbb{O})$, of all 3×3-matrices with entries in the algebra $\mathbb{O}$ of Cayley numbers, and endow it with the algebra involution $*$ consisting of transposing the matrix and taking standard involution in each entry. Then, $H_3(\mathbb{O}) := H(M_3(\mathbb{O}), *)$ is a Jordan algebra under the product $\bullet$ defined as in [4.2], and, endowed with a suitable norm, it becomes, in fact, a *JB*-algebra, which is not a *JC*-algebra (Hanche-Olsen and Stormer 1984, Corollary 2.8.5, Proposition 2.9.2, and Corollary 3.1.7).

Jordan algebras are *power associative*, i.e. all subalgebras generated by a single element are associative. In the case of *JB*-algebras, this is specified by assertion (ii) as follows.

PROPOSITION 4.1.– Let A be a *JB*-algebra. We have:

i) (Hanche-Olsen and Stormer 1984, Theorem 3.2.2) If A is associative, then A identifies with the *JB*-algebra $C_0^{\mathbb{R}}(E)$ of all real-valued continuous functions vanishing at infinity on a suitable locally compact Hausdorff topological space E.

ii) (Hanche-Olsen and Stormer 1984, Theorems 3.2.4 and 3.3.9) For $a \in A$, the closed subalgebra of A generated by a identifies with the *JB*-algebra $C_0^{\mathbb{R}}(E)$ for some locally compact bounded subset of $\mathbb{R}$.

iii) (Hanche-Olsen and Stormer 1984, Theorem 3.4.2) If M is any closed ideal of A, then A/M is a *JB*-algebra in the quotient norm.

iv) (Hanche-Olsen and Stormer 1984, Proposition 3.4.3) If B is any *JB*-algebra, and if $F : A \to B$ is an algebra homomorphism, then F is contractive. Moreover, if the algebra homomorphism F is injective, then F is an isometry.

An element u of a unital *JB*-algebra is said to be a *symmetry* if $u^2 = \mathbf{1}$. We note that symmetries of unital *JB*-algebras are norm-one elements. An element of an algebra A is said to be *central* when it commutes with any element of A and it associates with any two elements of A.

PROPOSITION 4.2.– (Cabrera and Rodríguez 2014, Propositions 3.1.9 and 3.1.15) Let A be a non-zero JB-algebra. Then A is unital if and only if the closed unit ball, $\mathbb{B}_A$, of A has extreme points. Moreover, if A is unital, then the extreme points of $\mathbb{B}_A$ are the symmetries in A, and the vertices of $\mathbb{B}_A$ are the central symmetries in A.

We recall that an element u of a normed space X is said to be a *vertex* of $\mathbb{B}_X$ if $\|u\| = 1$ and the set $D(X, u) := \{f \in \mathbb{B}_{X'} : f(u) = 1\}$ separates the points of X, and remark the general fact that vertices of $\mathbb{B}_X$ are extreme points of $\mathbb{B}_X$ (Cabrera and Rodríguez 2014, Lemma 2.1.25).

According to the definition of a JBW-algebra in (Hanche-Olsen and Stormer 1984, 4.1.1), and (Hanche-Olsen and Stormer 1984, Theorem 4.4.16), *JBW-algebras* can be introduced as those JB-algebras, which are dual Banach spaces. Combining the Banach–Alaoglu and Krein–Milman theorems with proposition 4.2, we realize that non-zero JBW-algebras are unital.

PROPOSITION 4.3.– (Hanche-Olsen and Stormer 1984, Proposition 4.2.3) Let A be a JBW-algebra, a be in A and $\varepsilon > 0$. Then there exist pairwise orthogonal idempotents $e_1, \ldots, e_n \in A$ and real numbers $\lambda_1, \ldots, \lambda_n$ such that $\|a - \sum_{i=1}^{n} \lambda_i e_i\| < \varepsilon$.

PROPOSITION 4.4.– Let A be a JB-algebra. Then:

i) (Hanche-Olsen and Stormer 1984, Theorem 4.4.3) The bidual, A'', of A, endowed with the Arens product, becomes a JBW-algebra.

ii) (Cabrera and Rodríguez 2014, Proposition 3.1.18) The set

$$M(A) := \{a'' \in A'' : a''A \subseteq A\}$$

becomes a closed subalgebra of A'' containing the unit of A'', and containing A as an ideal.

The set $M(A)$ above is called the *JB-algebra of multipliers* of A. We note that, by (Cabrera and Rodríguez 2014, Corollary 2.2.13), the equality $M(A) = A$ holds if and only if A has a unit.

Propositions 4.1–4.4 become the main ingredients in the proof of the following.

THEOREM 4.1.– (Cabrera and Rodríguez 2014, Theorem 3.1.21) Let A and B be JB-algebras, and let $F : A \to B$ be a mapping. Then F is a surjective linear isometry if and only if there are a central symmetry $u \in M(B)$ and a bijective algebra homomorphism $\Phi : A \to B$ such that $F(x) = u\,\Phi(x)$ for every $x \in A$.

As a straightforward consequence of theorem 4.1, we obtain the following.

COROLLARY 4.1.– Linearly isometric JB-algebras are isomorphic (as algebras).

Let X be a normed space over $\mathbb{K}$. We denote by I_X the identity mapping on X, and by $BL(X)$ the normed algebra of all bounded linear operators on X. Given a norm-one element u of X, and an element $x \in X$, the *numerical range* of x relative to (X, u) is defined as the non-empty compact convex subset of $\mathbb{K}$ given by $V(X, u, x) := \{f(x) : f \in D(X, u)\}$.

COROLLARY 4.2.– (Cabrera and Rodríguez 2014, Corollary 3.1.32) Let A be a non-zero JB-algebra, and let $T : A \to A$ be a linear mapping. Then T is a derivation of A if and only if T is bounded with $V(BL(A), I_A, T) = 0$.

We note that the above corollary contains one of the ingredients in its proof, namely that derivations on JB-algebras are automatically continuous (Cabrera and Rodríguez 2014, Corollary 3.1.31).

DEFINITION 4.1.– *By an approximate identity in a normed algebra A, we mean a net* $\{a_\lambda\}_{\lambda \in \Lambda}$ *in A such that* $\lim_\lambda a_\lambda a = \lim_\lambda a a_\lambda = a$ *for every* $a \in A$. *Now let A be a JB-algebra. Then, as in the particular case that A is the self-adjoint part of a C*-algebra (see example 4.1), the set* $A^+ := \{a^2 : a \in A\}$ *becomes a closed convex proper cone in A, giving rise to a natural order in A converting* A^+ *into the set of all positive elements of A (Hanche-Olsen and Stormer 1984, 3.3.3).*

PROPOSITION 4.5.– (Hanche-Olsen and Stormer 1984, Proposition 3.5.4) Every JB-algebra has an increasing approximate identity bounded by 1.

DEFINITION 4.2.– *Let A be a JB-algebra. A linear functional f on A is said to be positive if* $f(a) \geq 0$ *whenever a is a positive element of A. The JB-algebra A is said to be monotone complete if each bounded increasing net* a_λ *in A has a least upper bound in A. Suppose that A is monotone complete. A linear functional f on A is called normal if it is bounded and if* $f(a_\lambda) \to f(\sup_\lambda a_\lambda)$ *for each net* a_λ *as above.*

THEOREM 4.2.– (Hanche-Olsen and Stormer 1984, Theorem 4.4.16) A JB-algebra A is a JBW-algebra if and only if it is monotone complete and the set of all positive normal linear functionals on A separates the points of A. Moreover, if A is a JBW-algebra, then the predual of A is unique and consists of the normal linear functionals on A.

A classical definition of a *von Neumann algebra* is that of a self-adjoint algebra of bounded linear operators on a complex Hilbert space, which is closed in the weak operator topology. Since a JB-algebra cannot in general be represented as a Jordan algebra of bounded linear operators on a complex Hilbert space (see example 4.2), to say that a JB-algebra is "closed in the weak operator topology" makes no sense. However, we recall that non-spatial characterizations of von Neumann algebras are known. Indeed, Kadison (1956) (see also (Kadison and Ringrose 1997, Exercise 7.6.38)) showed that an abstract C^*-algebra A admits a faithful representation as a von Neumann algebra of operators if and only if A is monotone complete and the set of all positive normal linear functionals on A separates the points of A. Another characterization is that of (Sakai 1956) (see also (Sakai 1971, Theorem 1.16.7)) who showed that a C^*-algebra admits such a representation if and only if it is a dual Banach space. In this way, both the definition itself of a JBW-algebra (as a JB-algebra which is a dual Banach space) and theorem 4.2 show that JBW-algebras become the reasonable JB-algebra substitutes of von Neumann algebras.

Roughly speaking, JB-algebras can be "described" by means of the so-called representation theory, which essentially consists of theorems 4.3 and 4.4.

An algebra A over $\mathbb{K}$ is said to be *quadratic* if it is unital and, for every $a \in A$, a^2 lies in the linear hull of $\{\mathbf{1}, a\}$. By a $*$-*involution* on a $*$-algebra A over $\mathbb{K}$, we mean a linear involution τ on A commuting with $*$ and satisfying $\tau(ab) = \tau(b)\tau(a)$ for all $a, b \in A$. By a *JBW-factor* (respectively, a *W^*-factor*), we mean a prime JBW-algebra (respectively, a prime von Neumann algebra).

The following classification of JBW-factors follows from (Hanche-Olsen and Stormer 1984, Theorems 5.3.8, 5.3.9, and 7.3.5, and Proposition 7.3.3).

THEOREM 4.3.– The JBW-factors are as follows:

i) the JB-algebra $H_3(\mathbb{O})$ (see example 4.2);

ii) the simple quadratic JB-algebras;

iii) the JB-algebras of the form $H(A, *)$, where A is a W^*-factor;

iv) the JB-algebras of the form $H(A, *) \cap H(A, \tau)$, where A is a W^*-factor and τ is a $*$-involution on A.

The simple quadratic JB-algebras A different from $\mathbb{R}$ (called *JB-spin factors*) can be described from a real Hilbert space $(H, (\cdot|\cdot))$ of dimension at least 2, by taking $A := H \oplus \mathbb{R}\mathbf{1}$ with product

$$(a + \lambda\mathbf{1})(b + \mu\mathbf{1}) := (\mu a + \lambda b) + ((a|b) + \lambda\mu)\mathbf{1},$$

and norm $\|a + \lambda\mathbf{1}\| := \|a\| + |\lambda|$. According to (Hanche-Olsen and Stormer 1984, Chapter 6), all JB-spin factors are JC-algebras.

By a *JBW-factor representation* of a JB-algebra A, we mean a w^*-dense-range algebra homomorphism from A to some JBW-factor.

THEOREM 4.4.– (Hanche-Olsen and Stormer 1984, Proposition 4.6.4) Every JB-algebra has a faithful family of JBW-factor representations.

Since the closed subalgebra generated by two elements of a JB-algebra satisfies a certain identity, which is not satisfied by $H_3(\mathbb{O})$, theorems 4.3 and 4.4 imply the following.

COROLLARY 4.3.– (Hanche-Olsen and Stormer 1984, 7.2.5) Let A be a JB-algebra. Then the closed subalgebra of A generated by two elements is a JC-algebra.

The original sources for the material developed in this section are as follows: (Albert 1934; Alfsen *et al.* 1978; Arens 1951; Behncke 1979; Cohn 1954; Edwards 1980a; Hanche-Olsen 1980; Isidro and Rodríguez 1995; Jordan 1932; Jordan *et al.* 1934; Leung *et al.* 2009; Shirshov 1956; Shultz 1979; Stormer 1965; Topping 1965; Wright and Youngson 1978). This material and more has been fully re-elaborated in the books by (Alfsen and Shultz 2003; Cabrera and Rodríguez 2014, 2018; Hanche-Olsen and Stormer 1984). The reader is referred to the historical notes in these books for additional bibliography on the topic.

4.3. The non-associative Vidav–Palmer and Gelfand–Naimark theorems

An attractive approach to the non-associative generalizations of C^*-algebras consists of removing associativity in the abstract characterizations of unital (associative) C^*-algebras given either by the Gelfand–Naimark theorem or by the Vidav–Palmer theorem, and of studying (possibly non-unital) closed $*$-subalgebras of the Gelfand–Naimark or Vidav–Palmer algebras born after removing associativity.

To be more precise, for a norm-unital complete normed (possibly non-associative) complex algebra A, we may consider the following conditions:

(*GN*) (**GELFAND–NAIMARK AXIOM**). *There is a conjugate-linear vector space involution* $*$ *on* A *satisfying* $\mathbf{1}^* = \mathbf{1}$ *and* $\|a^*a\| = \|a\|^2$ *for every* $a \in A$.

(*VP*) (**VIDAV–PALMER AXIOM**). $A = H(A, \mathbf{1}) + iH(A, \mathbf{1})$.

In both conditions, $\mathbf{1}$ denotes the unit of A, whereas, in (*VP*), $H(A, \mathbf{1})$ stands for the closed real subspace of A consisting of those elements $h \in A$ such that

$$V(A, \mathbf{1}, h) \subseteq \mathbb{R}.$$

As is well known, if the norm-unital complete normed complex algebra A above is associative, then (*GN*) and (*VP*) are equivalent conditions, both providing nice characterizations of unital C^*-algebras (see (Doran and Belfi 1986)). In the general non-associative case, we are considering things that begin to be amusing. Indeed, it is easily seen that (*GN*) implies (*VP*) (Cabrera and Rodríguez 2014, Lemma 2.2.5), but the converse implication is not true (Cabrera and Rodríguez 2014, Example 2.3.65).

The amusing aspect of the non-associative consideration of the Vidav–Palmer and the Gelfand–Naimark axioms greatly increases due to the fact that condition (*VP*) (respectively, (*GN*)) on a norm-unital complete normed complex algebra A implies that A is "nearly" (respectively, "very nearly") associative. To specify this, let us recall some elementary concepts of non-associative algebra. *Alternative algebras* are defined as those algebras A satisfying $a^2b = a(ab)$ and $ba^2 = (ba)a$ for all $a, b \in A$. By Artin's theorem (Cabrera and Rodríguez 2014, Theorem 2.3.61), an algebra A is alternative (if and) only if, for all $a, b \in A$, the subalgebra of A generated by $\{a, b\}$ is associative. An algebra A is called

1) *flexible* if the identity $(ab)a = a(ba)$ holds for all $a, b \in A$;

2) *Jordan admissible* if the algebra A^{sym} (obtained by symmetrization of the product of A) is a Jordan algebra;

3) *non-commutative Jordan* if A is both flexible and Jordan admissible.

Basic references for non-commutative Jordan algebras are as follows: (Braun and Koecher 1966; Elduque and Myung 1994; Schafer 1995). Non-commutative Jordan algebras are power associative (Cabrera and Rodríguez 2014, Proposition 2.4.19) and, as a result of Artin's theorem, alternative algebras are non-commutative Jordan. Now we can advance that condition (*VP*) (respectively, (*GN*)) on a norm-unital complete normed complex algebra A implies that A is non-commutative Jordan (respectively, alternative).

Actually, the definitive version of the so-called "unital non-associative Gelfand–Naimark theorem" can be formulated by considering *alternative C^*-algebras* (namely those complete normed alternative complex algebras A endowed with a conjugate-linear algebra involution $*$ satisfying $\|a^*a\| = \|a\|^2$ for every $a \in A$), and by establishing the following.

THEOREM 4.5.– (Cabrera and Rodríguez 2014, Lemma 2.2.5 and Theorem 3.2.5) Norm-unital complete normed complex algebras fulfilling the Gelfand–Naimark axiom are nothing other than unital alternative C^*-algebras.

As the next example shows, not all alternative C^*-algebras are associative.

EXAMPLE 4.3.– The algebra of complex octonions $C(\mathbb{C}) := \mathbb{C}\otimes\mathbb{O}$, endowed with the projective norm and the involution obtained by tensorizing the standard involutions of $\mathbb{C}$ and $\mathbb{O}$, becomes a unital alternative C^*-algebra, which is not associative (Cabrera and Rodríguez 2014, Proposition 2.6.8).

The so-called "non-associative Bohnenblust–Karlin theorem" asserts that the unit of any norm-unital normed complex algebra A is a vertex of $\mathbb{B}_A$ (a consequence of (Cabrera and Rodríguez 2014, Proposition 2.1.11)). Moreover, according to (Cabrera and Rodríguez 2014, Corollary 2.1.13), the result just formulated is equivalent to the fact that $H(A, \mathbf{1}) \cap iH(A, \mathbf{1}) = 0$. Therefore, in the case that A is a Vidav–Palmer algebra, there is a unique conjugate-linear vector space involution $*$ on A (called the *natural involution* of A) such that $H(A, *) = H(A, \mathbf{1})$ (Cabrera and Rodríguez 2014, Definition 2.3.6).

The proof of Theorem 4.5 involves the facts that (GN) implies (*VP*) and that unit-preserving surjective linear isometries between unital JB-algebras are algebra homomorphisms (a consequence of Theorem 4.1), as well as the ones that the natural involution of any Vidav–Palmer algebra is an algebra involution (Cabrera and Rodríguez 2014, Theorem 2.3.8), and that Vidav–Palmer algebras are non-commutative Jordan (Cabrera and Rodríguez 2014, Theorem 2.4.11). These last two facts become also essential ingredients in the proof of the so-called "non-associative Vidav–Palmer theorem". Actually, the definitive version of this theorem can be formulated by considering *non-commutative JB^*-algebras* (namely those complete normed non-commutative Jordan complex algebras A endowed with a conjugate-linear algebra involution $*$ satisfying $\|U_a(a^*)\| = \|a\|^3$ for every $a \in A$, where $U_a(b) := a(ab + ba) - a^2b$ for all $a, b \in A$), and by establishing the following.

THEOREM 4.6.– (Cabrera and Rodríguez 2014, Lemma 2.2.5 and Theorem 3.3.11) Norm-unital complete normed complex algebras fulfilling the Vidav–Palmer axiom are nothing other than unital non-commutative JB^*-algebras.

Since, for elements a, b in an alternative algebra, the equality $U_a(b) = aba$ holds, it is not difficult to realize that alternative C^*-algebras become particular examples of non-commutative JB^*-algebras. Actually, alternative C^*-algebras are precisely those non-commutative JB^*-algebras which are alternative (Cabrera and Rodríguez 2014, Fact 3.3.2).

Now, keeping in mind theorems 4.5 and 4.6 above, together with the obvious fact that closed $*$-subalgebras of an alternative C^*-algebra (respectively, of a non-commutative JB^*-algebra) are alternative C^*-algebras (respectively, non-commutative JB^*-algebras), there is no doubt that, even in the non-unital case, both alternative C^*-algebras and non-commutative JB^*-algebras become reasonable non-associative generalizations (the latter containing the former) of classical C^*-algebras.

Before sketching the basic theory of alternative C^*-algebras and of non-commutative JB^*-algebras, let us establish the following "dual version" of theorem 4.6.

THEOREM 4.7.– (Cabrera and Rodríguez 2014, Corollary 3.3.26) Let A be a norm-unital complete normed complex algebra. Then A is a non-commutative JB^*-algebra (for some involution) if and only if $\mathrm{lin}_{\mathbb{R}}(D(A, \mathbf{1})) \cap i\,\mathrm{lin}_{\mathbb{R}}(D(A, \mathbf{1})) = 0$, where $\mathrm{lin}_{\mathbb{R}}(\cdot)$ means real linear hull.

The theory of non-commutative *JB**-algebras begins with the following proposition, which is an easy consequence of the fact, already pointed out, that non-commutative *JB**-algebras that are associative are actually C^*-algebras.

PROPOSITION 4.6.– (Cabrera and Rodríguez 2014, Proposition 3.4.1) Let A be a non-commutative *JB**-algebra. We have

i) every closed associative $*$-subalgebra of A is a C^*-algebra;

ii) for every $h \in H(A, *)$, the closed subalgebra of A generated by h (occasionally, by h and $\mathbf{1}$ if A is unital) is $*$-invariant, and is indeed a commutative C^*-algebra.

Now, example 4.1 can be generalized as follows:

PROPOSITION 4.7.– (Cabrera and Rodríguez 2014, Corollary 3.4.3) Let A be a non-commutative *JB**-algebra. Then $H(A, *)$ becomes a *JB*-algebra under the product of A^{sym} and the norm of A.

After the above proposition, the theory of *JB*-algebras has a relevant role in the development of the theory of non-commutative *JB**-algebras. Thus, proposition 4.1(iv) is one of the ingredients in the proof of the following.

PROPOSITION 4.8.– (Cabrera and Rodríguez 2014, Proposition 3.4.4) and (Cabrera and Rodríguez 2018, Proposition 5.9.3) Let A and B be non-commutative *JB**-algebras, and let $F : A \to B$ be an algebra homomorphism. We have:

i) F is contractive if and only if it preserves involutions;

ii) if F preserves involutions and is injective, then F is an isometry.

In its turn, corollary 4.3 is one of the ingredients in the proof of proposition 4.9 that follows. Non-commutative *JB**-algebras that are commutative are simply called *JB***-algebras*. We note that if A is a non-commutative *JB**-algebra, then A^{sym} becomes naturally a *JB**-algebra (Cabrera and Rodríguez 2014, Fact 3.3.4) such that $H(A, *) = H(A^{\mathrm{sym}}, *)$ as *JB*-algebras (see proposition 4.7). As a result, closed self-adjoint subalgebras of A^{sym} for some C^*-algebra A are *JB**-algebras. These particular examples of *JB**-algebras are called *JC^*-algebras*.

PROPOSITION 4.9.– (Cabrera and Rodríguez 2014, Proposition 3.4.6) Let A be a *JB**-algebra, and let x be in A. Then the closed $*$-subalgebra of A generated by x is a JC^*-algebra.

THEOREM 4.8.– (Cabrera and Rodríguez 2014, Theorem 3.4.8) Let B be a *JB*-algebra. Then there is a unique *JB**-algebra A such that $B = H(A, *)$.

In view of proposition 4.7 and the above theorem, JB^*-algebras and JB-algebras are in a one-to-one categorical correspondence (Cabrera and Rodríguez 2014, Fact 3.4.9).

PROPOSITION 4.10.– (Cabrera and Rodríguez 2014, Proposition 3.4.13 and Corollary 3.4.14) Let A be a non-commutative JB^*-algebra (respectively, an alternative C^*-algebra) and let M be a closed ideal of A. Then M is $*$-invariant, and A/M is a non-commutative JB^*-algebra (respectively, an alternative C^*-algebra) for the quotient norm and the quotient involution.

According to (Cabrera and Rodríguez 2014, Corollary 3.4.7), for every unital non-commutative JB^*-algebra A, the equality $\mathbb{B}_A = \overline{\text{co}}(E)$ holds, where

$$E := \{\exp(ih) : h \in H(A, *)\}.$$

This result is one of the main ingredients in the proof of the following.

PROPOSITION 4.11.– (Cabrera and Rodríguez 2014, Proposition 3.4.17) Let A be a non-commutative JB^*-algebra.

Then we have $\|U_{a,c}(b)\| \leq \|a\|\|b\|\|c\|$ for all $a, b, c \in A$.

Here, given an algebra A, $(a, b) \to U_{a,b}$ stands for the unique symmetric bilinear operator-valued mapping on $A \times A$ such that $U_{a,a} = U_a$ for every $a \in A$. As a straightforward consequence of theorem 4.8 and the above proposition, we obtain the following.

COROLLARY 4.4.– Let A be a JB-algebra. Then we have

$$\|U_{a,c}(b)\| \leq \|a\|\|b\|\|c\| \text{ for all } a, b, c \in A.$$

Let A and B be algebras over $\mathbb{K}$. By a *Jordan-homomorphism* from A to B, we mean an algebra homomorphism from A^{sym} to B^{sym}.

PROPOSITION 4.12.– (Cabrera and Rodríguez 2014, Proposition 3.4.25) Unit preserving surjective linear isometries between unital non-commutative JB^*-algebras are precisely the bijective Jordan-$*$-homomorphisms.

Nevertheless, we have the following.

THEOREM 4.9.– (Cabrera and Rodríguez 2014, Antitheorem 3.4.34) There exist linearly isometric unital JC^*-algebras, which are not $*$-isomorphic.

The above theorem shows that no reasonable version for non-commutative JB^*-algebras of Kadison's celebrated Theorem A in Kadison (1951) can be expected.

THEOREM 4.10.– (Cabrera and Rodríguez 2014, Theorem 3.4.75) Let A and B be non-commutative JB^*-algebras, and let F be a bijective algebra homomorphism. Then F can be written in a unique way as $F = G\exp(iD)$, where $G : A \to B$ is a bijective algebra $*$-homomorphism and D is a $*$-derivation of A.

The following relevant corollary is known as "the essential uniqueness of the JB^*-structure in non-commutative JB^*-algebras".

COROLLARY 4.5.– (Cabrera and Rodríguez 2014, Corollary 3.4.76) If two non-commutative JB^*-algebras are isomorphic, then they are $*$-isomorphic.

Now, keeping in mind example 4.3, we can assert that the alternative complex algebra of complex octonions can be structured as an alternative C^*-algebra in an essentially unique way. Analogously, keeping in mind example 4.2 and theorem 4.8, we are provided with the following.

EXAMPLE 4.4.– The complex Jordan algebra $H_3(C(\mathbb{C})) := \mathbb{C} \otimes H_3(\mathbb{O})$ can be structured as a JB^*-algebra in an essentially unique way (Cabrera and Rodríguez 2018, §6.1.38).

One of the ingredients in the proof of theorem 4.10 is the following.

PROPOSITION 4.13.– (Cabrera and Rodríguez 2014, Theorem 3.4.49) Let A be a complete normed complex algebra, and let Φ be a continuous algebra automorphism of A whose spectrum is contained in the open angle $\{z \in \mathbb{C} \setminus \{0\} : |\arg(z)| < \frac{2\pi}{3}\}$. Then there exists a continuous derivation D of A such that $\Phi = \exp(D)$.

The above proposition is the first result of the theory of associative normed algebras, which remains true in the non-associative setting without needing any change in its proof. It appears as lemma III.9.9 of Dixmier's (1969) book and, according to Dixmier (1969, p. 315), is essentially due to J.-P. Serre.

THEOREM 4.11.– Let A be a non-zero non-commutative JB^*-algebra. Then:

i) (Cabrera and Rodríguez 2014, Theorem 3.5.34) A'', endowed with the Arens product and the bitranspose of the involution of A, becomes a unital non-commutative JB^*-algebra satisfying all multilinear identities satisfied by A.

ii) (Cabrera and Rodríguez 2018, Theorem 5.10.90) The set

$$M(A) := \{a'' \in A'' : a''A + Aa'' \subseteq A\}$$

becomes a closed $*$-subalgebra of A'' containing the unit of A'', and containing A as an ideal.

The set $M(A)$ above is called the *non-commutative JB^*-algebra of multipliers* of A. We note that the equality $M(A) = A$ holds if and only if A has a unit. We also note that theorem 4.11(i) is involved in the proof of the "only if" part of proposition 4.8(i). As a straightforward consequence of theorem 4.11(i), we obtain the following.

COROLLARY 4.6.– (Cabrera and Rodríguez 2014, Corollary 3.5.35) Let A be a non-zero alternative C^*-algebra. Then A'', endowed with the Arens product and the bitranspose of the involution of A, becomes a unital alternative C^*-algebra.

By a Jordan derivation of an algebra A, we mean a derivation of A^{sym}.

THEOREM 4.12.– (Cabrera and Rodríguez 2018, Theorem 5.10.116) Let A be a non-zero non-commutative JB^*-algebra, and let $R : A \to A$ be a mapping. Then R lies in $H(BL(A), I_A)$ if and only if there exists $x \in H(M(A), *)$, and a Jordan $*$-derivation D of A, such that $R(a) = xa + iD(a)$ for every $a \in A$.

We note that the above theorem contains one of the ingredients in its proof, namely that derivations of non-commutative JB^*-algebras are automatically continuous (Cabrera and Rodríguez 2014, Lemma 3.4.26).

An element u of a unital alternative $*$-algebra A is said to be *unitary* if $uu^* = u^*u = \mathbf{1}$. The fact that unitary elements of a unital alternative C^*-algebra A are precisely the vertices of $\mathbb{B}_A$ (Cabrera and Rodríguez 2014, Proposition 3.4.31) is the key ingredient in the following proof.

THEOREM 4.13.– (Cabrera and Rodríguez 2018, Theorem 5.10.102) Let A be a non-zero alternative C^*-algebra, B be a non-commutative JB^*-algebra and $F : B \to A$ be a mapping. Then F is a surjective linear isometry if and only if there exists a bijective Jordan-$*$-homomorphism $G : B \to A$, and a unitary element u in the alternative C^*-algebra $M(A)$, satisfying $F(b) = uG(b)$ for every $b \in B$.

When the algebras A and B in the above theorem are in fact unital (associative) C^*-algebras, we obtain Kadison (1951) Theorem A. As a straightforward consequence of theorem 4.13, we obtain the following.

COROLLARY 4.7.– Linearly isometric alternative C^*-algebras are Jordan-$*$-isomorphic.

Keeping in mind theorem 4.9 and the above corollary, we realize that, concerning surjective linear isometries, the class of alternative C^*-algebras behaves better than the larger class of non-commutative JB^*-algebras.

Somehow the next theorem is a converse to theorem 4.11(i). Among others tools, its proof involves theorem 4.7 and the "only if" part of proposition 4.8(i).

THEOREM 4.14.– (Cabrera and Rodríguez 2018, Theorem 5.9.7) Let A be a complete normed complex algebra such that A'', endowed with the Arens product and a suitable involution $*$, is a non-commutative JB^*-algebra. Then A is a $*$-subalgebra of A'', and hence is a non-commutative JB^*-algebra.

As an easy consequence of propositions 4.5 and 4.7, we obtain the following.

PROPOSITION 4.14.– (Cabrera and Rodríguez 2014, Proposition 3.5.23) Every non-commutative JB^*-algebra has an approximate identity bounded by 1 and consisting of self-adjoint elements.

Now we can formulate a theorem containing the various sides of the so-called "unit-free version of the non-associative Gelfand–Naimark theorem".

THEOREM 4.15.– (Cabrera and Rodríguez 2014, Theorem 3.5.68) Let A be a complete normed complex algebra with a conjugate-linear algebra involution $*$. Then the following conditions are equivalent:

i) A has an approximate identity bounded by 1 and consisting of self-adjoint elements, and the equality $\|a^*a\| = \|a^*\|\|a\|$ holds for every $a \in A$;

ii) A has an approximate identity bounded by 1, and the equality $\|a^*a\| = \|a\|^2$ holds for every $a \in A$;

iii) A is alternative, and the equality $\|a^*a\| = \|a^*\|\|a\|$ holds for every $a \in A$;

iv) A is a non-commutative Jordan algebra, and the equality $\|a^*a\| = \|a^*\|\|a\|$ holds for every $a \in A$;

v) A is an alternative C^*-algebra.

Some relevant properties of a purely algebraic-topological nature are summarized in the following.

THEOREM 4.16.– Let A be a non-commutative JB^*-algebra. We have:

i) the topology of any algebra norm on A is stronger than that of the natural norm (Cabrera and Rodríguez 2014, Theorem 4.4.29);

ii) surjective algebra homomorphisms from complete normed complex algebras to A are continuous (Cabrera and Rodríguez 2014, Corollary 4.4.63).

Now the reader in encouraged to combine theorems 4.16(i), 4.11(i) and 4.6 with proposition 4.8(ii) in order to prove the following.

COROLLARY 4.8.– (Cabrera and Rodríguez 2014, Proposition 4.4.34) Let A be a non-commutative JB^*-algebra, and let $|||\cdot|||$ be an algebra norm on A such that $|||\cdot||| \leq \|\cdot\|$. Then $|||\cdot||| = \|\cdot\|$.

By a *non-commutative* JBW^**-algebra* (respectively, a JBW^**-algebra*, an *alternative* W^**-algebra*), we mean a non-commutative JB^*-algebra (respectively, a JB^*-algebra, an alternative C^*-algebra), which is a dual Banach space. Thus, JBW^*-algebras are precisely those non-commutative JBW^*-algebras which are commutative, and alternative W^*-algebras are precisely those non-commutative JBW^*-algebras which are alternative. Since a non-commutative JB^*-algebra is unital if and only if $\mathbb{B}_A$ has extreme points (Cabrera and Rodríguez 2014, Theorem 4.2.36), it follows that non-zero non-commutative JBW^*-algebras are unital (Cabrera and Rodríguez 2018, Fact 5.1.7).

The fundamental results on non-commutative JBW^*-algebras are summarized in theorem 4.17 that follows. Let A be a non-commutative JB^*-algebra. We recall that the JB-algebra $H(A, *)$ (see proposition 4.7) has a natural order (see definition 4.1), that A is said to be *monotone complete* whenever $H(A, *)$ is monotone complete (see definition 4.2), and that, when A is monotone complete, a linear functional f on A is called *normal* if it is bounded and if $f(a_\lambda) \to f(\sup_\lambda a_\lambda)$ whenever a_λ is any bounded increasing net in $H(A, *)$.

THEOREM 4.17.– (Cabrera and Rodríguez 2018, Theorem 5.1.29(ii), and Corollaries 5.1.30(iii) and 5.1.40) Let A be a non-commutative JB^*-algebra. Then the following conditions are equivalent:

i) A is a non-commutative JBW^*-algebra;

ii) the JB-algebra $H(A, *)$ is a JBW-algebra;

iii) A is monotone complete and the set of all normal linear functionals on A separates the points of A.

Moreover, if the above conditions are fulfilled, then the product of A is separately w^*-continuous, the involution of A is w^*-continuous and the predual of A is unique and consists of the normal linear functionals on A.

Combining theorem 4.8 with the implication (ii)⇒(i) in the above theorem, we obtain the following.

COROLLARY 4.9.– (Cabrera and Rodríguez 2018, Corollary 5.1.41) Let B be a JBW-algebra. Then there is a unique JBW^*-algebra A such that $B = H(A, *)$.

According to the implication (i)⇒(ii) in theorem 4.17 and the above corollary, JBW^*-algebras and JBW-algebras are in a one-to-one categorical correspondence (Cabrera and Rodríguez 2018, Fact 5.1.42).

Non-commutative JB^*-algebras enjoy a continuous functional calculus at each of their normal elements (Cabrera and Rodríguez 2014, Definition 3.4.20 and Corollary 4.1.72). In the particular case of non-commutative JBW^*-algebras, this functional calculus can be suitably enlarged (Cabrera and Rodríguez 2018, Proposition 5.10.7), allowing us to prove, in this case, a reasonable variant of Kadison (1951) Theorem A (see (Cabrera and Rodríguez 2018, Theorem 5.10.9)). As a result, we are provided with the following.

COROLLARY 4.10.– (Cabrera and Rodríguez 2018, Corollary 5.10.10) Linearly isometric non-commutative JBW^*-algebras are Jordan-$*$-isomorphic.

Before to deal with the representation theory of non-commutative JB^*-algebras, let us formulate the following.

THEOREM 4.18.– (Cabrera and Rodríguez 2014, Theorem 4.6.63) Let A be a unital non-commutative JB^*-algebra, D be a closed densely defined derivation of A, a be a self-adjoint element in $\mathrm{dom}(D) \setminus \mathbb{R}\mathbf{1}$, K denote the convex hull of the J-spectrum of a relative to A (Cabrera and Rodríguez 2014, p. 456) and f be of class C^2 on K. Then $f(a)$ (in the sense of the continuous functional calculus) lies in $\mathrm{dom}(D)$.

If the non-commutative JB^*-algebra A in the above theorem is in fact a commutative C^*-algebra, then the requirement that f is of class C^2 can be relaxed to the one that f is of class C^1 (Cabrera and Rodríguez 2014, Proposition 4.6.58). Nevertheless, this relaxation is not allowed in the general case, even if A is a C^*-algebra (see McIntosh (1978)).

As in the case of JB-algebras, alternative C^*-algebras and non-commutative JB^*-algebras enjoy a precise representation theory, which is summarized in theorems 4.19 and 4.21, proposition 4.16 and corollary 4.11.

By a *non-commutative JBW^*-factor* (respectively, a *JBW^*-factor*, or an *alternative W^*-factor*), we mean a prime non-commutative JBW^*-algebra (respectively, a prime JBW^*-algebra, or a prime alternative W^*-algebra). By a *non-commutative JBW^*-factor* (respectively, a *JBW^*-factor*, or an *alternative W^*-factor*) *representation* of a non-commutative JB^*-algebra (respectively, a JB^*-algebra, or an alternative C^*-algebra) A, we mean a w^*-dense-range algebra $*$-homomorphism from A to some non-commutative JBW^*-factor (respectively, JBW^*-factor, or alternative W^*-factor).

THEOREM 4.19.– (Cabrera and Rodríguez 2018, Corollaries 6.1.11 and 6.1.12) Every non-commutative JB^*-algebra (respectively, JB^*-algebra, or alternative C^*-algebra) has a faithful family of non-commutative JBW^*-factor (respectively, JBW^*-factor, or alternative W^*-factor) representations.

It is worthy to mention that, because of the separate w^*-continuity of the product of any non-commutative JBW^*-algebra, the bracketed version of the above theorem follows from the bracket-free version.

As a first application of theorem 4.19, the non-associative generalization of Kaplansky (1968) Theorem III.B can be proved. Indeed, we have the following.

PROPOSITION 4.15.– (Cabrera and Rodríguez 2018, Theorem 6.1.17 and Corollary 6.1.19) An alternative C^*-algebra (respectively, a non-commutative JB^*-algebra) is commutative (respectively, associative and commutative) if and only if it has no non-zero nilpotent element.

We note that, since alternative commutative algebras are associative (Cabrera and Rodríguez 2018, Fact 6.1.18), the bracket-free version of the result just formulated follows from the bracketed version.

Keeping in mind the categorical correspondence between JBW^*-algebras and JBW-algebras, theorem 4.3 implies the following.

PROPOSITION 4.16.– (Cabrera and Rodríguez 2018, Proposition 6.1.41) The JBW^*-factors are the following:

i) the JB^*-algebra $H_3(C(\mathbb{C}))$ (see example 4.4);

ii) the simple quadratic JB^*-algebras (see (Cabrera and Rodríguez 2014, Corollary 3.5.7));

iii) the JB^*-algebras of the form A^{sym}, where A is a W^*-factor;

iv) the JB^*-algebras of the form $H(A, \tau)$, where A is a W^*-factor with a $*$-involution τ.

Actually, combining theorem 4.19 and proposition 4.16 with Zelmanov's techniques (Zelmanov 1983; McCrimmon and Zel'manov 1988), all prime JB^*-algebras can be described. Indeed, we have the following.

THEOREM 4.20.– (Cabrera and Rodríguez 2018, Theorem 6.1.57) The prime JB^*-algebras are the following:

i) the JB^*-algebra $H_3(C(\mathbb{C}))$;

ii) the simple quadratic JB^*-algebras;

iii) the closed $*$-subalgebras of $M(A)^{\mathrm{sym}}$ containing A, where A is any prime C^*-algebra;

iv) the closed $*$-subalgebras of $M(A)^{\mathrm{sym}}$ contained in $H(M(A), \tau)$ and containing $H(A, \tau)$, where A is a prime C^*-algebra with a $*$-involution τ.

The above theorem becomes one of the main tools in the proof of the fact that, for every non-zero ideal I of a prime non-commutative JB^*-algebra A, and every linear mapping $f : I \to A$ such that $f(ax) = af(x)$ and $f(xa) = f(x)a$ for all $x \in I$

and $a \in A$, there exists $\lambda \in \mathbb{C}$ such that $f(x) = \lambda x$ for every $x \in I$ (Cabrera and Rodríguez 2018, Corollary 6.1.79).

Let A be an algebra over $\mathbb{K}$ and λ be in $\mathbb{K}$. The λ-*mutation* of A, denoted by $A^{(\lambda)}$, is defined as the algebra whose vector space is that of A, and whose product (say $\odot$) is defined by $a \odot b := \lambda ab + (1-\lambda)ba$. We note that the class of non-commutative Jordan algebras is closed under λ-mutations for arbitrary $\lambda \in \mathbb{K}$, and that, as a consequence, the class of non-commutative JB^*-algebras is closed under λ-mutations whenever $0 \leq \lambda \leq 1$ (see (Cabrera and Rodríguez 2018, §5.10.109) for details).

THEOREM 4.21.– (Cabrera and Rodríguez 2018, Theorem 6.1.112) The non-commutative JBW^*-factors are the JBW^*-factors (see proposition 4.16), the simple quadratic non-commutative JB^*-algebras (see Cabrera and Rodríguez 2014, Theorem 3.5.5; 2018, Corollary 6.1.3) and the non-commutative JB^*-algebras of the form $B^{(\lambda)}$ for some (associative) W^*-factor B and some $0 \leq \lambda \leq 1$.

Actually, combining theorems 4.19 and 4.21 with a normed refinement of an argument by Zelmanov (1985), all prime non-commutative JB^*-algebras can be described. Indeed, we have the following.

THEOREM 4.22.– (Cabrera and Rodríguez 2018, Theorem 6.2.27) The prime non-commutative JB^*-algebras are the prime JB^*-algebras (see Theorem 4.20), the simple quadratic non-commutative JB^*-algebras, and the non-commutative JB^*-algebras of the form $B^{(\lambda)}$ for some prime C^*-algebra B and some $0 \leq \lambda \leq 1$.

As another relevant application of theorems 4.19 and 4.21, let us cite that, for every element a in an arbitrary non-commutative JB^*-algebra, the inequality $0 \leq a^*a$ holds (Cabrera and Rodríguez 2018, Theorem 6.2.18), a fact that allows us to introduce the strong topology of a non-commutative JBW^*-algebra, and to establish its basic properties (see (Cabrera and Rodríguez 2018, Proposition 6.2.22 and Corollary 6.2.23)).

Since alternative commutative algebras are associative, and quadratic alternative C^*-algebras are either associative or equal to the alternative C^*-algebra $C(\mathbb{C})$ of complex octonions (Cabrera and Rodríguez 2014, Corollary 3.5.6), and alternative algebras that are mutations of some associative algebra are associative (Cabrera and Rodríguez 2018, Fact 6.1.114), it follows from Theorem 4.22 that every prime alternative C^*-algebra is either associative or equal to $C(\mathbb{C})$ (Cabrera and Rodríguez 2018, Corollary 6.2.28). In particular, we have the following.

COROLLARY 4.11.– (Cabrera and Rodríguez 2018, Corollary 6.1.115) Every alternative W^*-factor is either associative or the alternative C^*-algebra $C(\mathbb{C})$ of complex octonions.

The most relevant application of theorem 4.19 and corollary 4.11 is the following.

THEOREM 4.23.– (Cabrera and Rodríguez 2018, Corollary 6.2.9 and Theorem 6.2.16) Every alternative W^*-algebra A admits a unique decomposition $A = B \oplus_\infty C(E, C(\mathbb{C}))$ (direct sum of w^*-closed ideals), where B is a von Neumann algebra, E is a hyper-Stonean compact Hausdorff topological space, and $C(E, C(\mathbb{C}))$ stands for the alternative W^*-algebra of all $C(\mathbb{C})$-valued continuous functions on E.

The above theorem contains one of the ingredients in its proof, namely that, if the alternative W^*-algebra A is purely non-associative (Cabrera and Rodríguez 2018, Definition 6.2.7), then $A = C(E, C(\mathbb{C}))$ for E as above. The proof of this essential part of theorem 4.23 is not discussed in Cabrera and Rodríguez (2014, 2018). The reader is referred to Horn (1987c) for such a proof, which involves the whole paper Stacey (1982), as well as the version of Stacey (1982) as given in (Hanche-Olsen and Stormer 1984, Section 6.3).

The original sources for the material developed in this section, not previously quoted in it, are as follows: (Albert 1948; Alvermann and Janssen 1984; Arens 1951; Aupetit 1982; Bensebah 1992; Bohnenblust and Karlin 1955; Braun 1983, 1984; Braun *et al.* 1978; Cabrera and Rodríguez 1992, 1993, 2012; Cleveland 1963; Edwards 1980b; Essaleh *et al.* 2018; Fernández *et al.* 1992; Gelfand and Naimark 1943; Iochum *et al.* 1989; Jordan 1932; Kaidi *et al.* 1981, 2000, 2001b,c; Martínez *et al.* 1985; Mathieu 1989; McCrimmon 1969, 1971; Moore 1971; Okayasu 1968; Palmer 1968a,b; Paterson and Sinclair 1972; Payá *et al.* 1982, 1984; Perez *et al.* 1994; Rodríguez 1980, 1983, 1985, 2011; Russo and Dye 1966; Sinclair 1971; Upmeier 1976; Vidav 1956; Viola Devapakkiam 1971; Vowden 1967; Wright 1977; Wright and Youngson 1977, 1978; Youngson 1978, 1979b, 1981). This material and more has been fully re-elaborated in Cabrera and Rodríguez (2014, 2018), where the reader can find additional bibliography on the topic. Concerning associative forerunners, the reader is referred also to the books (Allan 2011; Ara and Mathieu 2003; Bonsall and Duncan 1971, 1973; Bratteli 1986; Doran and Belfi 1986; Palmer 1994, 2001; Sakai 1971, 1991).

4.4. *JB**-triples

By a *Jordan ∗-triple* over $\mathbb{K}$, we mean a vector space X over $\mathbb{K}$ endowed with a mapping $\{\cdots\} : X \times X \times X \to X$ (called the *triple product* of X), which is linear in the outer variables and conjugate-linear in the middle variable, and satisfies the commutative condition $\{xyz\} = \{zyx\}$ and the *Jordan triple identity* $\{uv\{xyz\}\} = \{\{uvx\}yz\} - \{x\{vuy\}z\} + \{xy\{uvz\}\}$. Given elements x, y in a Jordan ∗-triple X, we denote by $L(x, y)$ the linear operator on X defined by $L(x, y)(z) := \{xyz\}$.

By a *Banach Jordan ∗-triple* over $\mathbb{K}$, we mean a Jordan ∗-triple over $\mathbb{K}$ endowed with a complete norm making the triple product continuous. Let X be a Banach

Jordan $*$-triple over $\mathbb{K}$. We say that X is *hermitian* if $\mathbb{K} = \mathbb{C}$ and if $L(x,x)$ lies in $H(BL(X), I_X)$ for every $x \in X$. If, in addition, for every $x \in X$, the spectrum of $L(x,x)$ consists only of non-negative real numbers, then we say that X is *positive hermitian*.

By a *JB*-triple,* we mean a positive hermitian Banach Jordan $*$-triple X such that the equality $\|\{xxx\}\| = \|x\|^3$ holds for every $x \in X$.

As the next example shows, C^*-algebras are closely related to *JB**-triples.

EXAMPLE 4.5.– Let A be a C^*-algebra. Then A becomes a *JB**-triple under its own norm and the triple product $\{xyz\} := \frac{1}{2}(xy^*z + zy^*x)$ (Cabrera and Rodríguez 2014, Fact 4.1.41).

More generally, we have the following.

THEOREM 4.24.– (Cabrera and Rodríguez 2014, Theorem 4.1.45) Let A be a non-commutative *JB**-algebra. Then A becomes a *JB**-triple (called the *JB**-triple underlying A) under its own norm and the triple product $\{xyz\} := U_{x,z}(y^*)$.

The following characterization of *JB**-triples is very useful.

THEOREM 4.25.– (Cabrera and Rodríguez 2014, Corollary 4.1.51) *JB**-triples are precisely those positive hermitian Banach Jordan $*$-triples X satisfying $\|L(x,x)\| = \|x\|^2$ for every $x \in X$.

Jordan $*$-triples satisfying the identity $\{uv\{xyz\}\} = \{\{uvx\}yz\}$ are called *abelian*.

Now let us set $\mathbb{T} := \{z \in \mathbb{C} : |z| = 1\}$, and let E be a subset of a Hausdorff locally convex complex topological vector space such that $0 \notin E$, $E \cup \{0\}$ is compact, and $\mathbb{T}E \subseteq E$. Then E is a locally compact Hausdorff topological space. Let us write

$$C_0^{\mathbb{T}}(E) := \{x \in C_0^{\mathbb{C}}(E) : x(zt) = zx(t) \text{ for every } (z,t) \in \mathbb{T} \times E\},$$

and note that $C_0^{\mathbb{T}}(E)$ is a closed subtriple of the *JB**-triple underlying the C^*-algebra $C_0^{\mathbb{C}}(E)$ (see example 4.5), and hence $C_0^{\mathbb{T}}(E)$ becomes a *JB**-triple (Cabrera and Rodríguez 2014, Fact 4.1.40), which is clearly abelian. The main result on abelian *JB**-triples establishes that there are no more abelian *JB**-triples than those given by the construction just done (Cabrera and Rodríguez 2014, Theorem 4.2.7). Actually, as a result, *JB**-triples generated by a single element are in fact commutative C^*-algebras (regarded as *JB**-triples). Indeed, we have the following "continuous triple functional calculus" at each non-zero element of a *JB**-triple.

THEOREM 4.26.– (Cabrera and Rodríguez 2014, Theorem 4.2.9) Let X be a JB^*-triple, let x be a non-zero element of X, and let M stand for the closed subtriple of X generated by x. Then there exists a unique subset E of $\mathbb{R}^+$, such that $E \cup \{0\}$ is compact, in such a way that M identifies with the JB^*-triple underlying the C^*-algebra $C_0^{\mathbb{C}}(E)$, and x converts into the inclusion mapping $E \hookrightarrow \mathbb{C}$.

Let X and x be as in theorem 4.26. Then the locally compact subset E of $\mathbb{R}^+$, given by that theorem, is called the *triple spectrum* of x (relative to X).

Let A be a JB^*-algebra generated (as a normed $*$-algebra) by a non-self-adjoint idempotent e. Then, according to (Cabrera and Rodríguez 2014, Theorems 4.3.29 and 4.3.32), A can be described in terms of the triple spectrum of e relative to the JB^*-triple underlying A (see theorem 4.24). As a result, A is generated (as a normed algebra) by two self-adjoint idempotents (Cabrera and Rodríguez 2014, Corollary 4.3.34).

Let X be a Jordan $*$-triple, and let u be in X. We say that u is a *tripotent* if $\{uuu\} = u$. This is the case when u is a *unitary tripotent*, which means that the equality $\{uux\} = x$ holds for every $x \in X$.

THEOREM 4.27.– (Cabrera and Rodríguez 2014, Theorem 4.2.24) Let X be a JB^*-triple, and let u be a norm-one element of X. Then the following conditions are equivalent:

i) u is a unitary tripotent of X;

ii) X is the JB^*-triple underlying a JB^*-algebra with unit u;

iii) u is a vertex of the closed unit ball of X.

A tripotent e in a Jordan $*$-triple X is said to be *complete* if the conditions $x \in X$ and $\{eex\} = 0$ imply $x = 0$.

THEOREM 4.28.– (Cabrera and Rodríguez 2014, Theorem 4.2.34) An element u of a non-zero JB^*-triple X is a complete tripotent if and only if u is an extreme point of $\mathbb{B}_X$.

So far, we have summarized the part of the basic theory of JB^*-triples, which can be established without involving infinite-dimensional holomorphy. Actually, JB^*-triples were introduced by Kaup (1977, 1983) in the search of a functional-analytic solution to the problem of the classification (up to biholomorphic bijections) of all bounded symmetric domains in complex Banach spaces (see theorem 4.32).

Let X be a Banach space over $\mathbb{K}$, and let Ω be an open subset of X. By a *vector field* on Ω, we simply mean a mapping $\Lambda : \Omega \to X$. Given a point $x_0 \in \Omega$, a *local flow* of Λ at x_0 is a differentiable function $\varphi : I \to \Omega$, where I is an open interval of $\mathbb{R}$ containing 0, which is a solution of the initial value problem

$$\begin{cases} x'(t) = \Lambda(x(t)) & (t \in I) \\ x(0) = x_0. \end{cases}$$

If $\Lambda : \Omega \to X$ is in fact a locally Lipschitz vector field, then there exists a unique maximal local flow of Λ at x_0 (Cabrera and Rodríguez 2018, Lemma 5.4.6). Actually we have the following.

PROPOSITION 4.17.– (Cabrera and Rodríguez 2018, Theorem 5.4.9) Let X be a Banach space over $\mathbb{K}$, Ω be an open subset of X and $\Lambda : \Omega \to X$ be a locally Lipschitz vector field. Then there exist an open subset $\mathfrak{D}$ of $\mathbb{R} \times \Omega$ containing $\{0\} \times \Omega$, and a continuous mapping φ from $\mathfrak{D}$ to Ω which is a solution of the Cauchy problem

$$\begin{cases} \dfrac{\partial}{\partial t} f(t,x) = \Lambda(f(t,x)) & ((t,x) \in \mathfrak{D}) \\ f(0,x) = x & (x \in \Omega), \end{cases}$$

and satisfies the following properties:

i) if for each $x \in \Omega$ we consider the set $\mathfrak{D}_x := \{t \in \mathbb{R} : (t,x) \in \mathfrak{D}\}$ and the function $\varphi_x : \mathfrak{D}_x \to \Omega$ defined by $\varphi_x(t) := \varphi(t,x)$, then φ_x is the maximal local flow of Λ at x;

ii) if for each $t \in \mathbb{R}$, we consider the set $\mathfrak{D}_t := \{x \in \Omega : (t,x) \in \mathfrak{D}\}$ and the mapping $\varphi_t : \mathfrak{D}_t \to \Omega$ defined by $\varphi_t(x) := \varphi(t,x)$, then $\mathfrak{D}_t$ is open in X and, in the case that $\mathfrak{D}_t \neq \emptyset$, the mapping φ_t is a homeomorphism from $\mathfrak{D}_t$ onto $\mathfrak{D}_{-t}$ with $\varphi_t^{-1} = \varphi_{-t}$.

Given a locally Lipschitz vector field Λ on an open subset Ω of a Banach space X, the function $\varphi : \mathfrak{D} \to \Omega$ in the statement of proposition 4.17 is called the *flow* of Λ, and, when $\mathfrak{D}_1 \neq \emptyset$, the mapping $\varphi_1 : \mathfrak{D}_1 \to \Omega$ is called the *exponential* of Λ, and is denoted by $\exp(\Lambda)$. The vector field Λ is said to be *complete* whenever $\mathfrak{D} = \mathbb{R} \times \Omega$.

From now on until the discussion after proposition 4.22, X shall stand for a non-zero complex Banach space. Let Ω be a bounded domain in X. If Λ is a complete holomorphic vector field on Ω, then, denoting by $\varphi : \mathbb{R} \times \Omega \to \Omega$ the flow of Λ, it turns out that, for each $t \in \mathbb{R}$, the mapping $\varphi_t : \Omega \to \Omega$ is biholomorphic (Cabrera and Rodríguez 2018, Theorem 5.4.22). Therefore, denoting by $\mathrm{aut}(\Omega)$ the set of all complete holomorphic vector fields on Ω, and by $\mathrm{Aut}(\Omega)$ the set of all biholomorphic mappings from Ω onto Ω, we can consider the *exponential mapping*

$\exp : \text{aut}(\Omega) \to \text{Aut}(\Omega)$. Moreover, $\text{aut}(\Omega)$ is a real Banach Lie algebra in a natural way (Cabrera and Rodríguez 2018, Theorem 5.5.11), and the exponential mapping induces on $\text{Aut}(\Omega)$ a topology (called the *analytic topology*) (Cabrera and Rodríguez 2018, Theorem 5.5.21), as well as an analytic structure, converting $\text{Aut}(\Omega)$ into a real Banach Lie group, which acts analytically on Ω, and whose Banach Lie algebra is $\text{aut}(\Omega)$ (Cabrera and Rodríguez 2018, Theorem 5.5.45). One of the main ingredients in the proof of the results just reviewed is as follows.

THEOREM 4.29.– (Cabrera and Rodríguez 2018, Theorem 5.5.20) Let Ω be a bounded domain in X. Then there are an open neighborhood $\mathfrak{M}_0$ of the origin in $\text{aut}(\Omega)$ and a real analytic mapping $C : \mathfrak{M}_0 \times \mathfrak{M}_0 \to \text{aut}(\Omega)$ such that

$$\exp(C(\Lambda_1, \Lambda_2)) = \exp(\Lambda_1) \circ \exp(\Lambda_2) \text{ for all } \Lambda_1, \Lambda_2 \in \mathfrak{M}_0.$$

The proof of theorem 4.29 relies on the convergence of the Baker–Campbell–Hausdorff series in Banach Lie algebras (Beltita 2006, Proposition 1.33).

PROPOSITION 4.18.– (Cabrera and Rodríguez 2018, Fact 5.6.14) Let Ω be a bounded domain in X, endow $\text{Aut}(\Omega)$ with the analytic topology and denote by $\text{Aut}_0(\Omega)$ the connected component of the identity in $\text{Aut}(\Omega)$. Then $\text{Aut}_0(\Omega)$ is a closed normal subgroup of $\text{Aut}(\Omega)$ and

$$\text{Aut}_0(\Omega) = \{\exp(\Lambda_1) \circ \cdots \circ \exp(\Lambda_n) : n \in \mathbb{N},\ \Lambda_k \in \text{aut}(\Omega)\ (1 \le k \le n)\}.$$

We denote by Δ_X the open unit ball of X. The *symmetric part*, X_s, of X is defined by the equality

$$X_s := \{\Lambda(0) : \Lambda \in \text{aut}(\Delta_X)\}.$$

PROPOSITION 4.19.– (Cabrera and Rodríguez 2018, Fact 5.6.28) X_s is a closed complex subspace of X with open unit ball $\Delta_{X_s} = \text{Aut}(\Delta_X)(0) = \text{Aut}_0(\Delta_X)(0)$. As a result, both $\text{Aut}(\Delta_X)$ and $\text{Aut}_0(\Delta_X)$ act transitively on Δ_{X_s}.

According to (Cabrera and Rodríguez 2018, Fact 5.6.29(i)), for each $y \in X_s$ there exists a unique quadratic mapping q_y from X to X such that the restriction to Δ_X of the vector field $x \to y - q_y(x)$ on X belongs to $\text{aut}(\Delta_X)$. Consequently, we can define the so-called *partial triple product* of X as the mapping $\langle \cdots \rangle$ from $X \times X_s \times X$ to X given by

$$\langle xyz \rangle := q_y(x, z) \text{ for all } x, z \in X \text{ and } y \in X_s,$$

where, for $y \in X_s$, $q_y(\cdot, \cdot) : X \times X \to X$ stands for the unique continuous symmetric bilinear mapping such that $q_y(x, x) = q_y(x)$ for every $x \in X$.

PROPOSITION 4.20.– (Cabrera and Rodríguez 2018, Facts 5.6.29 and 5.6.30) The partial triple product of X is continuous, symmetric bilinear in the outer variables and conjugate-linear in the middle variable. Moreover, it satisfies the so-called "partial Jordan triple identity"

$$\langle uv\langle xyz\rangle\rangle = \langle\langle uvx\rangle yz\rangle - \langle x\langle vuy\rangle z\rangle + \langle xy\langle uvz\rangle\rangle$$

for all $u, v, y \in X_s$ and $x, z \in X$. Furthermore, for every $y \in X_s$, the holomorphic vector field $x \to y - \langle xyx\rangle$ is complete on Δ_X, and the operator $x \to \langle yyx\rangle$ on X lies in $H(BL(X), I_X)$.

The following statement expresses the fact that complete holomorphic vector fields on Δ_X are characterized by tangency to the unit sphere, $\mathbb{S}_X$, of X.

PROPOSITION 4.21.– (Cabrera and Rodríguez 2018, Fact 5.6.38) Let Λ be a holomorphic vector field on Δ_X. Then Λ is complete if and only if it has a bounded holomorphic extension (say $\widehat{\Lambda}$) to an open neighborhood of $\mathbb{B}_X$ satisfying

$$V(X, x, \widehat{\Lambda}(x)) \subseteq i\mathbb{R} \text{ for every } x \in \mathbb{S}_X.$$

A domain Ω in X is said to be *homogeneous* if the group $\text{Aut}(\Omega)$ acts transitively on Ω. On the other hand, if $X = X_s$, then the partial triple product of X works on the whole $X \times X \times X$, and consequently the term "partial" seems to us inadequate: it seems better to call it the *intrinsic triple product* of X.

THEOREM 4.30.– (Cabrera and Rodríguez 2018, Theorem 5.6.68) The following conditions are equivalent:

i) Δ_X is a homogeneous domain;

ii) $X_s = X$;

iii) X is a JB^*-triple for some triple product.

Moreover, if the above conditions are fulfilled, then the triple product of X as a JB^*-triple coincides with the intrinsic triple product of X as a complex Banach space.

The above theorem is known as "Kaup's holomorphic characterization of JB^*-triples". The equivalence (i)$\Leftrightarrow$(ii) follows straightforwardly from the equality $\Delta_{X_s} = \text{Aut}(\Delta_X)(0)$ in proposition 4.19. The proof of the implication (iii)$\Rightarrow$(ii) given in Cabrera and Rodríguez (2018) (which is actually done in (Cabrera and Rodríguez 2018, Proposition 5.6.56)) involves proposition 4.21 and the following proposition as main tools.

PROPOSITION 4.22.– (Cabrera and Rodríguez 2018, Corollary 5.6.47) Let $B(\mathbb{S}_X, X)$ denote the Banach space of all bounded functions from $\mathbb{S}_X$ to X, and let i_X denote the inclusion $\mathbb{S}_X \hookrightarrow X$. Suppose that X is separable. Then, for every bounded and uniformly continuous function $f : \mathbb{S}_X \to X$, the numerical range $V(B(\mathbb{S}_X, X), i_X, f)$ is equal to the closed convex hull of the set

$$\{\phi(f(x)) : x \in \mathbb{S}_X,\ X \text{ is smooth at } x, \text{ and } \phi \in D(X, x)\}.$$

We note that, by proposition 4.20, condition (ii) in theorem 4.30 implies that X is a hermitian Banach Jordan $*$-triple. Starting from this fact, the proof of the implication (ii)⇒(iii) given in Cabrera and Rodríguez (2018) follows Kaup's original arguments (Kaup 1983) to show that X can be equivalently renormed as a JB^*-triple in such a way that surjective linear isometries on X remain isometries on the renormed space (Cabrera and Rodríguez 2018, Fact 5.6.67) and concludes by showing that the new norm coincides with the initial one. In this concluding step, the already discussed implication (iii)⇒(ii) in the theorem, as well as propositions 4.18 and 4.19, are applied. Anyway, implicitly or explicitly, most arguments in the above proof of theorem 4.30 are given by (Kaup 1977, 1983). In particular, Kaup's theorems 4.25 and 4.26 are involved.

The first proof that the open unit ball of a JB^*-triple is a homogeneous domain appears in Kaup (1977). Later, Kaup (1983) gives a more intrinsic proof of this result. Indeed, he proves the following.

THEOREM 4.31.– Let X be a JB^*-triple, and let x be in Δ_X. Then the spectrum of the Bergmann operator on X

$$B(x, x) : y \to x - 2\{xxy\} + \{x\{xyx\}x\}$$

consists only of positive numbers, $I_X + L(z, x)$ is a bijective operator on X for each $z \in \Delta_X$, and the "Möbius transformation" g_x, defined by

$$g_x(z) := x + B(x, x)^{\frac{1}{2}}(I_X + L(z, x))^{-1}(z) \text{ for every } z \in \Delta_X,$$

is a biholomorphic automorphism of Δ_X satisfying $g_x(0) = x$, $g_x^{-1} = g_{-x}$, and $Dg_x(0) = B(x, x)^{\frac{1}{2}}$.

By a *symmetric domain* in a complex Banach space X, we mean a domain $\Omega \subseteq X$ such that, for each $x_0 \in \Omega$, there exists $S \in \text{Aut}(\Omega)$ such that $S^2 = I_\Omega$ and x_0 is an isolated fixed point for S. Keeping in mind the implication (iii)⇒(i) in theorem 4.30, we realize that the open unit ball of any JB^*-triple is a bounded symmetric domain. Actually, Kaup (1983) concluding theorem (called by Kaup "a Riemann mapping theorem for bounded symmetric domains in complex Banach spaces") reads as follows.

THEOREM 4.32.– A bounded domain in a complex Banach space is symmetric (if and) only if it is biholomorphically equivalent to the open unit ball of a JB^*-triple.

A holomorphy-free argument involving theorem 4.25 shows that bijective triple homomorphisms between JB^*-triples are isometries (Cabrera and Rodríguez 2014, Proposition 4.1.52). Therefore, invoking theorem 4.30 we obtain the following.

COROLLARY 4.12.– Surjective linear isometries between JB^*-triples are precisely the bijective triple homomorphisms.

As pointed out in Kaup (1983), the above corollary yields the following.

COROLLARY 4.13.– A bounded linear operator T on a JB^*-triple X lies in $H(BL(X), I_X)$ if and only if the equality

$$T(\{xyz\}) = \{T(x)yz\} - \{xT(y)z\} + \{xyT(z)\}$$

holds for all $x, y, z \in X$.

Keeping in mind an elementary fact on numerical ranges (Cabrera and Rodríguez 2014, Corollary 2.1.2(i)), proposition 4.21 and theorem 4.30 lead to the following "contractive projection theorem".

THEOREM 4.33.– (Cabrera and Rodríguez 2018, Theorem 5.6.59) Let X be a JB^*-triple, let $\pi : X \to X$ be a contractive linear projection and set $Y := \pi(X)$. Then we have:

i) Y becomes a JB^*-triple under the triple product $\{\cdots\}^\pi$ defined by

$\{xyz\}^\pi := \pi(\{xyz\})$ for all $x, y, z \in Y$.

ii) $\pi(\{xyz\}) = \pi(\{x\pi(y)z\})$ for every $y \in X$ and all $x, z \in Y$.

According to the version of the local reflexivity principle in terms of (normed) ultrapowers (Cabrera and Rodríguez 2018, Proposition 5.7.8), the bidual of any Banach space X can be seen as the range of a contractive projection on a suitable ultraproduct of X. Since every ultraproduct of a JB^*-triple is a JB^*-triple (Cabrera and Rodríguez 2018, Corollary 5.7.6), it follows from theorem 4.33(i) that the bidual of any JB^*-triple is a JB^*-triple (in a unique way, because of theorem 4.30). The above argument, due to Dineen (1986b), has been deeply refined by Barton and Timoney (1986) to prove assertion (i) in the following.

THEOREM 4.34.– Let X be a JB^*-triple. Then:

i) (Cabrera and Rodríguez 2018, Theorem 5.7.18) The bidual of X becomes a JB^*-triple under a triple product that extends that of X and is separately w^*-continuous.

ii) (Cabrera and Rodríguez 2018, Proposition 5.10.95) The set

$$\mathrm{Mult}(X) := \{x \in X'' : \{xXX\} \subseteq X\}$$

is a closed subtriple of X'' containing X as a triple ideal.

The set $\mathrm{Mult}(X)$ above is called the *JB*-triple of multipliers* of X.

PROPOSITION 4.23.– (Cabrera and Rodríguez 2018, Proposition 5.10.97) A *JB**-triple X underlies a non-commutative *JB**-algebra if and only if the *JB**-triple $\mathrm{Mult}(X)$ has a unitary tripotent.

The "if" part of the above proposition follows easily from theorems 4.27 and 4.34(ii). In turn, the "only if" part of proposition 4.23 relies on the fact that the non-commutative *JB**-algebra of multipliers of a non-commutative *JB**-algebra A (see theorem 4.11(ii)) coincides with the *JB**-triple of multipliers of the *JB**-triple underlying A (Cabrera and Rodríguez 2018, Proposition 5.10.96).

JBW-triples* are defined as those *JB**-triples, which are dual Banach spaces. The next lemma is crucial in the original proof of theorem 4.34(i) in Barton and Timoney (1986), as well as in theorem 4.35 given in Cabrera and Rodríguez (2018).

LEMMA 4.1.– (Cabrera and Rodríguez 2018, Lemma 5.7.14) Let X be a *JBW**-triple whose triple product is w^*-continuous in the middle variable. Then the triple product of X is separately w^*-continuous.

Actually, we have the following.

THEOREM 4.35.– (Cabrera and Rodríguez 2018, Theorems 5.7.20 and 5.7.38) Let X be a *JBW**-triple. Then the triple product of X is separately w^*-continuous, and the predual of X is unique.

The proof of the above theorem given in Cabrera and Rodríguez (2018) is not the original one, and, contrarily to what happens later, avoids any Banach space result on uniqueness of preduals. Actually, the proof of the separate w^*-continuity of the triple product given in Cabrera and Rodríguez (2018) only involves theorems 4.33 and 4.34(i), and (as we already commented) lemma 4.1. In its turn, in the proof of Cabrera and Rodríguez (2018), the uniqueness of the predual is derived from the uniqueness of preduals of *JBW**-algebras (assured by theorem 4.2) via the separate w^*-continuity of the triple product (previously established) and a consequence of theorem 4.27, see Cabrera and Rodríguez (2014, Corollary 4.2.30(iii)(b)). In this process, the fact that abelian *JBW**-triples are commutative von Neumann algebras (Cabrera and Rodríguez 2018, Proposition 5.7.26) is applied. (We note that, in view of the comments before theorem 4.26, abelian *JB**-triples need not be commutative C^*-algebras.)

As a first application of theorems 4.34(i) and 4.35, it can be shown that the quotient of a JB^*-triple by a closed triple ideal is a JB^*-triple (Cabrera and Rodríguez 2018, Proposition 5.7.40). The same references can be applied to prove proposition 4.24, which, keeping in mind theorem 4.24, becomes a partial generalization of proposition 4.8.

PROPOSITION 4.24.– (Cabrera and Rodríguez 2018, Proposition 5.7.41) Let X and Y be JB^*-triples, and let $F : X \to Y$ be a triple homomorphism. We have:

i) F is contractive;

ii) If F is injective, then F is isometric.

Another relevant application of theorem 4.35 allows us to prove the following.

PROPOSITION 4.25.– (Cabrera and Rodríguez 2018, Theorem 5.7.36) Let X be the predual of a JBW^*-triple. Then there is a subspace Y of X'' such that $X'' = X \oplus_1 Y$.

According to Harmand *et al.* (1993, Definition III.1.1(b) (see also Chapter IV)), those Banach spaces X fulfilling the conclusion in the above proposition are called *L-embedded*. Keeping in mind that the dual of a JB^*-triple is the predual of a JBW^*-triple (see Theorem 4.34(i)), the next theorem follows from proposition 4.25 and general results on L-embedded Banach spaces (Cabrera and Rodríguez 2018, Section 5.8).

THEOREM 4.36.– (Cabrera and Rodríguez 2018, Corollary 5.8.33 and Fact 5.8.39) Let X be a JB^*-triple. Then all bounded linear operators from X to X' are weakly compact. Equivalently, all continuous bilinear mappings from $X \times X$ to X are Arens regular.

Now we can formulate the following theorem, which can be seen as a unit-free version of theorem 4.6.

THEOREM 4.37.– (Cabrera and Rodríguez 2018, Theorem 5.9.9) A non-zero complete normed complex algebra A is a non-commutative JB^*-algebra (for some involution) if and only if A has an approximate identity bounded by one and the open unit ball of A is a homogeneous domain.

To show how most of the results reviewed until now come together, we provide the reader with a proof of the above theorem.

PROOF.– The "only if" part follows from proposition 4.14, and theorems 4.24 and 4.30. Suppose that A is a non-zero complete normed complex algebra having an approximate identity bounded by one and such that Δ_A is a homogeneous domain. Then, by Kaup's theorem 4.30, there exists a surjective linear isometry ϕ from A to a suitable JB^*-triple X. As a first consequence, it follows from theorem 4.36 that A is an Arens regular normed algebra, and hence the product of A'' is separately w^*-continuous. Therefore, since A has an approximate identity bounded by one, a straightforward verification shows that A'' has a unit $\mathbf{1}$ with $\|\mathbf{1}\| = 1$ (take a w^*-cluster point in A'' of the approximate identity of A). On the other hand, the bitranspose ϕ'' of ϕ becomes a surjective linear isometry from A'' to X'', and X'' is a *JB**-triple because of theorem 4.34(i). Then, by the non-associative Bohnenblust–Karlin theorem, $\phi''(\mathbf{1})$ is a vertex of $\mathbb{B}_{X''}$, and hence, by the implication (iii)⇒(ii) in theorem 4.27, X'' is the underlying Banach space of a *JB**-algebra with unit $\phi''(\mathbf{1})$. Therefore, by the easy part of theorem 4.6, we have $X'' = H(X'', \phi''(\mathbf{1})) + iH(X'', \phi''(\mathbf{1}))$, and hence $A'' = H(A'', \mathbf{1}) + iH(A'', \mathbf{1})$. Therefore, by the difficult part of theorem 4.6, A'' is a non-commutative *JB**-algebra. Finally, it follows from theorem 4.14 that A is a non-commutative *JB**-algebra. □

Keeping in mind the great amount of results involved in the above proof, and the large genealogy of each of them, we feel that theorem 4.37 becomes one of culminating points of the theory we are reviewing in this chapter.

The next two corollaries follow straightforwardly from theorem 4.37.

COROLLARY 4.14.– A complete normed associative complex algebra is a C^*-algebra if and only if it has an approximate identity bounded by one and its open unit ball is a homogeneous domain.

COROLLARY 4.15.– A normed associative complex algebra is a C^*-algebra if and only if it is linearly isometric to a C^*-algebra and has an approximate identity bounded by one.

Combining theorem 4.37 with the bracketed version of proposition 4.15, we obtain the following corollary.

COROLLARY 4.16.– A non-zero complete normed complex algebra is a commutative C^*-algebra if and only if A has an approximate identity bounded by one, the open unit ball of A is a homogeneous domain, and there is no non-zero element $a \in A$ such that $a^2 = 0$.

A different non-associative characterization of commutative C^*-algebras can be obtained by combining theorem 4.37 with proposition 3.5.44 in Cabrera and Rodríguez (2014).

To conclude this section, let us report on the representation theory of *JB**-triples and its applications.

Prime JBW^*-triples are called *JBW^*-triple factors*. A *Cartan factor* is a JBW^*-triple factor such that the closed unit ball of its predual has extreme points. According to Horn (1987b), Cartan factors come in the following six types:

1) $\mathbf{I}_{n,m} := BL(H,K)$ (the Banach space of all bounded linear mappings from H to K), where H, K are complex Hilbert spaces of hilbertian dimension n, m, respectively, with $1 \leq n \leq m$, and the triple product is defined by

$$\{xyz\} := \frac{1}{2}(xy^*z + zy^*x) \text{ for all } x, y, z \in BL(H,K).$$

2) $\mathbf{II}_n := \{x \in BL(H) : \sigma x^* \sigma = -x\}$ as closed subtriple of $BL(H)$, where H is a complex Hilbert space of hilbertian dimension $n \geq 5$ and σ is an isometric conjugate-linear involution on H.

3) $\mathbf{III}_n := \{x \in BL(H) : \sigma x^* \sigma = x\}$ as closed subtriple of $BL(H)$, where H is a complex Hilbert space of hilbertian dimension $n \geq 2$ and σ is an isometric conjugate-linear involution on H.

4) $\mathbf{IV}_n := H$, where H is a complex Hilbert space of hilbertian dimension $n \geq 3$, σ is an isometric conjugate-linear involution on H, and the triple product and the norm are given by

$$\{xyz\} := (x|y)z + (z|y)x - (x|\sigma(z))\sigma(y) \text{ and}$$

$$\|x\|^2 := (x|x) + \sqrt{(x|x)^2 - |(x|\sigma(x))|^2},$$

respectively, for all $x, y, z \in H$.

5) $\mathbf{V} := M_{12}(C(\mathbb{C}))$ the 1×2-matrices over the complex octonions $C(\mathbb{C})$, regarded as a closed subtriple of the type $\mathbf{VI}$ immediately below.

6) $\mathbf{VI} := H_3(C(\mathbb{C}))$ the *JB**-algebra of all hermitian 3×3-matrices over $C(\mathbb{C})$, regarded as a *JB**-triple.

Type $\mathbf{IV}_n$ Cartan factors are usually called *spin triple factors.*

By a *JBW***-factor representation* of a *JB**-triple X, we mean a w^*-dense-range triple homomorphism from X to some *JBW**-triple factor.

THEOREM 4.38.– (Cabrera and Rodríguez 2018, Theorem 7.1.5) Every *JB**-triple has a faithful family of Cartan factor representations.

Proposition 4.24(ii) and the above theorem lead to the following.

COROLLARY 4.17.– (Cabrera and Rodríguez 2018, Corollary 7.1.6) Let X be a *JB**-triple. Then there exists a *JB**-algebra A containing X as a closed subtriple.

The next result follows straightforwardly from proposition 4.11, theorem 4.24, and the above corollary.

COROLLARY 4.18.– Let X be a *JB**-triple. Then, for all $x, y, z \in X$, the inequality $\|\{xyz\}\| \leq \|x\|\|y\|\|z\|$ holds.

Combining theorem 4.38 with Zelmanov's techniques (Zelmanov 1984, 1985; D'Amour 1992; D'Amour and McCrimmon 2000), all prime JB^*-triples can be described. Indeed, we have the following.

THEOREM 4.39.– (Cabrera and Rodríguez 2018, Theorem 7.1.41) Let X be a prime JB^*-triple. Then one of the following assertions holds for X:

i) X is either the type **V** or the type **VI** Cartan factor;

ii) X is a spin triple factor;

iii) There exists a prime C^*-algebra A and a self-adjoint idempotent $e \in M(A)$ such that X can be regarded as a closed subtriple of the C^*-algebra $M(A)$ contained in $eM(A)(\mathbf{1} - e)$ and containing $eA(\mathbf{1} - e)$;

iv) There exists a prime C^*-algebra A, a self-adjoint idempotent $e \in M(A)$, and a $*$-involution τ on A with $e + e^\tau = \mathbf{1}$ such that X can be regarded as a closed subtriple of the C^*-algebra $M(A)$ contained in $H(eM(A)e^\tau, \tau)$ and containing $H(eAe^\tau, \tau)$.

With similar techniques, the next result can be proved.

THEOREM 4.40.– (Moreno and Rodríguez 2003, Theorem 3.8) Let X be a JBW^*-triple factor. Then one of the following assertions holds for X:

i) X is either the type **V** or the type **VI** Cartan factor;

ii) X is a spin triple factor;

iii) There exists a W^*-factor A and a self-adjoint idempotent $e \in A$ such that $X = eA(\mathbf{1} - e)$;

iv) There exists a W^*-factor A, a self-adjoint idempotent $e \in A$, and a $*$-involution τ on A with $e + e^\tau = \mathbf{1}$ such that $X = H(eAe^\tau, \tau)$.

The original sources for the material developed in this section, not previously quoted in it, are as follows: (Akemann 1967; Alvermann 1985; Arazy 1987; Barton *et al.* 1988; Becerra and Rodríguez 2007a,b; Bouhya and Fernández 1994; Braun *et al.* 1978; Bunce and Chu 1992; Burgos *et al.* 2010; Chu *et al.* 1989; Friedman and Russo 1985, 1986; Godefroy 1983; Godefroy and Iochum 1988; Godefroy and Saab 1986; Harris 1972; Heinrich 1980; Horn 1987a; Kaidi *et al.* 2001a,b,c; Kaup 1984; Kaup and Upmeier 1977; Lumer and Phillips 1961; Martínez 1980; Moreno and Rodríguez 2000; Pelczynski 1962; Peralta 2015; Pfitzner 1993, 2010; Rodríguez 1998, 2004; Spitkovsky 1994; Stachó 1982; Vigué 1976; Youngson 1981). This material and more has been fully re-elaborated in the books (Bourbaki 2006; Cabrera and Rodríguez 2014, 2018; Chu 2012; Isidro 2019; Isidro and Stachó 1985; Lang 1995; Loos 1977; Upmeier 1985, 1987).

4.5. Past, present and future of non-associative C^*-algebras

The birth of non-associative C^*-algebras can be dated in 1970, when, in the last section of the monograph (Kaplansky 1970), Kaplansky designed the basic lines of research in JB-algebras, JB^*-algebras and alternative C^*-algebras. The reader is referred to Cabrera and Rodríguez (2014, pp. xi–xv) for a full review of Kaplansky's words. There the reader can find the appropriate comments concerning how Kaplansky's predictions have come true, some even exceeding the original expectations.

Eight years after the publication of Kaplansky's monograph, Alfsen *et al.* (1978) established the foundations of the theory of JB-algebras. Shortly later, the work in Alfsen *et al.* (1978) was complemented by Shultz (1979) by considering JBW-algebras. The results in Alfsen *et al.* (1978); Shultz (1979), together with some simplifications in Behncke (1979); Hanche-Olsen (1980), early results on JC-algebras (like those in Stormer (1965) and Topping (1965)) and many other later results (like those in Stacey (1982)) were fully re-elaborated in Hanche-Olsen and Stormer (1984). Since its publication to date, this book has converted into the irreplaceable reference for the topic. Actually, most of the results reviewed in section 4.1 of this chapter have been taken from (Hanche-Olsen and Stormer 1984). The remaining results in section 4.1 have been taken from (Cabrera and Rodríguez 2014), and are rooted in the papers of Edwards (1980a) and Wright and Youngson (1978). The theory of JB-algebras enjoys an autonomous development. Only occasionally (as could be the case of corollary (4.4)) the theory of JB-algebras depends on that of JB^*-algebras.

JB^*-algebras were introduced by Kaplansky in his final lecture to the 1976 St. Andrews Colloquium of the Edinburgh Mathematical Society, and were first studied by Wright (1977), who, as main results, proved the unital versions of proposition 4.7 and theorem 4.8. Shortly later, unital JB^*-algebras were reconsidered by Braun *et al.* (1978), who proved theorems 4.9 and 4.27, as well as the unital version of theorem 4.24. Simultaneously, Wright and Youngson (1977, 1978) proved the Russo–Dye–Palmer type theorem reviewed before proposition 4.11, and the unital version of propositions 4.11 and 4.12. Non-unital JB^*-algebras and JBW^*-algebras were considered first by Youngson (1981) and Edwards (1980b), respectively.

Non-commutative JB^*-algebras arose in the literature as a necessity for a right formulation of the non-associative Vidav–Palmer theorem 4.6, a result whose paternity can be settled by combining the papers (Rodríguez 1983; Kaidi *et al.* 1981; Youngson 1978, 1979b) in this order. A systematic study of non-commutative JB^*-algebras (paying attention to alternative C^*-algebras, non-commutative JBW^*-algebras and alternative W^*-algebras) began in Payá *et al.* (1982, 1984), and continued in Alvermann and Janssen (1984); Braun (1983, 1984). The unital non-associative Gelfand–Naimark theorem 4.5 was proved in Rodríguez (1980), whereas example 4.3 was established in Kaidi *et al.* (1981).

JB^*-triples were introduced by Kaup (1977) under the name of C^*-triple systems. Among other results, in Kaup (1977) he proved the implication (iii)⇒(i) in theorem 4.30, corollaries 4.12 and 4.13, and proposition 4.24. The work done in Kaup (1977) culminated with Kaup's paper (1983) where he proved theorems 4.25, 4.26 and 4.32, as well as the remaining part of theorem 4.30. It is worthy mention that Kaup's proof of theorem 4.25 involves a general result in (Martínez 1980) on complete normed unital Jordan complex algebras (Cabrera and Rodríguez 2014, Theorem 4.1.10). The contractive projection theorem 4.33 was obtained in Kaup (1984) (see also Stachó (1982) for a forerunner).

Kaup's work just reviewed had remarkable finite-dimensional forerunners due to (Cartan 1935a, 1931, 1935b; Koecher 1969)) and (Loos 1977). It was Koecher who introduced Jordan triple systems as a vehicle for classifying finite-dimensional bounded symmetric domains. The infinite-dimensional version of Cartan's work on groups of holomorphic transformations was independently established by (Upmeier 1976; Vigué 1976). The connection between C^*-algebras and bounded symmetric domains was noticed first by (Harris 1974, 1981), who proved relevant forerunners of the implication (iii)⇒(i) in theorem 4.30. Kaup and Upmeier (1976) introduced the symmetric part and the partial triple product of a complex Banach space, and proved propositions 4.19 and 4.20. As the main result, they derive that complex Banach spaces with biholomorphically equivalent open unit balls are linearly isometric (Cabrera and Rodríguez 2018, Theorem 5.6.35).

So far, we have referred the reader to the pioneering works on JB-algebras, JB^*-algebras, non-commutative JB^*-algebras, alternative C^*-algebras and JB^*-triples. Thus, in a few years (1978–1984), the foundations of a large building were laid, over which later luminous floors would be built, like corollary 4.17 (Friedman and Russo 1986; Horn 1987b), and theorems 4.34(i) (Barton and Timoney 1986), 4.35 (Barton and Timoney 1986; Horn 1987a), 4.1 (Isidro and Rodríguez 1995), 4.37 (Kaidi *et al.* 2001b) and 4.15 (Cabrera and Rodríguez 2012; Rodríguez 2011).

Some of the main goals in the theory we are dealing with have been already reached. Thus, for example, the role played by JB^*-triples concerning the problem of the classification of symmetric bounded domains in complex Banach spaces was completely clarified by the categorical formulation of Theorem 4.32 given in Kaup (1983). Concerning our own goals, the main were (and have been until recently) that of obtaining non-associative characterizations of non-commutative JB^*-algebras and of alternative C^*-algebras, and that of establishing the Zelmanovian classifications of prime JB^*-algebras, prime non-commutative JB^*-algebras and prime JB^*-triples. Nevertheless, keeping in mind theorems 4.15, 4.37 (both already dated), 4.20 (Fernández *et al.* 1992), 4.22 (Kaidi *et al.* 2000) and 4.39 (Moreno and Rodríguez

2000, 2003), today it seems to us that these goals have no much future. We note that, according to (Rodríguez 2011) (see also (Cabrera and Rodríguez 2018, Corollary 6.2.3)) and corollary 4.16, even non-associative characterizations of C^*-algebras and of commutative C^*-algebras are also known.

Anyway, a honest objective in the theory we are dealing with was, is and shall be that of finding the appropriate generalization of a certain known result on C^*-algebras to the setting of JB-algebras, non-commutative JB^*-algebras and/or JB^*-triples (sometimes, according to Kaplansky's (1970) words, "in a suitably altered form"). Thus, for example, theorem 4.38 becomes the appropriate translation to the language of JB^*-triples of the fact that every C^*-algebra has a faithful family of irreducible representations on complex Hilbert spaces. To date, the line of research we are commented on has underlaid most works on the topic, and we feel that the same shall happen in the future. In many cases (as, for example, in that of Theorem 4.34(i)), it has encouraged the development of revolutionary new techniques. On the other hand, it is worthy to mention that, as Banach spaces, JB^*-triples strictly contain non-commutative JB^*-algebras (see theorem 4.24 and proposition 4.23), so in particular they contain C^*-algebras. Thus, in most cases, the establishment of Banach space properties of JB^*-triples entails not only generalizations but also new proofs of Banach space properties of C^*-algebras. As a outstanding example, we cite the theorem in Edwards *et al.* (2010) asserting that closed faces of the closed unit ball of any JB^*-triple are norm-semi-exposed, a result whose C^*-algebra forerunner is given by Akemann and Pedersen (1992).

In extending the theory of C^*-algebras to the non-associative substitutes, the researcher may encounter unpleasant surprises. As a sample, the fact that linearly isometric C^*-algebras are Jordan-$*$-isomorphic (see Paterson and Sinclair (1972)) survives for alternative C^*-algebras (corollary 4.7), but not for general non-commutative JB^*-algebras (theorem 4.9). On the other hand, the success does not always go along with the researcher. Indeed, according to Ptak (1972), complete normed associative complex $*$-algebras A, which are C^*-algebras in an equivalent norm can be characterized by Condition ♣ which follows:

♣ The set $\{(\exp -1)(ih) : h \in H(A, *)\}$ is bounded in A.

As a matter of fact, Condition ♣ also characterize complete normed Jordan (respectively, alternative) complex $*$-algebras A, which are JB^*-algebras (respectively, alternative C^*-algebras) in an equivalent norm (see (Aupetit and Youngson 1984); (Youngson 1979a); (Cabrera and Rodríguez 2014, Proposition 4.5.45)), but is not sufficient to characterize complete normed non-commutative Jordan complex $*$-algebras A, which are non-commutative JB^*-algebras in an equivalent norm (Cabrera and Rodríguez 2014, Example 4.5.43). Our conjecture is that a complete normed non-commutative Jordan complex $*$-algebra A is a non-commutative JB^*-algebra in an equivalent norm if (and only if) it fulfills

Condition ♣ and, for every normal element $a \in A$, the spectral radius of both left and right multiplication operators by a is equal to the spectral radius of a (Cabrera and Rodríguez 2014, Conjecture 4.5.52). Despite the setbacks like those just commented, sometimes the non-associative approach overtakes the associative one. For example, this is the case of the associative corollary 4.14, which was unknown before the establishment of the non-associative theorem 4.37 in Kaidi *et al.* (2001b).

In our opinion, one of the most interesting challenge of the theory we are dealing with is that of reorganizing all the material appeared in the literature (which is scattered along many hundreds of papers) in order to do it understandable by as many people as possible. This has been already partially done in various studies (Alfsen and Shultz 2003; Cabrera and Rodríguez 2014, 2018; Chu 2012; Hanche-Olsen and Stormer 1984; Isidro 2019; Isidro and Stachó 1985; Upmeier 1985, 1987).

Concerning our book (Cabrera and Rodríguez 2014, 2018), we would like to briefly illustrate the difficulty of the above sketched goal by commenting on successes and failures we have had, and on obstacles we have found along its writing. Concerning successes, let us say that we were able to develop the basic theory of non-commutative JB^*-algebras (including all results from theorem 4.5 to 4.18, and even theorem 4.37) without any explicit reference to the representation theory. Only the representation theory of JB-algebras was implicitly involved through corollary 4.3. Moreover, we have got remarkable simplifications of the original proofs of theorems 4.14, 4.26, and 4.28, by incorporating arguments in Essaleh *et al.* (2018); Burgos *et al.* (2010) and (Peralta 2015), respectively. Also we are happy with our new proof of theorem 4.35, which, contrarily to what happens in the original one, avoids any Banach space result on uniqueness of preduals. Concerning failures, let us discuss only one. Indeed, we have been unable to find a complete proof of theorem 4.23 which could fit reasonably in the framework of our book.

Concerning obstacles we have found along the writing of (Cabrera and Rodríguez 2014, 2018), let us say at first that theorem 4.29 appears in the literature without any hint for its proof, other than a vague reference to Bourbaki (2006). Actually, we were able to incorporate this theorem to our book thanks to H. Upmeier (reference [1113] of Cabrera and Rodríguez (2018)) who provided us with a complete proof. To conclude our discussion on obstacles, let us comment on theorem 4.36, which is originally given by Chu *et al.* (1989), who actually proved the deeper fact that each bounded linear operator from a JB^*-triple to its dual factors through a complex Hilbert space. The proof of the refinement of theorem 4.36 just reviewed is extremely involved. Therefore, we designed a proof of theorem 4.36 involving only proposition 4.25 and general results on Banach spaces. Looking for proofs of these results, we realized that the one of a theorem in Godefroy and Saab (1986) had been missing from the literature. We overcame this obstacle thanks to H. Pfitzner (reference [1047] of Cabrera and Rodríguez (2018)), who was able to rebuild the lost proof.

After concluding the writing of our book (Cabrera and Rodríguez 2014, 2018), we have continued thinking about the reorganization of results, obtaining some new advances. Indeed, according to (Becerra *et al.* 2003, Proposition 2.9) (see also (Cabrera and Rodríguez 2018, Corollary 5.6.70)) and its proof, if the symmetric part of the bidual X'' of a complex Banach space X contains X, then X'' becomes a JBW^*-triple whose triple product is w^*-continuous in the middle variable. This result, together with the key lemma 4.1, a forerunner of (Cabrera and Rodríguez 2018, Theorem 5.6.45) given by Harris (1971) and a simplification of the ultrapower techniques in Dineen (1986a) paper, allows us to quickly conclude the proof of Theorem 4.34(i) (see Cabrera and Rodríguez (2019) for details).

4.6. Acknowledgments

This chapter has been partially supported by the Junta de Andalucía and Spanish government grants FQM199 and MTMT2016-76327-C3-2-P.

4.7. References

Akemann, C.A. (1967). The dual space of an operator algebra. *Trans. Amer. Math. Soc.*, 126, 286–302.

Akemann, C.A., Pedersen, G.K. (1992). Facial structure in operator algebra theory. *Proc. London Math. Soc.*, 64, 418–448.

Albert, A.A. (1934). On a certain algebra of quantum mechanics. *Ann. Math.*, 35, 65–73.

Albert, A.A. (1948). Power-associative rings. *Trans. Amer. Math. Soc.*, 608, 552–593.

Alfsen, E.M., Shultz, F.W. (2003). *Geometry of State Spaces of Operator Algebras*. Birkhäuser Boston, Inc., Boston.

Alfsen, E.M., Shultz, F.W., Stormer, E. (1978). A Gelfand–Neumark theorem for Jordan algebras. *Adv. Math.*, 28, 11–56.

Allan, G.R. (2011). *Introduction to Banach Spaces and Algebras*. Oxford University Press, Oxford.

Alvermann, K. (1985). The multiplicative triangle inequality in noncommutative JB- and JB^*-algebras. *Abh. Math. Sem. Univ. Hamburg*, 55, 91–96.

Alvermann, K., Janssen, G. (1984). Real and complex non-commutative Jordan Banach algebras. *Math. Z.*, 185, 105–113.

Ara, P., Mathieu, M. (2003). *Local Multipliers of C^*-Algebras*. Springer, London.

Arazy, J. (1987). An application of infinite-dimensional holomorphy to the geometry of Banach spaces. In *Geometrical Aspects of Functional Analysis (1985/86)*, Lindenstrauss, J., Milman, V.D. (eds). Springer-Verlag, Berlin, 122–150.

Arens, R. (1951). The adjoint of a bilinear operation. *Proc. Amer. Math. Soc.*, 2, 839–848.

Aupetit, B. (1982). The uniqueness of the complete norm topology in Banach algebras and Banach Jordan algebras. *J. Funct. Anal.*, 47, 1–6.

Aupetit, B., Youngson, M.A. (1984). On symmetry of Banach Jordan algebras. *Proc. Amer. Math. Soc.*, 91, 364–366.

Barton, T.J., Dang, T., Horn, G. (1988). Normal representations of Banach Jordan triple systems. *Proc. Amer. Math. Soc.*, 102, 551–555.

Barton, T.J., Timoney, R.M. (1986). Weak*-continuity of Jordan triple products and its applications. *Math. Scand.*, 59, 177–191.

Becerra, J., Rodríguez, A. (2007a). Non self-adjoint idempotents in C^*- and JB^*-algebras. *Manuscripta Math.*, 124, 183–193.

Becerra, J., Rodríguez, A. (2007b). C^*- and JB^*-algebras generated by a non-self-adjoint idempotent. *J. Funct. Anal.*, 248, 107–127.

Becerra, J., Rodríguez, A., Wood, G.V. (2003). Banach spaces whose algebras of operators are unitary: A holomorphic approach. *Bull. London Math. Soc.*, 35, 218–224.

Behncke, H. (1979). Hermitian Jordan Banach algebras. *J. London Math. Soc.*, 20, 327–333.

Beltita, D. (2006). *Smooth Homogeneous Structures in Operator Theory*. Chapman & Hall/CRC, Boca Raton.

Bensebah, A. (1992). Weakness of the topology of a JB^*-algebra. *Canad. Math. Bull.*, 35, 449–454.

Bohnenblust, H.F., Karlin, S. (1955). Geometrical properties of the unit sphere of a Banach algebra. *Ann. Math.*, 62, 217–229.

Bonsall, F.F., Duncan, J. (1971). *Numerical Ranges of Operators on Normed Spaces and of Elements of Normed Algebras*. Cambridge University Press, Cambridge.

Bonsall, F.F., Duncan, J. (1973). *Numerical Ranges II*. Cambridge University Press, Cambridge.

Bouhya, K., Fernández, A. (1994). Jordan-$*$-triples with minimal inner ideals and compact JB^*-triples, *Proc. London Math. Soc.*, 68, 380–398.

Bourbaki, N. (2006). *Éléments de mathématique: groupes et algèbres de Lie*. Springer, Berlin.

Bratteli, O. (1986). *Derivations, Dissipations and Group Actions on C^*-algebras*. Springer, Berlin.

Braun, H., Koecher, M. (1966). *Jordan-Algebren*. Springer, Berlin.

Braun, R.B. (1983). Structure and representations of non-commutative C^*-Jordan algebras. *Manuscripta Math.*, 41, 139–171.

Braun, R.B. (1984). A Gelfand–Neumark theorem for C^*-alternative algebras. *Math. Z.*, 185, 225–242.

Braun, R.B., Kaup, W., Upmeier, H. (1978). A holomorphic characterization of Jordan C^*-algebras. *Math. Z.*, 161, 277–290.

Bunce, L.J., Chu, C.-H. (1992). Compact operations, multipliers and Radon-Nikodym property in JB^*-triples. *Pacific J. Math.*, 153, 249–265.

Burgos, M.J., Peralta, A.M., Ramírez, M.I., Ruiz, E.E. (2010). Von Neumann regularity in Jordan–Banach triples. In *Proceedings of Jordan Structures in Algebra and Analysis Meeting. Tribute to El Amin Kaidi for his 60th birthday*. Carmona, J., Morales, A., Peralta, A.M., and Ramírez, M.I. (eds). Editorial Círculo Rojo, Almería, 67–88.

Cabrera, M., Rodríguez, A. (1992). Nonassociative ultraprime normed algebras. *Quart. J. Math. Oxford*, 43, 1–7.

Cabrera, M., Rodríguez, A. (1993). New associative and nonassociative Gelfand–Naimark theorems. *Manuscripta Math.*, 79, 197–208.

Cabrera, M., Rodríguez, A. (2012). On the Gelfand–Naimark axiom $||a^*a|| = ||a^*||||a||$. *Quart. J. Math. Oxford*, 63, 855–860.

Cabrera, M., Rodríguez, A. (2014). *Non-Associative Normed Algebras. Volume 1: The Vidav–Palmer and Gelfand–Naimark Theorems.* Cambridge University Press, Cambridge.

Cabrera, M., Rodríguez, A. (2018). *Non-Associative Normed Algebras. Volume 2: Representation Theory and the Zel'manov Approach.* Cambridge University Press, Cambridge.

Cabrera, M., Rodríguez, A. (2019). A new proof of the Barton-Timoney theorem on the bidual of a JB^*-triple. *Math. Nachr.*, 292, 640–644.

Cartan, E. (1935a). Sur les domaines bornés homogènes de l'espace de n variables complexes. *Abh. Math. Sem. Univ. Hamburg*, 11, 116–162.

Cartan, H. (1931). Les fonctions de deux variables complexes et le problème de la représentation analytique. *J. Math. Pures Appl., 9e série*, 10, 1–114.

Cartan, H. (1935b). *Sur les groupes de transformations analytiques.* Hermann, Paris.

Chu, C.-H. (2012). *Jordan Structures in Geometry and Analysis.* Cambridge University Press, New York.

Chu, C.-H., Iochum, B., Loupias, G. (1989). Grothendieck's theorem and factorization of operators in Jordan triples. *Math. Ann.*, 284, 41–53.

Cleveland, S.B. (1963). Homomorphisms of non-commutative $*$-algebras. *Pacific J. Math.*, 13, 1097–1109.

Cohn, P.M. (1954). On homomorphic images of special Jordan algebras. *Canadian J. Math.*, 6, 253–264.

D'Amour, A. (1992). Quadratic Jordan systems of Hermitian type. *J. Algebra*, 149, 197–233.

D'Amour, A., McCrimmon, K. (2000). The structure of quadratic Jordan systems of Clifford type. *J. Algebra*, 234, 31–89.

Dineen, S. (1986a). Complete holomorphic vector fields on the second dual of a Banach space. *Math. Scand.*, 59, 131–142.

Dineen, S. (1986b). The second dual of a JB^*-triple system. In *Complex Analysis, Functional Analysis and Approximation Theory*, Mujica, J. (ed.). Notas de Matemática 110, North-Holland Publishing Co., Amsterdam, 67–69.

Dixmier, J. (1969). *Les algèbres d'opérateurs dans l'espace hilbertien (algèbres de von Neumann).* Second edition. Gauthier-Villars Éditeur, Paris.

Doran, R.S., Belfi, V.A. (1986). *Characterizations of C^*-Algebras: The Gelfand–Naimark Theorems.* Marcel Dekker, New York/Basel.

Edwards, C.M. (1980a). Multipliers of JB-algebras. *Math. Ann.*, 249, 265–272.

Edwards, C.M. (1980b). On Jordan W^*-algebras. *Bull. Sci. Math.*, 104, 393–403.

Edwards, C.M., Fernández-Polo, F.J., Hoskin, C.S., Peralta, A.M. (2010). On the facial structure of the unit ball in a JB^*-triple. *J. Reine Angew. Math.*, 641, 123–144.

Elduque, A., Myung, H.C. (1994). *Mutations of Alternative Algebras*. Academic Publishers Group, Dordrecht.

Essaleh, A.B.A., Peralta, A.M., Ramírez, M.I. (2018). Pointwise-generalized-inverses of linear maps between C^*-algebras and JB^*-triples. *Operators and Matrices*, 12, 369–391.

Fernández, A., García, E., Rodríguez, A. (1992). A Zel'manov prime theorem for JB^*-algebras. *J. London Math. Soc.*, 46, 319–335.

Friedman, Y., Russo, B. (1985). Structure of the predual of a JBW^*-triple. *J. Reine Angew. Math.*, 356, 67–89.

Friedman, Y., Russo, B. (1986). The Gelfand–Naimark theorem for JB^*-triples. *Duke Math. J.*, 53, 139–148.

Gelfand, I.M., Naimark, M.A. (1943). On the embedding of normed rings into the ring of operators in Hilbert space. *Mat. Sbornik*, 12, 197–213.

Godefroy, G. (1983). Parties admissibles d'un espace de Banach. Applications. *Ann. Sci. École Norm. Sup.*, 16(4), 109–122.

Godefroy, G., Iochum, B. (1988). Arens-regularity of Banach algebras and the geometry of Banach spaces. *J. Funct. Anal.*, 80, 47–59.

Godefroy, G., Saab, P. (1986). Quelques espaces de Banach ayant les propriétés (V) ou (V^*) de A. Peczyski. *C. R. Acad. Sci. Paris Sér. I Math.*, 303, 503–506.

Hanche-Olsen, H. (1980). A note on the bidual of a JB-algebra. *Math. Z.*, 175, 29–31.

Hanche-Olsen, H., Stormer, E. (1984). *Jordan Operator Algebras*. Pitman, Boston.

Harmand, P., Werner, D., Werner, W. (1993). *M-ideals in Banach Spaces and Banach Algebras*. Springer, Berlin.

Harris, L.A. (1971). The numerical range of holomorphic functions in Banach spaces. *Amer. J. Math.*, 93, 1005–1019.

Harris, L.A. (1972). Banach algebras with involution and Möbius transformations. *J. Funct. Anal.*, 11, 1–16.

Harris, L.A. (1974). Bounded symmetric homogeneous domains in infinite dimensional spaces. In *Proceedings on Infinite Dimensional Holomorphy (Internat. Conf., Univ. Kentucky, Lexington, Ky., 1973)*. Hayden, T.L. and Suffridge, T.J. (eds). Springer-Verlag, Berlin-New York, 13–40.

Harris, L.A. (1981). A generalization of C^*-algebras. *Proc. London Math. Soc.*, 42, 331–361.

Heinrich, S. (1980). Ultraproducts in Banach space theory. *J. Reine Angew. Math.*, 313, 72–104.

Horn, G. (1987a). Characterization of the predual and ideal structure of a JBW^*-triple. *Math. Scand.*, 61, 117–133.

Horn, G. (1987b). Classification of JBW^*-triples of type I. *Math. Z.*, 196, 271–291.

Horn, G. (1987c). Coordinatization theorem for JBW^*-triples. *Quart. J. Math. Oxford*, 38, 321–335.

Iochum, B., Loupias, G., Rodríguez, A. (1989). Commutativity of C^*-algebras and associativity of JB^*-algebras. *Math. Proc. Cambridge Phil. Soc.*, 106, 281–291.

Isidro, J.M. (2019). *Jordan Triple Systems in Complex and Functional Analysis.* American Mathematical Society, Providence.

Isidro, J.M., Rodríguez, A. (1995). Isometries of JB-algebras. *Manuscripta Math.*, 86, 337–348.

Isidro, J.M., Stachó, L.L. (1985). *Holomorphic Automorphism Groups in Banach spaces: An Elementary Introduction.* North-Holland Publishing Co., Amsterdam.

Jacobson, N. (1968). *Structure and Representations of Jordan Algebras.* American Mathematical Society Colloquium Publications. Providence, 39.

Jordan, P. (1932). Über eine klasse nichtassoziativer hyperkomplexer algebren. *Nachrichten von der Gesellschaft der Wissenschaften zu Göttingen, Mathematisch-Physikalische Klasse*, 569–575.

Jordan, P., von Neumann, J., Wigner, E. (1934). On an algebraic generalization of the quantum mechanical formalism. *Ann. Math.*, 35, 29–64.

Kadison, R.V. (1951). Isometries of operator algebras. *Ann. Math.*, 54, 325–338.

Kadison, R.V. (1956). Operator algebras with a faithful weakly-closed representation. *Ann. Math.*, 64, 175–181.

Kadison, R.V., Ringrose, J.R. (1997). *Fundamentals of the Theory of Operator Algebras. Vol. I. Elementary Theory, and Vol. II. Advanced Theory.* American Mathematical Society, Providence.

Kaidi, A., Martínez, J., Rodríguez, A. (1981). On a non-associative Vidav–Palmer theorem. *Quart. J. Math. Oxford*, 32, 435–442.

Kaidi, A., Morales, A., Rodríguez, A. (2000). Prime non-commutative JB^*-algebras. *Bull. London Math. Soc.*, 32, 703–708.

Kaidi, A., Morales, A., Rodríguez, A. (2001a). Geometrical properties of the product of a C^*-algebra. *Rocky Mountain J. Math.*, 31, 197–213.

Kaidi, A., Morales, A., Rodríguez, A. (2001b). A holomorphic characterization of C^*- and JB^*-algebras. *Manuscripta Math.*, 104, 467–478.

Kaidi, A., Morales, A., Rodríguez, A. (2001c). Non-associative C^*-algebras revisited. In *Recent Progress in Functional Analysis, Proceedings of the International Functional Analysis Meeting on the Occasion of the 70th Birthday of Professor Manuel Valdivia Valencia.* Bierstedt, K.D., Bonet, J., Maestre, M., Schmets, J. (eds), North-Holland Publishing Co., Amsterdam, 379–408.

Kaplansky, I. (1968). *Rings of Operators.* W. A. Benjamin, Inc., New York/Amsterdam.

Kaplansky, I. (1970). *Algebraic and Analytic Aspects of Operator Algebras.* American Mathematical Society, Providence.

Kaup, W. (1977). Algebraic characterization of symmetric complex Banach manifolds. *Math. Ann.*, 228, 39–64.

Kaup, W. (1983). A Riemann mapping theorem for bounded symmetric domains in complex Banach spaces. *Math. Z.*, 183, 503–529.

Kaup, W. (1984). Contractive projections on Jordan C^*-algebras and generalizations. *Math. Scand.*, 54, 95–100.

Kaup, W., Upmeier, H. (1976). Banach spaces with biholomorphically equivalent unit balls are isomorphic. *Proc. Amer. Math. Soc.*, 58, 129–133.

Kaup, W., Upmeier, H. (1977). Jordan algebras and symmetric Siegel domains in Banach spaces. *Math. Z.*, 157, 179–200.

Koecher, M. (1969). *An Elementary Approach to Bounded Symmetric Domains*. Rice University, Houston.

Lang, S. (1995). *Differential and Riemannian Manifolds*. Springer, New York.

Leung, C.-W., Ng, C.-K., Wong, N.-C. (2009). Geometric unitaries in JB-algebras. *J. Math. Anal. Appl.*, 360, 491–494.

Loos, O. (1977). *Bounded Symmetric Domains and Jordan Pairs*. University of California, Irvine.

Lumer, G., Phillips, R.S. (1961). Dissipative operators in a Banach space. *Pacific J. Math.*, 11, 679–698.

Martínez, J. (1980). JV-algebras. *Math. Proc. Cambridge Philos. Soc.*, 87, 47–50.

Martínez, J., Mena, J.F., Payá, R., Rodríguez, A. (1985). An approach to numerical ranges without Banach algebra theory. *Illinois J. Math.*, 29, 609–625.

Mathieu, M. (1989). Elementary operators on prime C^*-algebras, I. *Math. Ann.*, 284, 223–244.

McCrimmon, K. (1969). The radical of a Jordan algebra. *Proc. Nat. Acad. Sci.*, 62, 671–678.

McCrimmon, K. (1971). Noncommutative Jordan rings. *Trans. Amer. Math. Soc.*, 158, 1–33.

McCrimmon, K. (2004). *A Taste of Jordan Algebras*. Springer, New York.

McCrimmon, K., Zel'manov, E. (1988). The structure of strongly prime quadratic Jordan algebras. *Adv. Math.*, 69, 133–222.

McIntosh, A. (1978). Functions and derivations of C^*-algebras. *J. Funct. Anal.*, 30, 264–275.

Moore, R.T. (1971). Hermitian functionals on B-algebras and duality characterizations of C^*-algebras. *Trans. Amer. Math. Soc.*, 162, 253–265.

Moreno, A., Rodríguez, A. (2000). On the Zel'manovian classification of prime JB^*-triples. *J. Algebra*, 226, 577–613.

Moreno, A., Rodríguez, A. (2003). On the Zel'manovian classification of prime JB^*- and JBW^*-triples. *Comm. Algebra*, 31, 1301–1328.

Okayasu, T. (1968). A structure theorem of automorphisms of von Neumann algebras. *Tôhoku Math. J.*, 20, 199–206.

Palmer, T.W. (1968a). Characterizations of C^*-algebras. *Bull. Amer. Math. Soc.*, 74, 538–540.

Palmer, T.W. (1968b). Unbounded normal operators on Banach spaces. *Trans. Amer. Math. Soc.*, 133, 385–414.

Palmer, T.W. (1994). *Banach Algebras and the General Theory of $*$-Algebras. Volume I: Algebras and Banach algebras*. Cambridge University Press, Cambridge.

Palmer, T.W. (2001). *Banach Algebras and the General Theory of $*$-Algebras. Volume II: $*$-Algebras*. Cambridge University Press, Cambridge.

Paterson, A.L.T., Sinclair, A.M. (1972). Characterization of isometries between C^*-algebras. *J. London Math. Soc.*, 5, 755–761.

Payá, R., Pérez, J., Rodríguez, A. (1982). Non-commutative Jordan C^*-algebras. *Manuscripta Math.*, 37, 87–120.

Payá, R., Pérez, J., Rodríguez, A. (1984). Type I factor representations of non-commutative JB^*-algebras. *Proc. London Math. Soc.*, 48, 428–444.

Pelczynski, A. (1962). Banach spaces on which every unconditionally converging operator is weakly compact. *Bull. Acad. Polon. Sci.*, 10, 641–648.

Peralta, A.M. (2015). Positive definite hermitian mappings associated to tripotent elements. *Expo. Math.*, 33, 252–258.

Perez, J., Rico, L., Rodríguez, A. (1994). Full subalgebras of Jordan–Banach algebras and algebra norms on JB^*-algebras. *Proc. Amer. Math. Soc.*, 121, 1133–1143.

Pfitzner, H. (1993). L-summands in their biduals have Peczyski's property (V*). *Studia Math.*, 104, 91–98.

Pfitzner, H. (2010). Phillips' lemma for L-embedded Banach spaces. *Arch. Math. (Basel)*, 94, 547–553.

Ptak, V. (1972). Banach algebras with involution. *Manuscripta Math.*, 6, 245–290.

Rodríguez, A. (1980). A Vidav–Palmer theorem for Jordan C^*-algebras and related topics. *J. London Math. Soc.*, 22, 318–332.

Rodríguez, A. (1983). Non-associative normed algebras spanned by Hermitian elements. *Proc. London Math. Soc.*, 47, 258–274.

Rodríguez, A. (1985). The uniqueness of the complete algebra norm topology in complete normed nonassociative algebras. *J. Funct. Anal.*, 60, 1–15.

Rodríguez, A. (1998). Isometries and Jordan isomorphisms onto C^*-algebras. *J. Operator Theory*, 40, 71–85.

Rodríguez, A. (2004). Numerical ranges of uniformly continuous functions on the unit sphere of a Banach space. *J. Math. Anal. Appl.*, 297, 472–476.

Rodríguez, A. (2011). Approximately norm-unital products on C^*-algebras, and a non-associative Gelfand–Naimark theorem. *J. Algebra*, 347, 224–246.

Russo, B., Dye, H.A. (1966). A note on unitary operators in C^*-algebras. *Duke Math. J.*, 33, 413–416.

Sakai, S. (1956). A characterization of W^*-algebras. *Pacific J. Math.*, 6, 763–773.

Sakai, S. (1971). *C^*-Algebras and W^*-Algebras.* Springer, New York/Heidelberg.

Sakai, S. (1991). *Operator Algebras in Dynamical Systems. The Theory of Unbounded Derivations in C^*-Algebras.* Cambridge University Press, Cambridge.

Schafer, R.D. (1995). *An Introduction to Nonassociative Algebras* (Corrected reprint). Dover Publications, Inc., New York.

Shirshov, A.I. (1956). On special J-rings. *Mat. Sb.*, 38(80), 149–166.

Shultz, F.W. (1979). On normed Jordan algebras which are Banach dual spaces. *J. Funct. Anal.*, 31, 360–376.

Sinclair, A.M. (1971). The norm of a Hermitian element in a Banach algebra. *Proc. Amer. Math. Soc.*, 28, 446–450.

Spitkovsky, I. (1994). Once more on algebras generated by two projections. *Linear Algebra Appl.*, 208/209, 377–395.

Stacey, P.J. (1982). Type I_2 JBW-algebras. *Quart. J. Math. Oxford*, 33, 115–127.

Stachó, L.L. (1982). A projection principle concerning biholomorphic automorphisms. *Acta Sci. Math. (Szeged)*, 44, 99–124.

Stormer, E. (1965). On the Jordan structure of C^*-algebras. *Trans. Amer. Math. Soc.*, 120, 438–447.

Topping, D.M. (1965). Jordan algebras of self-adjoint operators. *Mem. Amer. Math. Soc.*, 53.

Upmeier, H. (1976). Über die automorphismengruppen von Banach-mannigfaltigkeiten mit invarianter Metrik. *Math. Ann.*, 223, 279–288.

Upmeier, H. (1985). *Symmetric Banach Manifolds and Jordan C^*-Algebras*. North-Holland, Amsterdam.

Upmeier, H. (1987). *Jordan Algebras in Analysis, Operator Theory, and Quantum Mechanics*. American Mathematical Society, Providence.

Vidav, I. (1956). Eine metrische kennzeichnung der selbstadjungierten operatoren. *Math. Z.*, 66, 121–128.

Vigué, J.-P. (1976). Le groupe des automorphismes analytiques d'un domaine borné d'un espace de Banach complexe. Application aux domaines bornés symétriques. *Ann. Sci. École Norm. Sup.*, 9(4), 203–281.

Viola Devapakkiam, C. (1971). Jordan algebras with continuous inverse. *Math. Japon.*, 16, 115–125.

Vowden, B.J. (1967). On the Gelfand–Neumark theorem. *J. London Math. Soc.*, 42, 725–731.

Wright, J.D.M. (1977). Jordan C^*-algebras. *Michigan Math. J.*, 24, 291–302.

Wright, J.D.M., Youngson, M.A. (1977). A Russo Dye theorem for Jordan C^*-algebras. In *Functional Analysis: Surveys and Recent Results (Proc. Conf., Paderborn, 1976)*, Bierstedt, K.D., Fuchssteiner, F. (eds). North-Holland Mathematics Studies, Vol. 27. Notas de Matemática, No. 63. North-Holland Publishing Co., Amsterdam-New York-Oxford. 279–282.

Wright, J.D.M., Youngson, M.A. (1978). On isometries of Jordan algebras. *J. London Math. Soc.*, 17, 339–344.

Youngson, M.A. (1978). A Vidav theorem for Banach Jordan algebras. *Math. Proc. Cambridge Philos. Soc.*, 84, 263–272.

Youngson, M.A. (1979a). Equivalent norms on Banach Jordan algebras. *Math. Proc. Cambridge Philos. Soc*, 86, 261–269.

Youngson, M.A. (1979b). Hermitian operators on Banach Jordan algebras. *Proc. Edinburgh Math. Soc.*, 22, 93–104.

Youngson, M.A. (1981). Nonunital Banach Jordan algebras and C^*-triple systems. *Proc. Edinburgh Math. Soc.*, 24, 19–29.

Zelmanov, E.I. (1983). On prime Jordan algebras II. *Siberian Math. J.*, 24, 89–104.

Zelmanov, E.I. (1984). Primary Jordan triple systems II. *Sibirsk. Mat. Zh.*, 25, 50–61.

Zelmanov, E.I. (1985). On prime Jordan triple systems III. *Siberian Math. J.*, 26, 71–82.

Zhevlakov, K.A., Slin'ko, A.M., Shestakov, I.P., Shirshov, A.I. (1982). *Rings That are Nearly Associative*. (Translated by H.F. Smith). Academic Press, Inc. New York–London.

5

Structure of $H^\star$-algebras

José Antonio CUENCA MIRA
Department of Algebra, Geometry and Topology, Malaga University, Spain

5.1. Introduction

$H^\star$-algebras were studied in the associative case by W. Ambrose and I. Kaplansky. Ambrose was interested in the topic by a particular case of $H^\star$-algebra appearing in a previous paper by I. Segal. Ambrose and Kaplansky prove analogues of the Wedderburn theorems for this kind of algebra. They found new algebra in infinite dimension playing a role similar to that of the matrix algebras of the classical Wedderburn–Artin theory. Later diverse authors deal with different classes of non-associative $H^\star$-algebras, with infinite-dimensional algebras appearing frequently, recalling, in some sense, the simple algebras of the finite-dimensional theory. This has been so in the case of the more classical classes of non-associative algebras, such as the Lie, Jordan and non-commutative Jordan algebras, among others. Abundant literature deals with $H^\star$-algebras. On the one hand, the discovery of the role played by the so-called topologically simple $H^\star$-algebras as the building blocks in the construction of all $H^\star$-algebras has been the origin of a wide program of determination of the topologically simple $H^\star$-algebras belonging to all the relevant classes of non-associative algebras described by algebraic identities. On the other hand, similarities with other classes of non-associative normed algebras have suggested the study of different properties related with the general theory of $H^\star$-algebras. This chapter is devoted to explaining the fundamental tools and results of the theory of $H^\star$-algebras, with a special emphasis on the structure theory. We also include modifications and new proofs of some old results.

Algebra and Applications 1,
coordinated by Abdenacer MAKHLOUF. © ISTE Ltd 2020.

5.2. Preliminaries: aspects of the general theory

Let $\mathfrak{A}$ be an $\mathbb{K}$-algebra, where $\mathbb{K} = \mathbb{R}$ or $\mathbb{C}$. A map $\star : \mathfrak{A} \longrightarrow \mathfrak{A}$ $(x \mapsto x^\star)$ is said to be an *involution* of $\mathfrak{A}$ if it satisfies: (i) $(x+y)^\star = x^\star + y^\star$ (ii) $(\lambda x)^\star = \bar{\lambda} x^\star$; (iii) $(x^\star)^\star = x$, for any x, $y \in \mathfrak{A}$ and $\lambda \in \mathbb{K}$. Obviously, $\star$ is a linear map if $\mathbb{K} = \mathbb{R}$. If, in addition, $(xy)^\star = y^\star x^\star$ for all elements x, $y \in \mathfrak{A}$, we will say that $\star$ is a *multiplicative involution* of $\mathfrak{A}$. Let $\mathfrak{A}$ be a $\mathbb{K}$-algebra provided with an involution $\star$, whose vector space is a Hilbert space over $\mathbb{K}$ with inner product denoted by $(\ |\ )$. $\mathfrak{A}$ is said to be a *semi-$H^\star$-algebra* if

$$(xy|z) = (x|zy^\star) = (y|x^\star z) \qquad [5.1]$$

for any x, y, $z \in \mathfrak{A}$. Semi-$H^\star$-algebras whose involution $\star$ is multiplicative are named $H^\star$-algebras.

Trivial examples of real $H^\star$-algebras are as follows: (i) $\mathbb{R}$ with involution identity and inner product given by $(\lambda, \mu) \mapsto \lambda\mu$; (ii) $\mathbb{C}$ with the involution the complex conjugation and taking as inner product the unique real inner product making $\{1, \mathrm{i}\}$ an orthonormal basis. It is also possible to see $\mathbb{C}$ as a complex $H^\star$-algebra with the same involution and with complex inner product given by $(\lambda, \mu) \mapsto \lambda\bar{\mu}$.

There are easy ways to obtain new $H^\star$-algebras from a given $H^\star$-algebra. We now describe several of them.

EXAMPLE 5.1.– *Change of the inner product*: If $\mathfrak{A}$ is an $H^\star$-algebra and λ a strictly positive real number, then $\mathfrak{A}$ is also an $H^\star$-algebra with the same involution and with inner product $\lambda(\ |\ )$.

EXAMPLE 5.2.– *Symmetrized and antisymmmetrized $H^\star$-algebras*: If $\mathfrak{A}$ is an $H^\star$-algebra, then the symmetrized (respectively, antisymmetrized) algebra $\mathfrak{A}^+$ (respectively, $\mathfrak{A}^-$) of product $x{\bullet}y = \frac{1}{2}(xy + yx)$ (respectively, $[x, y] = xy - yx$) is also an $H^\star$-algebra with the same inner product and involution. Similar assertions hold in the case of semi-$H^\star$-algebras. This is also so in many of the constructions and results that follow, although we will not remark on this fact in order to avoid repetitions.

EXAMPLE 5.3.– *Complex $H^\star$-algebras are real $H^\star$-algebras*: Obviously every complex algebra $\mathfrak{A}$ can be viewed as a real algebra with only real numbers considered in the multiplication by scalars. This algebra is named the *realified* of $\mathfrak{A}$. If $\mathfrak{A}$ is a complex $H^\star$-algebra, then the realified of $\mathfrak{A}$ is an $H^\star$-algebra with the same involution and the inner product defined by $(x, y) \mapsto \frac{1}{2}((x|y) + (y|x))$, for any x, $y \in \mathfrak{A}$.

EXAMPLE 5.4.– *Complexified of a real $H^\star$-algebra*: Let $\mathfrak{A}$ be a real algebra and $\mathfrak{A}_{\mathbb{C}} = \mathbb{C} \otimes \mathfrak{A}$ its complexified. If $\mathfrak{A}$ is a real $H^\star$-algebra, then $\mathfrak{A}_{\mathbb{C}}$ becomes a complex $H^\star$-algebra with the involution and inner product characterized by the equations

$$(\lambda \otimes x)^\star = \bar{\lambda} \otimes x^\star, \qquad (\lambda \otimes x | \mu \otimes y) = \lambda\bar{\mu}(x|y).$$

Here $\star$ and $(\ |\ )$ denote the involution and inner product of $\mathfrak{A}$ (respectively, $\mathfrak{A}_{\mathbb{C}}$) in the right (respectively, left) side of the above equations.

Let K be a field of characteristic $\neq 2$ and A a unital K-algebra endowed with a non-degenerate quadratic form. A is said to be a *composition algebra* if $q(xy) = q(x)q(y)$ for any x, $y \in A$. If A is a composition algebra and $q(\ ,\ )$ denotes the bilinear form associated with q, then the map $s : A \longrightarrow A$ given by $s : x \mapsto q(x, 1)1 - x$ is an involutive antiautomorphism: that is, s is linear satisfying $s^2 = \mathrm{Id}$ and $s(xy) = s(y)s(x)$ for all x, $y \in A$. Moreover, $xs(x) = q(x)1 = s(x)x$ and $x + s(x) \in K1$. The map s is named the *Cayley antiautomorphism* of A (see (Zhevlakov *et al.* 1982, p. 26)). The so-called *Cayley–Dickson process* allows us to obtain a new composition algebra from a given composition algebra A. It works in the following way. Being given a composition algebra A and a non-zero element γ of K, we consider the vector space $A \oplus vA$, which is the direct sum of the two isomorphic K-vector spaces A and vA. In this K-vector space, we introduce the product

$$(a + vb)(c + vd) = (ac + \gamma ds(b)) + v(s(a)d + cb).$$

If q denotes the quadratic form of A, then $A \oplus vA$ is a new composition algebra, denoted by $\mathrm{CD}(A, \gamma)$, whose non-degenerate quadratic form is $a + vb \mapsto q(a) - \gamma q(b)$. Furthermore, the Cayley antiautomorphism of $A \oplus vA$ is the map $\hat{s}. : a + vb \mapsto s(a) - vb$, the algebra A can be viewed as a subalgebra of $\mathrm{CD}(A, \gamma)$, and $\hat{s}$ extends the antiautomorphism s (for details, see (Zhevlakov *et al.* 1982) or (Jacobson 1974, pp. 417–427)). Moreover, it is well known that composition algebras are obtained from the base field by repeated application, at most three times, of the Cayley–Dickson process. This is an immediate consequence of the generalized Hurwitz theorem (see (Zhevlakov *et al.* 1982)). Whether two algebras $\mathrm{CD}(A, \gamma)$ and $\mathrm{CD}(A, \gamma')$ are isomorphic is something that depends on the arithmetic of the base field. So if K is an algebraically closed field, there exist exactly four non-isomorphic composition algebras, one in each of the dimensions 1, 2, 4 and 8. If $\mathbb{K} = \mathbb{R}$, there exist seven composition algebras: (i) the base field $\mathbb{R}$; (ii) two two-dimensional algebras isomorphic to $\mathbb{C}$ and $\mathbb{R} \oplus \mathbb{R}$; (iii) two four-dimensional algebras (the division quaternions algebra $\mathbb{H}$ and the so-called algebra of split quaternions $\mathbb{H}_s$); (iv) two eight-dimensional algebras (the division octonions algebra $\mathbb{O}$ and the algebra of split octonions $\mathbb{O}_s$). Furthermore, in the case that $K = \mathbb{R}$ all these algebras can be obtained by repeated application of the Cayley–Dickson process with $\gamma = 1$ or -1. So, for example, $\mathbb{C} = \mathrm{CD}(\mathbb{R}, -1)$, $\mathbb{R} \oplus \mathbb{R} \cong \mathrm{CD}(\mathbb{R}, 1)$, $\mathbb{H} = \mathrm{CD}(\mathbb{C}, -1)$, $\mathbb{H}_s = \mathrm{CD}(\mathbb{C}, 1)$, $\mathbb{O} = \mathrm{CD}(\mathbb{H}, -1)$ and $\mathbb{O}_s = \mathrm{CD}(\mathbb{H}, 1)$.

EXAMPLE 5.5.– *Real composition $H^\star$-algebras:* Let D be a real composition algebra with Cayley antiautomorphism s, which is also a real $H^\star$-algebra. Let $\mathrm{CD}(D, \gamma)$ be the real algebra $D \oplus vD$ obtained from D by the Cayley–Dickson process with $v^2 = \gamma$ and $\gamma \in \{1, -1\}$. Then $\mathrm{CD}(D, \gamma)$ becomes an $H^\star$-algebra with respect to the involution and inner product given by $x^\star = (a + vb)^\star = a^\star + \gamma v s(b^\star)$ and $(a + vb|c + vd) = (a|c) + (b|d)$, for all elements $x = a + vb$, $y = c + vd \in \mathrm{CD}(D, \gamma)$. This gives that all the real composition algebras are $H^\star$-algebras.

EXAMPLE 5.6.– *Complex composition $H^\star$-algebras:* Let D be a complex composition algebra. There then exists a real division composition algebra D_0, which is uniquely determined up of isomorphisms, such that $D = D_0 \otimes_{\mathbb{R}} \mathbb{C}$. Let s_0 be the Cayley antiautomorphism of D_0. Then D becomes a complex $H^\star$-algebra relative to the involution and inner product characterized by the conditions

$$(x_0 \otimes \lambda)^\star = s_0(x_0) \otimes \bar{\lambda}, \qquad (x_0 \otimes \lambda \,|\, y_0 \otimes \mu) = \lambda\bar{\mu}(x_0|y_0), \tag{5.2}$$

for any $x_0, y_0 \in D_0$ and $\lambda, \mu \in \mathbb{C}$. Obviously D can be also viewed as a real $H^\star$-algebra with inner product given as in [5.3] from that of [5.2].

EXAMPLE 5.7.– *Unital $H^\star$-algebras coming from anticommutative $H^\star$-algebras*: Let W be an anticommutative $H^\star$-algebra with product $\wedge$ and whose involution $\square$ is isometric. Then the direct sum $A = \mathbb{K}e \oplus W$ of the one-dimensional vector space $\mathbb{K}e$ with the vector space W endowed with the product, inner product and involution defined as

$$(\alpha e + v)(\beta e + w) = (\alpha\beta + (v|w^\square))e + \alpha w + \beta v + v \wedge w, \tag{5.3}$$

$$(\alpha e + v|\beta e + w) = \alpha\bar{\beta} + (v|w), \tag{5.4}$$

$$(\alpha e + v)^\star = \bar{\alpha} e + v^\square, \tag{5.5}$$

is an $H^\star$-algebra over $\mathbb{K}$. Furthermore, e is a unit of A and $\|e\| = 1$. In particular, this allows us to show the existence of commutative $H^\star$-algebras with unit element in all dimension ≥ 1. Indeed, if W is a real or complex Hilbert space, then W can be viewed as an anticommutative algebra whose product $\wedge$ is the zero product. Let now $\square : W \longrightarrow W$ be an arbitrary isometric map, which is linear if $\mathbb{K} = \mathbb{R}$ and conjugate-linear if $\mathbb{K} = \mathbb{C}$. Then W becomes an anticommutative $H^\star$-algebra with isometric involution $\square$. In consequence, $A = \mathbb{K}e \oplus W$ with the product, inner product and involution defined by [5.3]–[5.5] becomes a commutative $H^\star$-algebra with unit e such that $\|e\| = 1$.

As another particular case, we can see that the real algebras of real and complex numbers $\mathbb{R}$, $\mathbb{C}$, as well as the real division algebras $\mathbb{H}$ and $\mathbb{O}$ of quaternions and octonions are $H^\star$-algebras. Let us illustrate this in the case that the given division algebra D is $\mathbb{H}$. If $\{1, i, j, k\}$ is a standard basis of $\mathbb{H}$ and $W = \mathbb{R}i + \mathbb{R}j + \mathbb{R}k$, then we make W an Euclidean space with the unique inner product making $\{i, j, k\}$ an orthonormal basis. With the involution $\square$ yielding every $v \in W$ to $-v$, and the anticommutative product $\wedge$ characterized by the multiplication rules $i \wedge j = k$, $i \wedge k = -j$, $j \wedge k = i$, W is an anticommutative $H^\star$-algebra. So $\mathbb{H} = \mathbb{R}1 \oplus W$ is an $H^\star$-algebra with inner product and involution [5.4] and [5.5] and whose product agrees with [5.3].

EXAMPLE 5.8.– *$H^\star$-algebras of Hilbert–Schmidt operators*: Let D be one of the composition division algebras $\mathbb{R}$, $\mathbb{C}$ or $\mathbb{H}$. Real and complex Hilbert spaces appear frequently in all the mathematics. Although less frequent, it is also possible to consider quaternionic Hilbert spaces. So if V is a left vector space over D ($D = \mathbb{R}$, $\mathbb{C}$, or $\mathbb{H}$), we will say that V is a *pre-Hilbert space* if V is endowed with a map $[\,|\,] : V \times V \longrightarrow D$ satisfying the following conditions:

$$[v + v' \mid w] = [v \mid w] + [v' \mid w], \qquad [\lambda v|w] = \lambda[v|w]$$

$$\overline{[v|w]} = [w|v], \qquad [v|v] > 0 \text{ if } v \neq 0.$$

for any v, $w, v' \in V$ and $\lambda \in D$ and where $^-$ denotes the Cayley antiautomorphism of D. It can be defined a norm $\|\ \|$ in V given by $\|v\|^2 = [v|v]$ for every $v \in V$. Incidentally, we observe that $\|\ \|$ is a D-norm in the sense that $\|\mu v\| = |\mu|\,\|v\|$, for all $\mu \in D$ and $v \in V$, and where $|\mu| = \sqrt{\mu\bar{\mu}}$. If the pre-Hilbert space V is complete with respect to $\|\ \|$, then we say that V is a *Hilbert space* over D. In a similar way, Hilbert spaces can be defined in the case of right vector spaces over D. However, we will refer here only to left Hilbert spaces unless expressly stated otherwise. Some typical concepts and results of the theory of real and complex Hilbert spaces can be translated verbatim to this more general framework including quaternionic Hilbert spaces. So in a given D-Hilbert space V, there are *complete orthonormal systems*, that is, pairwise orthogonal norm-one vectors with a dense D-linear envelope. Moreover, every two of these complete orthonormal systems have the same cardinal, which is called the *Hilbert dimension* of V.

If V and W are D-Hilbert spaces, then the set $BL(V, W)$ of the continuous D-linear maps is a right D-vector space. As in the classical setting of the real and complex Hilbert spaces, it is possible to define the adjoint $G^\square$ of the continuous D-linear map $G : V \longrightarrow W$ as the unique continuous D-linear map $G^\square : W \longrightarrow V$, satisfying $[G(v)|w] = [v|G^\square(w)]$ for all $v \in V$ and $w \in W$.

A *Hilbert–Schmidt operator* $G : V \longrightarrow W$ is a D-linear continuous map such that there exists a complete orthonormal system $\{v_\gamma\}_{\gamma\in\Gamma}$, satisfying $\sum_{\gamma\in\Gamma} \|G(v_\gamma)\|^2 < \infty$. It is easily checked that the above sum is independent of the chosen complete orthonormal system. It is also possible to define a map $[\,|\,]_{\mathcal{HS}} : \mathcal{HS}(V,W) \times \mathcal{HS}(V,W) \longrightarrow D$ by $[\,G_1|G_2]_{\mathcal{HS}} = \sum_{\gamma\in\Gamma}[G_1(v_\gamma)\,|\,G_2(v_\gamma)]$ for any $G_1,\ G_2 \in \mathcal{HS}(V,W)$. The right vector space $\mathcal{HS}(V,W)$ with the map $[\,|\,]_{\mathcal{HS}}$ becomes a right Hilbert space over D. So it is a normed space over the center C of D, and the norm of an arbitrary Hilbert–Schmidt operator G is given by $\|G\|_{\mathcal{HS}} = \sqrt{\sum_{\gamma\in\Gamma}\|G(v_\gamma)\|^2}$. Moreover, the adjoint of a Hilbert–Schmidt operator G is also a Hilbert–Schmidt operator, satisfying $\|G^\square\|_{\mathcal{HS}} = \|G\|_{\mathcal{HS}}$.

In the cases that $V = W$, we will write $BL(V)$ and $\mathcal{HS}(V)$ instead of $BL(V,V)$ and $\mathcal{HS}(V,V)$. Since composition of Hilbert–Schmidt operators are Hilbert–Schmidt operators, we can check that if V is a Hilbert space over D, then the algebra $\mathcal{HS}(V)$ becomes a real $H^\star$-algebra with respect to the involution $^\square$ and the inner product given by $(G_1|G_2)_{\mathcal{HS}} = \frac{1}{2}\sum_{\gamma\in\Gamma}\big([G_1(v_\gamma)\,|\,G_2(v_\gamma)] + \overline{[G_1(v_\gamma)\,|\,G_2(v_\gamma)]}\big)$. Furthermore, if $D = \mathbb{C}$, then $\mathcal{HS}(V)$ is also a complex $H^\star$-algebra with the inner product $(G_1|G_2)_{\mathcal{HS}} = \sum_{\gamma\in\Gamma}\big([G_1(v_\gamma)\,|\,G_2(v_\gamma)]$.

The remainder paragraphs of this section are devoted to establish some properties of the general theory of $H^\star$-algebras. The common feature of all of them is their independence from any kind of algebraic identity. Nevertheless, very frequently all these properties have been first obtained under additional restrictions of this type.

If $\mathfrak{A}$ is a K-algebra and $x \in \mathfrak{A}$, we will denote by L_x and R_x to the linear operators $L_x;\ R_x : \mathfrak{A} \longrightarrow \mathfrak{A}$ given by $L_x : y \mapsto xy$ and $R_x : y \mapsto yx$. They are named *multiplication operators* of the element x. If $\mathfrak{A}$ is a semi-$H^\star$-algebra, then from [5.1] it follows that the adjoint of each L_x (respectively, R_x) is $L_{x^\star}$ (respectively, $R_{x^\star}$), which implies the continuity of these multiplication operators (see (Rudin 1973, Theorem 5.1)). As in (Cuenca and Rodríguez 1987, Proof of Proposition 2), taking in account the Banach–Steinhaus theorem we obtain:

PROPOSITION 5.1.– (Cuenca and Rodríguez 1987, Proposition 2) . The product of each semi-$H^\star$-algebra $\mathfrak{A}$ is continuous.

If $\mathfrak{A}$ is an $H^\star$-algebra, then the matrix algebra $\mathcal{M}_n(\mathfrak{A})$ of the $n \times n$-matrices with entries in $\mathfrak{A}$ is also an $H^\star$-algebra. This is generalized in our next example when the considered set $\mathcal{A}$ is infinite and the involution is continuous.

EXAMPLE 5.9.– *Matrix $H^\star$-algebras*: Let $\mathfrak{A}$ be an semi-$H^\star$-algebra over $\mathbb{K}$ ($\mathbb{K} = \mathbb{R}$ or $\mathbb{C}$), with continuous involution $\square$ and inner product $(\ ,\ )$. Let $\mathcal{A}$ be a non-empty set and $\mathcal{M}_{\mathcal{A}}(\mathfrak{A})$ the $\mathbb{K}$-vector space of the matrices (a_{ij}) $(i,\ j \in \mathcal{A},\ a_{ij} \in \mathfrak{A})$ such that $\sum_{i,j}\|a_{i,j}\|^2 < \infty$. Then $(a_{ij})(b_{ij}) = (c_{ij})$ with $c_{ij} = \sum_k a_{ik}b_{kj}$ is a well-defined product in $\mathcal{M}_{\mathcal{A}}(\mathfrak{A})$. With the inner product $(\,(a_{ij})|(b_{ij})\,) = \sum_{i,j}(a_{ij}|b_{ij})$ and involution $(a_{ij})^\star = (a_{ji}^\square)$, $\mathcal{M}_{\mathcal{A}}(\mathfrak{A})$ becomes an semi-$H^\star$-algebra, which is an $H^\star$-algebra if so is $\mathfrak{A}$.

Let V be a D-Hilbert space ($D = \mathbb{R}, \mathbb{C}$, or $\mathbb{H}$) where a complete orthonormal system has been chosen. It is not difficult to see that the $H^\star$-algebra $\mathcal{HS}(V)$ of the Hilbert–Schmidt operators can be identified in an obvious way with a matrix $H^\star$-algebra $\mathcal{M}_\Gamma(D)$, where Γ is a set with cardinal the Hilbert dimension of V.

In the same way that in the case of associative algebras, we can introduce the usual concepts of ideal, subalgebra, homomorphism, endomorphism, isomorphism, etc. Also the usual notations for operations with ideals will be used throughout this chapter. So if I and J are ideals of the K-algebra $\mathfrak{A}$, then by IJ we will denote the ideal of $\mathfrak{A}$ generated by the subset $\{xy : x \in I \text{ and } y \in J\}$. In the case that $I = J$, we will denote by I^2 to this ideal. If $\mathfrak{A}$ is a K-algebra and S a non-empty subset of $\mathfrak{A}$, we define the *left annihilator* $\operatorname{Lann} S$ and the *right annihilator* $\operatorname{Rann} S$ by $\operatorname{Lann} S = \{x \in \mathfrak{A} : \ xS = 0\}$ and $\operatorname{Rann} S = \{x \in \mathfrak{A} : \ Sx = 0\}$. So $\operatorname{Lann} S$ and $\operatorname{Rann} S$ are vector subspaces of $\mathfrak{A}$. The subspace $\operatorname{Ann} S$ defined as $\operatorname{Ann} S = \operatorname{Lann} S \cap \operatorname{Rann} S$ is named the *annihilator* of S. If $\mathfrak{A}$ is a semi-$H^\star$-algebra, then the subspaces $\operatorname{Lann} S$ and $\operatorname{Rann} S$ are closed, as well as $\operatorname{Ann} S$. A subset S of the semi-$H^\star$-algebra $\mathfrak{A}$ is named *self-adjoint* if $S^\star \subset S$ (so also $S^\star = S$). A self-adjoint closed subalgebra S of the semi-$H^\star$-algebra (respectively, $H^\star$-algebra) $\mathfrak{A}$ becomes a semi-$H^\star$-algebra (respectively, $H^\star$-algebra) with the inner product and involution inherited by those of $\mathfrak{A}$. We will say that S is a *semi-$H^\star$-subalgebra* (respectively, *$H^\star$-subalgebra*) of $\mathfrak{A}$. Since the general theory of $H^\star$-algebras has been carefully treated in the books by Cabrera and Rodríguez (2018), throughout the remainder of this section, in addition to the original papers, we will also include the appropriate reference (Cabrera and Rodríguez 2014, 2018), where complete proofs can be found.

PROPOSITION 5.2.– (Cuenca and Rodríguez 1987, Proposition 2; Cuenca and Sánchez 1994, Theorem 1; Cabrera and Rodríguez 2014, 2018, Theorem 8.1.10, Corollary 8.1.12 and Proposition 8.1.13). Let $\mathfrak{A}$ be a semi-$H^\star$-algebra. Then the following assertions holds:

1) $\operatorname{Lann} \mathfrak{A} = \operatorname{Rann} \mathfrak{A} = \operatorname{Ann} \mathfrak{A}$ and $\overline{A^2}^{\perp} = \operatorname{Ann} \mathfrak{A}$;

2) if $\operatorname{Ann} \mathfrak{A} = 0$, then $\mathfrak{A}$ has continuous involution, which is isometric if and only if $\mathfrak{A}$ is an $H^\star$-algebra;

3) if $\operatorname{Ann} \mathfrak{A} = 0$ and I is a closed ideal of $\mathfrak{A}$, then $I^\star = I$ and I is a semi-$H^\star$-subalgebra of $\mathfrak{A}$ with zero annihilator in itself. Furthermore, $\mathfrak{A}$ splits as an orthogonal direct sum of closed ideals $\mathfrak{A} = I \oplus I^\perp$ and $\overline{I^2} = I$.

We recall that a K-algebra $\mathfrak{A}$ is called a *semiprime algebra* if the unique ideal of $\mathfrak{A}$ with zero square is the zero ideal. As a result of proposition 5.2, we see that non-zero $H^\star$-algebras with zero annihilator are semiprime algebras.

PROPOSITION 5.3.– (Cuenca and Rodríguez 1987, Proposition 2; Cuenca and Sánchez 1994, Theorem 1; Cabrera and Rodríguez 2014, 2018, Theorem 8.1.10 and Corollary 8.1.12). Let $\mathfrak{A}$ be a semi-$H^\star$-algebra and $B = (\operatorname{Ann}\mathfrak{A})^\perp$. Then:

1) the $\star$-invariant closed ideal $\operatorname{Ann}\mathfrak{A}$ is a semi-$H^\star$-subalgebra of $\mathfrak{A}$, which is an $H^\star$-algebra with zero product;

2) the ideal B is a closed ideal of $\mathfrak{A}$, which is a semi-$H^\star$-algebra with the inner product inherited by that of $\mathfrak{A}$ and the involution $\sharp$ defined as $b^\sharp = \pi(b^\star)$, where π denotes the orthogonal projection of $\mathfrak{A}$ onto B;

3) if $\mathfrak{A}$ is an $H^\star$-algebra with continuous involution, then B is self-adjoint, and so an $H^\star$-subalgebra of $\mathfrak{A}$.

A first step in the reduction of the general theory of $H^\star$-algebras to the case of the $H^\star$-algebras with zero annihilator is provided by the next proposition, which is a result of proposition 5.3. A later reduction to the topologically simple case is given in theorem 5.1. Both results holds for semi-$H^\star$-algebras.

PROPOSITION 5.4.– Each semi-$H^\star$-algebra $\mathfrak{A}$ splits as an orthogonal direct sum of two closed ideals I_1 and I_2, which are semi-$H^\star$-algebras, with I_1 of zero product and I_2 with zero annihilator.

A closed ideal I of a semi-$H^\star$-algebra $\mathfrak{A}$ is said to be a *minimal closed ideal* of $\mathfrak{A}$ if I is minimal with respect to the inclusion in the set of the non-zero closed ideals of $\mathfrak{A}$. On the other hand, we will say that a semi-$H^\star$-algebra $\mathfrak{A}$ is *topologically simple* if $\mathfrak{A}^2 \neq 0$ and $\mathfrak{A}$ does not contain closed ideals different from 0 and $\mathfrak{A}$. We recall that an algebra $\mathfrak{A}$ is a *prime* algebra if $IJ = 0$ implies $I = 0$ or $J = 0$. Obviously every topologically simple $H^\star$-algebra is prime.

THEOREM 5.1.– (Cuenca and Rodríguez 1987, Theorem 1; Cuenca and Sánchez 1994, Theorem 1; Cabrera and Rodríguez 2014, 2018, Theorem 8.1.16). Let $\mathfrak{A} \neq 0$ be an $H^\star$-algebra. Then $\operatorname{Ann}\mathfrak{A} = 0$ iff $\mathfrak{A}$ is the closure of the orthogonal direct sum of its minimal closed ideals. Moreover, all of them are topologically simple $H^\star$-subalgebras of $\mathfrak{A}$.

Theorem 5.1, whose proof can be also found in Cabrera and Rodríguez (2014, 2018), reduces finally the structure theory of $H^\star$-algebras to the topologically simple case, because, jointly with proposition 5.4, this theorem allows us to see the topologically simple $H^\star$-algebras as the building blocks with which all the $H^\star$ algebras can be obtained.

The general theory of $H^\star$-algebras includes several interesting results on automatic continuity, as well as a careful study of isomorphisms between $H^\star$-algebras with zero annihilator. We first recall that if $\mathfrak{A}$ is a K-algebra and

$D : \mathfrak{A} \longrightarrow \mathfrak{A}$ is a K-linear map, then D is said to be a *derivation* if $D(xy) = D(x)y + xD(y)$ for all $x, y \in \mathfrak{A}$.

THEOREM 5.2.– (Villena 1994, Theorem 4; Cabrera and Rodríguez 2014, 2018, Theorem 8.1.41 and Corollary 8.1.83). Let $\mathfrak{A}$ be a semi-$H^\star$-algebra with zero annihilator. If $D : \mathfrak{A} \longrightarrow \mathfrak{A}$ is a derivation, then D is continuous.

THEOREM 5.3.– (Rodríguez 1995; Cabrera and Rodríguez 2014, 2018, Theorem 8.1.52 and Corollary 8.1.83). Let $\mathfrak{A}$ be a complete normed algebra, $\mathfrak{B}$ an $H^\star$-algebra with zero annihilator. If $G : \mathfrak{A} \longrightarrow \mathfrak{B}$ a dense-range homomorphism, then G is continuous.

THEOREM 5.4.– (Cuenca and Rodríguez 1985; Cabrera and Rodríguez 2014, 2018, Theorem 8.1.53 and Corollary 8.1.84). Let $\mathfrak{A}$ be a complete normed algebra and $\mathfrak{B}$ a semi-$H^\star$-algebra with zero annihilator. If $G : \mathfrak{A} \longrightarrow \mathfrak{B}$ is an algebra homomorphism with range $\star$-invariant and dense in $\mathfrak{B}$, then G is continuous.

Let $\mathfrak{A}$ be a K-algebra and $T : \mathfrak{A} \longrightarrow \mathfrak{A}$ a linear map. If $xT(y) = T(xy) = T(x)y$ for any $x, y \in \mathfrak{A}$, then T is named a *centralizer* of $\mathfrak{A}$. The set of the centralizers $Z(\mathfrak{A})$ is a subalgebra of the K-endomorphisms $\mathrm{End}_K \mathfrak{A}$ of the vector space $\mathfrak{A}$, which is named the *centroid* of $\mathfrak{A}$. Obviously, $K\,\mathrm{Id} \subset Z(\mathfrak{A})$. Moreover, if $\mathfrak{A}$ is a complete normed algebra, then the centralizers of $\mathfrak{A}$ are continuous maps (see (Cabrera and Rodríguez 2014, 2018, vol. 1, p. 4)). The next proposition shows that if $\mathfrak{A}$ is a topologically simple $H^\star$-algebra, then $Z(\mathfrak{A})$ is a field that determines in an important measure the structure of $\mathfrak{A}$.

PROPOSITION 5.5.– (Cuenca and Rodríguez 1985, Theorem 1; Cabrera and Rodríguez 2014, 2018, Lemma 8.1.29 and Corollary 8.1.89). Let $\mathfrak{A}$ be a topologically simple semi-$H^\star$-algebra over $\mathbb{K}$. Then we have the following possibilities for $Z(\mathfrak{A})$:

1) $Z(\mathfrak{A}) = \mathbb{C}\,\mathrm{Id}$ if $\mathbb{K} = \mathbb{C}$;

2) $Z(\mathfrak{A}) = \mathbb{R}\,\mathrm{Id}$ if $\mathbb{K} = \mathbb{R}$ and $\mathfrak{A}$ is not the realification of a complex semi-$H^\star$-algebra;

3) $Z(\mathfrak{A}) \cong \mathbb{C}$ if $\mathbb{K} = \mathbb{R}$ and $\mathfrak{A}$ is the realification of a complex semi-$H^\star$-algebra.

Let $G : \mathfrak{A} \longrightarrow \mathfrak{B}$ be a given continuous linear map between two $H^\star$-algebras over the same field. In example 5.8, we will denote by $G^\square : \mathfrak{B} \longrightarrow \mathfrak{A}$ to the adjoint map. It is well known that $G^\square$ satisfies $(G(x)|y) = (x|G^\square(y))$ for any $x \in \mathfrak{A}$ and $y \in \mathfrak{B}$. Furthermore, we denote by $G^\star$ the map from $\mathfrak{A}$ into $\mathfrak{B}$ given by $G^\star : x \mapsto G(x^\star)^\star$.

THEOREM 5.5.– (Cuenca and Rodríguez 1985, Theorem 3.1; Cabrera and Rodríguez 2014, 2018, Theorem 8.1.62). Let $\mathfrak{A}$ and $\mathfrak{B}$ be $H^\star$-algebras such that $\mathfrak{B}$ is topologically simple and $\mathfrak{A}$ has zero annihilator. If $G : \mathfrak{A} \longrightarrow \mathfrak{B}$ is a dense-range homomorphism, then there exists a positive real number λ such that $G^{\square} \circ G^{\star} = \lambda \operatorname{Id}_{\mathfrak{B}}$

THEOREM 5.6.– (Cuenca and Rodríguez 1985, Theorem 3.3 and Corollary 2.3; Cabrera and Rodríguez 2014, 2018, Theorem 8.1.64 and 8.1.92). Let $G : \mathfrak{A} \longrightarrow \mathfrak{B}$ be a bijective algebra homomorphism between $H^{\star}$-algebras with zero annihilator. Then G can be written in a unique way as $G = G_2 \circ G_1$ with $G_2 : \mathfrak{A} \longrightarrow \mathfrak{B}$ an algebra $\star$-isomorphism and $G_1 = \exp D$ with D a derivation of $\mathfrak{A}$ such that $D^{\square} = -D$.

COROLLARY 5.1.– If two $H^{\star}$-algebras with zero annihilator are isomorphic, then they are also $\star$-isomorphic.

THEOREM 5.7.– (Cuenca and Rodríguez 1985, Corollary 3.4; Cabrera and Rodríguez 2014, 2018, Corollary 8.1.65 and Theorem 8.1.95). Let $\mathfrak{A}$ and $\mathfrak{B}$ be topologically simple $H^{\star}$-algebras. Then there exists a positive real number ρ (with $\rho = 1$, if $\mathfrak{A} = \mathfrak{B}$) such that $G^{\square} = \rho(G^{\star})^{-1}$ for all bijective algebra homomorphisms $G : \mathfrak{A} \longrightarrow \mathfrak{B}$.

COROLLARY 5.2.– (Cuenca and Rodríguez 1985, Corollary 3.4; Cabrera and Rodríguez 2014, 2018, Corollaries 8.1.67 and 8.1.97). Let $\mathfrak{A}$ and $\mathfrak{B}$ be topologically simple $H^{\star}$-algebras. Then there exists a positive real number ρ such that ρG is an isometric map for every bijective $\star$-isomorphism $G : \mathfrak{A} \longrightarrow \mathfrak{B}$.

5.3. Ultraproducts of $H^{\star}$-algebras

In general, the ultraproduct of a family of algebras underlying to $H^{\star}$-algebras are not $H^{\star}$-algebras in a natural way. Nevertheless, under certain restrictions, it is possible to construct an $H^{\star}$-algebra living in this ultraproduct and containing a copy of every one of the $H^{\star}$-algebras of a given family. This construction works in the frame of the general theory of $H^{\star}$-algebras and we will see its utility in the determination of the topologically simple Lie $H^{\star}$- algebras, where this kind of construction was first introduced (Cuenca *et al.* 1990).

Let $\mathcal{M} \neq \emptyset$ be a partially ordered set whose partial order $\leq$ makes $\mathcal{M}$ a directed set. Let $\mathbb{K}$ be $\mathbb{R}$ or $\mathbb{C}$ and $(\{\mathfrak{A}_i\}_{i \in \mathcal{M}}, \{\nu_{ji}\}_{i \leq j})$ a pair where each $\mathfrak{A}_i$ is an $H^{\star}$-algebra over $\mathbb{K}$ and each ν_{ji} $(i \leq j)$ is an isometric $\star$-monomorphism $\nu_{ji} : \mathfrak{A}_i \longrightarrow \mathfrak{A}_j$. We will say that $(\{\mathfrak{A}_i\}_{\in \mathcal{M}}, \{\nu_{ji}\}_{i \leq j})$ is a *directed isometric $\star$-monomorphic system* of $H^{\star}$-algebras if $\nu_{ki} = \nu_{kj} \circ \nu_{ji}$ for all i, j, $k \in \mathcal{M}$ such that $i \leq j \leq k$. For each $i \in \mathcal{M}$, we denote by $[i, \infty)$ the subset defined by $[i, \infty) = \{j \in \mathcal{M} : i \leq j\}$. Since the family of subsets $\{[i, \infty) : i \in \mathcal{M}\}$ has the finite intersection property, there exists some ultrafilter $\mathcal{U}$ on $\mathcal{M}$ containing all the subsets $[i, \infty)$. Now we define an equivalence relation $\cong$ in the direct product $\prod_i \mathfrak{A}_i$ in the following way: $(x_i) \cong (x_i')$ iff $\{i \in \mathcal{M} : x_i = x_i'\} \in \mathcal{U}$. We will denote by $[(x_i)]$ the equivalence class of (x_i) and by $(\prod_i \mathfrak{A}_i)/\mathcal{U}$ the quotient set of the equivalence relation $\cong$. In $(\prod_i \mathfrak{A}_i)/\mathcal{U}$, it can be defined a sum, product, product by scalars and involution in the following way:

$$[(x_i)] + [(y_i)] = [(x_i + y_i)], \qquad [(x_i)]\,[(y_i)] = [(x_i y_i)]$$
$$\lambda[(x_i)] = [(\lambda x_i)], \qquad [(x_i)]^\star = [(x_i^\star)].$$

So $(\prod_i \mathfrak{A}_i)/\mathcal{U}$ becomes a $\mathbb{K}$-algebra with involution. This is a particular case of ultraproducts of algebras with involution (see (Eklof 1999)). The subalgebra with involution $\mathfrak{A}_{\mathcal{U}}$ of $(\prod_i \mathfrak{A}_i)/\mathcal{U}$ whose elements are the classes $[(x_i)]$ for which there exist some $i_0 \in \mathcal{M}$ and $x_{i_0} \in \mathfrak{A}_{i_0}$ satisfying $\{i \in [i_0, \infty) : x_i = \nu_{ii_0}(x_{i_0})\} \in \mathcal{U}$ becomes a pre-Hilbert space with the inner product defined by $([(x_i)] \,|\, [(y_i)]) = (x_{i_0} \,|\, y_{i_0})$, where $i_0 \in \mathcal{M}$ is chosen satisfying that the subset $\{i \in [i_0, \infty) : x_i = \nu_{ii_0}(x_{i_0})$ and $y_i = \nu_{ii_0}(y_{i_0})\}$ belongs to $\mathcal{U}$. Moreover, $([(x_i)]\,[(y_i)] \mid [(z_i)]\,) = ([(x_i)] \mid [(z_i)]\,[(y_i)]^\star\,) = ([(y_i)] \mid [(x_i)]^\star\,[(z_i)]\,)$. So $\mathfrak{A}_{\mathcal{U}}$ has now a structure very close to that of $H^\star$-algebra. Here, the only thing that is missing with respect to an $H^\star$-algebra is the possible completeness of the inner product. There are two different ways to avoid this difficulty. The first is to consider the completion of $\mathfrak{A}_{\mathcal{U}}$, which is an $H^\star$-algebra if the involutions of the $\mathfrak{A}_i$ are continuous. But there exists also a property on the directed set $\mathcal{M}$ implying the completeness of the above defined inner product. We will name this property as the countable directed property of the given directed set $\mathcal{M}$. So a directed set $\mathcal{M}$ is said to be *countable directed* if for each sequence $i_1, \ldots, i_n, \ldots$ of elements of $\mathcal{M}$ there exists some $\gamma \in \mathcal{M}$ such that $i_k \leq \gamma$ for all $k = 1, \ldots, n, \ldots$. Also we must point out that the algebra $\mathfrak{A}_{\mathcal{U}}$ has a good behavior in front of topological simplicity.

PROPOSITION 5.6.– (Cuenca *et al.* 1990, Proposition 1).

Let $(\mathcal{M}, \leq)$ be a countable directed set, $\mathcal{U}$ an ultrafilter on $\mathcal{M}$ containing all the subsets $[i, \infty)$ and $(\{\mathfrak{A}_i\}_{i \in \mathcal{M}}, \{\nu_{ji}\}_{i \leq j})$ a directed isometric $\star$-monomorphic system of $H^\star$-algebras. If $\mathfrak{A}_{\mathcal{U}}$ is the algebra with involution and inner product constructed above, then $\mathfrak{A}_{\mathcal{U}}$ is an $H^\star$-algebra and this $H^\star$-algebra is topologically simple if each $\mathfrak{A}_i$ is topologically simple.

It is also easy to prove the following proposition.

PROPOSITION 5.7.– Let $(\mathcal{M}, \leq)$ a countable directed set, $(\{\mathfrak{A}_i\}_{i \in \mathcal{M}}, \{\nu_{ji}\}_{i \leq j})$ a directed isometric $\star$-monomorphic system of $H^\star$-algebras and $\mathcal{U}$ an ultrafilter containing the subsets $[i, \infty)$. Then the following assertions hold:

1) if $(\{\mathfrak{B}_i\}_{i \in \mathcal{M}}, \{\nu'_{ji}\}_{i \leq j})$ is also a directed isometric $\star$-monomorphic system of $H^\star$-algebras and we suppose given for each $i \in \mathcal{M}$ an isometric $\star$-homomorphism $\eta_i : \mathfrak{A}_i \longrightarrow \mathfrak{B}_i$ such that ${\nu_{ji}}' \circ \eta_i = \eta_j \circ \nu_{ji}$ for every $j \geq i$, then the map $\eta : \mathfrak{A}_{\mathcal{U}} \longrightarrow \mathfrak{B}_{\mathcal{U}}$ given by $\eta : [(x_i)] \mapsto [(\eta_i(x_i))]$ is an isometric $\star$-homomorphism. Furthermore, if the η_i are $\star$-isomorphisms, then η is also a $\star$-isomorphism;

2) let $\mathcal{N}$ be a non-empty subset of $\mathcal{M}$ such that $\mathcal{N} \in \mathcal{U}$. Then with the restriction of $\leq$ the set $\mathcal{N}$ is countable directed and, in addition, the family of subsets $\{\mathcal{N} \cap U :$

$U \in \mathcal{M}\}$ is an ultrafilter on $\mathcal{N}$ containing all the subsets $[i, \infty) \cap \mathcal{N}$. Furthermore, if we denote this ultrafilter by $\mathcal{N}/\mathcal{U}$, then $\mathfrak{A}_{\mathcal{N}/\mathcal{U}}$ and $\mathfrak{A}_{\mathcal{U}}$ are isometrically $\star$-isomorphic;

3) if $\mathcal{M}_1, \ldots, \mathcal{M}_s$ are subsets of $\mathcal{M}$ such that $\mathcal{M} = \bigcup_{\ell=1}^{s} \mathcal{M}_\ell$, then there exists some ℓ_0 such that $\mathcal{M}_{\ell_0} \in \mathcal{U}$. In particular, $\mathcal{M}_{\ell_0}$ is countable directed with the inherited order.

Let $(\mathcal{M}, \leq)$ be a directed (respectively, countable directed) set and $(\{\mathfrak{A}_i\}_{i\in\mathcal{M}}, \{\mu_{ji}\}_{i\leq j})$ a directed isometric $\star$-monomorphic system of $H^\star$-algebras, where each $\mathfrak{A}_i$ is an $H^\star$-subalgebra of a given $H^\star$-algebra $\mathfrak{A}$ such that $\mathfrak{A}_i \subset \mathfrak{A}_j$ for all $j \geq i$ and where the monomorphisms μ_{ji} $(i \leq j)$ are the corresponding isometric $\star$-monomorphisms of inclusion. Then we say that $(\{\mathfrak{A}_i\}_{i\in\mathcal{M}}, \{\mu_{ji}\}_{i\leq j})$ is a *net* (respectively, *countable directed net*) of $H^\star$-subalgebras of $\mathfrak{A}$. If, in addition, $\mathfrak{A} = \bigcup_{i\in\mathcal{M}} \mathfrak{A}_i$, we will say that $(\{\mathfrak{A}_i\}_{i\in\mathcal{M}}, \{\mu_{ji}\}_{i\leq j})$ *covers* $\mathfrak{A}$.

PROPOSITION 5.8.– Let $(\{\mathfrak{A}\}_{i\in\mathcal{M}}), \{\mu_{ji}\})$ be a countable directed net of $H^\star$-subalgebras covering the $H^\star$-algebra $\mathfrak{A}$ and $\mathcal{U}$ an ultrafilter on $\mathcal{M}$ containing the subsets $[i, \infty)$. Then the following assertions hold:

1) $\mathfrak{A}_{\mathcal{U}}$ is isometrically $\star$-isomorphic to $\mathfrak{A}$;

2) if $\mathcal{N}$ is a non-empty subset of $\mathcal{M}$ such that $\mathcal{N} \in \mathcal{U}$, then $(\{\mathfrak{A}\}_{i\in\mathcal{N}}, \{\mu_{ji}\})$ is a countable directed net of $H^\star$-subalgebras covering $\mathfrak{A}$;

3) if $\mathcal{M}_1, \ldots, \mathcal{M}_s$ are subsets of $\mathcal{M}$ such that $\mathcal{M} = \bigcup_{\ell=1}^{s} \mathcal{M}_\ell$, then there exists some $\mathcal{M}_{\ell_0} \in \mathcal{U}$ such that $(\{\mathfrak{A}\}_{i\in\mathcal{M}_{\ell_0}}, \{\mu_{ji}\})$ is a countable directed net of $H^\star$-subalgebras covering $\mathfrak{A}$.

5.4. Quadratic $H^\star$-algebras

We recall that a K-algebra A with unit e is a *quadratic* K-algebra if for all $x \in A$ there exist λ_x, $\mu_x \in K$ such that $x^2 + \lambda_x x + \mu_x e = 0$. We can see that if $\mathbb{K} = \mathbb{R}$ or $\mathbb{C}$, then the $H^\star$-algebras in example 5.7 are quadratic $\mathbb{K}$-algebras. The next proposition asserts: (i) each quadratic $\mathbb{K}$-algebra A, which is an $H^\star$-algebra, can be endowed with an $H^\star$-algebra structure such as those in example 5.7; and (ii) with only one exception, which happens in the two-dimensional case, all the $H^\star$-structures in A agree up to a positive factor of the inner product.

PROPOSITION 5.9.– Let A be a quadratic $\mathbb{K}$-algebra with unit e, which is an $H^\star$-algebra over $\mathbb{K}$. Let $W = \{v \in A : \; v^2 \in \mathbb{K}e - \{0\}$ and $v \notin \mathbb{K} - \{0\}\}$. Then the following assertions hold:

1) W is a vector subspace of A such that $W^\star \subset W$, which is an anticommutative $H^\star$-algebra with respect to the product $\wedge$ defined by $v \wedge w = \frac{1}{2}(vw - wv)$, and the inner product and involution $\Box$ inherited by those of A;

2) $A = \mathbb{K}e \oplus W$ with the product [5.3], the inner product [5.4] and involution [5.5] is an $H^\star$-algebra. Furthermore, if A is not isomorphic to $\mathbb{K} \oplus \mathbb{K}$, then A is simple and the product and involution of A are given by [5.3] and [5.5] and its inner product is obtained from [5.4] by multiplying by $||e||^2$.

5.5. Associative $H^\star$-algebras

The structure theory of $H^\star$-algebras begins with (Ambrose 1945) in the complex associative case. He proposed proposition 5.4 and theorem 5.1 for complex associative $H^\star$-algebras, determining all the topologically simple complex associative $H^\star$-algebras. The associative real case was treated by (Kaplansky 1948) in the context of the dual rings. He determined all the topologically simple associative real $H^\star$-algebras (Kaplansky 1948, Theorem 13). This result has been also proved several times using different methods (Balachandran and Swaminathan 1986; Cabrera *et al.* 1988b; Castellón and Cuenca 1992b; Sánchez 1989). The principal aim of this section is to prove the Ambrose–Kaplansky theorem of characterization of the topologically simple associative $H^\star$-algebras.

We must point out that, in the associative frame, we do not need to consider semi-$H^\star$-algebras, because associative semi-$H^\star$-algebras with zero annihilator are $H^\star$-algebras (Cabrera and Rodríguez 2014, 2018, Vol. 1, pp. 492–493).

We recall that an element z of an associative $\mathbb{K}$-algebra A is said to be *right quasi-regular* (respectively, left quasi-regular) if $\mathrm{Id} - L_z : A \longrightarrow A$ (respectively, $\mathrm{Id} - R_z$) is a surjective map. We say that a right (respectively, left) ideal of A is *quasi-regular* if all its elements are right (respectively, left) quasi-regular. The existence of an ideal of A whose elements are right quasi-regular containing all quasi-regular right ideals of A is well known. This ideal is named the *Jacobson radical* of A. We will denote it as $\mathrm{Rad}_J\, A$, (see, for example, (Jacobson 1968b, p. 9; Kaplansky 1972, Theorems 8 and 10). Moreover, $\mathrm{Rad}_J\, A$ is also an ideal with all its elements left quasi-regular and containing all the quasi-regular left ideals of A (Jacobson 1968b, p. 9; Kaplansky 1972, Theorems 8 and 10).

The Jacobson radical of an associative algebra is closely related to sided modular ideals. First, we recall that a right ideal I is said to be a *modular* right ideal if there exists an element u of A, which is called a *left unit modulo* I, such that $ux - x \in I$ for every $x \in A$. Modular left ideals are defined in a similar way.

WELL-KNOWN FACTS 5.1.– (Jacobson 1968b; Kaplansky 1972; Herstein 1976) Let A be an associative algebra. Then the following assertions hold:

1) each proper modular right (respectively, left) ideal can be included in a modular maximal right (respectively, left) ideal;

2) $\mathrm{Rad}_J\, A$ is the intersection of all the modular maximal right ideals of A, which also agrees with the intersection of all its modular maximal left ideals, understanding that $\mathrm{Rad}_J\, A = A$ if there are no such ideals;

3) if I is a minimal right (respectively, left) ideal of A such that $I^2 \neq 0$, then there exists an idempotent f in A such that $I = fA$ (respectively, $I = Af$);

4) if $f \neq 0$ is an idempotent of A and $I = fA$ is a minimal right ideal or $J = Af$ is a minimal left ideal of A, then fAf is a division algebra. Furthermore, if A has non-zero nilpotent sided ideals and f is an idempotent such that fAf is a division algebra, then fA (respectively, Af) is a minimal right (respectively, left) ideal.

In the next lemma, we recall some properties of the general theory of $H^\star$-algebras relative to the closed sided ideals of the $H^\star$-algebras of zero annihilator.

LEMMA 5.1.– Let $\mathfrak{A}$ be an $H^\star$-algebra with zero annihilator, I a right ideal of $\mathfrak{A}$ and J a left ideal of $\mathfrak{A}$. Then the following assertions hold:

1) $I^\star$ (respectively, $J^\star$) is a left (respectively, right) ideal and $I^\perp$ (resp $J^\perp$) is a closed right (respectively, left) ideal of $\mathfrak{A}$.

2) If $x \in \mathfrak{A}$ and I and J are closed then $x\mathfrak{A} \subset I \Longleftrightarrow x \in I$ and $\mathfrak{A}x \subset J \Longleftrightarrow x \in J$.

3) If I and J are closed then $\mathrm{Lann}\, I = (I^\star)^\perp$ and $\mathrm{Rann}\, J = (J^\star)^\perp$.

PROOF.–

1) Follows from a straightforward verification;

2) It is sufficient to observe that if $y \in \mathfrak{A}$, $x\mathfrak{A} \subset I$ and $x = x_1 + x_2$, with $x_1 \in I$ and $x_2 \in I^\perp$, then $x_2 y \in I \cap I^\perp = \{0\}$. In consequence, $x_2 = 0$ and $x \in I$;

3) Since $I^\star$ and $I^{\star^\perp}$ are closed left ideals, $\overline{\mathfrak{A}I^\star} \subset I^\star$ and $\overline{\mathfrak{A}I^{\star\perp}} \subset I^{\star^\perp}$. So $\mathfrak{A} = \overline{\mathfrak{A}^2} = \overline{\mathfrak{A}I^\star + \mathfrak{A}I^{\star\perp}} \subset \overline{I^\star} + \overline{I^{\star\perp}}$. In consequence, $\overline{\mathfrak{A}I^\star} = I^\star$. Since $(xI|\mathfrak{A}) = (x|\mathfrak{A}I^\star) \subset (x|I^\star)$ for all $x \in \mathfrak{A}$, using part 2 we easily obtain $\mathrm{Lann}\, I = (I^\star)^\perp$. In a similar way, $\mathrm{Rann}\, J = (J^\star)^\perp$. □

LEMMA 5.2.– Let A be an associative $H^\star$-algebra with zero annihilator, I a closed right ideal and J a closed left ideal. Then the following assertions hold:

1) $\mathrm{Rann}\,\mathrm{Lann}\, I = I$ and $\mathrm{Lann}\,\mathrm{Rann}\, J = J$;

2) for any $z \in A$, let S_z and T_z be the linear maps defined by the equalities $S_z = \mathrm{Id} - L_z$ and $T_z = \mathrm{Id} - R_z$. Then $\mathrm{Rann}\,\mathrm{Im}\, T_z = \ker S_z$ and $\mathrm{Lann}\,\mathrm{Im}\, S_z = \ker T_z$;

3) if S_z and T_z are the linear maps previously defined, then S_z (respectively, T_z) is injective if and only if T_z (respectively, S_z) is onto;

4) $x^\star x \neq 0$ for all the non-zero elements of A;

5) $\operatorname{Rad}_J A = 0$;

6) if I is a minimal closed right ideal, then I is a minimal right ideal.

PROOF.–

1) It follows from lemma 5.1 taking into account that $\star$ is an isometric map.

2) We have $\operatorname{Rann}\operatorname{Im} T_z = \{x \in A : yx - yzx = 0 \text{ for all } y \in A\}$. So $x \in \operatorname{Rann}\operatorname{Im} T_z$ if and only if $x - zx \in \operatorname{Ann} A = 0$. This shows that $\operatorname{Rann}\operatorname{Im} T_z = \ker S_z$. A similar argument proves that $\operatorname{Lann}\operatorname{Im} S_z = \ker T_z$.

3) $\operatorname{Im} T_z$ (respectively, $\ker S_z$) is a left ideal (respectively, closed right ideal). Obviously if $\operatorname{Im} T_z = A$, then $0 = \operatorname{Rann}\operatorname{Im} T_z$. By (2), S_z is a injective map. Assume now $\operatorname{Im} T_z \neq A$. Since $\operatorname{Im} T_z$ is a z-modular left ideal, there exists some z-modular maximal left ideal M containing it. This left ideal is closed, because maximal modular left ideals of Banach algebras are closed $0 \neq \operatorname{Rann} M \subset \operatorname{Rann}(\operatorname{Im} T_z) = \ker S_z$. Thus, S_z is not injective.

4) Let x be an element of A such that $x^\star x = 0$. So for every $y \in A$ we have $0 = (x^\star x | yy^\star) = (x | xyy^\star) = \|xy\|^2$. Therefore, $x \in \operatorname{Lann} A = 0$.

5) It can be proved that $\operatorname{Rad}_J A = \{x \in A : x^\star x + xx^\star = 0\}$. This is so even in the more general setting of the non-commutative Jordan $H^\star$-algebras, as we will see later (theorem 5.11). Let us consider $x \in A$ such that $x^\star x + xx^\star = 0$. So $-\|x^\star x\|^2 = (x^\star x | xx^\star) = \|x^2\|^2$. In consequence, $x^\star x = 0$ and so $x = 0$.

6) By part (1), $M = \operatorname{Lann} I$ is a maximal closed left ideal of A and $I = \operatorname{Rann} M$. Let $N \neq 0$ be a right ideal of A contained in I. By (5), $\operatorname{Rad}_J A = 0$. So there exists some $z \in N$, $z \neq 0$, which is not right quasi-regular. Using the notations of part (3), $\operatorname{Im} S_z \neq A$. By (4), $\ker T_z$ is a non-zero left ideal of A. Furthermore, $R_z(M) = 0$ gives $T_z(M) = M$. Since $A = M + \ker T_z$, we obtain $\operatorname{Im} T_z = M$. In consequence, $I = \operatorname{Rann} M = \operatorname{Rann}\operatorname{Im} T_z = \ker S_z = \{x \in A : zx = x\} \subset N$. Therefore, $I = N$. □

An idempotent e of an associative algebra A is said to be *completely primitive* if the subalgebra eAe is a division subalgebra of A. If e is a self-adjoint idempotent of an associative $H^\star$-algebra A, then we say that e is a *projection* of A. If, in addition, e is completely primitive, we say that e is a *completely primitive projection*. Completely primitive projections are very abundant in associative $H^\star$-algebras with zero annihilator and they are closely related to the minimal sided ideals of the given $H^\star$-algebra. This is illustrated in the following proposition.

PROPOSITION 5.10.– Let $A \neq 0$ be an associative $H^\star$-algebra with zero annihilator. Then we have:

1) there exist minimal closed right (respectively, left) ideals in A;

2) if I is a minimal right ideal of A, then there exists a unique non-zero projection e such that $I = eA$. Furthermore, this idempotent is completely primitive;

3) minimal closed right (respectively, left) ideals agree with the minimal right (respectively, left) ideals of A;

4) the minimal closed right ideals of A are a family $\{I_i\}_{i\in\mathcal{M}}$ of pairwise orthogonal ideals whose orthogonal direct sum $\bigoplus_{i\in\mathcal{M}} I_i$ is a dense ideal of A. Furthermore, if each e_i is the unique non-zero projection such that $I_i = e_i A$, then the left ideals Ae_i are the minimal closed left ideals of A, which are pairwise orthogonal. Furthermore,

$$\bigoplus_{i\in\mathcal{M}} e_i A = \bigoplus_{i,j\in\mathcal{M}} e_i A e_j = \bigoplus_{i\in\mathcal{M}} Ae_i.$$

PROOF.–

1) Since $\operatorname{Rad}_J A = 0$, there exists some maximal modular left ideal M in A. Since M is closed, the ideal $I = \operatorname{Lann} M$ is a minimal closed right ideal of A. In a similar way, it can prove the existence of minimal closed left ideals in A.

2) By minimality of I, there exists a non-zero idempotent f such that $I = fA$. Moreover, fAf is a division subalgebra of A. By the Gelfand–Mazur theorem and its extension to the real case (see (Cabrera and Rodríguez 2014, 2018, Corollary 1.1.45 and Proposition 2.540)), fAf is isomorphic to $\mathbb{C}$, in the complex case; and to $\mathbb{R}$, $\mathbb{C}$ or $\mathbb{H}$, if $\mathbb{K} = \mathbb{R}$. In particular, fAf is a real quadratic algebra. So there exists λ, $\mu \in \mathbb{R}$ such that $(ff^\star f)^2 + \lambda ff^\star f + \mu f = 0$. In consequence, if $z = ff^\star$, then $z \neq 0$ and $z^3 + \lambda z^2 + \mu z = 0$. So $B = \mathbb{R}z + \mathbb{R}z^2$ is a non-zero commutative associative real $H^\star$-subalgebra of $A_{\mathbb{R}}$ with $\dim B \leq 2$ and identity involution. By lemma 5.2, B has no non-zero nilpotent elements, which implies the existence of some non-zero idempotent e in B. Obviously $B \subset I$ and $e^\star = e$. In consequence, $eA = I$. The projection e is completely primitive, because eAe is isomorphic to some of the division algebras $\mathbb{R}$, $\mathbb{C}$ or $\mathbb{H}$. If e and e' are now projections such that $eA = I = e'A$, then, by minimality of I, eAe and $e'Ae'$ are division $H^\star$-subalgebras of A that are isomorphic to $\mathbb{R}$, $\mathbb{C}$ or $\mathbb{H}$. We observe that $e' \in eAe$, because $e' \in eA$, $e' = e'^\star \in Ae$, and so $e' \in eA \cap Ae = eAe$. Since the $H^\star$-subalgebras $\mathbb{R}$, $\mathbb{C}$ and $\mathbb{H}$ have only one non-zero projection, $e' = e$.

3) It is a direct consequence of lemma 5.2 and part (2).

4) Orthogonality of the I_i is easy. Since $\operatorname{Rad}_J A$ is the intersection of the maximal modular left ideals of A, taking Lann in both sides of the equality $0 = \operatorname{Rad}_J A$ we obtain

$$A = \overline{\bigoplus_{i\in\mathcal{M}} I_i} = \overline{\bigoplus_{i\in\mathcal{M}} e_i A}.$$

Analogously, $A = \overline{\bigoplus_{i \in \mathcal{M}} Ae_i}$. The remainder of the assertions can be easily shown. □

THEOREM 5.8.– Ambrose (1945); Kaplansky (1948). Let A be a topologically simple associative real or complex $H^\star$-algebra. Then, up to a positive factor of the inner product, A is isometrically $\star$-isomorphic to the $H^\star$-algebra of the Hilbert–Schmidt operators $\mathcal{HS}(H)$, where H is a D-Hilbert space and $D = \mathbb{C}$ in the complex case, and $D = \mathbb{R}$, $\mathbb{C}$ or $\mathbb{H}$ in the real case.

PROOF.– Let $\{I_i\}_{i \in \mathcal{M}}$ be the family of the minimal closed right ideals of A and $\{e_i\}$ their completely primitive projections associated, as in proposition 5.10, satisfying $I_i = e_i A$. Let i and j be arbitrary in $\mathcal{M}$. The vector subspace $e_i A e_j$ will be denoted by A_{ij}. As in (Ambrose 1945, iii, p. 380), we can show that each A_{ij} is non-zero. Moreover, by minimality of the $e_i A$, the subspaces $e_i A e_i$ are division $H^\star$-subalgebras isomorphic to $\mathbb{R}$, $\mathbb{C}$ or $\mathbb{H}$.

Let $I = e_\omega A$ be a chosen minimal closed right ideal. Multiplying the inner product by a positive factor if it is necessary, we can assume $\|e_\omega\| = 1$. In particular, $D = e_\omega A e_\omega$ is a division $H^\star$-subalgebra, which is isometrically isomorphic to $\mathbb{R}$, $\mathbb{C}$ or $\mathbb{H}$. The right ideal I is a left vector space over D with the multiplication by scalars $D \times I \longrightarrow I$ given by $(d, e_\omega a) \mapsto d e_\omega a$. Each subspace $A_{\omega j}$ is also a non-zero left D-vector subspace of I. So we can choose a norm-one element $u_{\omega j}$ such that $u_{\omega j} \in A_{\omega j}$ for each $j \neq \omega$. In addition, we make $u_{\omega\omega} = e_\omega$. We define $u_{j\omega} = u_{\omega j}{}^\star$. Then $\|u_{j\omega}\| = 1$ and $u_{j\omega} u_{\omega j}$ and $u_{\omega j} u_{j\omega}$ are self-adjoint. We can check easily that $\|u_{\omega j} u_{j\omega}\| = 1 = \|u_{j\omega} u_{\omega j}\|$ and also we obtain $u_{\omega j} u_{j\omega} = e_\omega$ and $u_{j\omega} u_{\omega j} = e_j$. In consequence, the maps $\rho_{j\omega} : e_\omega A e_j \longrightarrow e_\omega A e_\omega$ and $\rho_{\omega j} : e_\omega A e_\omega \longrightarrow e_\omega A e_j$ given by $\rho_{j\omega} : e_\omega a e_j \mapsto e_\omega a e_j u_{j\omega}$ and $\rho_{\omega j} : e_\omega b e_\omega \mapsto e_\omega b e_\omega u_{\omega j}$ are inverse linear maps. This implies that $A_{\omega j}$ is a one-dimensional D-vector subspace. Moreover, it is easy to see that $e_i A e_j = u_{i\omega} A u_{\omega j} = u_{i\omega} D u_{\omega j}$. Now we define a map $[\ |\ ] : I \times I \longrightarrow D$, which associates each pair $(e_\omega a, e_\omega b)$ to the element $[e_\omega a | e_\omega b] = e_\omega a b^\star e_\omega$. It is easy to check that with this map I becomes a Hilbert space over D and that $\|v\|^2 = [v|v]$ for each $v \in I$. For every $x \in A$, we define the D-linear map $\varphi_x : I \longrightarrow I$ given by $\varphi_x : v \mapsto vx$. Then φ_x is continuous and $[\varphi_x(v)|w] = [v|\varphi_{x^\star}(w)]$ for any v, $w \in I$. The set $\{u_{\omega j}\}_{j \in \mathcal{M}}$ is a complete orthonormal system of the D-Hilbert space I and for each $d \in D$ the D-linear map $\varphi_{u_{i\omega} d u_{\omega j}}$ sends $u_{\omega i}$ to $d u_{\omega j}$, and $u_{\omega k}$ to 0 if $k \neq i$. So these maps are Hilbert–Schmidt operators and we can check easily that if d, $c \in D$, then

$$[\varphi_{u_{i\omega} d u_{\omega j}} \mid \varphi_{u_{i'\omega} c u_{\omega j'}}]_{\mathcal{HS}} = \begin{cases} d\bar{c} & \text{if } i = i' \text{ and } j = j' \\ 0 & \text{otherwise.} \end{cases}$$

In consequence, if $x = \sum_{i,j} u_{i\omega} d^{(i,j)} u_{\omega j}$ (with $d^{(i,j)} \in D$), then

$$\sum_{i,j,i',j'} [\varphi(u_{i\omega} d^{(i,j)} u_{\omega j}) \mid \varphi(u_{i'\omega} d^{(i',j')} u_{\omega j'})] = \sum_{i,j} d^{(i,j)} \overline{d^{(i,j)}} = \|x\|^2,$$

and so φ_x is a Hilbert–Schmidt operator and the map $\varphi : A \longrightarrow \mathcal{HS}(I)$ sending x to φ_x is an isometric linear map between the real or complex Hilbert space A and $\mathcal{HS}(I)$ such that $\varphi(x^\star)$ agrees with the adjoint of $\varphi(x)$. Furthermore, $\varphi(xx') = \varphi(x') \circ \varphi(x)$ for any x, $x' \in A$. Therefore, φ is an isometric $\star$-isomorphism, which completes the proof. □

Theorem 5.8 jointly with some elements of the theory of alternative algebras allows us to determine all the topologically simple alternative $H^\star$-algebras. We recall that an *alternative algebra* is a non-associative algebra satisfying the alternative laws given by $x^2 y = x(xy)$ and $(yx)x = yx^2$. Alternative algebras are algebras very close to associative algebras. In fact, the well-known Artin's theorem asserts that an algebra is alternative if and only if each of the two elements of A generates an associative algebra (see Zhevlakov *et al.* 1982, p. 36; Schafer 1966, p. 29). Nevertheless, there are alternative algebras that are not associative. These algebras are called *properly alternative*. The structure theory of alternative algebras has been deeply studied (Bruck and Kleinfeld 1951; Kleinfeld 1953b,a; Herstein 1964; Slater 1970, 1972). The next proposition is an easy consequence of the results obtained in Slater (1970) (see also (Zhevlakov *et al.* 1982, Chapter 9)).

PROPOSITION 5.11.– Let A be a prime properly alternative algebra over a field K of characteristic $\neq 3$. Assume that the centroid $Z(A)$ of A is a finite extension of the field K. Then $A_{\hat{K}}$ is an octonions algebra for some intermediate field $\hat{K}$ of the extension $Z(A)/K$. In particular, if the extension $Z(A)/K$ has prime degree, then either A or $A_{Z(A)}$ are octonions algebras.

Now taking into account proposition 5.5, we obtain the following consequence of proposition 5.11.

PROPOSITION 5.12.– If A is a topologically simple alternative $H^\star$-algebra, then either A is associative or A is some of the three octonions algebras $\mathbb{O}$, $\mathbb{O}_s$ or $\mathbb{O}_\mathbb{C}$.

REMARK 5.1.– Proposition 5.12 was originally derived in Pérez de Guzmán (1983) in the complex case, where a different proof has been given. We can also observe that a certain 'symmetric character' imposed by the existence of a multiplicative involution in an $H^\star$-algebra allows us to obtain the same conclusion of proposition 5.12 under the weaker assumption that A satisfies only one of the alternative laws. Indeed, if A is an $H^\star$-algebra satisfying the right alternative law, then taking involution on $(xy)y = xy^2$ we obtain $y^\star(y^\star x^\star) = {y^\star}^2 x^\star$. Since $\star$ is a bijective map, then the algebra A satisfies also the left alternative identity.

REMARK 5.2.– Smiley (1953) has introduced right $H^{\star}$-algebras in the case of associative algebras. These are associative algebras satisfying conditions similar to those of the $H^{\star}$-algebras but where only the first part of [5.1] is assumed. There are associative right $H^{\star}$-algebras that are not $H^{\star}$-algebras. In the topologically simple case, all of them can be obtained from a certain modification of the $H^{\star}$-algebras of Hilbert–Schmidt operators $\mathcal{HS}(H)$ (see (Smiley 1953) for details).

We close this section remembering an interesting theorem given by Rodríguez (1988).

THEOREM 5.9.– (Cabrera and Rodríguez 2014, 2018, Theorem 8.1.76 and Corollary 8.1.86 (ii)).

Let A be an associative semiprime $\mathbb{K}$-algebra such that A^{+} is an $H^{\star}$-algebra with respect to some inner product and involution. Then A is also an $H^{\star}$-algebra with the same inner product and involution.

5.6. Flexible $H^{\star}$-algebras

A K-algebra A is said to be *flexible* if

$$(xy)x = x(yx) \qquad [5.6]$$

for all elements x, $y \in A$. Flexible algebras generalize commutative as well as associative algebras. Flexible algebras also include many interesting classes of non-associative algebras (alternative, Jordan, non-commutative Jordan algebras, etc.). The algebraic identity [5.6] is named *flexible identity*.

Non-commutative Jordan $H^{\star}$-algebras will be treated in more detail in the next section. Now we consider $H^{\star}$-algebras of the wider class of the flexible $H^{\star}$-algebras. Theorem 5.10 states the way that the study of the flexible $H^{\star}$-algebras with zero annihilator can be reduced to those of the anticommutative $H^{\star}$-algebras and the flexible $H^{\star}$-algebras A with $\operatorname{Ann} A^{+} = 0$.

It is possible to characterize the flexible algebras in the following way.

WELL-KNOWN FACT 5.2.– Let A be an algebra over a field of characteristic $\neq 2$. Then A is a flexible algebra iff for each $x \in A$ the map $D_x : A \longrightarrow A$ defined by $D_x : y \mapsto xy - yx$ is a derivation of A^{+}.

From this, we can prove easily the following theorem (see (Cuenca 1984, Theorem 2.6.5) originally proved in the case of complex $H^{\star}$-algebras, but whose proof is also valid in the real case).

THEOREM 5.10.– Let A be a flexible semi-$H^\star$-algebra with zero annihilator. Then A is an orthogonal direct sum $A = A_1 \oplus A_2$ where A_1 and A_2 are self-adjoint closed ideals of A such that A_1 is an anticommutative algebra and $\operatorname{Ann} A_2^+ = 0 = \operatorname{Ann} A_1$.

A kind of flexible $H^\star$-algebras closely related to the associative $H^\star$-algebras are the quasi-associative $H^\star$-algebras. We say that a K-algebra A is an *scalar isotope* of the K-algebra D if A and D have the same underlying vector space and there exists some $\lambda \in K$ such the products (juxtaposition in D and $\underset{(\lambda)}{\cdot}$ in A) satisfy the relation $x \underset{(\lambda)}{\cdot} y = \lambda xy + (1-\lambda)yx$, for any $x, y \in D$. We say that A is the λ-*mutation* of D. A routine verification shows that if $A = D^{(\lambda)}$ and $\lambda \neq 1/2$, then $D = \left(A^{(\lambda)}\right)^{(\mu)}$, for $\mu = 1/(2\lambda - 1)$. It can be easily checked that if $\lambda \in \mathbb{R}$, then the λ-mutation of each $H^\star$-algebra is also an $H^\star$-algebra of same inner product and involution.

A K-algebra A is said to be *quasi-associative* if there exists an extension field $\hat{K}$ of K, an associative $\hat{K}$-algebra D and $\lambda \in \hat{K}$ such that the scalar extension $A_{\hat{K}}$ agrees with $D^{(\lambda)}$.

PROPOSITION 5.13.– (Albert 1948, p. 585). Let K be a field of characteristic $\neq 2$ and A a quasi-associative K-algebra, which is not commutative. Then there exists a field extension $\hat{K}$ of K of degree at most 2, an element $\lambda \in \hat{K}$, $\lambda \neq 1/2$ and an associative $\hat{K}$-algebra D such that $A_{\hat{K}} = D^{(\lambda)}$.

Now we dispose of the necessary elements to prove that the quasi-associative and non-commutative $H^\star$-algebras with zero annihilator agree with the real mutations of associative $H^\star$-algebras with zero annihilator. Using theorem 5.9, we give here a shorter proof than the original one (Cuenca and Rodríguez 1987; Cuenca and Sánchez 1994).

PROPOSITION 5.14.– (Cuenca and Rodríguez 1987, pp. 6–7), (Cuenca and Sánchez 1994, Theorem 3). Let A be a real (respectively, complex) quasi-associative $H^\star$-algebra, which is not commutative and with zero annihilator. Then there exists some real (respectively, complex) associative $H^\star$-algebra D and a real number $\lambda \neq 1/2$ such that A agrees with the $H^\star$-algebra $D^{(\lambda)}$.

PROOF.– The real case follows easily from the complex one (see (Cuenca and Sánchez 1994, p. 489) for details). We now assume A to be a complex quasi-associative $H^\star$-algebra. By proposition 5.13, there exists a non-commutative associative complex algebra D such that $A = D^{(\lambda)}$ for some $\lambda \in \mathbb{C}$, $\lambda \neq 1/2$. Since $D^+ = (D^{(\lambda)})^+ = A^+$, theorem 5.9 gives that D is an $H^\star$-algebra with the same inner product and involution as D^+ (so as A). If it is denoted, by juxtaposition, the associative product of D, then for $x, y, z \in A$ we have $\lambda(xy|z) + (1-\lambda)(yx|z) = (x \underset{(\lambda)}{\cdot} y|z) = (y|x^\star \underset{(\lambda)}{\cdot} z) = \bar{\lambda}(xy|z) + (1-\bar{\lambda})(yx|z)$. So $(\lambda - \bar{\lambda})(xy - yx|z) = 0$. This gives $\lambda \in \mathbb{R}$, because $\lambda \notin \mathbb{R}$ implies $xy = yx$, which contradicts the non-commutative character of A $(= D^{(\lambda)})$. □

5.7. Non-commutative Jordan $H^\star$-algebras

In this section, we give a classification theorem of the topologically simple non-commutative Jordan $H^\star$-algebras, which was originally established in (Cuenca and Rodríguez 1987) and (Cuenca and Sánchez 1994), where a different proof was given. Here, we will use a later theorem by Fernández and Rodríguez (1986, Thoerem 13). We recall that if K is a field of characteristic $\neq 2$ and A is a flexible K-algebra A, then A is a *non-commutative Jordan algebra* if it satisfies the identity

$$(x^2y)x = x^2(yx). \qquad [5.7]$$

Commutative K-algebras satisfying [5.7] are named *Jordan algebras* and they are a class of non-associative algebras deeply studied. It is not difficult to see that a flexible algebra is a non-commutative Jordan algebra if and only if the symmetrized K-algebra A^+ is a Jordan algebra (see (Schafer 1955, p. 473)). Standard references for Jordan algebras are as follows: (Braun and Koecher 1966; Jacobson 1968a, 1981; McCrimmon 2004). In these algebras, an important role is played by the quadratic operator U. It associates to each $x \in A$ the linear map $U_x = 2{L_x}^2 - L_{x^2}$. In the more general case of the non-commutative Jordan algebras, the U-operator is defined as $U_x = L_x \circ (L_x + R_x) - L_{x^2}$. If A is a non-commutative Jordan algebra, then the operator U_x agrees with the corresponding operator U_x^+ of the symmetrized Jordan algebra A^+. Non-commutative Jordan algebras are *power associative*, that is, the subalgebra generated by each element of A is an associative algebra. As in associative algebras, also in different classes of non-associative algebras it is possible to consider Peirce subspaces relative to a given idempotent e. So if A is a non-commutative Jordan algebra and e an idempotent of A, then the *Peirce subspaces* A_i ($i = 0, 1/2, 1$), of A relative to e are defined by $A_i = \{x \in A : ex + xe = 2ix\}$. The subspaces A_i are also denoted as $A_i(e)$ if the reference to the idempotent e cannot be understood. In (Albert 1948, pp. 559–562), it has been proved that A splits into a direct sum $A = A_1 \oplus A_{1/2} \oplus A_0$ and that A_1 and A_0 are subalgebras of A such that $A_1 = \{x \in A : ex = x = xe\}$ and $A_0 = \{x \in A : ex = 0 = xe\}$. So $U_e(A) = A_1$. Non-commutative Jordan algebras were introduced by Albert (1948) who focused his attention on the subclass of the standard algebras. The structure theory of the finite-dimensional semisimple non-commutative Jordan algebras over fields of characteristic 0 has been treated by Schafer (1955) and Oehmke (1958) in the case of arbitrary characteristic $\neq 2, 3$. Moreover, McCrimmon's papers (1966, 1971) complete the study of different aspects of the theory of non-commutative Jordan algebras, and, in Fernández and Rodríguez (1986), a classification theorem for the non-degenerate prime non-commutative Jordan algebras containing some completely primitive idempotent is established.

We recall that an element x of a unital non-commutative Jordan algebra A is said to be *invertible* if there exists some $y \in A$ satisfying $xy = 1 = yx$ and $x^2y = x = yx^2$.

This concept of invertibility has been previously introduced in the Jordan case (see (Jacobson 1968a, p. 52)). By the commutativity of A, in this case y must satisfy only the conditions $xy = 1$ and $x^2y = x$. It is well known that if A is a Jordan algebra, then x is invertible iff the Jordan operator U_x is a bijective linear map. Furthermore, if this is the case, the element y is uniquely determined and is called the *inverse* of x, moreover y is also invertible with inverse x (see (Jacobson 1968a, p. 52)). These assertions are also true for invertible elements of a non-commutative Jordan algebra, because in virtue of the fact (proved by K. McCrimmon) an element x of the unital non-commutative Jordan algebra A is invertible with inverse y if and only if x is invertible in A^+ with same inverse (McCrimmon 1965, p. 943).

A unital non-commutative Jordan algebra A is said to be a *division algebra* if every non-zero element of A is invertible. A non-zero idempotent e of a non-commutative Jordan algebra is said to be *completely primitive* if the Peirce subspace $A_1(e)$ $(= U_e^+(A))$ is a division non-commutative Jordan algebra. The concept of invertibility previously introduced allows us to define a Jacobson radical for non-commutative Jordan algebras. First, we proceed to define the *quasi-invertible* elements of the given algebra. So an element x in a non-commutative Jordan algebra A is a quasi-invertible element of A if $1 - x$ is invertible in the unitized algebra $\hat{A}$ of A. Ideals of A whose elements are all quasi-invertible are said to be *quasi-invertible ideals*. It can be prove the existence of a unique maximal quasi-invertible ideal of A, which is called the *Jacobson radical* of A (McCrimmon 1971, p. 673). We will denote it by $\mathrm{Rad}_J A$. Non-commutative Jordan algebras with zero Jacobson radical will be called *J-semisimple* algebras. The inclusion $\mathrm{Rad}_J A \subset \mathrm{Rad}_J A^+$ holds (McCrimmon 1971, Theorem 11), with equality in the case of complete normed algebras (Cabrera and Rodríguez 2014, 2018, Proposition 4.4.17). Elements z of a non-commutative Jordan algebra A such that $U_z = 0$ are named *absolute zero divisors*. Non-commutative Jordan algebras without non-zero absolute zero divisors are called *non-degenerate*. As per (McCrimmon 1969, Theorem 10), $\mathrm{Rad}_J A^+$ contains all the absolute zero divisors of A. For non-commutative Jordan $H^\star$-algebras, we also have the equality of both sets (see (Sánchez 1989, Proposition 2.15)). Taking into account (Cabrera and Rodríguez 2014, 2018, Proposition 8.1.145; Sánchez 1989, Corolario 2.16; Cuenca and Sánchez 1994, Lemma 2) and summarizing, we have the following theorem.

THEOREM 5.11.– Let A be a non-commutative Jordan $H^\star$-algebra and Z the set of its absolute zero divisors. Then Z is an ideal of A. Furthermore, $Z = \mathrm{Rad}_J A = \mathrm{Rad}_J A^+ = \mathrm{Ann} = A^+ = \{x \in A : x^\star x + x x^\star = 0\}$.

COROLLARY 5.3.– Each $H^\star$-subalgebra of a J-semisimple non-commutative Jordan $H^\star$-algebra is also J-semisimple.

As in the case of associative $H^\star$-algebras, we call *projection* to each self-adjoint idempotent of a given non-commutative Jordan $H^\star$-algebra A. By theorem 5.11, the closed subalgebra generated by each non-zero self-adjoint or skew-adjoint element of a J-semisimple non-commutative Jordan $H^\star$-algebra is a commutative associative $H^\star$-algebra with zero annihilator. Taking into account proposition 5.10, we obtain the following.

COROLLARY 5.4.– Let $A \neq 0$ be a J-semisimple non-commutative Jordan $H^\star$-algebra. Then A contains non-zero projections.

By standard methods in the theory of $H^\star$-algebras, every non-zero projection of a non-commutative Jordan $H^\star$-algebra can be decomposed into a finite sum $e = \sum_{i=1}^{n} e_i$ of pairwise orthogonal completely primitive projections (see (Devapakkiam 1975, Proposition 4.2; Cuenca 1984, Corollary 3.122; Sánchez 1989, Proposition 3.7)). We recall that here the pairwise orthogonality of the e_i means that $e_ie_j = 0$ for any $i \neq j$ (so also $e_je_i = 0$ and $(e_i|e_j) = 0$). In particular, we have the following.

COROLLARY 5.5.– If $A \neq 0$ is a J-semisimple non-commutative Jordan $H^\star$-algebra, then A contains completely primitive projections.

As we have pointed out, we will use the following theorem in the determination of the topologically simple non-commutative Jordan $H^\star$-algebras.

THEOREM 5.12.– (Fernández and Rodríguez 1986, Theorem 13). Let A be a non-commutative Jordan K-algebra. If A is a non-degenerate prime algebra containing some completely primitive idempotent, then some of the following possibilities occurs:

1) A is a non-commutative Jordan division K-algebra;

2) A is a simple flexible quadratic algebra over a field $\hat{K}$ extension of K;

3) A is a non-degenerate prime Jordan K-algebra containing a completely primitive idempotent;

4) A is a quasi-associative algebra that is not commutative.

In order to use the previous theorem in the determination of the topologically simple non-commutative Jordan $H^\star$-algebras, we do the following observations:

FACTS 5.1.–

1) let A be a normed non-commutative Jordan division algebra over $\mathbb{K}$. Then A is a quadratic algebra over $\mathbb{K}$ if $\mathbb{K} = \mathbb{R}$, or A is isomorphic to $\mathbb{C}$ in the complex case (see (Cabrera and Rodríguez 2014, 2018, Proposition 2.5.49));

2) every field contained in an $H^\star$-algebra is isomorphic to $\mathbb{R}$ or $\mathbb{C}$ (consequence of the Gelfand–Mazur theorem). In particular, if A is an $H^\star$-algebra over $\mathbb{K}$ that is quadratic over a field extension $\hat{\mathbb{K}}$ of $\mathbb{K}$, then $\hat{\mathbb{K}} = \mathbb{R}$ or $\mathbb{C}$.

THEOREM 5.13.– (Cuenca and Rodríguez 1987, Theorem 2; Cuenca and Sánchez 1994, Theorem 3). Let A be a topologically simple non-commutative Jordan $H^{\star}$-algebra over $\mathbb{K}$ ($\mathbb{K} = \mathbb{R}$ or $\mathbb{C}$). Then at least one of the following assertions hold:

1) A is anticommutative;

2) A is a flexible quadratic algebra over $\hat{K}$ not isomorphic to $\hat{K} \times \hat{K}$, with $\hat{K} = \mathbb{C}$ if $\mathbb{K} = \mathbb{C}$, and $\hat{\mathbb{K}} = \mathbb{R}$ or $\mathbb{C}$ if $\mathbb{K} = \mathbb{R}$ (see proposition 5.9);

3) A is a Jordan $H^{\star}$-algebra (see section 5.8);

4) A is a quasi-associative $H^{\star}$-algebra, which is not commutative (see proposition 5.14).

PROOF.– By theorem 5.10, A is anticommutative or $\operatorname{Ann} A^{+} = 0$. So we only need to consider the case that $\operatorname{Ann} A^{+} = 0$. Then by corollary 5.5, A has some completely primitive projection. Moreover, the topological simplicity of A implies that A is a prime non-commutative Jordan algebra. So A satisfies some of the alternatives of theorem 5.12. By fact 5.1.1, if A is a non-commutative Jordan division algebra, then A is a quadratic algebra over $\mathbb{K}$. By fact 5.1.2, $H^{\star}$-algebras quadratic as $\hat{K}$-algebras are excluded for extension fields $\hat{\mathbb{K}}$ of $\mathbb{K}$ different from $\mathbb{R}$ and $\mathbb{C}$. □

5.8. Jordan $H^{\star}$-algebras

Jordan $H^{\star}$-algebras have been first treated by Devapakkiam (1975) and Devapakkiam and Rema (1976). They determined the topologically simple complex Jordan $H^{\star}$-algebras under certain additional restrictions including the separable character of the underlying Hilbert space. The structure theory of Jordan $H^{\star}$-algebras has been accomplished in Cuenca and Rodríguez (1987) in the complex case (see also (Cuenca 1984)), and in Cuenca and Sánchez (1994) for real Jordan $H^{\star}$-algebras. In (Cuenca and Rodríguez 1987) and (Cuenca 1984), the determination of the topologically simple complex Jordan $H^{\star}$-algebras was achieved by an extension of the Jacobson strong coordinatization theorem to the framework of the complex Jordan $H^{\star}$-algebras with zero annihilator (see (Cuenca and Rodríguez 1987, Theorem 3)). Here we will follow a different way by using the classification, based on (Zelmanov 1979, 1983), that P.N. Áhn has provided of the simple Jordan algebras over fields of characteristic $\neq 2$ and containing some minimal quadratic ideal. We recall that a vector subspace Q of a Jordan algebra A is a *quadratic ideal* of A if $U_Q(A) \subset Q$.

THEOREM 5.14.– (Áhn 1986). Let K be a field of characteristic $\neq 2$. For each simple Jordan K-algebra A containing some minimal quadratic ideal, some of the following alternative occurs:

1) there exists a division associative K-algebra D such that each finite subset of A is contained in a quadratic ideal of A, which is isomorphic to a matrix algebra $\mathcal{M}_n(D)^+$;

2) there exists a division associative K-algebra D such that each finite subset of A is contained in some quadratic ideal of A, which is isomorphic to a Jordan algebra of symmetric elements $\mathrm{Sym}(\mathcal{M}_n(D), \sigma_n)$ of $\mathcal{M}_n(D)$ relative to an involutive antiautomorphism σ_n of $\mathcal{M}_n(D)$;

3) A is isomorphic to the quadratic Jordan $\hat{K}$-algebra of a non-degenerate symmetric bilinear form, where $\hat{K}$ is an extension of the base field K;

4) A is isomorphic to a simple exceptional Jordan algebra of dimension 27 over its center.

We will use theorem 5.14 in order to determine all the topologically simple real Jordan $H^\star$-algebras with identity involution. As a consequence, the classification theorem of the topologically simple complex $H^\star$-algebras is obtained (see remark 5.3). Now the next lemma can be easily proved.

LEMMA 5.3.– If A is a J-semisimple non-commutative Jordan $H^\star$-algebra and S a simple dense subalgebra of A containing some completely primitive projection, then A is topologically simple.

Let $\mathfrak{A}$ be a real algebra. We recall that $\mathfrak{A}$ is *formally real* if for every one of its finite sequences $x_1, \ldots, x_n$ the relation $\sum_{i=1}^{n} x_i^2 = 0$ implies $x_i = x_2 = \cdots = x_n = 0$. Formally, real Jordan algebras were introduced by Jordan *et al.* (1934) in the pioneering paper, where a complete structure theory of the formally real Jordan algebras of finite dimension has been given.

LEMMA 5.4.– Let A be a formally real normed Jordan algebra such that $A = \mathrm{Sym}(\mathcal{M}_n(D), \sigma)$, where $n \geq 3$, D is a real associative division algebra, and σ is an involutive antiautomorphism of $\mathcal{M}_n(D)$. Then we have:

1) D is isomorphic to $\mathbb{R}$, $\mathbb{C}$ or $\mathbb{H}$;

2) A is isomorphic to $\mathrm{Sym}(\mathcal{M}_n(D), \tau)$, where $D = \mathbb{R}$, $\mathbb{C}$ or $\mathbb{H}$ and τ is the involution of the $H^\star$-algebra $\mathcal{M}_n(D)$ given by $\tau : M \mapsto \bar{M}^t$, with $^-$ denoting the Cayley antiautomorphism of D.

PROOF.– Let $A = \mathrm{Sym}(\mathcal{M}_n(D), \sigma)$ be a formally real normed Jordan algebra. The well-known classification of involutive antiautomorphism as described in (Jacobson 1969, p. 9) implies the existence of an involutive antiautomorphism s of D and elements $\gamma_1, \ldots, \gamma_n \in D$ satisfying $0 \neq \gamma_i = s(\gamma_i)$ such that A is isomorphic to $\mathrm{Sym}(\mathcal{M}_n(D), \rho)$, with ρ given by $\rho : M \mapsto \Gamma(M^s)^t\Gamma^{-1}$ and Γ the diagonal matrix $\Gamma = \mathrm{diag}(\gamma_1, \ldots, \gamma_n)$ (we can observe that the symplectic case must be excluded by the formally real character of A). Let E_{ij} $(i,\, j = 1, \dots, n)$ be the corresponding unit (i, j)-matrices. Since $D_1 = \{aE_{12} + \gamma_2 s(a)\gamma_1^{-1}E_{21} : a \in D\}$ is a real algebra relative to the continuous product $\square$ defined as $c\square d = (c \bullet (E_{23} + E_{32}))\bullet((E_{23} + E_{32}) \bullet d)$ and this algebra is isomorphic to D, then D can be viewed as a real normed division associative algebra. By the Arens–Gelfand–Mazur theorem, D is isomorphic to $\mathbb{R}$, $\mathbb{C}$ or $\mathbb{H}$. The formally real character of A implies assertion 2. $\square$

The next theorem reformulates in an equivalent way (Cuenca and Sánchez 1994, Theorem 4).

THEOREM 5.15.– (Cuenca and Sánchez 1994, Theorem 4). Let A be a topologically simple real Jordan $H^\star$-algebra with identity involution. Then, up to a positive factor of the inner product, A is isometrically isomorphic to some of the following $H^\star$-algebras:

1) the $H^\star$-subalgebra $\mathrm{Sym}(\mathcal{M}_{\mathcal{A}}(D), \star)$ of the $H^\star$-algebra of $\mathcal{A} \times \mathcal{A}$-matrices $\mathcal{M}_{\mathcal{A}}(D)$, where $D = \mathbb{R}, \mathbb{C}$ or $\mathbb{H}$ and $|\mathcal{A}| \geq 3$;

2) a real Jordan $H^\star$-algebra of identity involution, which is quadratic over $\mathbb{R}$ and not isomorphic to $\mathbb{R} \oplus \mathbb{R}$ (see proposition 5.9);

3) the exceptional Jordan $H^\star$-algebra of conjugate-symmetric 3×3-matrices with entries in the real division $H^\star$-algebra of octonions $\mathbb{O}$.

PROOF.– Using (Osborn and Racine 1979, Theorem 17), we obtain easily the existence of a simple dense ideal B of A containing all the completely primitive projections of A. B holds some of the different possibilities of theorem 5.14. In particular, each finite subset $F = \{x_1, \dots, x_n\}$ of B is contained in an unital subalgebra B_F of B and, in addition, B_F can be chosen as B $(= A)$ in the cases that B is isomorphic to some of the algebras of the cases 3 or 4 of the cited theorem. Let e_F be the unity of B_F. So we have $(\sum_{i=1}^n x_i^2 | e_F) = \sum_{i=1}^n \|x_i\|^2$. Thus, B_F is a formally real algebra. As a result, we obtain the following: (i) if B is a quadratic Jordan algebra over $\hat{\mathbb{K}}$, then $\hat{\mathbb{K}} = \mathbb{R}$; (ii) if B is exceptional, then B is 27-dimensional over $\mathbb{R}$ and even isomorphic to the Jordan algebra of conjugate-symmetric 3×3-matrices with entries in the division algebra of octonions $\mathbb{O}$; (iii) if B is as in case 1 of theorem 5.14, then B is one-dimensional. Assume now B as in case 2 of theorem 5.14. Without loss in generality, we can assume that B has at least 3 pairwise orthogonal completely primitive idempotents, because if the greater number of pairwise orthogonal completely primitive idempotents of B is 2 and if two of these idempotents are e_{ω_1} and e_{ω_2}, then B $(= A)$ is unital with $e_{\omega_1} + e_{\omega_2}$ as unity and so an argument as that in (Jacobson 1968a, p. 202) shows that A is a quadratic Jordan algebra with $\dim A \geq 3$. Let now $e_{\omega_1}, e_{\omega_2}$ and e_{ω_3} be three pairwise orthogonal completely primitive idempotents of B. Let B_{i_0} be a quadratic ideal of B containing $e_{\omega_1}, e_{\omega_2}$ and e_{ω_3}, which is isomorphic to an algebra of the type $\mathrm{Sym}(\mathcal{M}_{n_0}(D), \sigma_{i_0})$. Let now $\{B_i\}_{i \in \mathcal{M}}$ be the family of the quadratic ideals B_i of B

containing B_{i_0} and such that there exists a positive integer n_i and an involutive antiautomorphism σ_i of $\mathcal{M}_{n_i}(D)$ such that $B_i \cong \text{Sym}(\mathcal{M}_{n_i}(D), \sigma_i)$. For all i, we will denote as η_i one chosen isomorphism from B_i onto $\text{Sym}(\mathcal{M}_{n_i}(D), \sigma_i)$. By Lemma 5.4, we can suppose that $D = \mathbb{R}$, $\mathbb{C}$ or $\mathbb{H}$ and σ_i is given by $\sigma_i : M \mapsto \bar{M}^t$. In particular, each $\text{Sym}(\mathcal{M}_{n_i}(D), \sigma_i)$ is a real $H^\star$-algebra of identity involution, which is an $H^\star$-subalgebra of the symmetrized $H^\star$-algebra $\mathcal{M}_{n_i}(D)^+$ (with the conjugate-transposition σ_i as involution). By multiplying the inner products of each $H^\star$-algebra $\mathcal{M}_{n_i}(D)^+$ by a real positive factor, if it were necessary, we can assume that the unity elements of each B_i and $\mathcal{M}_{n_i}(D)^+$ have the same norm. Since the $\star$-isomorphisms η_i multiply the norms of the elements of B_i by the same positive factor (corollary 5.2), we obtain that the η_i are also isometric maps.

In $\mathcal{M}$, a partial order can be defined by declaring $i \leq j$ iff B_i is a subalgebra of B_j. If for any $i, j \in \mathcal{M}$ such that $i \leq j$ we denote by μ_{ji} the monomorphism of inclusion $\mu_{ji} : B_i \longrightarrow B_j$, then $(\{B_i\}_{i \in \mathcal{M}}, \mu_{ji})$ is a net of $H^\star$-subalgebras with union B. For $i, j \in \mathcal{M}$ such that $i \leq j$, let $\nu_{ji} : \text{Sym}(\mathcal{M}_{n_i}(D), \sigma_i) \longrightarrow \text{Sym}(\mathcal{M}_{n_j}(D), \sigma_j)$ be the composition $\eta_j \circ \mu_{ji} \circ \eta_i^{-1}$. Obviously the ν_{ji} are isometric $\star$-monomorphisms. It is possible to show that each ν_{ji} can be extended in a unique way to an isometric associative $\star$-monomorphism $\hat{\nu}_{ji} : \mathcal{M}_{n_i}(D) \longrightarrow \mathcal{M}_{n_j}(D)$. So $(\{\mathcal{M}_{n_i}(D)\}_{i \in \mathcal{M}}, \{\hat{\nu}_{ji}\}_{i \leq j})$ is a directed isometric $\star$-monomorphic system of $H^\star$-algebras. Let $\mathcal{U}$ be an ultrafilter in $\mathcal{M}$ containing all the subsets $[i, \infty)$. Now from the B_i and μ_{ji} we construct, as in section 5.3, the algebra with identity involution and inner product $B_{\mathcal{U}}$, and the algebra C with inner product and isometric involution σ obtained from the directed isometric monomorphism system $(\{\mathcal{M}_{n_i}(D)\}_{i \in \mathcal{M}}, \{\hat{\nu}_{ji}\}_{i \leq j})$. Then $B_{\mathcal{U}}$ is isometrically isomorphic to B and the completion C' of C is an associative $H^\star$-algebra, which is topologically simple (lemma 5.3), whose involution $\star$ is the unique extension by continuity of σ to an involutive antiautomorphism of C'. Furthermore, $B_{\mathcal{U}}$ is isometrically isomorphic to $\text{Sym}(C, \sigma)$ and the extension by continuity of this isomorphism provides an isometric isomorphism from A onto $\text{Sym}(C', \star)$. □

REMARK 5.3.–

1) Theorem 5.15 allows the determination of all the topologically simple complex Jordan $H^\star$-algebras given in (Cuenca and Rodríguez 1987, Theorem 4) using completely different methods. We only need take into account that the self-adjoint elements of a complex Jordan $H^\star$-algebra is a real $H^\star$-algebra with identity involution. On the other hand, the complex case can be used as in (Cuenca and Sánchez 1994) in order to describe all the topologically simple real Jordan $H^\star$-algebras (Cuenca and Sánchez 1994).

2) We observe that in the proof of theorem 5.15, we have found that each topologically simple Jordan $H^\star$-algebra with identity involution has a net of unital $H^\star$-subalgebras $\{A_\gamma\}$ such that $\bigcup_\gamma A_\gamma$ is a dense subalgebra of A. It is not difficult to see that this is also true for J-semisimple non-commutative Jordan $H^\star$-algebras

(see (Cuenca and Rodríguez 1987, pp. 4–5), where this was originally proved in the complex case).

5.9. Moufang $H^\star$-algebras

It is well known that every alternative algebra A satisfies the following identities:

$$y((xz)x) = ((yx)z)x, \qquad [5.8]$$

$$(xy)(zx) = (x(yz))x, \qquad [5.9]$$

$$(x(yx))z = x(y(xz)), \qquad [5.10]$$

(see (Zhevlakov *et al.* 1982; Schafer 1966)). These identities are called *right, middle* and *left Moufang* identity, respectively.

In this section, we will prove that if A is an $H^\star$-algebra that satisfies some of the Moufang identities, then A satisfies all of them. We show that every such $H^\star$-algebra A is alternative. Since alternative $H^\star$-algebras have been completely described in proposition 5.12, this allows us to conclude the structure theory of the Moufang $H^\star$-algebras.

In order to determinate the $H^\star$-algebras satisfying some of the Moufang identities, we first need to establish some previous results on algebras endowed with a non-degenerate associative symmetric bilinear form. We recall that if $\mathfrak{A}$ is a non-associative algebra over a field K and $g : \mathfrak{A} \times \mathfrak{A} \longrightarrow K$ a symmetric bilinear form, then g is said to be *associative* if $g(xy, z) = g(x, yz)$, for any x, y, $z \in \mathfrak{A}$. It is well known that the presence of a non-degenerate associative symmetric bilinear form in a given algebra $\mathfrak{A}$ that satisfies some algebraic identity frequently implies stronger algebraic identities in $\mathfrak{A}$. This occurs in the case of the power-associative algebras. So following some ideas of (Albert 1949, pp. 320–321), it can be proved that each power-associative algebra over a field of characteristic $\neq 2, 3$ and 5 provided with a non-degenerate associative symmetric bilinear form is a non-commutative Jordan algebra (see (Cabrera and Rodríguez 2014, 2018, Proposition 8.1.7; Cuenca 1984, Theorem 2.1.7)). As a result, we have the following proposition.

PROPOSITION 5.15.– (Cuenca 1984, Theorem 2.1.8; Cabrera and Rodríguez 2014, 2018, Theorem 8.1.9). The power-associative $H^\star$-algebras are non-commutative Jordan algebras.

Also the presence of a non-degenerate associative symmetric bilinear form in algebras satisfying some of the Moufang identities implies that the given algebra must satisfy additionally other algebraic identities. This fact can be illustrated in the following lemma.

LEMMA 5.5.– (Cuenca 2002, Lemma 1.1). Let $\mathfrak{A}$ be a non-associative algebra over a field K and $g : \mathfrak{A} \times \mathfrak{A} \longrightarrow K$ a non-degenerate associative symmetric bilinear form. Then $\mathfrak{A}$ satisfies the right Moufang identity if and only if the middle Moufang identity holds in $\mathfrak{A}$.

PROOF.– For any $x, y, z, u \in \mathfrak{A}$, we have $g\Big(y((xz)x) - ((yx)z)x, u\Big) = g((xz)x, uy) - g(z, (xu)(yx)) = g\Big(z, (x(uy))x - (xu)(yx)\Big)$ which, together to the non-degenerate character of g, completes the proof. □

LEMMA 5.6.– Let A be an $H^\star$-algebra with zero annihilator. Then the following assertions are equivalent:

1) A satisfies the right Moufang identity;

2) A satisfies the middle Moufang identity;

3) A satisfies the left Moufang identity.

Furthermore, if the $H^\star$-algebra A satisfies some of the Moufang identities, then A is a flexible algebra, which is anticommutative only in the case that $A = 0$.

PROOF.– Let $g : A \times A \longrightarrow \mathbb{K}$ be the symmetric bilinear form defined by $g(x, y) = (x|y^\star) + (y|x^\star)$ for any $x, y \in A$. A straightforward verification shows that g is associative and non-degenerate. In consequence, lemma 5.5 gives the equivalence of assertions 1 and 2. Since $\star$ is bijective, the equivalence of 1 and 3 is obtained taking $\star$ in both sides of [5.8]. Now, taking $\star$ in both sides of [5.9], we found that A satisfies the middle Moufang identity if and only if $((xz)x)u = x(z(xu))$ holds for any $x, z, u \in A$. Since A has zero annihilator, this implies that A is a flexible algebra. Finally, assume that A is an anticommutative $H^\star$-algebra satisfying some of the Moufang identities. Let x be an arbitrary self-adjoint or skew-adjoint element of A. So $x^\star = \varepsilon x$ for $\varepsilon \in \{+, -\}$. Anticommutativity and [5.8] yield to $R_x^3 = 0$. Therefore, $0 = (R_x^4(y)|y) = (R_x^2(y)|R_x^2(y))$ for all $y \in A$. So $R_x^2(y) = 0$ and $0 = (R_x^2(y)|y) = \varepsilon(R_x(y)|R_x(y))$. Thus, $x \in \operatorname{Ann} A = \{0\}$. Since A or its realified (in the complex case) agrees with the direct sum $\operatorname{Sym}(A, \star) \oplus \operatorname{Sk}(A, \star)$ of the real vector subspaces $\operatorname{Sym}(A, \star)$ and $\operatorname{Sk}(A, \star)$ of self-adjoint and skew-adjoint elements, then we have $A = 0$. □

THEOREM 5.16.– (Cuenca 2002). Let A be an $H^\star$-algebra. Then A satisfies some of the Moufang identities if and only if A is an alternative algebra.

PROOF.– Without loss in generality, we assume $\operatorname{Ann} A = \{0\}$. By lemma 5.6, A is a flexible algebra satisfying all the Moufang identities. Now we will prove that A is an alternative $H^\star$-algebra. First we observe that [5.9] implies $x^2x^2 = (xx^2)x$. On the other hand, flexible algebras satisfy $x^2x = xx^2$. It is well known that the algebraic identities $x^2x^2 = (xx^2)x$ and $x^2x = xx^2$ imply that A is a power-associative algebra (in fact, they characterize power-associativity in the cases that the base field has zero characteristic). By proposition 5.15, A is a non-commutative Jordan $H^\star$-algebra. By lemma 5.6 and theorems 5.10 and 5.11, A is J-semisimple. By remark 5.3, there exists a net $\{A_\gamma\}$ of unital $H^\star$-subalgebras such that $A = \overline{\bigcup_\gamma A_\gamma}$. Let e_γ be the unity of every A_γ. Making $z = e_\gamma$ in [5.8] and $y = e_\gamma$ in [5.10], we see that every A_γ is an alternative algebra. Since $A = \overline{\bigcup_\gamma A_\gamma}$, then A is also an alternative algebra. □

5.10. Lie $H^\star$-algebras

We recall that an anticommutative algebra $\mathfrak{L}$ is said to be a *Lie algebra* if its product, usually denoted by $[\ ,\ ]$, satisfies $[[x,y],z]+[[y,z],x]+[[z,x],y]=0$ for any x, y, $z \in \mathfrak{L}$. This identity is called the *Jacobi identity*. Associative $H^\star$-algebras provides examples of Lie $H^\star$-algebras. So if A is an associative $H^\star$-algebra, then the antisymmetrized $H^\star$-algebra A^- is a Lie $H^\star$-algebra. The self-adjoint closed subalgebras of the antisymmetrized $H^\star$-algebra A^- of an associative $H^\star$-algebra A are also Lie $H^\star$-algebras. In particular, if $\sigma : A \longrightarrow A$ is an isometric involutive $\star$-antiautomorphism of the associative $H^\star$-algebra A, then the vector subspace $\mathrm{Sk}(A,\sigma)$ of the elements x of A such that $x = -\sigma(x)$ is a Lie $H^\star$-subalgebra of A^-. A way to obtain isometric involutive $\star$-antiautomorphisms of the associative complex $H^\star$-algebra of Hilbert–Schmidt operators $\mathcal{HS}(\mathcal{H})$ is through conjugations and anticonjugations of the complex Hilbert space $\mathcal{H}$. We recall that if $\mathcal{H}$ is a complex Hilbert space and $J : \mathcal{H} \longrightarrow \mathcal{H}$ is an isometric conjugate-linear map, then J is said to be a *conjugation* (respectively, *anticonjugation*) of the Hilbert space $\mathcal{H}$ if $J^2 = \mathrm{Id}$ (respectively, $J^2 = -\mathrm{Id}$). As in the above sections, we will denote by $T^\square$ the adjoint of a given linear operator T of the Hilbert space $\mathcal{H}$. Now if J is a conjugation or an anticonjugation of $\mathcal{H}$, then the map $\sigma_J : \mathcal{HS}(\mathcal{H}) \longrightarrow \mathcal{HS}(\mathcal{H})$ given by $\sigma_J : T \mapsto JT^\square J^{-1}$ is an isometric involutive $\star$-antiautomorphism. So we now dispose of three different kind of examples of Lie complex $H^\star$-algebras:

– Type A: The antisymmetrized Lie $H^\star$-algebra $\mathcal{HS}(\mathcal{H})^-$ of the associative $H^\star$-algebra $\mathcal{HS}(\mathcal{H})$ of the Hilbert–Schmidt operators of a complex Hilbert space $\mathcal{H}$.

– Type B: The Lie $H^\star$-subalgebra of $\mathcal{HS}(\mathcal{H})^-$ of the Hilbert–Schmidt operators T such that $JT^\square J^{-1} = -T$, where $\mathcal{H}$ is a complex Hilbert space and $J : \mathcal{H} \longrightarrow \mathcal{H}$ is a conjugation of $\mathcal{H}$.

– Type C: The Lie $H^\star$-subalgebra of $\mathcal{HS}(\mathcal{H})^-$ of the Hilbert–Schmidt operators T such that $JT^\square J^{-1} = -T$, where $\mathcal{H}$ is a complex Hilbert space and $J : \mathcal{H} \longrightarrow \mathcal{H}$ is an anticonjugation of $\mathcal{H}$.

So the $H^\star$-algebras of types B and C are always $H^\star$-subalgebras of antisymmetric elements $\mathrm{Sk}(\mathcal{HS}(\mathcal{H}),\sigma_J)$ of $\mathcal{HS}(\mathcal{H})^-$ with respect to the previously introduced isometric involutive $\star$-antiautomorphism σ_J (with J a conjugation in case

B, and an anticonjugation in case C). Usually, a Lie complex algebra is called a *Lie $H^\star$-algebra of classical type* if it is of some of the above types A, B or C. Lie complex $H^\star$-algebras were initially studied by Schue (1960, 1961) under the name of *$L^\star$-algebras*. With these papers, Schue opens the theory of $H^\star$-algebras to the non-associative world. He shows that the finite-dimensional simple complex Lie algebras are $H^\star$-algebras. He also determines all the topologically simple infinite-dimensional complex Lie algebras in the cases that the underlying Hilbert space is separable. He obtains that, up to a positive factor of the inner product, they are isometrically $\star$-isomorphic to some of the Lie $H^\star$-algebras of classical type, but with $\mathcal{H}$ a separable infinite-dimensional complex Hilbert space in all the cases. Balachandran (1969) studies the Lie $H^\star$-algebras of classical type. He shows that infinite-dimensional Lie complex $H^\star$-algebras of different types cannot be isomorphic, as well as several questions relative to the subalgebras of Cartan of these Lie $H^\star$-algebras. Furthermore, he poses explicitly the problem of knowing if all the topologically simple infinite-dimensional complex Lie $H^\star$-algebras are of classical type. An affirmative answer to this problem was given in Cuenca *et al.* (1990). We will now introduce the necessary elements to understand the way that this answer was obtained.

We recall that subalgebras of zero product of a given Lie algebra $\mathfrak{L}$ are named *abelian subalgebras* of $\mathfrak{L}$. Let now $\mathfrak{L}$ be a complex Lie $H^\star$-algebra with zero annihilator and H a subalgebra of $\mathfrak{L}$. As in Schue (1960), we will say that H is a *Cartan subalgebra* of $\mathfrak{L}$ if H is a maximal self-adjoint abelian subalgebra. By Zorn's lemma, $\mathfrak{L}$ has some Cartan subalgebra H. Moreover, Cartan subalgebras of $\mathfrak{L}$ are closed subalgebras. If $\alpha : H \longrightarrow \mathbb{C}$ is a linear map, we say that α is a root of $\mathfrak{L}$ relative to H if the subspace $V_\alpha = \{x \in \mathfrak{L} : [h,x] = \alpha(h)x \text{ for any } h \in H\}$ is non-zero. In this case, V_α is named the *root space* of α. In particular, 0 is a root and $V_0 = H$. If α is a root, then the continuity of the bracket product implies the continuity of α. So V_α is closed, and if $\alpha \neq 0$ there exists some $h_\alpha \in H$ such that $\alpha(h) = (h|h_\alpha)$ for all $h \in H$. Furthermore, each h_α is self-adjoint (see (Schue 1960, p. 711) for details). In consequence, if α is a root, then $-\alpha$ is also a root and $V_{-\alpha} = V_{\alpha}{}^\star$. Moreover, if α and β are different roots, then $(V_\alpha|V_\beta) = 0$ and $[V_\alpha, V_\beta] \subset V_{\alpha+\beta}$ understanding that $V_{\alpha+\beta} = 0$ if $\alpha + \beta$ is not a root. Let us to denote by Δ the set of non-zero roots of $\mathfrak{L}$ relative to H. We say that $\mathfrak{L}$ *admits a Cartan decomposition relative to H* if $\mathfrak{L}$ is the closure of the orthogonal direct sum $\bigoplus_{\alpha\in\Delta\cup\{0\}} V_\alpha$. It has been proved in (Schue 1961, p. 348) that every Lie $H^\star$-algebra with zero annihilator admits a Cartan decomposition relative to each of its Cartan subalgebras.

Using (Balachandran 1972, Remark 2 and Propositions 2 and 3), we can prove the next proposition (see (Cuenca *et al.* 1990, Proposition 1)).

PROPOSITION 5.16.– (Cuenca *et al.* 1990). Let $\mathfrak{L}$ be a topologically simple infinite-dimensional complex Lie $H^\star$-algebra and H a Cartan subalgebra of $\mathfrak{L}$. Then there exists a countable directed net $(\{\mathcal{L}_i\}_{i\in\mathcal{M}}, \{\mu_{ji}\}_{i\leq j})$ of infinite-dimensional topologically simple separable $H^\star$-subalgebras covering $\mathcal{L}$ such that: (i) each $H \cap \mathcal{L}_i$ is a Cartan subalgebra of $\mathcal{L}_i$; (ii) if $i \leq j$, then each root space of $\mathcal{L}_i$ relative to $H \cap \mathcal{L}_i$ and different from $H \cap \mathcal{L}_i$ is a root space of $\mathcal{L}_j$ relative to $H \cap \mathcal{L}_j$.

Let $\mathcal{L}_1$ (respectively, $\mathcal{L}_2$) be a Lie $H^\star$-algebra with zero annilator and H_1 (respectively, H_2) a Cartan subalgebra of $\mathcal{L}_1$ (respectively, $\mathcal{L}_2$). Let $G : \mathcal{L}_1 \longrightarrow \mathcal{L}_2$ be a $\star$-monomorphism. We say that G *preserves the root spaces* relative to H_1 and H_2 if $G(H_1) \subset H_2$ and the image of the root spaces of non-zero roots of $\mathcal{L}_1$ is root spaces of non-zero roots of $\mathcal{L}_2$. Let $\mathcal{L}_1$ and $\mathcal{L}_2$ be infinite-dimensional complex classical Lie $H^\star$-algebras, both of the same type. Using arguments similar to those of (Balachandran 1969, Proofs of Theorems 5 and 6), we can determine all the $\star$-monomorphisms from $\mathcal{L}_1$ into $\mathcal{L}_2$ that are positive multiples of isometries and preserve the root spaces relative to two given Cartan subalgebras H_1 and H_2. As a result, we can state the following proposition.

PROPOSITION 5.17.– Let $\mathcal{L}_1$ and $\mathcal{L}_2$ be two infinite-dimensional complex Lie $H^\star$-algebras with Cartan subalgebras H_1 and H_2 and $G : \mathcal{L}_1 \longrightarrow \mathcal{L}_2$ a $\star$-monomorphism preserving the root spaces relative to H_1 and H_2. Assume the existence of a real positive number ρ such that $\|G(x)\| = \rho\|x\|$ for all $x \in \mathcal{L}_1$. Then we have:

1) if $\mathcal{L}_1 = \mathcal{HS}(\mathcal{H}_1)^-$ and $\mathcal{L}_2 = \mathcal{HS}(\mathcal{H}_2)^-$ are both of type A, then either G is an associative $\star$-monomorphism from $\mathcal{HS}(\mathcal{H}_1)$ into $\mathcal{HS}(\mathcal{H}_2)$ or $-G$ is a $\star$-antimonomorphism between these associative $H^\star$-algebras;

2) if $\mathcal{L}_1$ and $\mathcal{L}_2$ are of the same type B or C with $\mathcal{L}_1 = \mathrm{Sk}(\mathcal{HS}(\mathcal{H}_1), \sigma_{J_1})$ and $\mathcal{L}_2 = \mathrm{Sk}(\mathcal{HS}(\mathcal{H}_2), \sigma_{J_2})$, then there exists a $\star$-monomorphism $G' : \mathcal{HS}(\mathcal{H}_1) \longrightarrow \mathcal{HS}(\mathcal{H}_2)$ extending G and satisfying $\sigma_{J_2} \circ G' = G' \circ \sigma_{J_1}$ and $\|G'(x)\| = \rho\|x\|$ for all $x \in \mathcal{L}_1$.

THEOREM 5.17.– (Cuenca *et al.* 1990). An infinite-dimensional complex Lie $H^\star$-algebra $\mathcal{L}$ is topologically simple if and only if there exist a $\star$-isomorphism G from $\mathcal{L}$ onto some of the Lie $H^\star$ algebras of classical type A, B or C with complex Hilbert space $\mathcal{H}$ of infinite dimension, and a strictly positive real number ρ such that ρG is isometric.

PROOF.– Let $\mathcal{L}$ be a topologically simple complex Lie $H^\star$-algebra of infinite dimension. By propositions 5.16 and 5.8, there exists a countable directed net $(\{\mathcal{L}_i\}_{i\in\mathcal{M}}, \{\mu_{ji}\}_{i\leq j})$ of infinite-dimensional topologically simple separable $H^\star$-subalgebras covering $\mathcal{L}$, with all the $\mathcal{L}_i$, up to a positive factor, isometrically $\star$-isomorphic to classical Lie $H^\star$-algebras of the same type A, B or C and with the $\star$-monomorphisms of inclusion $\mu_{ji} : \mathcal{L}_i \longrightarrow \mathcal{L}_j$ $(i \leq j)$ preserving the root spaces relative to $H \cap \mathcal{L}_i$ and $H \cap \mathcal{L}_j$. First, we suppose that all the $\mathcal{L}_i$ are $\star$-isomorphic to Lie $H^\star$-algebras A_i^- of type A. So for each $i \in \mathcal{M}$, there exists a separable complex

Hilbert space $\mathcal{H}_i$ of infinite dimension such that $A_i = \mathcal{HS}(\mathcal{H}_i)$, a positive real number ρ_i and a $\star$-isomorphism of Lie $H^\star$-algebras $\eta_i : \mathfrak{L}_i \longrightarrow A_i^-$ such that $\|\eta_i(x)\| = \rho_i\|x\|$ for all $x \in \mathfrak{L}_i$. Changing the inner product $(\mid)_i$ of A_i by $(\mid)_i/\rho_i^2$ we can suppose the η_i isometric $\star$-isomorphisms. So for each $i \leq j$, the monomorphism $e_{ji} : A_i^- \longrightarrow A_j^-$ given by $e_{ji} = \eta_j \circ \mu_{ji} \circ \eta_i^{-1}$ is an isometric $\star$-monomorphism preserving root spaces relative to $\eta_i(H \cap \mathfrak{L}_i)$ and $\eta_j(H \cap \mathfrak{L}_j)$. By proposition 5.17, each e_{ji} is either an isometric monomorphism or the opposite of an isometric antimonomorphism between the associative $H^\star$-algebras A_i and A_j. Fix $i_0 \in \mathcal{M}$. By proposition 5.8, there exists an ultrafilter $\mathcal{U}$ on the partial ordered set $\mathcal{N} = \{i \in \mathcal{M} : i \geq i_0\}$ such that $\mathcal{U}$ contains all the subsets $[i, \infty)$ $(i \in \mathcal{N})$, and a countable directed net of $H^\star$-subalgebras $(\{\mathfrak{L}_{i\in\mathcal{N}}\}, \mu_{ji})$ covering $\mathfrak{L}$. Furthermore, one and only one of the subsets $\mathcal{N}_1 = \{i \in \mathcal{N} : e_{ii_0} \text{ is a monomorphism}\}$ or $\mathcal{N}_2 = \{i \in \mathcal{N} : -e_{ii_0} \text{ is an antimonomorphism}\}$ belongs to $\mathcal{U}$ and, in addition, it gives a new countable directed net of $H^\star$-subalgebras covering $\mathfrak{L}$. Assume now that the first possibility occurs. Then $(\{\mathfrak{L}_i\}_{i\in\mathcal{N}_1}, \{e_{ji}\})$ is a countable directed net of $H^\star$-subalgebras covering $\mathfrak{L}$ and it is not difficult to see that for any $i, j \in \mathcal{N}_1$ such that $i \leq j$ the Lie monomorphism e_{ji} is an associative monomorphism. In consequence, $(\{A_i\}_{i,j\in\mathcal{N}_1}, \{e_{ji}\})$ is a directed isometric $\star$-monomorphic system of associative $H^\star$-algebras and $\mathcal{N}_1/\mathcal{U}$ is an ultrafilter on $\mathcal{N}_1$ containing all the subsets $[i, \infty)$ $(i \in \mathcal{N}_1)$. So $A_{\mathcal{N}_1/\mathcal{U}}$ is a topologically simple associative $H^\star$-algebra (see proposition 5.6), and there exists a complex Hilbert space $\mathcal{H}$ such that, up to a positive factor of the inner product, $A_{\mathcal{N}_1/\mathcal{U}}$ is $\star$-isometrically isomorphic to $\mathcal{HS}(\mathcal{H})$. Since $(A_{\mathcal{N}_1/\mathcal{U}})^-$ and $(A^-)_{\mathcal{N}_1/\mathcal{U}}$ are also isometrically $\star$-isomorphic, taking into account propositions 5.7 and 5.8, we obtain that $\mathfrak{L}$ is isometrically $\star$-isomorphic to $\mathcal{HS}(\mathcal{H})^-$. The case that $\mathcal{N}_2 \in \mathcal{U}$ can be reduced to the previous one (indeed, take an isometric involutive $\star$-antiautomorphism τ of A_{i_0}, substitute η_{i_0} by $\hat{\eta}_{i_0} = -\tau \circ \eta_{i_0}$ and each e_i such that $i \in \mathcal{N}_2$ and $i_0 \leq i$ by $\hat{e}_{ii_0} = -e_{ii_0} \circ \tau$ leaving the remaining η_i and e_{ij} unchanged).

If there exists a countable directed net $(\{\mathfrak{L}_i\}_{i\in\mathcal{M}}, \{\mu_{ji}\}_{i\leq j})$ of infinite-dimensional topologically separable Lie $H^\star$-subalgebras of the same type B or C with the inclusion $\star$-monomorphisms μ_{ji} $(i \leq j)$ preserving root spaces, we can argue in a similar way using part 2 of proposition 5.17 (see (Cuenca *et al.* 1990)). □

REMARK 5.4.– The topologically simple real Lie $H^\star$-algebras were described independently by Balachandran, de la Harpe and Unsain under the additional assumption of separability of its Hilbert space (Balachandran 1972; de la Harpe 1971; Unsain 1972). Furthermore, Balachandran (1972) has proved that the topologically simple real Lie $H^\star$-algebras can be completely described in the case that each topologically simple complex Lie $H^\star$-algebra of infinite dimension were of classical type. But, in virtue of theorem 5.17, this is so. In consequence, theorem 5.17 also concludes the structure theory of Lie $H^\star$-algebras in the real case.

5.11. Topics closely related to Lie $H^\star$-algebras

Let K be a given field. A *Malcev K-algebra* is an anticommutative algebra satisfying $J(x, y, xz) = J(x, y, z)x$ for any x, y, $z \in A$, where J is a trilinear map named *jacobiano* and defined as $J(x, y, z) = (xy)z + (yz)x + (zx)y$ for each 3-uple of A. Antisymmetrized of alternative algebras are Malcev algebras; in particular, Lie algebras are Malcev algebras. As was pointed out by Malcev, this kind of algebras are closely related with the theory of analytic loops. Many authors have been interested in the structure theory of Malcev algebras (Sagle 1962; Loos 1966; Kuz'min 1968b,a). Finally, Filippov (1977) characterizes the simple central non-Lie Moufang algebras by dropping the finite-dimension assumption of (Kuz'min 1968a). Proposition 5.5 and the Filippov theorem allow us to prove the next theorem, which was originally proved in Cabrera *et al.* (1988a) (see also (Cabrera and Rodríguez 1990) where the authors establish previously a generalization of proposition (5.5), part 1 relative to the extended centroid).

THEOREM 5.18.– (Cabrera *et al.* 1988a). Every topologically simple non-Lie Malcev $H^\star$-algebra A over $\mathbb{K}$ is up to a positive factor of the inner product isometrically $\star$-isomorphic to some of the following:

1) *Case* $\mathbb{K} = \mathbb{C}$: The anticommutative $H^\star$-algebra of the quadratic $H^\star$-algebra of complex octonions.

2) *Case* $\mathbb{K} = \mathbb{R}$: Either the above anticommutative $H^\star$-algebra of the complex octonions viewed as a real $H^\star$-algebra, or the anticommutative real $H^\star$-algebra of some of the two possible (split or division) real octonions $H^\star$-algebras.

Let A be an algebra endowed with an involutive antiautomorphism τ, whose subspace of symmetric (resp.skew-symmetric) elements is denoted by $\mathrm{Sym}(A; \tau)$ (respectively, $\mathrm{Sk}(A, \tau)$). If $(\,,\, ,\,)$ denotes the associator of A and the ground field is of characteristic $\neq 2, 3$, then we will say that A is a *structurable algebra* if the following conditions are satisfied:

1) $(s, x, y) = -(x, s, y) = (x, y, s)$;

2) $(a, b, c) - (c, a, b) = (b, a, c) - (c, b, a)$;

3) $\frac{2}{3}[\,[a^2, a], b] = (b, a^2, a) - (b, a, a^2)$;

for any $s \in \mathrm{Sk}(A, \tau)$, a, $b \in \mathrm{Sym}(A, \tau)$ and x, $y \in A$. Structurable algebras were introduced by Allison (1978) in the unital case by a different set of conditions, but equivalent to the above if A has a unit element. Examples of structurable algebras are alternative algebras with arbitrary involutive antiautomorphism, as well as Jordan algebras with $\tau = \mathrm{Id}$. The structure theory of the finite-dimensional structurable algebras over fields of characteristic other than 2 and 3 has been accomplished in (Schafer 1985; Smirnov 1990). Since every finite-dimensional τ-simple structurable complex algebra (A, τ) becomes an $H^\star$-algebra in an appropriate way with τ an isometric map (see (Cabrera *et al.* 1990)), the following theorem of (Cabrera *et al.* 1990) can be viewed as a proper extension of the classification theorem of the finite-dimensional complex τ-simple structurable algebras.

THEOREM 5.19.– (Cabrera *et al.* 1990). Let (A, τ) be a topologically τ-simple structurable complex $H^\star$-algebra with τ isometric. Then some of the following alternatives holds:

1) A has finite dimension;

2) A is associative;

3) A is a topologically simple $H^\star$-algebra with $\tau = \mathrm{Id}$ (Jordan algebra);

4) (A, τ) is a structurable $H^\star$-algebra constructed from an involutive Hilbert module over a topologically τ-simple associative complex $H^\star$-algebra $\mathcal{E}$ with isometric involutive antiautomorphism t (see (Cabrera *et al.* 1990) for a detailed description).

Structurable $H^\star$-algebras are linked to Lie $H^\star$-algebras through a certain construction in Cabrera *et al.* (1994) where to each structurable complex $H^\star$-algebra with zero annihilator and isometric involutive antiautomorphism a complex Lie $H^\star$-algebra with zero annihilator $\mathcal{K}(A, \tau)$ associates. This construction is an infinite-dimensional version of the Allison extension of the Kantor–Tits–Koecher construction and it has nice properties, as shown in the following theorem.

THEOREM 5.20.– (Cabrera *et al.* 1994). The above-mentioned construction of a complex Lie $H^\star$-algebras with zero annihilator $\mathcal{K}(A)$ from each structurable complex $H^\star$-algebra (A, τ) with zero annihilator and isometric involutive antiautomorphism τ induces an order preserving bijection between the τ-invariant closed ideals of everyone of these structurable $H^\star$-algebras (A, τ) and the closed ideals of $\mathcal{K}(A, \tau)$. In particular, (A, τ) is τ-topologically simple iff $\mathcal{K}(A, \tau)$ is topologically simple. Furthermore, all the complex topologically simple Lie $H^\star$-algebras arise in this construction.

An interesting application of the theory of Lie $H^\star$-algebras is provided by the equivariant monotone operators (see (Beltita 2006)). We will briefly outline the way in which the connection between equivariant monotone operators and Lie $H^\star$-algebras arises. First, we will introduce some concepts and notations.

Let $\mathfrak{g}$ be a real or complex complete normed algebra endowed with a continuous multiplicative involution $\star$. Often $\mathfrak{g}$ is called an *involutive Banach algebra.* Let $\mathfrak{g}^\sharp$ be the topological dual of $\mathfrak{g}$ and, in a similar way to page 163, for each $f \in \mathfrak{g}^\sharp$ we denote by $f^\star$ the linear form defined by $f^\star(a) = \overline{f(a^\star)}$ for all $a \in \mathfrak{g}$. Let now assume that $\mathfrak{g}$ is a real involutive Banach algebra. Paraphrasing (Beltita 2006, pp. 147–148) a continuous linear map $\iota : \mathfrak{g}^\sharp \longrightarrow \mathfrak{g}$ is said to be an *equivariant monotone operator* of $\mathfrak{g}$ if everyone of the following assertions hold:

1) $\iota(f^\star) = \iota(f)^\star$ for all $f \in \mathfrak{g}^\sharp$;

2) $L_a \circ \iota = \iota \circ (L_{a^\star})^\sharp$ and $R_a \circ \iota = \iota \circ (R_{a^\star})^\sharp$ for every $a \in \mathfrak{g}$ and where L_a (respectively, R_a) is the corresponding left (respectively, right) multiplication operator of a and $(L_{a^\star})^\sharp : \mathfrak{g}^\sharp \longrightarrow \mathfrak{g}$ and $(R_{a^\star})^\sharp : \mathfrak{g}^\sharp \longrightarrow \mathfrak{g}$ are the continuous linear maps given by $(L_{a^\star})^\sharp : f \mapsto f \circ L_{a^\star}$ and $(R_{a^\star})^\sharp : f \mapsto f \circ R_{a^\star}$;

3) $\hat{} \circ \iota = \iota^\sharp$, where $\hat{} : \mathfrak{g} \longrightarrow (\mathfrak{g}^\sharp)^\sharp$ is the linear map transforming each $a \in \mathfrak{g}$ into the map $\hat{a}$ given by $\hat{a}(f) = f(a)$ for any $f \in \mathfrak{g}^\sharp$ and $\iota^\sharp : \mathfrak{g}^\sharp \longrightarrow (\mathfrak{g}^\sharp)^\sharp$ is given by $\iota^\sharp : f \mapsto f \circ \iota$;

4) $f(\iota(f)) \geq 0$ for all $f \in \mathfrak{g}^\sharp$.

We will denote by $\mathcal{E}^+(\mathfrak{g})$ the set of the equivariant monotone operators of $\mathfrak{g}$ and by $\mathcal{E}_0^+(\mathfrak{g})$ the subset of those $\iota \in \mathcal{E}^+(\mathfrak{g})$ such that $\ker \iota = 0$. Equivariant monotone operators are closely related to the so-called Lie $H^\star$-ideals (or $L^\star$-ideals). Let $\mathfrak{g}$ be an involutive Banach algebra over $\mathbb{K}$ ($\mathbb{K} = \mathbb{R}$ or $\mathbb{C}$). Following (Beltita 2006, p. 152), an ideal I of $\mathfrak{g}$ is said to be an *$H^\star$-ideal* if there exist an $H^\star$-algebra with continuous involution A and a $\star$-homomorphism $\varphi : A \longrightarrow \mathfrak{g}$ such that $I = \operatorname{Im}\varphi$. If, in addition, A can be chosen being a Lie $H^\star$-algebra, then we say that I is an *$L^\star$-ideal* of $\mathfrak{g}$. By considering an appropriate $H^\star$-algebra structure on the orthogonal subspace $(\ker\varphi)^\perp$ of $\ker\varphi$ in A and by restricting φ to $(\ker\varphi)^\perp$, we see that we only need to consider $\star$-monomorphisms of $H^\star$-algebras with continuous involution into $\mathfrak{g}$ in order to obtain all the $H^\star$-ideals of $\mathfrak{g}$ (Cabrera and Rodríguez 2014, 2018, Proof of Theorem 8.1.101). Lie algebras, which are involutive Banach algebras, are called *involutive Banach–Lie algebras*. In the case that $\mathfrak{g}$ is a real involutive Banach–Lie algebra, it is not difficult to see that, under certain additional conditions, $L^\star$-ideals of $\mathfrak{g}$ allows to construct elements of $\mathcal{E}^+(\mathfrak{g})$ (see (Beltita 2006, Proposition 8.1)). In (Beltita 2006, Theorem 8.8), this situation is reversed establishing the following theorem that we reformulate in the following way.

THEOREM 5.21.– Let $\mathfrak{g}$ be a real involutive Banach–Lie algebra and $\iota : \mathfrak{g}^\sharp \longrightarrow \mathfrak{g}$ an equivariant monotone operator. Then there exist a real Lie $H^\star$-algebra $\mathfrak{L}$ with isometric involution and an injective $\star$-homomorphism $\varphi : \mathfrak{L} \longrightarrow \mathfrak{g}$ such that $\iota = \varphi \circ \theta \circ \varphi^\sharp$, where $\theta : \mathfrak{L}^\sharp \longrightarrow \mathfrak{L}$ is the linear map that for each linear form g is given by $\theta(g) = x$, where x is the unique element of $\mathfrak{L}$ such that $g(y) = (x|y)$ for all $y \in \mathfrak{L}$. Furthermore, the $L^\star$-ideal $\operatorname{Im}\varphi$ depends exclusively on ι.

5.12. Two-graded $H^\star$-algebras

Two-graded $H^\star$-algebras have been studied in the associative and Jordan cases. We recall that a K-algebra $\mathfrak{A}$ is said to be a *two-graded* algebra if there exist two vector subspaces $\mathfrak{A}_0$ and $\mathfrak{A}_1$ such that $\mathfrak{A} = \mathfrak{A}_0 \oplus \mathfrak{A}_1$ and satisfying for all $\alpha, \beta \in \{0,1\}$ that $\mathfrak{A}_\alpha \mathfrak{A}_\beta \subset \mathfrak{A}_{\alpha+\beta}$, where the sum of indices must be understood as modulo 2. Usual algebraic concepts such as ideals, homomorphisms, etc. are considered in a two-graded sense. So if $\mathfrak{A}$ and $\mathfrak{B}$ are two-graded algebras and $G : \mathfrak{A} \longrightarrow \mathfrak{B}$ a linear map, we say that G is a *homomorphism of two-graded algebras* if G is an homomorphism of algebras in the ungraded sense and, in addition, $G(\mathfrak{A}_\alpha) \subset \mathfrak{B}_\alpha$ for all $\alpha = 0,\ 1$. On the other hand, an *ideal* I of the two-graded algebra $\mathfrak{A}$ is an ideal of $\mathfrak{A}$ in the ungraded sense that can be decomposed as a direct sum of vector subspaces $I = I_0 \oplus I_1$ with $I_0 \subset \mathfrak{A}_0$ and $I_1 \subset \mathfrak{A}_1$. A *two-graded $H^\star$-algebra* is a real or complex two-graded algebra, which is also an $H^\star$-algebra such that $\mathfrak{A}_0$ and $\mathfrak{A}_1$ are orthogonal self-adjoint closed subspaces of $\mathfrak{A}$.

A two-graded $H^\star$-algebra $\mathfrak{A}$ is *topologically simple* if $\mathfrak{A}^2 \neq 0$ and $\mathfrak{A}$ has not closed (graded) ideals different from 0 and $\mathfrak{A}$. Also the minimal closed ideals, as in the ungraded case, must be understood as minimal in the set of the non-zero closed (graded) ideals of $\mathfrak{A}$.

The structure theory of the two-graded $H^\star$-algebras begins with the following theorem, which formulates the analogous of proposition 5.3 and theorem 5.1 in this context.

THEOREM 5.22.– (Castellón *et al.* 1993, Theorem 3). Let $\mathfrak{A}$ be a two-graded $H^\star$-algebra. Then:

1) $\mathfrak{A}$ is an orthogonal direct sum $\mathfrak{A} = \operatorname{Ann}\mathfrak{A} \oplus \overline{\mathfrak{A}^2}$ where, with respect to the restriction of the inner product and a suitable involution the closed ideal $\overline{\mathfrak{A}^2}$ becomes a two-graded $H^\star$-algebra with zero annihilator;

2) if $\mathfrak{A}$ has zero annihilator, then the orthogonal direct sum $\bigoplus I_\rho$ of the minimal closed ideals (in the graded sense) is dense in $\mathfrak{A}$. Furthermore, each I_ρ is a topologically simple two-graded $H^\star$-algebra.

The determination of the topologically simple two-graded $H^\star$-algebras have been achieved in the associative and Jordan cases by different methods. The way followed in the associative case was strongly inspired by the treatment of the prime associative two-graded algebras with non-zero socle of Cuenca *et al.* (1994). So using the results in Castellón *et al.* (1992), the following classification theorem of the topologically simple associative complex two-graded $H^\star$-algebras is obtained.

THEOREM 5.23.– (Castellón *et al.* 1993, Theorem 5). Let A be a topologically simple associative complex two-graded $H^\star$-algebra. Then up to a positive factor of the inner product A is isometrically $\star$-isomorphic to some of the following two-graded $H^\star$-algebras:

1) the $H^\star$-algebra $\mathcal{M}_{\mathcal{A}\cup\mathcal{B}}(\mathbb{C})$ with $\mathcal{A}$ and $\mathcal{B}$ two sets such that $\mathcal{A}\cap\mathcal{B}=\emptyset$ and $|\mathcal{A}\cup\mathcal{B}|\geq 1$ and the grading below

$$A_0=\begin{pmatrix}\mathcal{M}_{\mathcal{A}}(\mathbb{C}) & 0\\ 0 & \mathcal{M}_{\mathcal{B}}(\mathbb{C})\end{pmatrix},\qquad A_1=\begin{pmatrix}0 & \mathcal{M}_{\mathcal{A},\mathcal{B}}(\mathbb{C})\\ \mathcal{M}_{\mathcal{B},\mathcal{A}}(\mathbb{C}) & 0\end{pmatrix};$$

2) the two-graded $H^\star$-algebra $A=\mathcal{M}_{\mathcal{A}}(\mathbb{C})\oplus\mathcal{M}_{\mathcal{A}}(\mathbb{C})$ that is orthogonal direct sum of two copies of the Hilbert space $\mathcal{M}_{\mathcal{A}}(\mathbb{C})$, with product and involution $(a,b)(c,d)=(ac+bd,ad+bc)$, $((a_{ij}),(b_{ij}))^\star=((\overline{a_{ji}}),(\overline{b_{ji}})$ and grading $A_0=(\mathcal{M}_{\mathcal{A}}(\mathbb{C}),0)$ and $A_1=(0,\mathcal{M}_{\mathcal{A}}(\mathbb{C}))$.

The classification of the topologically simple associative real two-graded $H^\star$-algebras is also given in Castellón *et al.* (1993), a greater number of different possibilities now appearing.

The determination of the topologically simple Jordan two-graded $H^\star$-algebras is based on an easy dichotomy rule of the general theory of two-graded $H^\star$-algebras similar to that of (Cabrera *et al.* 1988b, Proof of Theorem 1), which can be established in the following way: If $\mathfrak{A}$ is a topologically simple two-graded $H^\star$-algebra, then one of the two following possibilities occurs:

1) there exists a topologically simple (ungraded) $H^\star$-algebra $\mathfrak{B}$ such that $\mathfrak{A}$ can be written as an orthogonal direct sum of Hilbert spaces $\mathfrak{A}=\mathfrak{B}\oplus\mathfrak{B}$ and $\mathfrak{A}$ has the graduation $\mathfrak{A}_0=(\mathfrak{B},0)$ and $\mathfrak{A}_1=(0,\mathfrak{B})$ and the product and involution given by

$$(b_0,b_1)(c_0,c_1)=(b_0c_0+b_1c_1,b_0c_1+b_1c_0)\ \text{and}\ (b_0,b_1)^\star=({b_0}^\star,{b_1}^\star) \qquad [5.11]$$

2) $\mathfrak{A}$ is a topologically simple as ungraded $H^\star$-algebra in which case the graduation is determined through an isometric involutive $\star$-automorphism G of $\mathfrak{A}$ so that $\mathfrak{A}_0=\mathrm{Sym}(\mathfrak{A},G)$ and $\mathfrak{A}_1=\mathrm{Sk}(\mathfrak{A},G)$.

The above dichotomy focuses the problem of the classification of the topologically simple two-graded $H^\star$-algebras in a given class $\mathcal{C}$ in the determination of the classes of the involutive isometric $\star$-automorphisms between ungraded topologically simple $H^\star$-algebras in $\mathcal{C}$ under the equivalence relation $G\equiv H$ iff there exists some isometric $\star$-isomorphism φ such that $G\circ\varphi=\varphi\circ H$. In the Jordan case, this classification was given in (Cuenca and Martín 1992). As a result, we obtain the determination of the topologically simple Jordan two-graded $H^\star$-algebras in the real and complex cases. In order to give an idea of these classifications, we give a detailed description when $\mathbb{C}$ is the ground field.

THEOREM 5.24.– (Cuenca and Martín 1992, Theorem 1). Up to a positive factor of the inner product each topologically simple complex Jordan two-graded $H^\star$-algebra A is isometrically $\star$-isomorphic to some of the following:

1) A is an orthogonal direct sum $A = B \oplus B$, where B is a topologically simple Jordan $H^{\star}$-algebra and A has the grading $A_0 = (B, 0)$, $A_1 = (0, B)$ and product and involution given as in [5.11];

2) A is the $H^{\star}$-algebra $\mathcal{M}_{\mathcal{A}}(\mathbb{C})^{+}$, with $|\mathcal{A}| \neq 2$ and grading $A_0 = \{M \in A : M^t = M\}$ and $A_1 = \{M \in A : M^t = -M\}$;

3) $A = \mathcal{M}_{\mathcal{A}}(\mathbb{C})^{+}$, with $|\mathcal{A}| > 2$, $A_0 = \{M : SM^t = MS\}$, $A_1 = \{M : SM^t = -MS\}$, and where S is the symplectic matrix and $|\mathcal{A}|$ is even if $|\mathcal{A}|$ is finite;

4) $A = \mathcal{M}_{\mathcal{A}\cup\mathcal{B}}(\mathbb{C})^{+}$, with $|\mathcal{A} \cup \mathcal{B}| > 2$ and grading

$$A_0 = \begin{pmatrix} \mathcal{M}_{\mathcal{A}}(\mathbb{C})^{+} & 0 \\ 0 & \mathcal{M}_{\mathcal{B}}(\mathbb{C})^{+} \end{pmatrix}, \quad A_1 = \begin{pmatrix} 0 & \mathcal{M}_{\mathcal{A},\mathcal{B}}(\mathbb{C}) \\ \mathcal{M}_{\mathcal{B},\mathcal{A}}(\mathbb{C}) & 0 \end{pmatrix};$$

5) the quadratic Jordan $H^{\star}$-algebra associated with a complex Hilbert space W of dimension > 1 provided with an isometric involution $\square$, where A has the grading given by $A_0 = \mathbb{C} \oplus W_0$ and $A_1 = W_1$, with W_0 and W_1 two orthogonal vector subspaces such that $W = W_0 \oplus W_1$ and satisfying $W_0^{\square} \subset W_0$ and $W_1^{\square} \subset W_1$;

6) the Jordan $H^{\star}$-algebra A of the symmetric $\mathcal{A} \times \mathcal{A}$-matrices with entries in $\mathbb{C}$, with $|\mathcal{A}| \geq 4$ and even in the finite case, provided with the grading $A_0 = \{M \in A : SM^t = MS\}$ and $A_1 = \{M \in A : SM^t = -MS\}$ for S the symplectic matrix;

7) the Jordan $H^{\star}$-algebra A of the symmetric $\mathcal{A} \times \mathcal{B}$-matrices with entries in $\mathbb{C}$ and $|\mathcal{A} \cup \mathcal{B}| \geq 3$, viewed as a two-graded $H^{\star}$-subalgebra of the two-graded $H^{\star}$-algebra $\mathcal{M}_{\mathcal{A}\cup\mathcal{B}}(\mathbb{C})^{+}$ of case 4;

8) the Jordan $H^{\star}$-algebra A of the $(\mathcal{A} \cup \mathcal{B}) \times (\mathcal{A} \cup \mathcal{B})$-matrices M with entries in $\mathbb{C}$ satisfying $SM^t = MS$ for S the symplectic matrix, with $|\mathcal{A}| \geq 6$ and even in the finite case, provided with the grading $A_0 = \{M \in A : M^t = M\}$ and $A_1 = \{M \in A : M^t = -M\}$;

9) the $H^{\star}$-algebra of the $(\mathcal{A} \cup \mathcal{B}) \times (\mathcal{A} \cup \mathcal{B})$-matrices M with entries in $\mathbb{C}$ satisfying $SM^t = MS$ for S the symplectic matrix, with $|\mathcal{A} \cup \mathcal{B}| \geq 6$ and even in the finite case, viewed as Jordan two-graded $H^{\star}$-subalgebra of the two-graded $H^{\star}$-algebra $\mathcal{M}_{\mathcal{A}\cup\mathcal{B}}(\mathbb{C})^{+}$ of case 4;

10) the 27-dimensional exceptional complex Jordan $H^{\star}$-algebra with some of its three possible gradings.

By realifying, all the Jordan two-graded $H^{\star}$-algebras appearing in the above theorem give topologically simple real two-graded Jordan $H^{\star}$-algebras. Nevertheless, the class of the topologically simple real two-graded Jordan $H^{\star}$-algebras is very much wider that is obtained by realifying theorem 5.24. A complete description of this that occurs in the real case can be seen in (Cuenca and Martín 1992, Theorems 2 and 3). The class of two-graded Lie $H^{\star}$-algebras have also been treated. So, as an easy consequence of proposition 5.17, the classification of the infinite-dimensional topologically simple complex Lie two-graded $H^\star$-algebras is obtained in (Calderón and Martín 2001, Theorem 1.1).

5.13. Other topics: beyond the $H^\star$-algebras

We have seen that Hilbert modules appear in the description of the topologically simple structurable $H^\star$-algebras. They are treated in Cabrera *et al.* (1995) (see also (Cabrera *et al.* 1993) for additional information, and (Cabrera *et al.* 1992) for its connection with ternary structures). $H^\star$-algebras admitting an absolute value has been studied in (Cuenca and Rodríguez 1995). The notable result asserting that topologically simple semi-$H^\star$-algebras are totally multiplicative prime is given by (Cabrera and Mohammed 2002, Theorem 4.1).

Furthermore, different classes of $H^\star$-triple systems, a ternary counterpart of $H^\star$-algebras, have been extensively studied (Kaup 1983; Neher 1987; Castellón and Cuenca 1992b,a, 1990; Castellón *et al.* 1994, 2000). In fact, the structure theory of the more relevant classes of $H^\star$-triple systems can be accomplished (associative, alternative and Jordan). An exception is the Lie case, where the problem of the classification of the topologically simple Lie $H^\star$-triple systems remains open. Ternary $H^\star$-algebras are another ternary structure related to associative $H^\star$-algebras (Castellón *et al.* 1992). As for $H^\star$-algebras, there exists also a well-developed general theory of $H^\star$-triple systems (Castellón and Cuenca 1992c, 1993; Villena *et al.* 1995; Villena and Zhory 1996), which also deals with problems of automatic continuity. Furthermore, the Krein counterparts of $H^\star$-algebras and $H^\star$-triple systems have been introduced by P.S. Rema, S. Sribala and N. Vijayarangan.

5.14. Acknowledgments

The author wishes to acknowledge professors Miguel Cabrera García and Ángel Rodríguez Palacios for their careful reading of the manuscript and their valuable comments.

5.15. References

Áhn, P.N. (1986). Simple Jordan algebras with minimal inner ideals. *Commun. Algebra*, 14(3), 489–492.

Albert, A.A. (1948). Power associative rings. *Trans. Amer. Math. Soc.*, 64, 552–593.

Albert, A.A. (1949). A theory of trace-admissible algebras. *Proc. Nat. Acad. Sci.*, 35(6), 317–322.

Allison, B.N. (1978). A class of nonassociative algebras with involution containing the class of Jordan algebras. *Math. Ann.*, 237, 133–1356.

Ambrose, W. (1945). Structure theorems for a special class of Banach algebras. *Trans. Amer. Math. Soc.*, 57, 364–386.

Balachandran, V.K. (1969). Simple $L^\star$-algebras of classical type. *Math. Ann.*, 180, 205–219.

Balachandran, V.K. (1972). Real $L^\star$-algebras. *Indian J. Pure Appl. Math.*, 3, 1224–1246.

Balachandran, V.K., Swaminathan, N. (1986). Real $H^\star$-algebras. *J. Funct. Anal.*, 65, 64–75.

Beltita, D. (2006). *Smooth Homogeneous Structures in Operator Theory*. Chapman & Hall/CRC, Boca Raton.

Braun, H., Koecher, M. (1966). *Jordan-Algebren*. Springer, Berlin/New York.

Bruck, R.H., Kleinfeld, E. (1951). The structure of alternative division rings. *Proc. Amer. Math Soc.*, 2, 878–890.

Cabrera, M., Rodríguez, A. (1990). Extended centroid and central closure of semiprime normed algebras: A first approach. *Comm. Algebra*, 18, 2293–2326.

Cabrera, M., Rodríguez, A. (2014). *Non-associative Normed Algebras. Volume 1: The Vidav-Palmer Theorem and Gelfand-Naimark Theorems*. Cambridge University Press, Cambridge.

Cabrera, M., Rodríguez, A. (2018). *Non-associative Normed Algebras. Volume 2: Representation Theory and the Zel'manov Approach*. Cambridge University Press, Cambridge.

Cabrera, M., Martínez, J., Rodríguez, A. (1988a). Malcev $H^\star$-algebras. *Math. Proc. Camb. Phil. Soc.*, 103, 463–471.

Cabrera, M., Martínez, J., Rodríguez, A. (1988b). Nonassociative real $H^\star$-algebras. *Publ. Mat.*, 32, 267–274.

Cabrera, M., Martínez, J., Rodríguez, A. (1990). Structurable $H^\star$-algebras. *J. Algebra*, 147, 19–62.

Cabrera, M., Martínez, J., Rodríguez, A. (1992). Hilbert modules over $H^\star$-algebras in relation with Hilbert ternary rings. In *Nonassociative Algebraic Models*, González, S., Myung, H.C. (eds). Nova Science Publishers, New York, 33–44.

Cabrera, M., Martínez, J., Rodríguez, A. (1993). Bounded differential operators on Hilbert modules and derivations of structurable $H^\star$-algebras. *Comm. Algebra*, 21, 2905–2945.

Cabrera, M., Marrackchi, A.E., Martínez, J., Rodríguez, A. (1994). A Tits-Koecher-Allison-Kantor construction for Lie $H^\star$-algebras. *J. Algebra*, 164(2), 361–408.

Cabrera, M., Martínez, J., Rodríguez, A. (1995). Hilbert modules revisited: Orthonormal bases and Hilbert-Schmidt operators. *Glasgow Math. J.*, 37, 45–54.

Cabrera, M., Mohammed, A.A. (2002). Totally multiplicatively prime algebras. *Proc. Roy. Soc, Edinbugh Sect A*, 132(11), 5241–5252.

Calderón, A.J., Martín, C. (2001). Hilbert space methods in the theory of Lie triple systems. In *Recent Progress in Functional Analysis*, Bierstedt, K.D., Bonet, J., Maestre, M., Schmets, J. (eds). Elsevier, Amsterdam, 309–319.

Castellón, A., Cuenca, J.A. (1990). Compatibility in Jordan $H^\star$-triple systems. *Bollettino U.M.I.*, 4-B, 433–447.

Castellón, A., Cuenca, J.A. (1992a). Alternative $H^\star$-triple systems. *Comm. Algebra*, 11(20), 3191–3260.

Castellón, A., Cuenca, J.A. (1992b). Associative $H^{\star}$-triple systems. In *Nonassociative Algebraic Models*, González, S., Myung, H.C. (eds). Nova Science Publishers, New York, 45–67.

Castellón, A., Cuenca, J.A. (1992c). Isomorphisms of $H^{\star}$-triple systems. *Annali Della Scuola Normale Superiore di Pisa. Ser. IV*, XIX(4), 507–514.

Castellón, A., Cuenca, J.A. (1993). Centroid and metacentroid of an $H^{\star}$-triple system. *Bulletin de la Societé Mathématique de Belgique*, 45(1-2 A), 85–91.

Castellón, A., Cuenca, J.A., Martín, C. (1992). Ternary $H^{\star}$-algebras. *Bolletino U.M.I.*, 6-B(7), 217–228.

Castellón, A., Cuenca, J.A., Martín, C. (1993). Applications of ternary $H^{\star}$-algebras to associative $H^{\star}$-superalgebras. *Algebras, Groups and Geometries*, 10, 181–190.

Castellón, A., Cuenca, J., Martín, C. (1994). Jordan $H^{\star}$-triple systems. In *Nonassociative Algebras and its Applications*, González, S. (ed.). Kluwer, Amsterdam, 66–72.

Castellón, A., Cuenca, J., Martín, C. (2000). Special Jordan $H^{\star}$-triple systems. *Comm. Algebra*, 28(10), 4699–4706.

Cuenca, J.A. (1984). Sobre $H^{\star}$-álgebras no asociativas. Teoría de estructura de las $H^{\star}$-álgebras de Jordan no conmutativas semisimples. Secretariado de Publicaciones de la Universidad de Málaga.

Cuenca, J.A. (2002). Moufang $H^{\star}$-algebras. *Extr. Math.*, 17(2), 239–246.

Cuenca, J.A., Martín, C. (1992). Jordan two-graded $H^{\star}$-algebras. In *Nonassociative Algebraic Models*, Gonzélez, S., Myung, H.C. (eds). Nova Science Publishers, New York, 91–100.

Cuenca, J.A., Rodríguez, A. (1987). Structure theory for noncommutative Jordan $H^{\star}$-algebras. *J. Algebra*, 106, 1–14.

Cuenca, J.A., Rodríguez, A. (1995). Absolute values on $H^{\star}$-algebras. *Comm. Algebra*, 23(5), 2595–2610.

Cuenca, J.A., Rodríguez, A. (1985). Isomorphisms of $H^{\star}$-algebras. *Math. Proc. Camb. Phil. Soc.*, 97, 93–99.

Cuenca, J.A., Sánchez, A. (1994). Structure theory for real noncommutative Jordan $H^{\star}$-algebras. *J. Algebra*, 164, 481–499.

Cuenca, J.A., García, A., Martín, C. (1990). Structure theory for $L^{\star}$-algebras. *Math. Proc. Camb. Phil. Soc.*, 107, 361–365.

Cuenca, J.A., García, A., Martín, C. (1994). Prime associative superalgebras with nonzero socle. *Algebras, Groups and Geometries*, 11, 359–369.

de la Harpe, P. (1971). Classification des $L^{\star}$-algèbres semi-simples réelles séparables. *C.R. Acad. Sci*, A(272), 1559–1561.

Devapakkiam, C.V. (1975). Hilbert space methods in the theory of Jordan algebras I. *Math. Proc. Camb. Phil. Soc.*, 78, 293–300.

Devapakkiam, C.V., Rema, P. (1976). Hilbert space methods in the theory of Jordan algebras II. *Math. Proc. Camb. Phil. Soc.*, 79, 307–319.

Eklof, P.C. (1999). Ultraproducts for algebraists. In *Handbook of Mathematical Logic*, Barwise, J. (ed.). North-Holland Publishing Corporation, Amsterdam, 106–137.

Fernández, A., Rodríguez, A. (1986). Primitive noncommutative Jordan algebras with nonzero socle. *Proc. Amer. Math. Soc.*, 96(2), 199–206.

Filippov, V.T. (1977). Central simple Malcev algebras. *Algebra Logic*, 15, 147–151.

Herstein, I.N. (1964). Anelli alternativi ed algebre di composizione. *Rend. Mat. e Appl.*, 23, 364–393.

Herstein, I.N. (1976). *Rings with Involution*. Chicago University Press, Chicago.

Jacobson, N. (1968a). *Structure and Representations of Jordan Algebras*. Amer. Math. Soc. Providence.

Jacobson, N. (1968b). *Structure of Rings*. Amer. Math. Soc. Providence.

Jacobson, N. (1969). *Lectures on Quadratic Jordan Algebras*. Tata Institute of Fundamental Research, Bombay.

Jacobson, N. (1974). *Basic Algebra I*. Freeman, San Francisco.

Jacobson, N. (1981). *Structure Theory of Jordan Algebras*. University of Arkansas, Fayetteville.

Jodan, P., von Neumann, J., Wigner, E. (1934). On an algebraic generalization of the quantum mechanical formalism. *Ann. Math.*, 35(1), 29–64.

Kaplansky, I. (1948). Dual rings. *Ann. Math.*, 49(3), 689–701.

Kaplansky, I. (1972). *Fields and Rings*, 2nd edition. Chicago University Press, Chicago.

Kaup, W. (1983). Über die klassifikation der symmetrischen hermiteschen mannigfaltigkeiten unendlicher dimension II. *Math. Ann.*, 262, 57–75.

Kleinfeld, E. (1953a). On simple alternative rings. *Amer. J. Math.*, 75(1), 98–104.

Kleinfeld, E. (1953b). Simple alternative rings. *Proc. Amer. Math. Soc.*, 58(3), 545–547.

Kuz'min, E.N. (1968a). Malcev algebras and their representations. *Algebra i Logika*, 7, 235–244.

Kuz'min, E.N. (1968b). Simple Malcev algebras over a field of characteristic zero. *Soviet Math. Dokl.*, 9, 1034–1036.

Loos, O. (1966). Uber eine Beziehung zwischen MalcevAlgebren und Lie-tripelsysteme. *Pacific J. Math.*, 18, 553–562.

McCrimmon, K. (1965). Norms and noncommutative Jordan algebras. *Pacific J. Math.*, 15, 925–956.

McCrimmon, K. (1966). Structure and representations of noncommutative Jordan algebras. *Trans. Amer. Math. Soc.*, 121(1), 187–199.

McCrimmon, K. (1969). The radical of a Jordan algebra. *Proc. Nat. Acad. Sci.*, 62, 671–678.

McCrimmon, K. (1971). Noncommutative Jordan rings. *Trans. Amer. Math. Soc.*, 158, 1–33.

McCrimmon, K. (2004). *A Taste of Jordan Algebras*. Springer, New York.

Neher, E. (1987). *Jordan Triple Systems by the Grid Approach*. Springer, Berlin/Heidelberg/New York.

Oehmke, R.H. (1958). On flexible algebras. *Ann. Math*, 68(2), 221–230.

Osborn, J.M., Racine, M.L. (1979). Jordan rings with nonzero socle. *Trans. Amer. Math. Soc.*, 251, 375–387.

Pérez de Guzmán, I. (1983). Structure theorems for alternative $H^\star$-algebras. *Math. Proc. Camb. Phil. Soc.*, 94, 437–446.

Rodríguez, A. (1988). Jordan axioms for $C^\star$-algebras. *Manuscripta Math.*, 61, 279–314.

Rodríguez, A. (1995). Continuity of densely valued homomorphisms into $H^\star$-algebras. *Quart. J. Math. Oxford*, 46, 107–118.

Rudin, W. (1973). *Functional Analysis*. McGraw-Hill, New York.

Sagle, A.A. (1962). Simple Malcev algebras over fields of characteristic zero. *Pacific J. Math.*, 12, 1057–1078.

Sánchez, A. (1989). $H^\star$-álgebras de Jordan no conmutativas reales. PhD Thesis. University of Málaga.

Schafer, R.D. (1955). Noncommutative Jordan algebras of characteristic 0. *Proc. Amer. Math. Soc.*, 172, 472–475.

Schafer, R.D. (1966). *An Introduction to Nonassociative Algebras*. Academic Press, New York.

Schafer, R.D. (1985). On structurable algebras. *J. Algebra*, 92, 400–412.

Schue, J.R. (1960). Hilbert space methods in the theory of Lie algebras. *Trans. Amer. Math. Soc.*, 95, 68–80.

Schue, J.R. (1961). Cartan decompositions for L*-algebras. *Trans. Amer. Math. Soc.*, 98, 334–349.

Slater, M. (1970). Prime alternative rings I and II. *J. Algebra*, 15, 229–251.

Slater, M. (1972). Prime alternative rings III. *J. Algebra*, 21(3), 394–409.

Smiley, M.F. (1953). Right $H^\star$-algebras. *Proc. Amer. Math. Soc.*, 4(1), 1–4.

Smirnov, O.N. (1990). Simple and semisimple structurable algebras. *Algebra and Logic*, 29, 377–394.

Unsain, I. (1972). Classification of the simple separable real $L^\star$-algebras. *J. Differential Geom.*, 7(106), 423–451.

Villena, A. (1994). Continuity of derivations of $H^\star$-algebras. *Proc. Amer. Math. Soc.*, 122, 821–826.

Villena, A., Zhory, M. (1996). Continuity of homomorphism pairs into $H^\star$-triple systems. *Proc. (Math. Sci.) I. A. Sc.*, 106, 79–90.

Villena, A.R., Zalar, B., Zohry, M. (1995). Continuity of derivations pairs on Hilbert triple systems. *Bollettino U.M.I.*, 7(9-B), 459–477.

Zelmanov, E.I. (1979). Prime Jordan algebras. *Algebra i Logika*, 18, 162–175.

Zelmanov, E.I. (1983). On prime Jordan algebras II. *Siberian Math. J.*, 24, 89–104.

Zhevlakov, K.A., Slin'ko, A.M., Shestakov, I.P., Shirshov, A.I. (1982). *Rings That are Nearly Associative*. Academic Press, New York.

6

Krichever–Novikov Type Algebras: Definitions and Results

Martin SCHLICHENMAIER
Department of Mathematics, Luxembourg University, Luxembourg

6.1. Introduction

The study of infinite dimensional (associative or Lie) algebras and their representations is a huge and rather involved field. Additional structures like a grading or additional information about their origin will be indispensable to obtain insights and results.

Krichever–Novikov (KN)-type algebras are an important class of infinite dimensional algebras. Roughly speaking, they are defined as algebras of meromorphic objects on compact Riemann surfaces, or equivalently, on projective curves. The non-holomorphicity is controlled by a fixed finite set of points where poles are allowed. Splitting of this set of possible points of poles into two disjoint subsets will induce an "almost-grading" (see definition 6.5). It is a weaker concept than a grading but still powerful enough to act as a basic tool in representation theory. For example, highest weight representations can still be defined. Of course, central extensions of these algebras are also needed. They are forced by representation theory and quantization.

Examples of KN-type algebras are the well-known algebras of conformal field theory (CFT), the Witt algebra, the Virasoro algebra, the affine Lie algebras (affine Kac–Moody algebras), etc. They appear when the geometric setting consists of the

Algebra and Applications 1,
coordinated by Abdenacer MAKHLOUF. © ISTE Ltd 2020.

Riemann Sphere, i.e. the genus zero Riemann surface, and the points of possible poles are $\{0\}$ and $\{\infty\}$. The almost-grading is now a honest grading.

Historically, starting from these well-known genus zero algebras, Krichever and Novikov (1987, 1989) suggested a global operator approach via KN objects. Still they only considered two possible points where poles are allowed and were dealing with the vector field and the function algebra. For work on affine algebras, Sheinman (1990) should be mentioned.

From the applications in CFT (e.g. string theory) but also from purely mathematical reasons, the need for a multipoint theory is evident. In 1990, the author of the current review developed a systematic theory valid for all genera (including zero) and any fixed finite set of points where poles are allowed (Schlichenmaier 1990b,c,a,d). These extensions were not at all straightforward. The main point was to introduce a replacement of the graded algebra structure present in the "classical" case. Krichever and Novikov found that the already mentioned almost-grading (definition 6.5) will be enough to allow for the standard constructions in representation theory. In Schlichenmaier (1990a,d), it was realized that a splitting of the set A of points where poles are allowed into two disjoint non-empty subsets $A = I \cup O$ is crucial for introducing an almost-grading. The corresponding almost-grading was explicitly given. In contrast to the classical situation, where there is only one grading, we will have a finite set of non-equivalent gradings and new interesting phenomena show up. This is already true for the genus zero case (i.e. the Riemann sphere case) with more than two points where poles are allowed. These algebras will only be almost-graded (e.g. Schlichenmaier (1993); Fialowski and Schlichenmaier (2003, 2005); Schlichenmaier (2017)).

Other (Lie) algebras were also introduced. In fact, most of them come from a *Mother Poisson Algebra*, the algebra of meromorphic form (see section 6.4.2). This algebra carries a (weak) almost-grading that gives the almost-grading for the other algebras. For the relevant algebras, almost-graded central extensions are constructed and classified. In the case of genus zero, universal central extensions are obtained.

KN-type algebras have a lot of interesting applications. They show up in the context of deformations of algebras, moduli spaces of marked curves, Wess–Zumino–Novikov–Witten (WZNW) models, Knizhnik–Zamolodchikov (KZ) equations, integrable systems, quantum field theories, symmetry algebras, and in many more domains of mathematics and theoretical physics. The KN-type algebras carry a very rich representation theory. We have Verma modules, highest weight representations, Fermionic and Bosonic Fock representations, semi-infinite wedge forms, $b - c$ systems, Sugawara representations and vertex algebras.

In 2014, the author published the book *KN-Type Algebras. Theory and Applications* (Schlichenmaier (2014a)), which collects all the results, proofs and some applications of the multipoint KN algebras. Quite an extensive list of references can also be found, including articles published by physicists on applications in the field-theoretical context. For some applications in the context of integrable systems, see also Sheinman (2012).

Recently, a revived interest in the theory of KN-type algebras appeared again in mathematics. The goal of this review is to give a gentle introduction to the KN-type algebras in the multipoint setting, and to collect the basic definitions and results so that they are accessible for an interested audience not yet familiar with them. For the proof and more material, we have to refer to the original articles and the corresponding parts of Schlichenmaier (2014a). There is a certain overlap with a previous survey of mine: Schlichenmaier (2016).

6.2. The Virasoro algebra and its relatives

These algebras supply examples of non-trivial infinite dimensional Lie algebras. They are widely used in CFT. For the convenience of the reader, we will start by recalling their conventional algebraic definitions.

The *Witt algebra* $\mathcal{W}$, sometimes also called Virasoro[1] algebra without central term, is the Lie algebra generated as a vector space over $\mathbb{C}$ by the basis elements $\{e_n \mid n \in \mathbb{Z}\}$ with the Lie structure

$$[e_n, e_m] = (m - n)e_{n+m}, \quad n, m \in \mathbb{Z}. \tag{6.1}$$

The algebra $\mathcal{W}$ is more than just a Lie algebra. It is a graded Lie algebra. If we set for the degree $\deg(e_n) := n$, then

$$\mathcal{W} = \bigoplus_{n\in\mathbb{Z}} \mathcal{W}_n, \qquad \mathcal{W}_n = \langle e_n \rangle_{\mathbb{C}}. \tag{6.2}$$

Obviously, $\deg([e_n, e_m]) = \deg(e_n) + \deg(e_m)$.

Algebraically $\mathcal{W}$ can also be given as the Lie algebra of derivations of the algebra of Laurent polynomials $\mathbb{C}[z, z^{-1}]$.

REMARK 6.1.– In the purely algebraic context, our field of definition $\mathbb{C}$ can be replaced by an arbitrary field $\mathbb{K}$ of characteristics 0.

1 For some remarks this would be a correct name, see Guieu and Roger (2007).

For the Witt algebra, the universal one-dimensional central extension is the *Virasoro algebra* $\mathcal{V}$. As a vector space, it is the direct sum $\mathcal{V} = \mathbb{C} \oplus \mathcal{W}$. If we set for $x \in \mathcal{W}$, $\hat{x} := (0, x)$, and $t := (1, 0)$, then its basis elements are $\hat{e}_n$, $n \in \mathbb{Z}$ and t with the Lie product [2].

$$[\hat{e}_n, \hat{e}_m] = (m-n)\hat{e}_{n+m} + \frac{1}{12}(n^3 - n)\delta_n^{-m}\, t, \qquad [\hat{e}_n, t] = [t, t] = 0, \quad [6.3]$$

for all $n, m \in \mathbb{Z}$. By setting $\deg(\hat{e}_n) := \deg(e_n) = n$ and $\deg(t) := 0$, the Lie algebra $\mathcal{V}$ becomes a graded algebra. The algebra $\mathcal{W}$ will only be a subspace, not a subalgebra of $\mathcal{V}$. But it will be a quotient. Up to equivalence of central extensions and rescaling the central element t, this is beside the trivial (splitting) central extension, the only central extension of $\mathcal{W}$.

Given a finite-dimensional Lie algebra $\mathfrak{g}$ (e.g. a finite-dimensional simple Lie algebra), the tensor product of $\mathfrak{g}$ with the associative algebra of Laurent polynomials $\mathbb{C}[z, z^{-1}]$ carries a Lie algebra structure via

$$[x \otimes z^n, y \otimes z^m] := [x, y] \otimes z^{n+m}. \quad [6.4]$$

This algebra is called *current algebra* or *loop algebra* and denoted by $\overline{\mathfrak{g}}$. Again we consider central extensions. For this, let β be a symmetric, bilinear form for $\mathfrak{g}$ which is invariant (i.e. $\beta([x, y], z) = \beta(x, [y, z])$ for all $x, y, z \in \mathfrak{g}$). Then a central extension is given by

$$[\widehat{x \otimes z^n}, \widehat{y \otimes z^m}] := \widehat{[x, y] \otimes z^{n+m}} - \beta(x, y) \cdot m\, \delta_n^{-m} \cdot t. \quad [6.5]$$

This algebra is denoted by $\widehat{\mathfrak{g}}$ and called *affine Lie algebra*. With respect to the classification of Kac–Moody Lie algebras, in the case of a simple $\mathfrak{g}$ they are exactly the Kac–Moody algebras of untwisted affine type (Kac 1968, 1990; Moody 1969).

To complete the description, let me introduce the Lie superalgebra of Neveu–Schwarz type. The centrally extended superalgebra has as a basis (we drop the $\hat{}$)

$$e_n,\ n \in \mathbb{Z}, \qquad \varphi_m,\ m \in \mathbb{Z} + \frac{1}{2}, \qquad t \quad [6.6]$$

2 Here δ_k^l is the Kronecker delta, which is equal to 1 if $k = l$, otherwise zero.

with structure equations

$$
\begin{aligned}
[e_n, e_m] &= (m-n)e_{m+n} + \frac{1}{12}(n^3-n)\,\delta_n^{-m}\,t,\\
[e_n, \varphi_m] &= (m-\frac{n}{2})\,\varphi_{m+n},\\
[\varphi_n, \varphi_m] &= e_{n+m} - \frac{1}{6}\,(n^2-\frac{1}{4})\,\delta_n^{-m}\,t.
\end{aligned}
\tag{6.7}
$$

By "setting $t = 0$", we obtain the non-extended superalgebra. The elements e_n (and t) are a basis of the subspace of even elements, the elements φ_m are a basis of the subspace of odd elements.

These algebras are Lie superalgebras. For completeness, I recall their definition here.

REMARK 6.2.– (Definition of a Lie superalgebra) Let $\mathcal{S}$ be a vector space that is decomposed into even and odd elements $\mathcal{S} = \mathcal{S}_{\bar{0}} \oplus \mathcal{S}_{\bar{1}}$, i.e. $\mathcal{S}$ is a $\mathbb{Z}/2\mathbb{Z}$-graded vector space. Furthermore, let $[.,.]$ be a $\mathbb{Z}/2\mathbb{Z}$-graded bilinear map $\mathcal{S}\times\mathcal{S}\to\mathcal{S}$ such that for elements x, y of pure parity

$$
[x,y] = -(-1)^{\bar{x}\bar{y}}[y,x]. \tag{6.8}
$$

Here $\bar{x}$ is the parity of x, etc. These conditions state that

$$
[\mathcal{S}_{\bar{0}},\mathcal{S}_{\bar{0}}] \subseteq \mathcal{S}_{\bar{0}}, \qquad [\mathcal{S}_{\bar{0}},\mathcal{S}_{\bar{1}}] \subseteq \mathcal{S}_{\bar{1}}, \qquad [\mathcal{S}_{\bar{1}},\mathcal{S}_{\bar{1}}] \subseteq \mathcal{S}_{\bar{0}}, \tag{6.9}
$$

and that $[x,y]$ is symmetric for x and y odd, otherwise anti-symmetric. Now $\mathcal{S}$ is a *Lie superalgebra* if in addition the *super-Jacobi identity* (for x, y, z of pure parity)

$$
(-1)^{\bar{x}\bar{z}}[x,[y,z]] + (-1)^{\bar{y}\bar{x}}[y,[z,x]] + (-1)^{\bar{z}\bar{y}}[z,[x,y]] = 0 \tag{6.10}
$$

is valid. As long as the type of the arguments is different from *(even, odd, odd)*, all signs can be put to $+1$ and we obtain the form of the usual Jacobi identity. In the remaining cases, we get

$$
[x,[y,z]] + [y,[z,x]] - [z,[x,y]] = 0. \tag{6.11}
$$

By the definitions, $\mathcal{S}_0$ is a Lie algebra.

6.3. The geometric picture

In the previous section, I gave the Virasoro algebra and its relatives by purely algebraic means, i.e. by basis elements and structure equations. The full importance and strength will become more visible in a geometric context. Also from this geometric realization, the need for a generalization, as obtained via the KN-type algebras, will become evident.

6.3.1. *The geometric realizations of the Witt algebra*

A geometric description of the Witt algebra over $\mathbb{C}$ can be given as follows.

Let W be the algebra of those meromorphic vector fields on the Riemann sphere $S^2 = \mathbb{P}^1(\mathbb{C})$, which are holomorphic outside $\{0\}$ and $\{\infty\}$. Its elements can be given as

$$v(z) = \tilde{v}(z)\frac{d}{dz} \qquad [6.12]$$

where $\tilde{v}$ is a meromorphic function on $\mathbb{P}^1(\mathbb{C})$, which is holomorphic outside $\{0, \infty\}$. Those are exactly the Laurent polynomials $\mathbb{C}[z, z^{-1}]$. Consequently, this subalgebra has the set $\{e_n, n \in \mathbb{Z}\}$ with $e_n = z^{n+1}\frac{d}{dz}$ as a vector space basis. The Lie bracket of vector fields is calculated as

$$[v, u] = \left(\tilde{v}\frac{d}{dz}\tilde{u} - \tilde{u}\frac{d}{dz}\tilde{v}\right)\frac{d}{dz}. \qquad [6.13]$$

By evaluating for the basis elements e_n, this gives [6.1] and the algebra can be identified with the Witt algebra, defined purely algebraically.

The subalgebra of global holomorphic vector fields is the three-dimensional subspace $\langle e_{-1}, e_0, e_1\rangle_{\mathbb{C}}$. It is isomorphic to the Lie algebra $\mathfrak{sl}(2, \mathbb{C})$.

Similarly, the algebra $\mathbb{C}[z, z^{-1}]$ can be given as the algebra of meromorphic functions on $S^2 = \mathbb{P}^1(\mathbb{C})$ holomorphic outside of $\{0, \infty\}$.

6.3.2. *Arbitrary genus generalizations*

In the geometric setup for the Virasoro algebra the objects are defined on the Riemann sphere and might have poles at two fixed points, at most. For a global operator approach to CFT and its quantization, this is not sufficient. One needs Riemann surfaces of arbitrary genus. Moreover, one needs more than two points where singularities are allowed[3]. Such generalizations were initiated by Krichever and Novikov (1987, 1989), who considered arbitrary genus and the two-point case. As far as the current algebras are concerned, see also Sheinman (1990, 1992, 1993, 1995). The multipoint case was systematically examined by Schlichenmaier (1990b,c,a,d, 1993, 1996, 2003b,a). For a related approach, see also Sadov (1991).

For the whole contribution, let Σ be a compact Riemann surface without any restriction for the genus $g = g(\Sigma)$. Furthermore, let A be a finite subset of Σ. Later we will need a splitting of A into two non-empty disjoint subsets I and O, i.e. $A = I \cup O$. Set $N := \#A$, $K := \#I$, $M := \#O$, with $N = K + M$. More precisely, let

$$I = (P_1, \ldots, P_K), \quad \text{and} \quad O = (Q_1, \ldots, Q_M) \qquad [6.14]$$

be disjoint ordered tuples of distinct points ("marked points", "punctures") on the Riemann surface. In particular, we assume $P_i \neq Q_j$ for every pair (i, j). The points in I are called the *in-points*, and the points in O are the *out-points*. Occasionally, we consider I and O simply as sets.

Sometimes we refer to the classical situation. By this, we get

$$\Sigma = \mathbb{P}^1(\mathbb{C}) = S^2, \quad I = \{z = 0\}, \quad O = \{z = \infty\}, \qquad [6.15]$$

and the situation considered in section 6.3.1.

The figures should indicate the geometric picture. Figure 6.1 shows the classical situation.

Figure 6.2 is genus 2, but still a two-point situation.

Finally, in Figure 6.3 the case of a Riemann surface of genus 2 with two incoming points and one outgoing point is explained.

REMARK 6.3.– We stress the fact that these generalizations are also needed in the case of genus zero if one considers more than two points. Even in the case of genus zero and three points, interesting algebras show up (see also Schlichenmaier (2017)).

3 The singularities correspond to points where free fields are entering the region of interaction or leaving it. In particular, from the very beginning there is a natural decomposition of the set of points into two disjoint subsets.

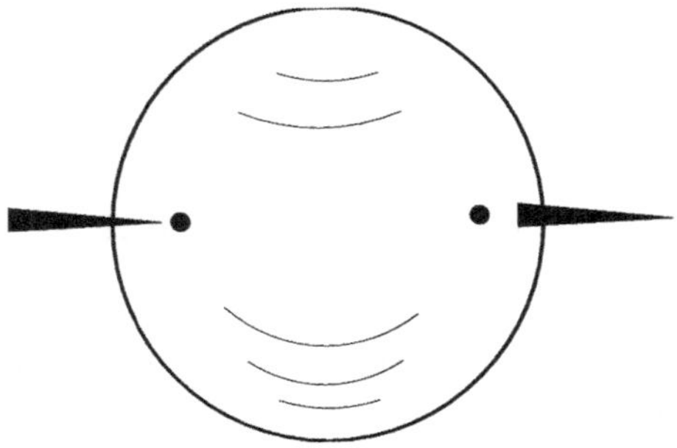

Figure 6.1. *Riemann surface of genus zero with one incoming and one outgoing point*

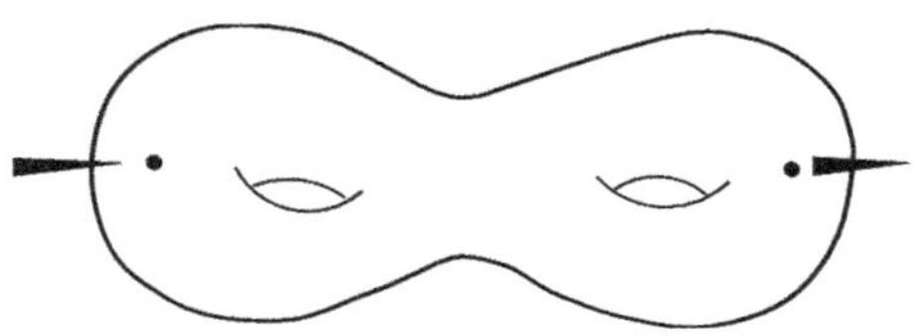

Figure 6.2. *Riemann surface of genus two with one incoming and one outgoing point*

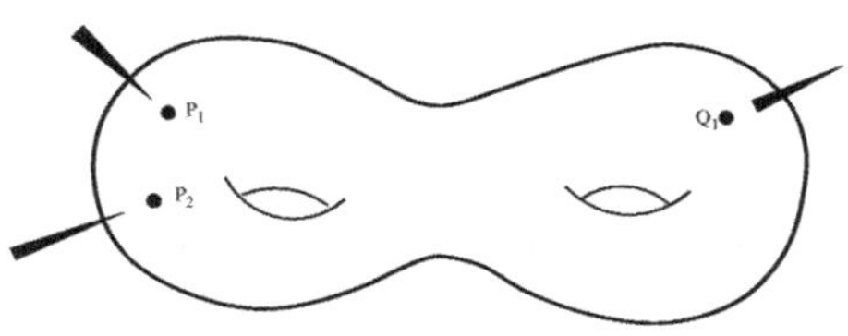

Figure 6.3. *Riemann surface of genus two with two incoming points and one outgoing point*

6.3.3. *Meromorphic forms*

To introduce the elements of the generalized algebras (later called KNtype algebras), we first have to discuss forms of certain (conformal) weights. Recall that Σ is a compact Riemann surface of genus $g \geq 0$. Let A be a fixed finite subset of Σ. In fact, we could allow, for this and the following sections (as long as we do not discuss almost-grading), that A is an arbitrary subset. This includes the extremal cases $A = \emptyset$ or $A = \Sigma$.

Let $\mathcal{K} = \mathcal{K}_\Sigma$ be the canonical line bundle of Σ. Its local sections are the local holomorphic differentials. If $P \in \Sigma$ is a point and z a local holomorphic coordinate at P, then a local holomorphic differential can be written as $f(z)dz$ with a local holomorphic function f defined in a neighborhood of P. A global holomorphic section can be described locally in coordinates $(U_i, z_i)_{i\in J}$ by a system of local holomorphic functions $(f_i)_{i\in J}$, which are related by the transformation rule induced by the coordinate change map $z_j = z_j(z_i)$ and the condition $f_i dz_i = f_j dz_j$. This yields

$$f_j = f_i \cdot \left(\frac{dz_j}{dz_i}\right)^{-1}. \qquad [6.16]$$

A meromorphic section of $\mathcal{K}$, i.e. a *meromorphic differential* is given as a collection of local meromorphic functions $(h_i)_{i\in J}$ with respect to a coordinate covering for which the transformation law [6.16] remains true. We will not make any distinction between the canonical bundle and its sheaf of sections, which is a locally free sheaf of rank 1.

In the following, λ is either an integer or a half-integer. If λ is an integer, then

1) $\mathcal{K}^\lambda := \mathcal{K}^{\otimes\lambda}$ for $\lambda > 0$;

2) $\mathcal{K}^0 := \mathcal{O}$, the trivial line bundle;

3) $\mathcal{K}^\lambda := (\mathcal{K}^*)^{\otimes(-\lambda)}$ for $\lambda < 0$.

Here, $\mathcal{K}^*$ denotes the dual line bundle of the canonical line bundle. This is the holomorphic tangent line bundle, whose local sections are the holomorphic tangent vector fields $f(z)(d/dz)$.

If λ is a half-integer, then we first have to fix a "square root" of the canonical line bundle, sometimes called a *theta characteristic*. This means we fix a line bundle L for which $L^{\otimes 2} = \mathcal{K}$. After L has been chosen, we set $\mathcal{K}^\lambda := \mathcal{K}^\lambda_L := L^{\otimes 2\lambda}$. In most cases, we will not mention L, but we have to keep the choice in mind. The fine structure of the algebras we are about to define will depend on the choice, but the main properties will remain the same.

REMARK 6.4.– A Riemann surface of genus g has exactly 2^{2g} non-isomorphic square roots of $\mathcal{K}$. For $g = 0$, we have $\mathcal{K} = \mathcal{O}(-2)$, and $L = \mathcal{O}(-1)$, the tautological bundle, is the unique square root. Already for $g = 1$, we have four non-isomorphic ones. As in this case $\mathcal{K} = \mathcal{O}$, one solution is $L_0 = \mathcal{O}$. But we have also other bundles L_i, $i = 1, 2, 3$. Note that L_0 has a non-vanishing global holomorphic section, whereas this is not the case for L_1, L_2 and L_3. In general, depending on the parity of the dimension of the space of globally holomorphic sections, i.e. of $\dim \mathrm{H}^0(\Sigma, L)$, one distinguishes even and odd theta characteristics L. For $g = 1$, the bundle $\mathcal{O}$ is an odd, the others are even theta characteristics. The choice of a theta characteristic is also called a spin structure on Σ (Atiyah 1971).

We set

$$\mathcal{F}^{\lambda}(A) := \{f \text{ is a global meromorphic section of } \mathcal{K}^{\lambda} \mid \\ f \text{ is holomorphic on } \Sigma \setminus A\}. \qquad [6.17]$$

Obviously this is a $\mathbb{C}$-vector space. To avoid a cumbersome notation, we will often drop the set A in the notation if A is fixed and clear from the context. Recall that in the case of half-integer λ, everything depends on the theta characteristic L.

DEFINITION 6.1.– *The elements of the space* $\mathcal{F}^{\lambda}(A)$ *are called* meromorphic forms of weight λ *(with respect to the theta characteristic* L*).*

REMARK 6.5.– In the two extremal cases for the set A, we obtain $\mathcal{F}^{\lambda}(\emptyset)$ the global holomorphic forms, and $\mathcal{F}^{\lambda}(\Sigma)$ all meromorphic forms. By compactness, each $f \in \mathcal{F}^{\lambda}(\Sigma)$ will have only finitely many poles. In the case that $f \not\equiv 0$, it will also have only finitely many zeros.

If f is a meromorphic λ-form, it can be represented locally by meromorphic functions f_i via $f = f_i (dz_i)^{\otimes \lambda}$. If $f \not\equiv 0$, the local representing functions have only finitely many zeros and poles. Whether a point P is a zero or a pole of f does not depend on the coordinate z_i chosen. We can define for $P \in \Sigma$ the *order*

$$\mathrm{ord}_P(f) := \mathrm{ord}_P(f_i), \qquad [6.18]$$

where $\mathrm{ord}_P(f_i)$ is the lowest non-vanishing order in the Laurent series expansion of f_i in the variable z_i around P. It will not depend on the coordinate z_i chosen.

The order $\mathrm{ord}_P(f)$ is (strictly) positive if and only if P is a zero of f. It is negative if and only if P is a pole of f. Moreover, its value gives the order of the zero and pole, respectively. By compactness, our Riemann surface Σ can be covered by finitely many coordinate patches. Hence, f can only have finitely many zeros and poles. We define the *(sectional) degree* of f to be

$$\mathrm{sdeg}(f) := \sum_{P \in \Sigma} \mathrm{ord}_P(f). \qquad [6.19]$$

PROPOSITION 6.1.– Let $f \in \mathcal{F}^{\lambda}$, $f \not\equiv 0$ then

$$\mathrm{sdeg}(f) = 2\lambda(g-1). \qquad [6.20]$$

For this and related results, see Schlichenmaier (2007b).

6.4. Algebraic structures

Next we introduce algebraic operations on the vector space of meromorphic forms of arbitrary weights. This space is obtained by summing over all weights

$$\mathcal{F} := \bigoplus_{\lambda \in \frac{1}{2}\mathbb{Z}} \mathcal{F}^{\lambda}. \qquad [6.21]$$

The basic operations will allow us to finally introduce the algebras we are heading for.

6.4.1. *Associative structure*

In this section, A is still allowed to be an arbitrary subset of points in Σ. We will drop the subset A in the notation. The natural map of the locally free sheaves of rank 1

$$\mathcal{K}^{\lambda} \times \mathcal{K}^{\nu} \to \mathcal{K}^{\lambda} \otimes \mathcal{K}^{\nu} \cong \mathcal{K}^{\lambda+\nu}, \quad (s,t) \mapsto s \otimes t, \qquad [6.22]$$

defines a bilinear map

$$\cdot : \mathcal{F}^{\lambda} \times \mathcal{F}^{\nu} \to \mathcal{F}^{\lambda+\nu}. \qquad [6.23]$$

With respect to local trivializations, this corresponds to the multiplication of the local representing meromorphic functions

$$(s\, dz^{\lambda}, t\, dz^{\nu}) \mapsto s\, dz^{\lambda} \cdot t\, dz^{\nu} = s \cdot t\; dz^{\lambda+\nu}. \qquad [6.24]$$

If there is no danger of confusion, then we will mostly use the same symbol for the section and for the local representing function.

PROPOSITION 6.2.– The space $\mathcal{F}$ is an associative and commutative graded (over $\frac{1}{2}\mathbb{Z}$) algebra. Moreover, $\mathcal{A} = \mathcal{F}^0$ is a subalgebra and the $\mathcal{F}^{\lambda}$ are modules over $\mathcal{A}$.

Of course, $\mathcal{A}$ is the algebra of those meromorphic functions on Σ which are holomorphic outside of A. In the case that $A = \emptyset$, it is the algebra of global holomorphic functions. By compactness, these are only the constants, hence $\mathcal{A}(\emptyset) = \mathbb{C}$. In the case that $A = \Sigma$, it is the field of all meromorphic functions $\mathcal{M}(\Sigma)$.

6.4.2. *Lie and Poisson algebra structure*

Next we define a Lie algebra structure on the space $\mathcal{F}$. The structure is induced by the map

$$\mathcal{F}^{\lambda} \times \mathcal{F}^{\nu} \to \mathcal{F}^{\lambda+\nu+1}, \qquad (e, f) \mapsto [e, f], \qquad [6.25]$$

which is defined in local representatives of the sections by

$$(e\, dz^{\lambda}, f\, dz^{\nu}) \mapsto [e\, dz^{\lambda}, f\, dz^{\nu}] := \left((-\lambda)e\frac{df}{dz} + \nu f\frac{de}{dz}\right) dz^{\lambda+\nu+1}, \qquad [6.26]$$

and bilinearly extended to $\mathcal{F}$. Of course, we have to show the following.

PROPOSITION 6.3.– (Schlichenmaier 2014a, Proposition 2.6 and 2.7) The prescription $[.,.]$ given by [6.26] is well defined and defines a Lie algebra structure on the vector space $\mathcal{F}$.

PROPOSITION 6.4.– (Schlichenmaier 2014a, Proposition 2.8) The subspace $\mathcal{L} = \mathcal{F}^{-1}$ is a Lie subalgebra, and the $\mathcal{F}^{\lambda}$'s are Lie modules over $\mathcal{L}$.

DEFINITION 6.2.– *An algebra* $(\mathcal{B}, \cdot, [.,.])$ *such that* $\cdot$ *defines the structure of an associative algebra on* $\mathcal{B}$ *and* $[.,.]$ *defines the structure of a Lie algebra on* $\mathcal{B}$ *is called a* Poisson algebra *if and only if the Leibniz rule is true, i.e.*

$$\forall e, f, g \in \mathcal{B} : [e, f \cdot g] = [e, f] \cdot g + f \cdot [e, g]. \qquad [6.27]$$

In other words, via the Lie product $[.,.]$ the elements of the algebra act as derivations on the associative structure.

THEOREM 6.1.– (Schlichenmaier 2014a, Theorem 2.10) The space $\mathcal{F}$ with respect to $\cdot$ and $[.,.]$ is a Poisson algebra.

Next we consider important substructures. We already encountered the subalgebras $\mathcal{A}$ and $\mathcal{L}$, but there are more structures around.

6.4.3. *The vector field algebra and the Lie derivative*

First, we look again at the Lie subalgebra $\mathcal{L} = \mathcal{F}^{-1}$. Here, the Lie action respects the homogeneous subspaces $\mathcal{F}^{\lambda}$. As forms of weight -1 are vector fields, it could also be defined as the Lie algebra of those meromorphic vector fields on the Riemann surface Σ, which are holomorphic outside of A. For vector fields, we have the usual

Lie bracket and the usual Lie derivative for their actions on forms. For the vector fields, we have (again naming the local functions with the same symbol as the section) for $e, f \in \mathcal{L}$

$$[e,f]_{|} = [e(z)\frac{d}{dz}, f(z)\frac{d}{dz}] = \left(e(z)\frac{df}{dz}(z) - f(z)\frac{de}{dz}(z)\right)\frac{d}{dz}\,. \qquad [6.28]$$

For the Lie derivative, we get

$$\nabla_e(f)_{|} = L_e(g)_{|} = e\,.\,g_{|} = \left(e(z)\frac{df}{dz}(z) + \lambda f(z)\frac{de}{dz}(z)\right)\frac{d}{dz}\,. \qquad [6.29]$$

Obviously, these definitions coincide with the definitions already given in [6.26], but now we have obtained a geometric interpretation.

6.4.4. *The algebra of differential operators*

If we look at $\mathcal{F}$, considered as a Lie algebra, more closely, we see that $\mathcal{F}^0$ is an abelian Lie subalgebra and the vector space sum $\mathcal{F}^0 \oplus \mathcal{F}^{-1} = \mathcal{A} \oplus \mathcal{L}$ is also a Lie subalgebra. In an equivalent way, it can also be constructed as semidirect sum of $\mathcal{A}$ considered as a abelian Lie algebra and $\mathcal{L}$ operating on $\mathcal{A}$ by taking the derivative.

DEFINITION 6.3.– *The* Lie algebra of differential operators of degree ≤ 1 *is defined as the semidirect sum of $\mathcal{A}$ with $\mathcal{L}$ and is denoted by $\mathcal{D}^1$.*

In terms of elements the Lie product is

$$[(g,e),(h,f)] = (e\,.\,h - f\,.\,g\,,\ [e,f]). \qquad [6.30]$$

Using the fact that $\mathcal{A}$ is an abelian subalgebra in $\mathcal{F}$, this is exactly the definition for the Lie product given for this algebra. Hence, $\mathcal{D}^1$ is a Lie algebra.

The projection on the second factor $(g,e) \mapsto e$ is a Lie homomorphism and we obtain a short exact sequences of Lie algebras

$$0 \longrightarrow \mathcal{A} \longrightarrow \mathcal{D}^1 \longrightarrow \mathcal{L} \longrightarrow 0\,. \qquad [6.31]$$

Hence, $\mathcal{A}$ is an (abelian) Lie ideal of $\mathcal{D}^1$ and $\mathcal{L}$ a quotient Lie algebra. Obviously, $\mathcal{L}$ is also a subalgebra of $\mathcal{D}^1$.

PROPOSITION 6.5.– The vector space $\mathcal{F}^\lambda$ becomes a Lie module over $\mathcal{D}^1$ by the operation

$$(g,e).f := g \cdot f + e.f, \quad (g,e) \in \mathcal{D}^1(A), f \in \mathcal{F}^\lambda(A). \qquad [6.32]$$

6.4.5. *Differential operators of all degrees*

In the following, we want to also consider differential operators of arbitrary degree acting on $\mathcal{F}^\lambda$. This is obtained via some universal constructions. First, we consider the universal enveloping algebra $U(\mathcal{D}^1)$. We denote its multiplication by $\odot$ and its unit by $\mathbf{1}$.

The universal enveloping algebra contains many elements which act in the same manner on $\mathcal{F}^\lambda$. For example, if h_1 and h_2 are functions different from constants, then $h_1 \cdot h_2$ and $h_1 \odot h_2$ are different elements of $U(\mathcal{D}^1)$. Nevertheless, they act in the same way on $\mathcal{F}^\lambda$.

Hence, we will divide out further relations

$$\mathcal{D} = U(\mathcal{D}^1)/J, \quad \text{respectively} \quad \mathcal{D}_\lambda = U(\mathcal{D}^1)/J_\lambda \qquad [6.33]$$

with the two-sided ideals

$$J := (\ a \odot b - a \cdot b,\ \mathbf{1} - 1 \mid a, b \in \mathcal{A}\ ),$$
$$J_\lambda := (\ a \odot b - a \cdot b,\ \mathbf{1} - 1,\ a \odot e - a \cdot e + \lambda e\,.\,a \mid a, b \in \mathcal{A}, e \in L\ ).$$

We can show that for all λ, the $\mathcal{F}^\lambda$ are modules over $\mathcal{D}$, and for a fixed λ, the space $\mathcal{F}^\lambda$ is a module over $\mathcal{D}_\lambda$.

We denote by $\mathrm{Diff}(\mathcal{F}^\lambda)$ the associative algebra of algebraic differential operators, as defined in Grothendieck and Dieudonné (1971, IV,16.8,16.11) and Bernstein *et al.* (1975). Let $D \in \mathcal{D}$ and assume that D is one of the generators

$$D = a_0 \odot e_1 \odot a_1 \odot e_2 \odot \cdots \odot a_{n-1} \odot e_n \odot a_n \qquad [6.34]$$

with $e_i \in \mathcal{L}$ and $a_i \in \mathcal{A}$ (written as elements in $U(\mathcal{D}^1)$), then we have the following.

PROPOSITION 6.6.– (Schlichenmaier 2014a, Proposition 2.14) Every element $D \in \mathcal{D}$ of $\mathcal{D}_\lambda$ of the form [6.34] operates as an (algebraic) differential operator of degree $\leq n$ on $\mathcal{F}^\lambda$.

In fact, we get (associative) algebra homomorphisms

$$\mathcal{D} \to \mathrm{Diff}(\mathcal{F}^\lambda), \qquad \mathcal{D}_\lambda \to \mathrm{Diff}(\mathcal{F}^\lambda)\,. \qquad [6.35]$$

In the case that the set A of points where poles are allowed is finite and non-empty, the complement $\Sigma \setminus A$ is affine (Hartshorne 1977, p.297). Hence, as shown in Grothendieck and Dieudonné (1971) every differential operator can be obtained by successively applying first-order operators, i.e. by applying elements from $U(\mathcal{D}^1)$. In other words, the homomorphisms [6.35] are surjective.

6.4.6. *Lie superalgebras of half forms*

Recall from remark 6.2 the definition of a Lie superalgebra.

With the help of our associative product [6.22], we will obtain examples of Lie superalgebras. First, we consider

$$\cdot \; \mathcal{F}^{-1/2} \times \mathcal{F}^{-1/2} \to \mathcal{F}^{-1} = \mathcal{L} \,, \qquad [6.36]$$

and introduce the vector space $\mathcal{S}$ with the product

$$\mathcal{S} := \mathcal{L} \oplus \mathcal{F}^{-1/2}, \quad [(e,\varphi),(f,\psi)] := ([e,f] + \varphi \cdot \psi, e \,.\, \varphi - f \,.\, \psi). \qquad [6.37]$$

The elements of $\mathcal{L}$ are denoted by $e, f, \ldots$, and the elements of $\mathcal{F}^{-1/2}$ by $\varphi, \psi, \ldots$

The definition [6.37] can be reformulated as an extension of $[.,.]$ on $\mathcal{L}$ to a super-bracket (denoted by the same symbol) on $\mathcal{S}$ by setting

$$[e,\varphi] := -[\varphi, e] := e \,.\, \varphi = (e\frac{d\varphi}{dz} - \frac{1}{2}\varphi\frac{de}{dz})(dz)^{-1/2} \qquad [6.38]$$

and

$$[\varphi, \psi] := \varphi \cdot \psi \,. \qquad [6.39]$$

We call the elements of $\mathcal{L}$ elements of even parity, and the elements of $\mathcal{F}^{-1/2}$ elements of odd parity. For such elements as x, we denote their parity by $\bar{x} \in \{\bar{0}, \bar{1}\}$.

The sum [6.37] can also be described as $\mathcal{S} = \mathcal{S}_{\bar{0}} \oplus \mathcal{S}_{\bar{1}}$, where $\mathcal{S}_{\bar{i}}$ is the subspace of elements of parity $\bar{i}$.

PROPOSITION 6.7.– (Schlichenmaier 2014a, Proposition 2.15) The space $\mathcal{S}$ with the above introduced parity and product is a Lie superalgebra.

REMARK 6.6.– The choice of the theta characteristics corresponds to choosing a *spin structure* on Σ. For the relation of the Neveu–Schwarz superalgebra to the geometry of graded Riemann surfaces, (see Bryant (1990)).

6.4.7. *Jordan superalgebra*

Leidwanger and Morier-Genoud (2012) introduced a *Jordan superalgebra* in our geometric setting. They put

$$\mathcal{J} := \mathcal{F}^0 \oplus \mathcal{F}^{-1/2} = \mathcal{J}_{\bar{0}} \oplus \mathcal{J}_{\bar{1}}. \qquad [6.40]$$

Recall that $\mathcal{A} = \mathcal{F}^0$ is the associative algebra of meromorphic functions. They define the (Jordan) product $\circ$ via the algebra structures for the spaces $\mathcal{F}^\lambda$ by

$$\begin{aligned} f \circ g &:= f \cdot g \quad &\in \mathcal{F}^0, \\ f \circ \varphi &:= f \cdot \varphi \quad &\in \mathcal{F}^{-1/2}, \\ \varphi \circ \psi &:= [\varphi, \psi] \quad &\in \mathcal{F}^0. \end{aligned} \qquad [6.41]$$

By rescaling the second definition with the factor 1/2, one obtains a *Lie anti-algebra*, as introduced by Ovsienko (2011). See Leidwanger and Morier-Genoud (2012) for more details and additional results on representations.

6.4.8. *Higher genus current algebras*

We fix an arbitrary finite-dimensional complex Lie algebra $\mathfrak{g}$. Our goal is to generalize the classical current algebra to a higher genus. For this, let (Σ, A) be the geometrical data consisting of the Riemann surface Σ and the subset of points A used to define $\mathcal{A}$, the algebra of meromorphic functions, which are holomorphic outside of the set $A \subseteq \Sigma$.

DEFINITION 6.4.– *The* higher genus current algebra *associated with the Lie algebra* $\mathfrak{g}$ *and the geometric data* (Σ, A) *is the Lie algebra* $\overline{\mathfrak{g}} = \overline{\mathfrak{g}}(A) = \overline{\mathfrak{g}}(\Sigma, A)$ *given as vector space by* $\overline{\mathfrak{g}} = \mathfrak{g} \otimes_{\mathbb{C}} \mathcal{A}$ *with the Lie product*

$$[x \otimes f, y \otimes g] = [x, y] \otimes f \cdot g, \qquad x, y \in \mathfrak{g}, \quad f, g \in \mathcal{A}. \qquad [6.42]$$

PROPOSITION 6.8.– $\overline{\mathfrak{g}}$ is a Lie algebra.

As usual we will suppress the mention of (Σ, A) if not needed. The elements of $\overline{\mathfrak{g}}$ can be interpreted as meromorphic functions $\Sigma \to \mathfrak{g}$ from the Riemann surface Σ to the Lie algebra $\mathfrak{g}$, which are holomorphic outside of A.

Later, we will introduce central extensions for these current algebras. They will generalize affine Lie algebras, and affine Kac–Moody algebras of untwisted type.

For some applications, it is useful to extend the definition by considering differential operators (of degree ≤ 1) associated with $\overline{\mathfrak{g}}$. We define $\mathcal{D}^1_{\mathfrak{g}} := \overline{\mathfrak{g}} \oplus \mathcal{L}$ and take for the summands the Lie product defined there and add

$$[e, x \otimes g] := -[x \otimes g, e] := x \otimes (e.g). \qquad [6.43]$$

This operation can be described as the semidirect sum of $\overline{\mathfrak{g}}$ with $\mathcal{L}$ and we get the following proposition.

PROPOSITION 6.9.– (Schlichenmaier 2014a, Proposition 2.15) $\mathcal{D}^1_{\mathfrak{g}}$ is a Lie algebra.

6.4.9. *KN-type algebras*

Above, the set A of points where poles are allowed was arbitrary. In the case that A is finite and moreover $\#A \geq 2$, the constructed algebras are called KN-type algebras. In this way, we get the KN vector field algebra, the function algebra, the current algebra, the differential operator algebra, the Lie superalgebra, etc. The reader may question what is so special about this situation that these algebras deserve special names. In fact, in this case we can endow the algebra with a (strong) almost-graded structure. This will be discussed in the next section. The almost-grading is a crucial tool for extending the classical result to a higher genus. Recall that in the classical case, we have genus zero and $\#A = 2$.

To be precise, a KN-type algebra should be considered to be one of the above algebras for $2 \leq \#A < \infty$ together with a fixed chosen almost-grading which is induced by splitting $A = I \cup O$ into two disjoint non-empty subsets (see definition 6.5).

6.5. Almost-graded structure

6.5.1. *Definition of almost-gradedness*

In the classical situation discussed in section 6.2, the algebras introduced in the last section are graded algebras. In the higher genus case and even in the genus zero case with more than two points where poles are allowed, there is no non-trivial grading anymore. As stated by Krichever and Novikov (1987), there is a weaker concept, an almost-grading, which to a large extent is a valuable replacement of an honest grading. Such an almost-grading is induced by a splitting of the set A into two non-empty and disjoint sets I and O. The (almost-)grading is fixed by exhibiting certain basis elements in the spaces $\mathcal{F}^\lambda$ as homogeneous.

DEFINITION 6.5.– *Let $\mathcal{L}$ be a Lie or an associative algebra such that $\mathcal{L} = \oplus_{n\in\mathbb{Z}}\mathcal{L}_n$ is a vector space direct sum, then $\mathcal{L}$ is called an* almost-graded *(Lie-) algebra if*

i) $\dim \mathcal{L}_n < \infty$*;*

ii) there exist constants $L_1, L_2 \in \mathbb{Z}$ such that

$$\mathcal{L}_n \cdot \mathcal{L}_m \subseteq \bigoplus_{h=n+m-L_1}^{n+m+L_2} \mathcal{L}_h, \qquad \forall n, m \in \mathbb{Z}.$$

The elements in $\mathcal{L}_n$ are called homogeneous elements of degree n, and $\mathcal{L}_n$ is called homogeneous subspace *of degree n.*

If $\dim \mathcal{L}_n$ *is bounded with a bound independent of n, we call $\mathcal{L}$* strongly almost-graded. *If we drop the condition that* $\dim \mathcal{L}_n$ *is finite dimensional, we call $\mathcal{L}$* weakly almost-graded.

In a similar manner, almost-graded modules over almost-graded algebras are defined. We can extend the definition to superalgebras, in an obvious way, even to more general algebraic structures. Note that this definition makes complete sense also for more general index sets $\mathbb{J}$. In fact, we will consider the index set $\mathbb{J} = (1/2)\mathbb{Z}$ in the case of superalgebras. The even elements (with respect to the super-grading) will have integer degree, the odd elements half-integer degree.

6.5.2. *Separating cycle and KN pairing*

Before we explain the almost-grading, we introduce an important structure. Let C_i be positively oriented (deformed) circles around the points P_i in I, $i = 1, \ldots, K$ and C_j^* positively oriented circles around the points Q_j in O, $j = 1, \ldots, M$.

A cycle C_S is called a separating cycle if it is smooth, positively oriented of multiplicity one and if it separates the in-points from the out-points. It might have more than one component. In the following, we will integrate meromorphic differentials on Σ without poles in $\Sigma \setminus A$ over closed curves C. Hence, we might consider C and C' as equivalent if $[C] = [C']$ in $\mathrm{H}_1(\Sigma \setminus A, \mathbb{Z})$. In this sense, we write for every separating cycle

$$[C_S] = \sum_{i=1}^{K} [C_i] = -\sum_{j=1}^{M} [C_j^*]. \qquad [6.44]$$

The minus sign appears due to the opposite orientation. Another way for giving such a C_S is via level lines of a "proper time evolution", for which I refer to in Schlichenmaier (2014a, section 3.9).

Given such a separating cycle C_S (respectively cycle class), we define a linear map

$$\mathcal{F}^1 \to \mathbb{C}, \qquad \omega \mapsto \frac{1}{2\pi \mathrm{i}} \int_{C_S} \omega. \qquad [6.45]$$

The map will not depend on the separating line C_S chosen, as two of such will be homologous and the poles of ω are only located in I and O.

Consequently, the integration of ω over C_S can also be described over the special cycles C_i or equivalently over C_j^*. This integration corresponds to calculating residues

$$\omega \quad \mapsto \quad \frac{1}{2\pi \mathrm{i}} \int_{C_S} \omega \;=\; \sum_{i=1}^{K} \mathrm{res}_{P_i}(\omega) \;=\; -\sum_{l=1}^{M} \mathrm{res}_{Q_l}(\omega). \qquad [6.46]$$

DEFINITION 6.6.– *The pairing*

$$\mathcal{F}^\lambda \times \mathcal{F}^{1-\lambda} \to \mathbb{C}, \quad (f,g) \mapsto \langle f,g\rangle := \frac{1}{2\pi \mathrm{i}} \int_{C_S} f \cdot g, \qquad [6.47]$$

between λ *and* $1-\lambda$ *forms is called a KN pairing.*

Note that the pairing depends not only on A (as $\mathcal{F}^\lambda$ depends on it) but also critically on the splitting of A into I and O, as the integration path will depend on it. Once the splitting is fixed, the pairing will be fixed too.

In fact, there exist dual basis elements (see equation [6.52]), hence the pairing is non-degenerate.

6.5.3. *The homogeneous subspaces*

Given the vector spaces $\mathcal{F}^\lambda$ of forms of weight $\mathcal{L}$, we will now single out subspaces $\mathcal{F}^\lambda_m$ of degree m by giving a basis of these subspaces. This has been done in the two-point case by Krichever and Novikov (1987) and in the multipoint case by Schlichenmaier (1990b,c,a,d); see also Sadov (1991). See, in particular, Schlichenmaier (2014a, Chapters 3, 4 and 5) for a complete treatment. All proofs of the statements that follow can be found there.

Depending on whether the weight λ is an integer or half-integer, we set $\mathbb{J}_\lambda = \mathbb{Z}$ or $\mathbb{J}_\lambda = \mathbb{Z} + 1/2$. For $\mathcal{F}^\lambda$, we introduce for $m \in \mathbb{J}_\lambda$ subspaces $\mathcal{F}^\lambda_m$ of dimension K, where $K = \#I$, by exhibiting certain elements $f^\lambda_{m,p} \in \mathcal{F}^\lambda$, $p = 1, \ldots, K$, which constitute a basis of $\mathcal{F}^\lambda_m$. The elements are the elements of degree m. As explained in the following, the degree is, in an essential way, related to the zero orders of the elements at the points in I.

Let $I = (P_1, P_2, \ldots, P_K)$, then we require for the zero-order at the point $P_i \in I$ of the element $f^\lambda_{n,p}$

$$\mathrm{ord}_{P_i}(f^\lambda_{n,p}) = (n+1-\lambda) - \delta^p_i, \quad i = 1, \ldots, K\,. \qquad [6.48]$$

The prescription at the points in O is made in such a way that the element $f^\lambda_{m,p}$ is essentially uniquely given. Essentially unique means up to multiplication with a constant[4]. After fixing as additional geometric data, a system of coordinates z_l centered at P_l for $l = 1, \ldots, K$ and requiring that

$$f^\lambda_{n,p}(z_p) = z_p^{n-\lambda}(1 + O(z_p))(dz_p)^\lambda \qquad [6.49]$$

4 Strictly speaking, there are some special cases where some constants have to be added such that the Krichever–Novikov duality [6.52] is valid.

the element $f_{n,p}$ is uniquely fixed. In fact, the element $f^{\lambda}_{n,p}$ only depends on the first-order jet of the coordinate z_p.

EXAMPLE 6.1.– Here we will not give the general recipe for the prescription at the points in O. Just to give an example that is also an important special case, assume $O = \{Q\}$ is a one-element set. If either the genus $g = 0$, or $g \geq 2$, $\lambda \neq 0,\ 1/2,\ 1$ and the points in A are in generic position, then we require

$$\mathrm{ord}_Q(f^{\lambda}_{n,p}) = -K \cdot (n + 1 - \lambda) + (2\lambda - 1)(g - 1). \qquad [6.50]$$

In the other cases (e.g. for $g = 1$), there are some modifications at the point in O necessary for finitely many n.

THEOREM 6.2.– (Schlichenmaier 2014a, Theorem 3.6) Set

$$\mathcal{B}^{\lambda} := \{\, f^{\lambda}_{n,p} \mid n \in \mathbb{J}_{\lambda},\ p = 1, \ldots, K \,\}. \qquad [6.51]$$

Then (a) $\mathcal{B}^{\lambda}$ is a basis of the vector space $\mathcal{F}^{\lambda}$.

b) The introduced basis $\mathcal{B}^{\lambda}$ of $\mathcal{F}^{\lambda}$ and $\mathcal{B}^{1-\lambda}$ of $\mathcal{F}^{1-\lambda}$ are dual to each other with respect to the KN pairing [6.47], i.e.

$$\langle f^{\lambda}_{n,p}, f^{1-\lambda}_{-m,r} \rangle = \delta^{r}_{p}\, \delta^{m}_{n}, \quad \forall n, m \in \mathbb{J}_{\lambda}, \quad r, p = 1, \ldots, K. \qquad [6.52]$$

In particular, from part (b) of the theorem it follows that the KN pairing is non-degenerate. Moreover, any element $v \in \mathcal{F}^{1-\lambda}$ acts as a linear form on $\mathcal{F}^{\lambda}$ via

$$\Phi_v : \mathcal{F}^{\lambda} \mapsto \mathbb{C}, \quad w \mapsto \Phi_v(w) := \langle v, w \rangle. \qquad [6.53]$$

Through this pairing, $\mathcal{F}^{1-\lambda}$ can be considered as the restricted dual of $\mathcal{F}^{\lambda}$. The identification depends on the splitting of A into I and O as the KN pairing depends on it. The full space $(\mathcal{F}^{\lambda})^*$ can even be described with the help of the pairing in a "distributional interpretation" via the distribution $\Phi_{\hat{v}}$ associated with the formal series

$$\hat{v} := \sum_{m \in \mathbb{J}_{\lambda}} \sum_{p=1}^{K} a_{m,p} f^{1-\lambda}_{m,p}, \quad a_{m,p} \in \mathbb{C}\,. \qquad [6.54]$$

The dual elements of $\mathcal{L}$ will be given by the formal series [6.54] with basis elements from $\mathcal{F}^2$, the quadratic differentials, the dual elements of $\mathcal{A}$ correspondingly from $\mathcal{F}^1$, the differentials and the dual elements of $\mathcal{F}^{-1/2}$ correspondingly from $\mathcal{F}^{3/2}$.

It is quite convenient to use special notations for elements of some important weights:

$$\begin{aligned} e_{n,p} &:= f^{-1}_{n,p}, \quad \varphi_{n,p} := f^{-1/2}_{n,p}, \quad A_{n,p} := f^{0}_{n,p}, \\ \omega^{n,p} &:= f^{1}_{-n,p}, \quad \Omega^{n,p} := f^{2}_{-n,p}. \end{aligned} \qquad [6.55]$$

In view of [6.52] for the forms of weight 1 and 2, we invert the index n and write it as superscript.

REMARK 6.7.– The existence of the basis elements is shown by using Riemann–Roch type arguments. It is also possible (and for certain applications necessary) to write down explicitly the basis elements $f^{\lambda}_{n,p}$ in terms of "usual" objects defined on the Riemann surface Σ. For genus zero, they can be given with the help of rational functions in the quasi-global variable z. For genus one (i.e. the torus case), representations with the help of Weierstraß σ and Weierstraß $\wp$ functions exist. For genus $\geq$ 1, there exist expressions in terms of theta functions (with characteristics) and prime forms. Here, the Riemann surface has first to be embedded into its Jacobian via the Jacobi map. See Schlichenmaier (1990c, 1993, 2014a, Chapter 5) for more details.

6.5.4. *The algebras*

THEOREM 6.3.– (Schlichenmaier 2014a, Theorem 3.8) There exist constants R_1 and R_2 (depending on the number and splitting of the points in A and on the genus g) independent of λ and ν and $n, m \in \mathbb{J}$ such that for the basis elements

$$\begin{aligned} f^{\lambda}_{n,p} \cdot f^{\nu}_{m,r} =& f^{\lambda+\nu}_{n+m,r}\delta^{r}_{p} \\ &+ \sum_{h=n+m+1}^{n+m+R_1} \sum_{s=1}^{K} a^{(h,s)}_{(n,p)(m,r)} f^{\lambda+\nu}_{h,s}, \quad a^{(h,s)}_{(n,p)(m,r)} \in \mathbb{C}, \\ [f^{\lambda}_{n,p}, f^{\nu}_{m,r}] =& (-\lambda m + \nu n)\, f^{\lambda+\nu+1}_{n+m,r}\,\delta^{r}_{p} \\ &+ \sum_{h=n+m+1}^{n+m+R_2} \sum_{s=1}^{K} b^{(h,s)}_{(n,p)(m,r)} f^{\lambda+\nu+1}_{h,s}, \quad b^{(h,s)}_{(n,p)(m,r)} \in \mathbb{C}. \end{aligned} \qquad [6.56]$$

In generic situations and for $N = 2$ points, one obtains $R_1 = g$ and $R_2 = 3g$.

The theorem states, in particular, that with respect to both the associative and Lie structure, the algebra $\mathcal{F}$ is weakly almost-graded. The reason why we only have weakly almost-gradedness is that

$$\mathcal{F}^\lambda = \bigoplus_{m \in \mathbb{J}_\lambda} \mathcal{F}^\lambda_m, \qquad \text{with} \quad \dim \mathcal{F}^\lambda_m = K, \tag{6.57}$$

and if we add up for a fixed m all λ we get that our homogeneous spaces are infinite dimensional.

In the definition of our KN-type algebra, only finitely many λs are involved, hence the following is immediate.

THEOREM 6.4.– The KN-type vector field algebras $\mathcal{L}$, function algebras $\mathcal{A}$, differential operator algebras $\mathcal{D}^1$, Lie superalgebras $\mathcal{S}$ and Jordan superalgebras $\mathcal{J}$ are all (strongly) almost-graded algebras and the corresponding modules $\mathcal{F}^\lambda$ are almost-graded modules.

We obtain with $n \in \mathbb{J}_\lambda$

$$\begin{aligned} \dim \mathcal{L}_n &= \dim \mathcal{A}_n = \dim \mathcal{F}^\lambda_n = K, \\ \dim \mathcal{S}_n &= \dim \mathcal{J}_n = 2K, \quad \dim \mathcal{D}^1_n = 3K\,. \end{aligned} \tag{6.58}$$

If $\mathcal{U}$ is any of these algebras, with the product denoted by $[\, , \,]$, then

$$[\mathcal{U}_n, \mathcal{U}_m] \subseteq \bigoplus_{h=n+m}^{n+m+R_i} \mathcal{U}_h, \tag{6.59}$$

with $R_i = R_1$ for $\mathcal{U} = \mathcal{A}$ and $R_i = R_2$ otherwise.

For further reference, let us specify the lowest degree term component in [6.56] for certain special cases.

$$\begin{aligned} A_{n,p} \cdot A_{m,r} &= A_{n+m,r}\, \delta^p_r \; + \text{ h.d.t.} \\ A_{n,p} \cdot f^\lambda_{m,r} &= f^\lambda_{n+m,r}\, \delta^p_r \; + \text{ h.d.t.} \\ &= (m-n) \cdot e_{n+m,r}\, \delta^p_r \; + \text{ h.d.t.} \\ e_{n,p} \cdot f^\lambda_{m,r} &= (m+\lambda n) \cdot f^\lambda_{n+m,r}\, \delta^p_r \; + \text{ h.d.t.} \end{aligned} \tag{6.60}$$

Here, h.d.t. denote linear combinations of basis elements of degree between $n + m + 1$ and $n + m + R_i$,

Finally, the almost-grading of $\mathcal{A}$ induces an almost-grading of the current algebra $\overline{\mathfrak{g}}$ by setting $\overline{\mathfrak{g}}_n = \mathfrak{g} \otimes \mathcal{A}_n$. We obtain

$$\overline{\mathfrak{g}} = \bigoplus_{n \in \mathbb{Z}} \overline{\mathfrak{g}}_n, \quad \dim \overline{\mathfrak{g}}_n = K \cdot \dim \mathfrak{g}. \qquad [6.61]$$

6.5.5. *Triangular decomposition and filtrations*

Let $\mathcal{U}$ be one of the above introduced algebras (including the current algebra). On the basis of the almost-grading, we obtain a triangular decomposition of the algebras

$$\mathcal{U} = \mathcal{U}_{[+]} \oplus \mathcal{U}_{[0]} \oplus \mathcal{U}_{[-]}, \qquad [6.62]$$

where

$$\mathcal{U}_{[+]} := \bigoplus_{m>0} \mathcal{U}_m, \quad \mathcal{U}_{[0]} = \bigoplus_{m=-R_i}^{m=0} \mathcal{U}_m, \quad \mathcal{U}_{[-]} := \bigoplus_{m<-R_i} \mathcal{U}_m. \qquad [6.63]$$

By the almost-gradedness, the $[+]$ and $[-]$ subspaces are in (infinite dimensional) subalgebras. The $[0]$ spaces in general are not. Sometimes we call them *critical strips*.

With respect to the almost-grading of $\mathcal{F}^\lambda$, we introduce a filtration

$$\begin{gathered}\mathcal{F}^\lambda_{(n)} := \bigoplus_{m \geq n} \mathcal{F}^\lambda_m, \\ \ldots \supseteq \mathcal{F}^\lambda_{(n-1)} \supseteq \mathcal{F}^\lambda_{(n)} \supseteq \mathcal{F}^\lambda_{(n+1)} \ldots \end{gathered} \qquad [6.64]$$

PROPOSITION 6.10.– (Schlichenmaier 2014a, Proposition 3.15)

$$\mathcal{F}^\lambda_{(n)} = \{\, f \in \mathcal{F}^\lambda \mid \operatorname{ord}_{P_i}(f) \geq n - \lambda, \forall i = 1, \ldots, K \,\}. \qquad [6.65]$$

6.6. Central extensions

Central extensions of our algebras appear naturally in the context of quantization and regularization of actions. Of course, they are also of independent mathematical interest.

6.6.1. *Central extensions and cocycles*

For the convenience of the reader, let us repeat the relation between central extensions and the second Lie algebra cohomology with values in the trivial module. A central extension of a Lie algebra W is a special Lie algebra structure on the vector space direct sum $\widehat{W} = \mathbb{C} \oplus W$. If we denote $\hat{x} := (0, x)$ and $t := (1, 0)$, then the Lie structure is given by

$$[\hat{x}, \hat{y}] = \widehat{[x, y]} + \psi(x, y) \cdot t, \quad [t, \widehat{W}] = 0, \quad x, y \in W, \qquad [6.66]$$

with bilinear form ψ. The map $x \mapsto \hat{x} = (0, x)$ is a linear splitting map. $\widehat{W}$ will be a Lie algebra, e.g. will fulfill the Jacobi identity, if and only if ψ is an antisymmetric bilinear form and fulfills the Lie algebra 2-cocycle condition

$$0 = d_2\psi(x, y, z) := \psi([x, y], z) + \psi([y, z], x) + \psi([z, x], y). \qquad [6.67]$$

Two central extensions are equivalent if they essentially correspond only to the choice of different splitting maps. A 2-cochain ψ is a coboundary if there exists a linear form $\varphi : W \to \mathbb{C}$ such that

$$\psi(x, y) = \varphi([x, y]). \qquad [6.68]$$

Every coboundary is a cocycle. The second Lie algebra cohomology $\mathrm{H}^2(W, \mathbb{C})$ of W with values in the trivial module $\mathbb{C}$ is defined as the quotient of the space of 2-cocycles modulo coboundaries. Moreover, two central extensions are equivalent if and only if the difference of their defining 2-cocycles ψ and ψ' is a coboundary. In this way, the second Lie algebra cohomology $\mathrm{H}^2(W, \mathbb{C})$ classifies equivalence classes of central extensions. The class $[0]$ corresponds to the trivial central extension. In this case, the splitting map is a Lie homomorphism. We construct central extensions of our algebras by exhibiting such Lie algebra 2-cocycles.

Clearly, equivalent central extensions are isomorphic. The opposite is not true. In our case, we can always rescale the central element by multiplying it with a non-zero scalar. This is an isomorphism but not an equivalence of central extensions. Nevertheless, it is an irrelevant modification. Hence, we will be mainly interested in central extensions modulo equivalence and rescaling. They are classified by $[0]$ and the elements of the projectivized cohomology space $\mathbb{P}(\mathrm{H}^2(W, \mathbb{C}))$.

In the classical case, we have $\dim \mathrm{H}^2(\mathcal{W}, \mathbb{C}) = 1$, hence there are only two essentially different central extensions, the splitting one given by the direct sum $\mathbb{C} \oplus \mathcal{W}$ of Lie algebras and the up to equivalence and rescaling unique non-trivial one, the Virasoro algebra $\mathcal{V}$.

6.6.2. *Geometric cocycles*

The cocycle of the Witt algebra

$$\frac{1}{12}(n^3 - n)\delta_n^{-m} \qquad [6.69]$$

to define the Virasoro algebra is very special. Obviously, it does not make any sense in the higher genus and/or multipoint case. We need to find a geometric description. For this, we have first to introduce connections.

6.6.2.1. *Projective and affine connections*

Let $(U_\alpha, z_\alpha)_{\alpha\in J}$ be a covering of the Riemann surface by holomorphic coordinates with transition functions $z_\beta = f_{\beta\alpha}(z_\alpha)$.

DEFINITION 6.7.–

a) A system of local (holomorphic, meromorphic) functions $R = (R_\alpha(z_\alpha))$ *is called a (holomorphic, meromorphic) projective connection if it transforms as*

$$R_\beta(z_\beta)\cdot(f'_{\beta,\alpha})^2 = R_\alpha(z_\alpha) + S(f_{\beta,\alpha}), \quad \textit{with} \quad S(h) = \frac{h'''}{h'} - \frac{3}{2}\left(\frac{h''}{h'}\right)^2, \qquad [6.70]$$

the Schwartzian derivative. Here, $'$ *denotes differentiation with respect to the coordinate* z_α.

b) A system of local (holomorphic, meromorphic) functions $T = (T_\alpha(z_\alpha))$ *is called a (holomorphic, meromorphic) affine connection if it transforms as*

$$T_\beta(z_\beta)\cdot(f'_{\beta,\alpha}) = T_\alpha(z_\alpha) + \frac{f''_{\beta,\alpha}}{f'_{\beta,\alpha}}. \qquad [6.71]$$

Every Riemann surface admits a holomorphic projective connection (Hawley and Schiffer 1966; Gunning 1966). Given a point P, there then always exists a meromorphic affine connection holomorphic outside of P and having maximally a pole of order one (Schlichenmaier 1990d).

From their very definition, it follows that the difference of two affine (projective) connections will be a (quadratic) differential. Hence, after fixing one affine (projective) connection, all others are obtained by adding (quadratic) differentials.

Next we introduce in a geometric way Lie algebra 2-cocycles by integrating certain differentials, associated with pairs of Lie algebra elements, over arbitrary smooth curves. For the proofs that the following expressions are indeed 2-cocycles, the reader can refer to Schlichenmaier (1990d) and Schlichenmaier (2014a).

6.6.2.2. *The function algebra $\mathcal{A}$*

We consider it as an abelian Lie algebra. Let C be an arbitrary smooth but not necessarily connected curve. We set

$$\psi_C^1(g,h) := \frac{1}{2\pi \mathrm{i}} \int_C g dh, \quad g,h \in \mathcal{A}. \qquad [6.72]$$

6.6.2.3. *The current algebra $\overline{\mathfrak{g}}$*

For $\overline{\mathfrak{g}} = \mathfrak{g} \otimes \mathcal{A}$, we fix a symmetric, invariant, bilinear form β on $\mathfrak{g}$ (not necessarily non-degenerate). Recall that invariance means that we have $\beta([x,y],z) = \beta(x,[y,z])$ for all $x,y,z \in \mathfrak{g}$. Then a cocycle is given as

$$\psi_{C,\beta}^2(x \otimes g, y \otimes h) := \beta(x,y) \cdot \frac{1}{2\pi \mathrm{i}} \int_C g dh, \quad x,y \in \mathfrak{g},\ g,h \in \mathcal{A}. \qquad [6.73]$$

6.6.2.4. *The vector field algebra $\mathcal{L}$*

Here, it is a little bit more delicate. First, we have to choose a (holomorphic) projective connection R. We define

$$\psi_{C,R}^3(e,f) := \frac{1}{24\pi \mathrm{i}} \int_C \left(\frac{1}{2}(e'''f - ef''') - R \cdot (e'f - ef') \right) dz\,. \qquad [6.74]$$

Only by the term with the projective connection will it be a well-defined differential, i.e. independent of the coordinate chosen. Another choice of a projective connection will result in a cohomologous one. Hence, the equivalence class of the central extension will be the same.

6.6.2.5. *The differential operator algebra $\mathcal{D}^1$*

For the differential operator algebra, the cocycles of type [6.72] for $\mathcal{A}$ can be extended by zero on the subspace $\mathcal{L}$. The cocycles for $\mathcal{L}$ can be pulled back. In addition, there is a third type of cocycle mixing $\mathcal{A}$ and $\mathcal{L}$:

$$\psi_{C,T}^4(e,g) := \frac{1}{24\pi \mathrm{i}} \int_C (eg'' + Teg')dz, \qquad e \in \mathcal{L}, g \in \mathcal{A}, \qquad [6.75]$$

with an affine connection T, with, at most, a pole of order one at a fixed point in O. Again, a different choice of connection will not change the cohomology class.

6.6.2.6. *The Lie superalgebra $\mathcal{S}$*

Here, we have to take into account that it is not a Lie algebra. Hence, the Jacobi identity has to be replaced by the super-Jacobi identity. The conditions for being a cocycle for the superalgebra cohomology will change too. Recall the definition of the algebra from section 6.4.6, in particular that the even elements (parity 0) are the vector fields and the odd elements (parity 1) are the half-forms. A bilinear form c is a cocycle if the following is true. The bilinear map c will be symmetric if both x and y are odd, otherwise it will be antisymmetric:

$$c(x,y) = -(-1)^{\bar{x}\bar{y}} c(x,y). \qquad [6.76]$$

The super-cocycle condition reads as

$$(-1)^{\bar{x}\bar{z}} c(x,[y,z]) + (-1)^{\bar{y}\bar{x}} c(y,[z,x]) + (-1)^{\bar{z}\bar{y}} c(z,[x,y]) = 0. \qquad [6.77]$$

With the help of c, we can define central extensions in the Lie superalgebra sense. If we put the condition that the central element is even, then the cocycle c has to be an even map and c vanishes for pairs of elements of different parity.

By convention, we denote vector fields by $e, f, g, ...$ and -1/2-forms by $\varphi, \psi, \chi, ..$ and get

$$c(e,\varphi) = 0, \quad e \in \mathcal{L},\ \varphi \in \mathcal{F}^{-1/2}. \qquad [6.78]$$

The super-cocycle conditions for the even elements is just the cocycle condition for the Lie subalgebra $\mathcal{L}$. The only other non-vanishing super-cocycle condition is for the *(even,odd,odd)* elements and reads as

$$c(e,[\varphi,\psi]) - c(\varphi, e\,.\,\psi) - c(\psi, e\,.\,\varphi) = 0. \qquad [6.79]$$

Here, the definition of the product $[e,\psi] := e\,.\,\psi$ was used.

If we have a cocycle c for the algebra $\mathcal{S}$, we obtain, by restriction, a cocycle for the algebra $\mathcal{L}$. For the mixing term, we know that $c(e,\psi) = 0$. A naive try to put just anything for $c(\varphi,\psi)$ (for example 0) will not work as [6.79] relates the restriction of the cocycle on $\mathcal{L}$ with its values on $\mathcal{F}^{-1/2}$.

PROPOSITION 6.11 (Schlichenmaier (2013)).– Let C be any closed (differentiable) curve on Σ not meeting the points in A, and let R be any (holomorphic) projective connection, then the bilinear extension of

$$\begin{aligned} \Phi_{C,R}(e,f) &:= \frac{1}{24\pi\mathrm{i}}\int_C \left(\frac{1}{2}(e'''f - ef''') - R\cdot(e'f - ef')\right)dz \\ \Phi_{C,R}(\varphi,\psi) &:= -\frac{1}{24\pi\mathrm{i}}\int_C (\varphi''\cdot\psi + \varphi\cdot\psi'' - R\cdot\varphi\cdot\psi)\,dz \\ \Phi_{C,R}(e,\varphi) &:= 0 \end{aligned} \qquad [6.80]$$

gives a Lie superalgebra cocycle for $\mathcal{S}$, hence defines a central extension of $\mathcal{S}$. A different projective connection will yield a cohomologous cocycle.

Note that the $\Phi_{C,R}$ restricted to $\mathcal{L}$ gives $\Psi^3_{C,R}$.

A similar formula was given by Bryant (1990). By adding the projective connection in the second part of [6.80], he corrected some formulas appearing in Bonora *et al.* (1988). He only considered the two-point case and only the integration over a separating cycle. See also Kreusch (2013) for the multipoint case, where still only the integration over a separating cycle is considered.

In contrast to the differential operator algebra case, the two parts cannot be prescribed independently. Only with the same integration path (more precisely, homology class) and the given factors in front of the integral, will it work. The reason for this is that [6.79] relates both.

6.6.3. *Uniqueness and classification of central extensions*

The above-introduced cocycles depend on the choice of the connections R and T. Different choices will not change the cohomology class. Hence, this ambiguity does not concern us. What really matters is that they depend on the integration curve C chosen.

In contrast to the classical situation, for the higher genus and/or multipoint situation there are many essentially different closed curves and hence many non-equivalent central extensions defined by the integration.

But we should take into account that we want to extend the almost-grading from our algebras to the centrally extended ones. This means we take $\deg\hat{x} := \deg x$ and assign a degree $deg(t)$ to the central element t, and still we want to obtain almost-gradedness.

This is possible if and only if our defining cocycle ψ is *local* in the following sense (the name was introduced in the two point case by Krichever and Novikov (1987)). There exists $M_1, M_2 \in \mathbb{Z}$ such that

$$\forall n, m: \quad \psi(W_n, W_m) \neq 0 \implies M_1 \leq n + m \leq M_2. \qquad [6.81]$$

Here, W stands for any of our algebras (including the supercase). Most importantly, "local" is defined in terms of the almost-grading, and the almost-grading itself depends on the splitting $A = I \cup O$. Hence, what is "local" depends on the splitting too.

We will call a cocycle *bounded* (from above) if there exists $M \in \mathbb{Z}$ such that

$$\forall n, m: \quad \psi(W_n, W_m) \neq 0 \implies n + m \leq M. \qquad [6.82]$$

Similarly bounded from below can be defined. Locality means bounded from above and from below.

Given a cocycle class we call it *bounded* (respectively, *local*) if and only if it contains a representing cocycle, which is bounded (respectively, local). Not all cocycles in a bounded class have to be bounded. If we choose as integration path a separating cocycle C_S or one of the C_i, then the above-introduced geometric cocycles are local, respectively, bounded. Recall that in this case integration can be done by calculating residues at the in-points or at the out-points. All of these cocycles are cohomologically non-trivial. The theorems in the following concern the opposite direction. They were treated in Schlichenmaier (2003b,a, 2013). See also Schlichenmaier (2014a) for a complete and common treatment.

The following result for the vector field algebra $\mathcal{L}$ gives the principal structure of the classification results.

THEOREM 6.5.– Schlichenmaier (2003b), (Schlichenmaier 2014a, Theorem 6.41) Let $\mathcal{L}$ be the KN vector field algebra with a given almost-grading induced by the splitting $A = I \cup O$.

a) The space of bounded cohomology classes is K-dimensional ($K = \#I$). A basis is given by setting the integration path in [6.74] to C_i, $i = 1, \ldots, K$ the little (deformed) circles around the points $P_i \in I$.

b) The space of local cohomology classes is one-dimensional. A generator is given by integrating [6.74] over a separating cocycle C_S, i.e.

$$\psi^3_{C_S,R}(e,f) = \frac{1}{24\pi \mathrm{i}} \int_{C_S} \left(\frac{1}{2}(e'''f - ef''') - R \cdot (e'f - ef') \right) dz\,. \qquad [6.83]$$

c) Up to equivalence and rescaling, there is only one non-trivial one-dimensional central extension $\widehat{\mathcal{L}}$ of the vector field algebra $\mathcal{L}$, which allows an extension of the almost-grading.

REMARK 6.8.– In the classical situation, part c) also shows that the Virasoro algebra is the unique non-trivial central extension of the Witt algebra (up to equivalence and rescaling). This result extends to the more general situation under the condition that one fixes the almost-grading, hence the splitting $A = I \cup O$. Here, I would like to repeat the fact that for $\mathcal{L}$ depending on the set A and its possible splittings into two disjoint subsets there are different almost-gradings. Hence, the "unique" central extension finally obtained will also depend on the splitting. Only in the two-point case, is there only one splitting possible. In the case that the genus $g \geq 1$, there are even integration paths possible in the definition of [6.74], which are not homologous to a separating cycle of any splitting. Hence, there are other central extensions possible not corresponding to any almost-grading.

The above theorem is a model for all other classification results. We will always obtain a statement about the bounded (from above) cocycles and then for the local cocycles.

If we consider the function algebra $\mathcal{A}$ as an abelian Lie algebra, then every skew-symmetric bilinear form will be a non-trivial cocycle. Hence, there is no hope of uniqueness. But if we add the condition of $\mathcal{L}$-invariance, which is given as

$$\psi(e.g, h) + \psi(g, e.h) = 0, \quad \forall e \in \mathcal{L},\ g, h \in \mathcal{A} \qquad [6.84]$$

things will change.

Let us denote the subspace of local cohomology classes by H^2_{loc}, and the subspace of local and $\mathcal{L}$-invariant cohomology classes by $\mathrm{H}^2_{\mathcal{L},loc}$. Note that the conditions are only required for at least one representative in the cohomology class. We collect a part of the results for the cocycle classes of the other algebras in the following theorem.

THEOREM 6.6.– (Schlichenmaier 2014a, Corollary 6.48)

a) $\dim \mathrm{H}^2_{\mathcal{L},loc}(\mathcal{A}, \mathbb{C}) = 1$;

b) $\dim \mathrm{H}^2_{loc}(\mathcal{L}, \mathbb{C}) = 1$;

c) $\dim \mathrm{H}^2_{loc}(\mathcal{D}^1, \mathbb{C}) = 3$;

d) $\dim \mathrm{H}^2_{loc}(\overline{\mathfrak{g}}, \mathbb{C}) = 1$ for $\mathfrak{g}$ a simple finite-dimensional Lie algebra;

e) $\dim \mathrm{H}^2_{loc}(\mathcal{S}, \mathbb{C}) = 1$.

A basis of the cohomology spaces are given by taking the cohomology classes of the cocycles [6.72], [6.74], [6.75], [6.73] and [6.80] obtained by integration over a separating cycle C_S.

Consequently, we also obtain for these algebras the corresponding result about uniqueness of almost-graded central extensions. For the differential operator algebra, we get three independent cocycles. This generalizes results of Arbarello *et al.* (1988) for the classical case.

For results on the bounded cocycle classes, we have to multiply the dimensions above by $K = \#I$. For the supercase, with odd central elements, the bounded cohomology vanishes.

For $\mathfrak{g}$ a reductive Lie algebra and if the cocycle is $\mathcal{L}$-invariant if restricted to the abelian part, a complete classification of local cocycle classes for both $\overline{\mathfrak{g}}$ and $\mathcal{D}^1_{\mathfrak{g}}$ can be found in Schlichenmaier (2003a), Schlichenmaier (2014a, Chapter 9).

I would like to mention that in all the applications I know of, the cocycles coming from representations, regularizations, etc., are local. Hence, the uniqueness or classification result above can be used.

6.7. Examples and generalizations

6.7.1. *The genus zero and three-point situation*

To illustrate, let us consider the three-point KN-type algebras of genus zero. We consider the Riemann sphere $S^2 = \mathbb{P}^1$ and a set A consisting of three points. Given any triple of points, there always exists an analytic automorphism of $\mathbb{P}^1$ mapping this triple to $\{a, -a, \infty\}$, with $a \neq 0$. In fact, $a = 1$ would suffice. Without restriction, we can take

$$I := \{\infty\}, \quad O := \{a, -a\} .$$

Due to the symmetry of the situation, it is more convenient to take a symmetrized basis of $\mathcal{A}$ (with $k \in \mathbb{Z}$)

$$A_{2k} := (z-a)^k(z+a)^k, \qquad A_{2k+1} := z(z-a)^k(z+a)^k ,$$

for $\mathcal{L}$ (with $k \in \mathbb{Z}$)

$$V_{2k} := z(z-a)^k(z+a)^k \frac{d}{dz}, \qquad V_{2k+1} := (z-a)^{k+1}(z+a)^{k+1}\frac{d}{dz} ,$$

and for the $-1/2$-forms

$$\varphi_{2k-1/2} := (z-a)^k(z+a)^k(\frac{d}{dz})^{-1/2}, \quad \varphi_{2k+1/2} := z(z-a)^k(z+a)^k(\frac{d}{dz})^{-1/2} .$$

We also inverted the grading. By straight-forward calculations, we obtain for the algebras the following structures.

The function algebra:

$$A_n \cdot A_m = \begin{cases} A_{n+m}, & n \text{or } m \text{even}, \\ A_{n+m} + a^2 \otimes A_{n+m-2}, & n \text{and } m \text{odd}. \end{cases} \qquad [6.85]$$

The vector field algebra:

$$[V_n, V_m] = \begin{cases} (m-n)V_{n+m}, & n, m \text{ odd}, \\ (m-n)\big(V_{n+m} + a^2 V_{n+m-2}\big), & n, m \text{ even}, \\ (m-n)V_{n+m} + (m-n-1)a^2 V_{n+m-2}, & n \text{ odd}, \ m \text{ even}. \end{cases} \qquad [6.86]$$

The current algebra:

$$[x \otimes A_n, y \otimes A_m] = \begin{cases} [x, y] \otimes A_{n+m}, & n \text{or } m \text{even}, \\ \otimes A_{n+m} + a^2 [x, y] \otimes A_{n+m-2}, & n \text{and } m \text{odd}. \end{cases} \qquad [6.87]$$

The structure equations for the superalgebra look similar and can be easily calculated.

The central extensions can be given by determining the cocycle values by calculating the residues of the integrand at ∞. For example, the local cocycle $\psi^1_{C_S}$ for the function algebra is calculated as (see Fialowski and Schlichenmaier (2005, A.13 and A.14))

$$\frac{1}{2\pi \mathrm{i}} \int_{C_S} A_n dA_m = \begin{cases} -n\delta_m^{-n}, & n, m \text{even}, \\ 0, & n, m \text{different parity}, \\ -n\delta_m^{-n} + a^2(-n+1)\delta_m^{-n+2}, & n, m \text{odd}. \end{cases}$$

The **affine algebra** is now given as the almost-graded central extension $\widehat{\mathfrak{g}}_{\beta,S}$ of the current algebra given by the cocycle

$$\psi^2_{C_S,\beta}(x \otimes A_n, y \otimes A_m) = \beta(x, y) \cdot \frac{1}{2\pi \mathrm{i}} \int_{C_S} A_n dA_m = \beta(x, y) \cdot \psi^1_{C_S}(A_n, A_m).$$

Three-point $\mathfrak{sl}(2,\mathbb{C})$-current algebra for genus 0:

Given a simple Lie algebra $\mathfrak{g}$ with generators and structure equations, the relations above can be written in these terms. An important example is $\mathfrak{sl}(2,\mathbb{C})$ with the standard generators

$$h := \begin{pmatrix} 1 & 0 \\ 0 & -1 \end{pmatrix}, \quad e := \begin{pmatrix} 0 & 1 \\ 0 & 0 \end{pmatrix}, \quad f := \begin{pmatrix} 0 & 0 \\ 1 & 0 \end{pmatrix}.$$

We set $e_n := e \otimes A_n$, $n \in \mathbb{Z}$ and in the same way f_n and h_n. Recall that the invariant bilinear form $\beta(x,y) = \mathrm{tr}(x \cdot y)$. We calculate

$$[e_n, f_m] = \begin{cases} h_{n+m}, & n \text{ or } m \text{ even}, \\ h_{n+m} + a^2 h_{n+m-2}, & n \text{ and } m \text{ odd}, \end{cases} \quad [6.88]$$

$$[h_n, e_m] = \begin{cases} 2e_{n+m}, & n \text{ or } m \text{ even}, \\ 2e_{n+m} + 2a^2 e_{n+m-2}, & n \text{ and } m \text{ odd}, \end{cases} \quad [6.89]$$

$$[h_n, f_m] = \begin{cases} -2f_{n+m}, & n \text{ or } m \text{ even}, \\ -2f_{n+m} - 2a^2 f_{n+m-2}, & n \text{ and } m \text{ odd}. \end{cases} \quad [6.90]$$

For the central extension, we obtain

$$[e_n, f_m] = \begin{cases} h_{n+m} - n\delta_m^{-n}, & n \text{ or } m \text{ even}, \\ h_{n+m} + a^2 h_{n+m-2} - n\delta_m^{-n} - a^2(n-1)\delta_m^{-n+2}, & n \text{ and } m \text{ odd}, \end{cases}$$

and

$$[h_n, h_m] = \begin{cases} -2n\delta_m^{-n}, & n, m \text{ even}, \\ 0, & n, m \text{ different parity}, \\ -2n\delta_m^{-n} + 2a^2(-n+1)\delta_m^{-n+2}, & n, m \text{ odd}. \end{cases}$$

For the other commutators, we do not have contributions to the center.

6.7.2. *Genus zero multipoint algebras – integrable systems*

Already the Witt and Virasoro algebra, in genus zero with two points where poles are allowed, are mathematically highly interesting objects which have a non-trivial representation theory. If we remain on the Riemann sphere but now allow more than two poles, we obtain an even more demanding mathematical theory. For the multipoint case, the related systems are important. For example, the classical KZ models of CFT are of this type (see, for example, Knizhnik and Zamolodchikov (1984)). Integrable systems show up.

Due to the connection between CFT and statistical mechanics, it is no surprise that the genus zero multipoint Krichever–Novikov algebras turn out to be related to algebras appearing in statistical mechanics. For example, the *Onsager algebra* appears as subalgebra of the three-point, $g = 0$, $\mathfrak{sl}(2, \mathbb{C})$-KN algebra. In this context, see the work of Hartwig and Terwilliger (2007); Benkart and Terwilliger (2007); Ito and Terwilliger (2008).

For the genus zero multipoint situation, a number of publications appeared. Some references are as follows: (Schlichenmaier 1990c; Fialowski and Schlichenmaier 2003, 2005, 2007; Bremner 1991, 1995; Schlichenmaier 2007a; Anzaldo-Meneses 1992; Cox 2008). Recently, Schlichenmaier (2017) gave a thorough and unified treatment of universal central extensions of the genus zero algebras.

From the point of view of symmetries of integrable systems, the concept of *automorphic Lie algebras* shows up. It was developed by Lombardo and Mikhailov (2005a,b); Lombardo and Sanders (2010). Invariant objects under finite subgroups of $PGL(2, \mathbb{C})$, the symmetry group of the Riemann sphere, are studied. Of course, there are relations to the $g = 0$, multipoint KN-type algebras. Chopp (2011) obtained some results for the genus one multipoint setting.

6.7.3. *Deformations*

As the second Lie algebra cohomology of the Witt and Virasoro algebra in their adjoint module vanishes (Schlichenmaier 2011; Fialowski 2012, 1990), both are formally and infinitesimally rigid. This means that those formal (and infinitesimal) families, which have these algebras as special fiber, are equivalent to the trivial one. If we consider the examples of section 6.7.1 parameterized by a variable a, then they are non-trivial (even locally non-trivial) families that have themselves as special elements for $a = 0$ the classical algebras. The geometric context is clear: the two points a and $-a$ move together. By Fialowski and Schlichenmaier (2003, 2005, 2007) the above algebras and similar families of algebras on tori were used to exhibit the fact that, for example, the Witt and Virasoro algebra allows non-trivial algebraic-geometric deformations, despite their formal rigidity. This is an effect that cannot appear in the finite-dimensional algebra setting. For families on tori, see the above quoted results (Schlichenmaier 2014a, Chapter 12); (see also Schlichenmaier (1990c); Bremner (1990, 1994); Deck (1990); Ruffing *et al.* (1992)).

6.8. Lax operator algebras

Recently, a new class of current type algebras appeared: the Lax operator algebras. As the name indicates, they are related to integrable systems (Sheinman 2011). The algebras were introduced by Krichever (2002), and Krichever and Sheinman (2007). Here, I will report on their definition; see Sheinman (2012) for more details.

Compared to the KN current type algebra, we will allow additional singularities that will play a special role. The points where these singularities are allowed are called *weak singular points*. The set of such points is denoted by

$$W = \{\gamma_s \in \Sigma \setminus A \mid s = 1, \dots, R\}. \qquad [6.91]$$

Let $\mathfrak{g}$ be one of the classical matrix algebras $\mathfrak{gl}(n)$, $\mathfrak{sl}(n)$, $\mathfrak{so}(n)$, $\mathfrak{sp}(2n)$. We assign to every point γ_s a vector $\alpha_s \in \mathbb{C}^n$ (respectively, $\in \mathbb{C}^{2n}$ for $\mathfrak{sp}(2n)$). The system

$$\mathcal{T} := \{(\gamma_s, \alpha_s) \in \Sigma \times \mathbb{C}^n \mid s = 1, \dots, R\} \qquad [6.92]$$

is called *Tyurin data*.

REMARK 6.9.– In the case that $R = n \cdot g$ and for generic values of (γ_s, α_s) with $\alpha_s \neq 0$, the tuples of pairs $(\gamma_s, [\alpha_s])$ with $[\alpha_s] \in \mathbb{P}^{n-1}(\mathbb{C})$ parametrize semi-stable rank n and degree $n\,g$ framed holomorphic vector bundles, as shown by Tjurin (1965), hence, it is named Tyurin data.

We consider $\mathfrak{g}$-valued meromorphic functions[5]

$$L : \Sigma \to \mathfrak{g}, \qquad [6.93]$$

which are holomorphic outside $W \cup A$, have at most poles of order one (respectively, of order two for $\mathfrak{sp}(2n)$) at the points in W, and fulfill certain conditions at W depending on $\mathcal{T}$. To describe them, let us fix local coordinates w_s centered at γ_s, $s = 1, \dots, R$. For $\mathfrak{gl}(n)$, the conditions are as follows. For $s = 1, \dots, R$, we require that there exist $\beta_s \in \mathbb{C}^n$ and $\kappa_s \in \mathbb{C}$ such that the matrix-valued function L has the following expansion at $\gamma_s \in W$

$$L(w_s) = \frac{L_{s,-1}}{w_s} + L_{s,0} + \sum_{k>0} L_{s,k} w_s^k, \qquad [6.94]$$

with

$$L_{s,-1} = \alpha_s \beta_s^t, \quad \mathrm{tr}(L_{s,-1}) = \beta_s^t \alpha_s = 0, \quad L_{s,0}\, \alpha_s = \kappa_s \alpha_s. \qquad [6.95]$$

5 To be precise, the interpretation as functions is a little bit misleading, as they behave under differentiation like operators on trivialized sections of vector bundles.

In particular, if $L_{s,-1}$ is non-vanishing, then it is a rank 1 matrix, and if $\alpha_s \neq 0$, then it is an eigenvector of $L_{s,0}$. The requirements [6.95] are independent of the chosen coordinates w_s.

It is not at all clear that the commutator of two such matrix functions fulfills these conditions again. But it is shown in Krichever and Sheinman (2007) that they indeed close to a Lie algebra (in fact in the case of $\mathfrak{gl}(n)$ they constitute an associative algebra under the matrix product). If $\alpha_s = 0$, then the conditions at the point γ_s correspond to the fact that L has to be holomorphic there. If all α_s's are zero or $W = \emptyset$, we obtain back the current algebra of KN-type. For the algebra under consideration here, in some sense the Lax operator algebras generalize them. In the bundle interpretation of the Tyurin data, the KN case corresponds to the trivial rank n bundle.

For $\mathfrak{sl}(n)$, the only additional condition is that in [6.94] all matrices $L_{s,k}$ are traceless. The conditions [6.95] remain unchanged.

In the case of $\mathfrak{so}(n)$, one requires that all $L_{s,k}$ in [6.94] are skew-symmetric. In particular, they are traceless. Following Krichever and Sheinman (2007), the set-up has to be slightly modified. First only those Tyurin parameters α_s are allowed that satisfy $\alpha_s^t \alpha_s = 0$. Then, [6.95] is changed in the following way:

$$L_{s,-1} = \alpha_s \beta_s^t - \beta_s \alpha_s^t, \quad \operatorname{tr}(L_{s,-1}) = \beta_s^t \alpha_s = 0, \quad L_{s,0}\, \alpha_s = \kappa_s \alpha_s. \qquad [6.96]$$

For $\mathfrak{sp}(2n)$, we consider a symplectic form $\hat{\sigma}$ for $\mathbb{C}^{2n}$ given by a non-degenerate skew-symmetric matrix σ. The Lie algebra $\mathfrak{sp}(2n)$ is the Lie algebra of matrices X such that $X^t \sigma + \sigma X = 0$. The condition $\operatorname{tr}(X) = 0$ will be automatic. At the weak singularities, we have the expansion

$$L(w_s) = \frac{L_{s,-2}}{w_s^2} + \frac{L_{s,-1}}{w_s} + L_{s,0} + L_{s,1} w_s + \sum_{k>1} L_{s,k} w_s^k. \qquad [6.97]$$

The condition [6.95] is modified as follows (see Krichever and Sheinman (2007)): there exist $\beta_s \in \mathbb{C}^{2n}$, $\nu_s, \kappa_s \in \mathbb{C}$ such that

$$L_{s,-2} = \nu_s \alpha_s \alpha_s^t \sigma,\ L_{s,-1} = (\alpha_s \beta_s^t + \beta_s \alpha_s^t)\sigma,\ {\beta_s}^t \sigma \alpha_s = 0,\ L_{s,0}\, \alpha_s = \kappa_s \alpha_s. \qquad [6.98]$$

Moreover, we require $\alpha_s^t \sigma L_{s,1} \alpha_s = 0$. Again under the point-wise matrix commutator, the set of such maps constitutes a Lie algebra.

It is possible to introduce an almost-graded structure for these Lax operator algebras induced by a splitting of the set $A = I \cup O$. This is done for the two-point case in Krichever and Sheinman (2007) and for the multipoint case in Schlichenmaier (2014b). From the applications, there is again a need to classify almost-graded central extensions.

The author obtained this jointly Sheinman in Schlichenmaier and Sheinman (2008) for the two-point case. For the multipoint case, see Schlichenmaier (2014b). For the Lax operator algebras associated with the simple algebras $\mathfrak{sl}(n), \mathfrak{so}(n), \mathfrak{sp}(n)$, it will be unique (given a splitting of A, there is an almost-grading and with respect to this there is up to equivalence and rescaling of only one non-trivial almost-graded central extension). For $\mathfrak{gl}(n)$, we obtain two independent local cocycle classes if we assume $\mathcal{L}$-invariance on the reductive part. Both in the definition of the cocycle and in the definition of $\mathcal{L}$-invariance a connection shows up.

REMARK 6.10.– Recently, Sheinman extended the set-up to G_2 (Sheinman 2014) and moreover gave a recipe for all semi-simple Lie algebras (Sheinman 2015).

6.9. Fermionic Fock space

6.9.1. *Semi-infinite forms and fermionic Fock space representations*

Our KN vector field algebras $\mathcal{L}$ have as Lie modules the spaces $\mathcal{F}^\lambda$. These representations are not of the type physicists are usually interested in. There are neither annihilation nor creation operators that can be used to construct the full representation out of a vacuum state.

To obtain representation with the required properties, the almost-grading again comes into play. First, using the grading of $\mathcal{F}^\lambda$, it is possible to construct starting from $\mathcal{F}^\lambda$, the forms of weight $\lambda \in 1/2\mathbb{Z}$, the *semi-infinite wedge forms* $\mathcal{H}^\lambda$s.

The vector space $\mathcal{H}^\lambda$ is generated by basis elements, which are formal expressions of the type

$$\Phi = f^\lambda_{(i_1)} \wedge f^\lambda_{(i_2)} \wedge f^\lambda_{(i_3)} \wedge \cdots, \qquad [6.99]$$

where $(i_1) = (m_1, p_1)$ is a double index indexing our basis elements. The indices are in strictly increasing lexicographical order. They are stabilizing in the sense that they will increase exactly by one starting from a certain index that depends on Φ. The action of $\mathcal{L}$ should be extended by the Leibniz rule from $\mathcal{F}^\lambda$ to $\mathcal{H}^\lambda$, but a problem arises. For elements of the critical strip $\mathcal{L}_{[0]}$ of the algebra $\mathcal{L}$, it might happen that they produce infinitely many contributions. The action has to be regularized (as physicists like to call it), which is a well-defined mathematical procedure.

Here, the almost-grading has another appearance. By the (strong) almost-graded module structure of $\mathcal{F}^\lambda$, the algebra $\mathcal{L}$ can be embedded into the Lie algebra of both sided infinite matrices

$$\overline{gl}(\infty) := \{A = (a_{ij})_{i,j\in\mathbb{Z}} \mid \exists r = r(A), \text{ such that } a_{ij} = 0 \text{ if } |i-j| > r\},$$

with "finitely many diagonals". The embedding will depend on the weight λ. For $\overline{gl}(\infty)$, there exists a procedure for the regularization of the action on the semi-infinite wedge product (Date *et al.* 1982; Kac and Peterson 1981) (see also Kac and Raina (1987)). In particular, there is a unique non-trivial central extension $\widehat{gl}(\infty)$. If we pull-back the defining cocycle for the extension, we obtain a central extension $\widehat{\mathcal{L}}_\lambda$ of $\mathcal{L}$ and the required regularization of the action of $\widehat{\mathcal{L}}_\lambda$ on $\mathcal{H}^\lambda$. As the embedding of $\mathcal{L}$ depends on the weight λ, the cocycle will also depend. The pull-back cocycle will be local. Hence, by the classification results of section 6.6.3, it is the unique central extension class defined by [6.74] integrated over C_S (up to a rescaling).

In $\mathcal{H}^\lambda$ there are invariant subspaces, which are generated by a certain "vacuum vector". The subalgebra $\mathcal{L}_{[+]}$ annihilates the vacuum, the central element and the other elements of degree zero act by multiplication with a constant and the whole representation space is generated by $\mathcal{L}_{[-]} \oplus \mathcal{L}_{[0]}$ from the vacuum.

As the function algebra $\mathcal{A}$ operates as multiplication operators on $\mathcal{F}^\lambda$, the above representation can be extended to the algebra $\mathcal{D}^1$ (see details in Schlichenmaier (1990d, 2014a)) after one passes to central extensions. The cocycle again is local and hence, up to coboundary, it will be a certain linear combination of the three generating cocycles for the differential operator algebra. In fact, its class will be

$$c_\lambda \cdot [\psi^3_{C_S}] + \frac{2\lambda - 1}{2}[\psi^4_{C_S}] - [\psi^1_{C_S}], \qquad c_\lambda := -2(6\lambda^2 - 6\lambda + 1). \qquad [6.100]$$

Recall that ψ^3 is the cocycle for the vector field algebra, ψ^1 the cocycle for the function algebra and ψ^4 the mixing cocycle. Note that the expression for c_λ appears also in Mumford's formula (Schlichenmaier 2007b) relating divisors on the moduli space of curves.

For $\mathcal{L}$, we could rescale the central element. Hence, the central extension $\widehat{\mathcal{L}}$ did not depend on the weight. Here, this is different. The central extension $\widehat{\mathcal{D}^1}_\lambda$ depends on it. Furthermore, the representation on $\mathcal{H}^\lambda$ gives a projective representation of the algebra of $\mathcal{D}_\lambda$ of differential operators of all orders. It is exactly the combination [6.100], which lifts to a cocycle for $\mathcal{D}_\lambda$ and gives a central extension $\widehat{\mathcal{D}}_\lambda$.

For the centrally extended algebras $\widehat{\mathfrak{g}}$ in a similar way, fermionic Fock space representations can be constructed (see Sheinman (2001); Schlichenmaier and Sheinman (1999)).

6.9.2. $b - c$ *systems*

As discussed previously, there are other quantum algebra systems that can be realized on $\mathcal{H}^{\lambda}$. On the space $\mathcal{H}^{\lambda}$, the forms $\mathcal{F}^{\lambda}$ act by wedging elements $f^{\lambda} \in \mathcal{F}^{\lambda}$ in front of the semi-infinite wedge form, i.e.

$$\Phi \mapsto f^{\lambda} \wedge \Phi. \qquad [6.101]$$

Using the KN duality pairing [6.47] and by contracting the elements in the semi-infinite wedge forms, the forms $f^{1-\lambda} \in \mathcal{F}^{1-\lambda}$ will also act on them. For Φ a basis element [6.99] of $\mathcal{H}^{\lambda}$, the contraction is defined via

$$i(f^{1-\lambda})\Phi = \sum_{l=1}^{\infty}(-1)^{l-1}\langle f^{1-\lambda}, f^{\lambda}_{i_l}\rangle \cdot f^{\lambda}_{(i_1)} \wedge f^{\lambda}_{(i_2)} \wedge \cdots \check{f}^{\lambda}_{(i_l)}. \qquad [6.102]$$

Here $\check{f}^{\lambda}_{(i_l)}$ indicates as usual that this element will not be there anymore.

Both operations create a Clifford algebra like structure, which is sometimes called a $b - c$ system (see Schlichenmaier (2014a, Chapters 7 and 8)).

6.10. Sugawara representation

In the classical set-up, the (two-dimensional) Sugawara construction relates to a representation of the classical affine Lie algebra $\widehat{\mathfrak{g}}$ a representation of the Virasoro algebra (e.g. Kac (1990); Kac and Raina (1987)). In joint work with Sheinman, the author succeeded in extending it to arbitrary genus and the multipoint setting (Schlichenmaier and Scheinman 1998). For an updated treatment, also incorporating the uniqueness results of central extensions, see Schlichenmaier (2014a, Chapter 10). Here, we will give a very rough sketch.

We start with an admissible representation V of a centrally extended current algebra $\widehat{\mathfrak{g}}$. Admissible means that the central element operates as constant $\times$ identity, and that every element v in the representation space will be annihilated by the elements in $\widehat{\mathfrak{g}}$ of sufficiently high degree (which depends on the element v).

For simplicity, let $\mathfrak{g}$ be either abelian or simple and β the non-degenerate symmetric invariant bilinear form used to construct $\widehat{\mathfrak{g}}$ (now we need that it is non-degenerate). Let $\{u_i\}$, $\{u^j\}$ be a system of dual basis elements for $\mathfrak{g}$ with respect to β, i.e. $\beta(u_i, u^j) = \delta_i^j$. Note that the Casimir element of $\mathfrak{g}$ can be given by

$\sum_i u_i u^i$. For $x \in \mathfrak{g}$, we consider the family of operators $x(n,p)$ given by the operation of $x \otimes A_{n,p}$ on V. We group them together in a formal sum

$$\widehat{x}(Q) := \sum_{n\in\mathbb{Z}} \sum_{p=1}^{K} x(n,p)\omega^{n,p}(Q), \quad Q \in \Sigma. \qquad [6.103]$$

Such a formal sum is called a field if applied to a vector $v \in V$ and it again gives a formal sum (now of elements from V), which is bounded from above. By the condition of admissibility, $\widehat{x}(Q)$ is a field. It is of conformal weight one, as the one-differentials $\omega^{n,p}$ show up.

The current operator fields are defined as 6

$$J_i(Q) := \widehat{u_i}(Q) = \sum_{n,p} u_i(n,p)\omega^{n,p}(Q). \qquad [6.104]$$

The Sugawara operator field $T(Q)$ is defined by

$$T(Q) := \frac{1}{2}\sum_i :J_i(Q)J^i(Q): \ . \qquad [6.105]$$

Here :...: denotes some normal ordering which is needed to make the product of two fields again a field. The *standard normal ordering* is defined as

$$:x(n,p)y(m,r): \quad := \begin{cases} x(n,p)y(m,r), & (n,p) \leq (m,r) \\ y(m,r)x(n,p), & (n,p) > (m,r) \end{cases} \qquad [6.106]$$

where the indices (n,p) are lexicographically ordered. By this prescription, the annihilation operators, i.e. the operators of positive degree, are brought as much as possible to the right so that they act first.

The current operators are fields of conformal weight one, hence the Sugawara operator field is a field of weight two. Consequently, we write it as

$$T(Q) = \sum_{k\in\mathbb{Z}} \sum_{p=1}^{K} L_{k,p} \cdot \Omega^{k,p}(Q) \qquad [6.107]$$

6 For simplicity, we drop mentioning the range of summation here and in the following when it is clear.

with certain operators $L_{k,p}$. The $L_{k,p}$ are called *modes of the Sugawara field* T or simply Sugawara operators.

Let 2κ be the eigenvalue of the Casimir operator in the adjoint representation. For $\mathfrak{g}$ abelian, $\kappa = 0$. For $\mathfrak{g}$ simple and β normalized such that the longest roots have square length 2, then κ is the dual Coxeter number. Recall that the central element t acts on the representation space V as $c \cdot id$ with a scalar c. This scalar is called the *level* of the representation. The key result is (where $x(g)$ denotes the operator corresponding to the element $x \otimes g$) as follows.

PROPOSITION 6.12.– (Schlichenmaier 2014a; Schlichenmaier and Scheinman 1998, Proposition 10.8) Let $\mathfrak{g}$ be either an abelian or a simple Lie algebra. Then

$$[L_{k,p}, x(g)] = -(c+\kappa) \cdot x(e_{k,p} \cdot g)\,. \qquad [6.108]$$

$$[L_{k,p}, \widehat{x}(Q)] = (c+\kappa) \cdot (e_{k,p} \,.\, \widehat{x}(Q))\,. \qquad [6.109]$$

Recall that $e_{k,p}$ are the KN basis elements for the vector field algebra $\mathcal{L}$.

In the next step, the commutators of the operators $L_{k,p}$ can be calculated. In the case that $c + \kappa = 0$, called the *critical level*, these operators generate a subalgebra of the center of $\mathfrak{gl}(V)$. If $c + \kappa \neq 0$ (i.e. at a non-critical level), the $L_{k,p}$ can be replaced by rescaled elements $L^*_{k,p} = \frac{-1}{c+\kappa} L_{k,p}$ and we denote by $T[..]$ the linear representation of $\mathcal{L}$ induced by

$$T[e_{k,p}] = \mathcal{L}^*_{k,p}. \qquad [6.110]$$

The result is that T defines a projective representation of $\mathcal{L}$ with a local cocycle. This cocycle is up to rescaling our geometric cocycle $\psi^3_{C_S,R}$ with a suitable projective connection[7] R. In detail,

$$T[[e,f]] = [T[e], T[f]] + \frac{c \dim \mathfrak{g}}{c+\kappa}\, \psi^3_{C_S,R}(e,f) id. \qquad [6.111]$$

Consequently, by setting

$$T[\hat{e}] := T[e], \quad T[t] := \frac{c \dim \mathfrak{g}}{c+\kappa}\, id\,. \qquad [6.112]$$

7 The projective connection takes care of the "up to coboundary". It is induced by the normal ordering prescription.

we obtain an honest Lie representation of the centrally extended vector field algebra $\widehat{\mathcal{L}}$ given by this local cocycle. For the general reductive case, see Schlichenmaier (2014a, section 10.2.1).

6.11. Application to moduli space

This application deals with *WZNW models* and the *KZ connection*. Despite the fact that it is a very important application, the following description is very condensed. More details can be found in Schlichenmaier and Sheinman (1999, 2004) and see also Schlichenmaier (2014a) and Sheinman (2012).

WZNW models are defined on the basis of a fixed finite-dimensional simple (or semi-simple) Lie algebra $\mathfrak{g}$. One considers families of representations of the affine algebras $\widehat{\mathfrak{g}}$ (which is an almost-graded central extension of $\overline{\mathfrak{g}}$) defined over the moduli space of Riemann surfaces of genus g with $K+1$ marked points and splitting of type $(K, 1)$. The single point in O will be a reference point. The data of the moduli of the Riemann surface and the marked points enter the definition of the algebra $\widehat{\mathfrak{g}}$ and the representation. The construction of certain co-invariants yields a special vector bundle of finite rank over moduli space called the vector bundle of conformal blocks, or Verlinde bundle. With the help of the KN vector field algebra, and using the Sugawara construction, the *KZ connection* is given. It is projectively flat. An essential fact is that certain elements in the critical strip $\mathcal{L}_{[0]}$ correspond to infinitesimal deformations of the moduli and to moving the marked points. This gives a global operator approach in contrast to the semi-local approach of Tsuchiya *et al.* (1989).

6.12. Acknowledgments

I acknowledge partial support by the Internal Research Project GEOMQ15, University of Luxembourg, and in the frame of the OPEN scheme of the *Fonds National de la Recherche* (FNR) with the project QUANTMOD O13/570706.

6.13. References

Anzaldo-Meneses, A. (1992). Krichever-Novikov algebras on Riemann surfaces of genus zero and one with N punctures. *J. Math. Phys.*, 33(12), 4155–4163.

Arbarello, E., De Concini, C., Kac, V.G., Procesi, C. (1988). Moduli spaces of curves and representation theory. *Comm. Math. Phys.*, 117(1), 1–36.

Atiyah, M.F. (1971). Riemann surfaces and spin structures. *Ann. Sci. École Norm. Sup.*, 4, 47–62.

Benkart, G., Terwilliger, P. (2007). The universal central extension of the three-point $\mathfrak{sl}_2$ loop algebra. *Proc. Amer. Math. Soc.*, 135(6), 1659–1668.

Bernstein, I.N., Gelfand, I.M., Gelfand, S.I. (1975). Differential operators on the base affine space and a study of $\mathfrak{g}$-modules. In *LieGroups and Their Representations (Proc. Summer School, Bolyai János Math. Soc., Budapest, 1971)*. Halsted, New York.

Bonora, L., Martellini, M., Rinaldi, M., Russo, J. (1988). Neveu-Schwarz- and Ramond-type superalgebras on genus-g Riemann surfaces. *Phys. Lett. B*, 206(3), 444–450.

Bremner, M. (1994). Universal central extensions of elliptic affine Lie algebras. *J. Math. Phys.*, 35(12), 6685–6692.

Bremner, M. (1995). Four-point affine Lie algebras. *Proc. Amer. Math. Soc.*, 123(7), 1981–1989.

Bremner, M.R. (1990). On a Lie algebra of vector fields on a complex torus. *J. Math. Phys.*, 31(8), 2033–2034.

Bremner, M.R. (1991). Structure of the Lie algebra of polynomial vector fields on the Riemann sphere with three punctures. *J. Math. Phys.*, 32(6), 1607–1608.

Bryant, P. (1990). Graded Riemann surfaces and Krichever-Novikov algebras. *Lett. Math. Phys.*, 19(2), 97–108.

Chopp, M. (2011). Lie-admissible structures on Witt type algebras and automorphic algebras. PhD Thesis, University of Luxembourg and University of Metz.

Cox, B. (2008). Realizations of the four point affine Lie algebra $\mathfrak{sl}(2, R) \oplus (\Omega_R/dR)$. *Pacific J. Math.*, 234(2), 261–289.

Date, E., Jimbo, M., Kashiwara, M., Miwa, T. (1982). Transformation groups for soliton equations. Euclidean Lie algebras and reduction of the KP hierarchy. *Publ. Res. Inst. Math. Sci.*, 18(3), 1077–1110.

Deck, T. (1990). Deformations from Virasoro to Krichever-Novikov algebras. *Phys. Lett. B*, 251(4), 535–540.

Fialowski, A. (1990). Deformations of some infinite-dimensional Lie algebras. *J. Math. Phys.*, 31(6), 1340–1343.

Fialowski, A. (2012). Formal rigidity of the Witt and Virasoro algebra. *J. Math. Phys.*, 53(7), 073501.

Fialowski, A., Schlichenmaier, M. (2003). Global deformations of the Witt algebra of Krichever-Novikov type. *Commun. Contemp. Math.*, 5(6), 921–945.

Fialowski, A., Schlichenmaier, M. (2005). Global geometric deformations of current algebras as Krichever-Novikov type algebras. *Comm. Math. Phys.*, 260(3), 579–612.

Fialowski, A., Schlichenmaier, M. (2007). Global geometric deformations of the Virasoro algebra, current and affine algebras by Krichever-Novikov type algebras. *Internat. J. Theoret. Phys.*, 46(11), 2708–2724.

Grothendieck, A., Dieudonné, J.A. (1971). *Eléments de géométrie algébrique I*, (French). Grundlehren der Mathematischen Wissenschaften [Fundamental Principles of Mathematical Sciences], vol. 166. Springer, Berlin.

Guieu, L., Roger, C. (2007). *L'algèbre et le groupe de Virasoro: aspects géométriques et algébriques, généralisations*. CRM Publications. Montreal.

Gunning, R.C. (1966). *Lectures on Riemann Surfaces*. Princeton Mathematical Notes. Princeton University Press, Princeton.

Hartshorne, R. (1977). *Algebraic Geometry*. Springer, New York, Heidelberg.

Hartwig, B., Terwilliger, P. (2007). The tetrahedron algebra, the Onsager algebra, and the $\mathfrak{sl}_2$ loop algebra. *J. Algebra*, 308(2), 840–863.

Hawley, N.S., Schiffer, M. (1966). Half-order differentials on Riemann surfaces. *Acta Math.*, 115, 199–236.

Ito, T., Terwilliger, P. (2008). Finite-dimensional irreducible modules for the three-point $\mathfrak{sl}_2$ loop algebra. *Comm. Algebra*, 36(12), 4557–4598.

Kac, V.G. (1968). Simple irreducible graded Lie algebras of finite growth. *Izv. Akad. Nauk SSSR Ser. Mat.*, 32, 1323–1367.

Kac, V.G. (1990). *Infinite-dimensional Lie Algebras*, 3rd edition. Cambridge University Press, Cambridge.

Kac, V.G., Peterson, D.H. (1981). Spin and wedge representations of infinite-dimensional Lie algebras and groups. *Proc. Nat. Acad. Sci.*, 78(6, part 1), 3308–3312.

Kac, V.G., Raina, A.K. (1987). *Bombay Lectures on Highest Weight Representations of Infinite-Dimensional Lie Algebras*. World Scientific Publishing Co., Inc., Teaneck.

Knizhnik, V.G., Zamolodchikov, A.B. (1984). Current algebra and Wess-Zumino model in two dimensions. *Nuclear Phys. B*, 247(1), 83–103.

Kreusch, M. (2013). Extensions of superalgebras of Krichever-Novikov type. *Lett. Math. Phys.*, 103(11), 1171–1189.

Krichever, I.M. (2002). Vector bundles and Lax equations on algebraic curves. *Comm. Math. Phys.*, 229(2), 229–269.

Krichever, I.M., Novikov, S.P. (1987). Algebras of Virasoro type, Riemann surfaces and strings in Minkowski space. *Funktsional. Anal. i Prilozhen.*, 21(4), 47–61, 96.

Krichever, I.M., Novikov, S.P. (1989). Algebras of Virasoro type, the energy-momentum tensor, and operator expansions on Riemann surfaces. *Funktsional. Anal. i Prilozhen.*, 23(1), 24–40.

Krichever, I.M., Sheinman, O.K. (2007). Lax operator algebras. *Funktsional. Anal. i Prilozhen.*, 41(4), 284–294.

Leidwanger, S., Morier-Genoud, S. (2012). Superalgebras associated to Riemann surfaces: Jordan algebras of Krichever-Novikov type. *Int. Math. Res. Not.*, 19, 4449–4474.

Lombardo, S., Mikhailov, A.V. (2005a). Reduction groups and automorphic Lie algebras. *Comm. Math. Phys.*, 258(1), 179–202.

Lombardo, S., Mikhailov, A.V. (2005b). Reductions of integrable equations and automorphic Lie algebras. In *SPT 2004—Symmetry and Perturbation Theory*. World Scientific Publishing, Hackensack.

Lombardo, S., Sanders, J.A. (2010). On the classification of automorphic Lie algebras. *Comm. Math. Phys.*, 299(3), 793–824.

Moody, R.V. (1969). Euclidean Lie algebras. *Canad. J. Math.*, 21, 1432–1454.

Ovsienko, V. (2011). Lie antialgebras: Premices. *J. Algebra*, 325, 216–247.

Ruffing, A., Deck, T., Schlichenmaier, M. (1992). String branchings on complex tori and algebraic representations of generalized Krichever-Novikov algebras. *Lett. Math. Phys.*, 26(1), 23–32.

Sadov, V.A. (1991). Bases on multipunctured Riemann surfaces and interacting strings amplitudes. *Comm. Math. Phys.*, 136(3), 585–597.

Schlichenmaier, M. (1990a). Central extensions and semi-infinite wedge representations of Krichever-Novikov algebras for more than two points. *Lett. Math. Phys.*, 20(1), 33–46.

Schlichenmaier, M. (1990b). Krichever-Novikov algebras for more than two points. *Lett. Math. Phys.*, 19(2), 151–165.

Schlichenmaier, M. (1990c). Krichever-Novikov algebras for more than two points: Explicit generators. *Lett. Math. Phys.*, 19(4), 327–336.

Schlichenmaier, M. (1990d). Verallgemeinerte Krichever-Novikov Algebren und deren Darstellungen. PhD Thesis, University of Mannheim.

Schlichenmaier, M. (1993). Degenerations of generalized Krichever-Novikov algebras on tori. *J. Math. Phys.*, 34(8), 3809–3824.

Schlichenmaier, M. (1996). Zwei Anwendungen algebraisch-geometrischer Methoden in der theoretischen Physik: Berezin-Toeplitz-Quantisierung und globale Algebren der zweidimensionalen konformen Feldtheorie. Habilitation Thesis, University of Mannheim.

Schlichenmaier, M. (2003a). Higher genus affine algebras of Krichever-Novikov type. *Mosc. Math. J.*, 3(4), 1395–1427.

Schlichenmaier, M. (2003b). Local cocycles and central extensions for multipoint algebras of Krichever-Novikov type. *J. Reine Angew. Math.*, 559, 53–94.

Schlichenmaier, M. (2007a). Higher genus affine Lie algebras of Krichever-Novikov type. In *Difference Equations, Special Functions and Orthogonal Polynomials*. World Scientific Publishing, Hackensack.

Schlichenmaier, M. (2007b). *An Introduction to Riemann Surfaces, Algebraic Curves and Moduli Spaces*. Theoretical and Mathematical Physics, 2nd edition. Springer, Berlin.

Schlichenmaier, M. (2011). An elementary proof of the vanishing of the second cohomology of the Witt and Virasoro algebra with values in the adjoint module. *Forum Math.*, 26(3), 913–929.

Schlichenmaier, M. (2013). Lie superalgebras of Krichever-Novikov type and their central extensions. *Anal. Math. Phys.*, 3(3), 235–261.

Schlichenmaier, M. (2014a). *Krichever-Novikov Type Algebras: Theory and Applications*. De Gruyter, Berlin.

Schlichenmaier, M. (2014b). Multipoint Lax operator algebras: Almost-graded structure and central extensions. *Mat. Sb.*, 205(5), 117–160.

Schlichenmaier, M. (2016). Krichever-Novikov type algebras: An introduction. In *Lie Algebras, Lie Superalgebras, Vertex Algebras and Related Topics – Proceedings of Symposia in Pure Mathematics*, American Mathematical Society, Providence, 92, 181–220.

Schlichenmaier, M. (2017). N-point Virasoro algebras are multipoint Krichever-Novikov-type algebras. *Comm. Algebra*, 45(2), 776–821.

Schlichenmaier, M., Scheinman, O.K. (1998). The Sugawara construction and Casimir operators for Krichever-Novikov algebras. *J. Math. Sci.* 1, 92(2), 3807–3834.

Schlichenmaier, M., Sheinman, O.K. (1999). The Wess-Zumino-Witten-Novikov theory, Knizhnik-Zamolodchikov equations, and Krichever-Novikov algebras. *Uspekhi Mat. Nauk*, 54(1(325)), 213–250.

Schlichenmaier, M., Sheinman, O.K. (2004). The Knizhnik-Zamolodchikov equations for positive genus, and Krichever-Novikov algebras. *Uspekhi Mat. Nauk*, 59(4(358)), 147–180.

Schlichenmaier, M., Sheinman, O.K. (2008). Central extensions of Lax operator algebras. *Uspekhi Mat. Nauk*, 66(1(2011)), 145–171.

Sheinman, O.K. (1990). Elliptic affine Lie algebras. *Funktsional. Anal. i Prilozhen.*, 24(3), 51–61, 96.

Sheinman, O.K. (1992). Highest weight modules of some quasigraded Lie algebras on elliptic curves. *Funktsional. Anal. i Prilozhen.*, 26(3), 65–71.

Sheinman, O.K. (1993). Affine Lie algebras on Riemann surfaces. *Funktsional. Anal. i Prilozhen.*, 27(4), 54–62, 96.

Sheinman, O.K. (1995). Highest-weight modules for affine Lie algebras on Riemann surfaces. *Funktsional. Anal. i Prilozhen.*, 29(1), 56–71, 96.

Sheinman, O.K. (2001). A fermionic model of representations of affine Krichever-Novikov algebras. *Funktsional. Anal. i Prilozhen.*, 35(3), 60–72, 96.

Sheinman, O.K. (2011). Lax operator algebras and Hamiltonian integrable hierarchies. *Uspekhi Mat. Nauk*, 66(1(397)), 151–178.

Sheinman, O.K. (2012). *Current Algebras on Riemann Surfaces: New Results and Applications*, vol. 58 of De Gruyter Expositions in Mathematics. Walter de Gruyter GmbH & Co. KG, Berlin.

Sheinman, O.K. (2014). Lax operator algebras of type G_2. In *Topology, Geometry, Integrable Systems, and Mathematical Physics, Amer. Math. Soc. Transl. Ser. 2*. American Mathematical Society, Providence, 373–392.

Sheinman, O.K. (2015). Lax operator algebras and gradings on semisimple Lie algebras. *Dokl. Akad. Nauk*, 461(2), 143–145.

Tjurin, A.N. (1965). The classification of vector bundles over an algebraic curve of arbitrary genus. *Izv. Akad. Nauk SSSR Ser. Mat.*, 29, 657–688.

Tsuchiya, A., Ueno, K., Yamada, Y. (1989). Conformal field theory on universal family of stable curves with gauge symmetries. In *Integrable Systems in Quantum Field Theory and Statistical Mechanics, vol. 19 of Adv. Stud. Pure Math.*. Academic Press, Boston, 459–566.

7

An Introduction to Pre-Lie Algebras

Chengming BAI
Chern Institute of Mathematics, Nankai University, Tianjin, China

7.1. Introduction

In this chapter, we give a brief introduction to pre-Lie algebras, with emphasis on their connections with some related structures.

DEFINITION 7.1.– A *pre-Lie algebra* A is a vector space A with a binary operation $(x, y) \to xy$ satisfying

$$(xy)z - x(yz) = (yx)z - y(xz), \quad \forall x, y, z \in A. \tag{7.1}$$

7.1.1. *Explanation of notions*

Pre-Lie algebras have several other names as follows:

1) *Left-symmetric algebra*: define the associator as

$$(x, y, z) = (xy)z - x(yz), \forall x, y, z \in A. \tag{7.2}$$

Then equation [7.1] is exactly the following identity:

$$(x, y, z) = (y, x, z), \forall x, y, z \in A, \tag{7.3}$$

Algebra and Applications 1,
coordinated by Abdenacer MAKHLOUF.

that is, the associator [7.2] is symmetric in the left two variables x, y. The notion of left-symmetric algebra was given by Vinberg (1963) in the study of convex homogeneous cones. Such a notion was used in many studies related to geometry.

2) *Right-symmetric algebra*: for example, the associator [7.2] is symmetric in the right two variables y, z, that is, the following identity is satisfied:

$$(xy)z - x(yz) = (xz)y - x(zy), \quad \forall x, y, z \in A. \qquad [7.4]$$

Note that a vector space with a binary operation $(x, y) \to x \cdot y$ is a left-symmetric algebra (pre-Lie algebra) if and only if its opposite algebra $(A, \cdot^{\mathrm{opp}})$ is a right-symmetric algebra, where $x \cdot^{\mathrm{opp}} y = y \cdot x$ for any $x, y \in A$. In this sense, the study of right-symmetric algebras is completely parallel to the study of left-symmetric algebras. Thus, we only need to consider the case of left-symmetric algebras.

3) The notion of pre-Lie algebra is due to its close relations with Lie algebras, which will be seen in the following sections. This notion was given by Gerstenhaber (1963) in the study of deformations and cohomology theory of associative algebras. The original form was given as a (graded) right-symmetric algebra. In the above sense and in order to be consistent, we use left-symmetry uniformly to denote a pre-Lie algebra.

4) *Quasi-associative algebra*: pre-Lie algebras include associative algebras whose associators are zero. So in this sense, pre-Lie algebras can be regarded as a kind of generalization of associative algebras. The notion of quasi-associative algebra was given by Kupershmidt (1994) in the study of phase spaces of Lie algebras.

5) *Vinberg algebra* or *Koszul algebra* or *Koszul–Vinberg algebra*: these notions are due to the pioneering work of Koszul (1961) and Vinberg (1963), whereas the former is in the study of affine manifolds and affine structures on Lie groups.

7.1.2. *Two fundamental properties*

Let A be a pre-Lie algebra. For any $x, y \in A$, let $L(x)$ and $R(x)$ denote the left and right multiplication operators, respectively, that is, $L(x)(y) = xy,\ R(x)(y) = yx$. Let $L : A \to gl(A)$ with $x \to L(x)$ and $R : A \to gl(A)$ with $x \to R(x)$ (for every $x \in A$) be two linear maps.

One of the close relationships between pre-Lie algebras and Lie algebras is given as follows.

PROPOSITION 7.1.– Let A be a pre-Lie algebra.

1) The commutator

$$[x, y] = xy - yx, \quad \forall x, y \in A, \qquad [7.5]$$

defines a Lie algebra $\mathfrak{g}(A)$, which is called the sub-adjacent Lie algebra of A and A is also called a compatible pre-Lie algebra structure on the Lie algebra $\mathfrak{g}(A)$.

2) Equation [7.1] is expressed as

$$[L(x), L(y)] = L([x, y]), \quad \forall x, y \in A, \qquad [7.6]$$

which means that $L : \mathfrak{g}(A) \to gl(A)$ with $x \to L(x)$ gives a representation of the Lie algebra $\mathfrak{g}(A)$.

REMARK 7.1.– Recall that a *Lie-admissible algebra* is a vector space with a binary operation $(x, y) \to xy$ whose commutator [7.5] defines a Lie algebra. It is equivalent to the following identity:

$$(x, y, z) + (y, z, x) + (z, x, y) = (y, x, z) + (z, y, x) + (x, z, y), \forall x, y, z \in A. \quad [7.7]$$

So a pre-Lie algebra is a special Lie-admissible algebra whose left multiplication operators give a representation of the associated commutator Lie algebra.

A direct consequence is that if a Lie algebra $\mathfrak{g}$ has a compatible pre-Lie algebra structure, then there are two representations of the Lie algebra $\mathfrak{g}$ on the underlying vector space of $\mathfrak{g}$ itself: one is given by the adjoint representation ad and another is given by L induced from the compatible pre-Lie algebra. Many interesting structures related to geometry are obtained with this approach.

7.1.3. *Some subclasses*

Some subclasses of pre-Lie algebras are very interesting. Some of them were even introduced and then developed independently.

1) *Associative algebra*: needless to say more.

2) *Transitive left-symmetric algebra* or *complete left-symmetric algebra*: a left-symmetric algebra A is called *transitive* or *complete* if for any $x \in A$, the right multiplication operator $R(x)$ is nilpotent. In affine geometry, real transitive left-symmetric algebras correspond to the complete affine connections (Kim 1986). There are several equivalent conditions (Segal 1992). They play important roles in the study of structures of pre-Lie algebras (Burde 2006).

3) *Left-symmetric derivation algebra* and *left-symmetric inner derivation algebra*: a left-symmetric algebra A is called a *derivation algebra* (an *inner derivation algebra* respectively) if for any $x \in A$, $L(x)$ or $R(x)$ is a derivation (an inner derivation, respectively) of the sub-adjacent Lie algebra $\mathfrak{g}(A)$. These two notions were introduced in Medina (1981) to study the left-invariant affine connections adapted to the (inner) automorphism structure of a Lie group.

4) *Novikov algebra*: a *Novikov algebra* A is a pre-Lie algebra satisfying an additional identity:

$$(xy)z = (xz)y, \quad \forall x, y, z \in A. \qquad [7.8]$$

In other words, a Novikov algebra is a pre-Lie algebra whose right multiplication operators are commutative. Novikov algebras were introduced in connection with the Hamiltonian operators in the formal variational calculus (Gel'fand and Dorfman 1979) and Poisson brackets of hydrodynamic type (Balinskii and Novikov 1985).

5) *Bi-symmetric algebra* or *assosymmetric algebra*: a *bi-symmetric algebra* is a pre-Lie algebra that is also a right-symmetric algebra with the same product. Such structures were introduced under the notion of *assosymmetric algebra* by Kleinfeld from the pure algebraic point of view in order to study the so-called near-associative algebras (Kleinfeld 1957). Note that the study of assosymmetric algebras began a little earlier than the study of pre-Lie algebras.

7.1.4. *Organization of this chapter*

Pre-Lie algebras have relations with many fields in mathematics and mathematical physics. As was pointed out by Chapoton and Livernet (2001), pre-Lie algebra "deserves more attention than it has been given". In particular, it has become a very active topic since the end of the last century due to the role in the quantum field theory (Connes and Kreimer 1998). A survey article was proposed by Burde (2006). However, there are a lot of results in the study of pre-Lie algebras not mentioned in Burde (2006) and, more importantly, this topic has been developing very quickly so many new studies have been appearing. So it is impossible to list every result or progress and mention every reference here. Alternatively, we only choose some materials to give a brief introduction to pre-Lie algebras. We would like to emphasize that these materials cannot provide a complete picture, but we hope that they might help readers to know why pre-Lie algebras are interesting. For this aim, we pay attention to interpret **the relationships between pre-Lie algebras and some related structures**.

This chapter is organized as follows:

– in section 7.2, we introduce some background information and different motivations of introducing the notion of pre-Lie algebra;

– in section 7.3, we introduce some basic properties of pre-Lie algebras including some comments on the studies of structure theory and representation theory and some classification results. We also give the constructions of pre-Lie algebras from some known structures such as commutative associative algebras, Lie algebras, associative algebras and linear functions;

– in section 7.4, we interpret the close relationship between pre-Lie algebras and the classical Yang–Baxter equation (CYBE), where the former are regarded as the underlying algebraic structures of the latter;

– in section 7.5, we put pre-Lie algebras into a bigger framework as one of the algebraic structures of the Lie analogues of Loday algebras.

Throughout this chapter, unless otherwise specified, all vector spaces and algebras are finite-dimensional over the complex field $\mathbb{C}$, although many results still hold over other fields or in the infinite-dimensional case.

7.2. Some appearances of pre-Lie algebras

In this section, we illustrate some appearances of pre-Lie algebras in different topics. We hope that with the introduction of these appearances, one can know some background and different motivations of introducing the notion of pre-Lie algebra. We emphasize the appearance of "left-symmetry". Some materials can be found in Burde (2006) and the references therein. We would like to point out that for these materials appearing in Burde (2006) or this chapter, we might express them in a "short version" since it seems enough to exhibit only the appearances of "left-symmetry" in those often long and important theories.

7.2.1. *Left-invariant affine structures on Lie groups: a geometric interpretation of "left-symmetry"*

Let G be a Lie group with a left-invariant affine structure: there is a flat torsion-free left-invariant affine connection ∇ on G, namely, for all left-invariant vector fields $X, Y, Z \in \mathfrak{g} = T(G)$,

$$R(X,Y)Z = \nabla_X \nabla_Y Z - \nabla_Y \nabla_X Z - \nabla_{[X,Y]} Z = 0, \qquad [7.9]$$

$$T(X,Y) = \nabla_X Y - \nabla_Y X - [X,Y] = 0. \qquad [7.10]$$

This means both the curvature $R(X,Y)$ and torsion $T(X,Y)$ are zero for the connection ∇. If we define

$$\nabla_X Y = XY, \qquad [7.11]$$

then the identity [7.1] for a pre-Lie algebra exactly amounts to equations [7.9] and [7.10].

See Koszul 1961; Vinberg 1963; Medina 1981 and Kim 1986 for more details.

7.2.2. Deformation complexes of algebras and right-symmetric algebras

Let V be a vector space. Denote by $C^m(V,V)$ the space of all m-multilinear maps from $V^{\otimes m}$ to V. For $f \in C^p(V,V)$ and $g \in C^q(V,V)$, define the product

$$\circ : C^p(V,V) \times C^q(V,V) \to C^{p+q-1}(V,V), \ \ (f,g) \mapsto f \circ g$$

as

$$f \circ g(x_1, \cdots, x_{p+q-1}) = \qquad [7.12]$$
$$\hat{E} \sum_{i=1}^{p} f(x_1, \cdots, x_{i-1}, g(x_i, \cdots, x_{i+q-1}), x_{i+q}, \cdots, x_{p+q-1}).$$

PROPOSITION 7.2.– The algebra $(C^\bullet(V,V), \circ)$ is a right-symmetric algebra.

When $V = A$, where A is an associative algebra, the role of pre-Lie algebra is necessary for the construction of cohomology theory. The product given by equation [7.12] should be modified to be a "graded version":

$$f \circ g(x_1, \cdots, x_{p+q-1}) = \qquad [7.13]$$
$$\sum_{i=1}^{p} (-1)^{(q-1)(i-1)} f(x_1, \cdots, x_{i-1}, g(x_i, \cdots, x_{i+q-1}), x_{i+q}, \cdots, x_{p+q-1}).$$

It satisfies the *graded right-symmetry*

$$(xy)z - x(yz) = (-1)^{|y||z|}((xz)y - x(zy)), \qquad [7.14]$$

for any x, y, z in a graded vector space and $|x|$ denotes the degree.

For the complex $C^\bullet(A,A)$, the key is to define the coboundary operator $d : C^p(A,A) \to C^{p+1}(A,A)$ such that $d^2 = 0$. In fact, the operator d is given as

$$d(f) = -\mu \circ f + (-1)^{|\mu||f|} f \circ u, \forall f \in C^p(A,A), \qquad [7.15]$$

where $\mu \in C^2(A,A)$ is the multiplication map of A.

See Gerstenhaber (1963) for more details.

7.2.3. *Rooted tree algebras: free pre-Lie algebras*

A *rooted tree* is a finite, connected oriented graph without loops in which every vertex has exactly one coming edge, except one (called the root) that has no incoming but only outgoing edges.

Let $\mathbb{T}$ be the vector space spanned by all rooted trees. One can introduce a bilinear product $\curvearrowright$ on $\mathbb{T}$ as follows. Let τ_1 and τ_2 be two rooted trees.

$$\tau_1 \curvearrowright \tau_2 = \sum_{s \in \mathrm{Vertices}(\tau_2)} \tau_1 \circ_s \tau_2,$$

where $\tau_1 \circ_s \tau_2$ is the rooted tree obtained by adding to the disjoint union of τ_2 and τ_1 an edge going from the vertex s of τ_2 to the root vertex of τ_1.

PROPOSITION 7.3.– $(\mathbb{T}, \curvearrowright)$ is a free pre-Lie algebra on one generator.

REMARK 7.2.– The pre-Lie algebra $(\mathbb{T}, \curvearrowright)$ is isomorphic to the pre-Lie algebra (in the sense of left-symmetry) given by Connes and Kreimer (1998) in the study of quantum field theory. Note that in (Connes and Kreimer 1998), the corresponding action (the so-called "glue" action of rooted trees obtained as the opposite of the "cut" action) is not the same as the above $\curvearrowright$ in the expressing form, whereas in fact the two algebras are isomorphic.

See Cayley 1890; Connes and Kreimer 1998; Chapoton and Livernet 2001 and Dzhumadil'daev and Lofwall 2002 for more details.

7.2.4. *Complex structures on Lie algebras*

DEFINITION 7.2.– Let $\mathfrak{g}$ be a real Lie algebra. A *complex structure* on $\mathfrak{g}$ is a linear endomorphism $J : \mathfrak{g} \to \mathfrak{g}$ satisfying $J^2 = -\mathrm{id}$ and the integrable condition:

$$J[x, y] = [Jx, y] + [x, Jy] + J[Jx, Jy], \quad \forall x, y \in \mathfrak{g}. \qquad [7.16]$$

NOTATION.– Let $\rho : \mathfrak{g} \to gl(V)$ be a representation of a Lie algebra $\mathfrak{g}$. On the vector space $\mathfrak{g} \oplus V$, there is a natural Lie algebra structure (denoted by $\mathfrak{g} \ltimes_\rho V$) given as follows:

$$[x_1 + v_1, x_2 + v_2] = [x_1, x_2] + \rho(x_1)v_2 - \rho(x_2)v_1, \qquad [7.17]$$

for any $x_1, x_2 \in \mathfrak{g}, v_1, v_2 \in V$.

PROPOSITION 7.4.– Let A be a real left-symmetric algebra A. Define a linear map $J : A \oplus A \rightarrow A \oplus A$ by

$$J(x,y) = (-y,x), \quad \forall\, x,y \in A. \qquad [7.18]$$

Then J is a complex structure on the Lie algebra $\mathfrak{g}(A) \ltimes_L A$, where L is the representation of the sub-adjacent Lie algebra $\mathfrak{g}(A)$ induced by the left multiplication operators of A.

REMARK 7.3.– In fact, there is a correspondence between pre-Lie algebras and complex product structures on Lie algebras (Andrada and Salamon 2005), where a complex product structure is a pair of a complex structure J and a product structure E satisfying $JE = -EJ$.

See Andrada and Salamon 2005 and Bai 2006 for more details.

7.2.5. *Symplectic structures on Lie groups and Lie algebras, phase spaces of Lie algebras and Kähler structures*

A *symplectic Lie group* is a Lie group G with a left-invariant symplectic form ω^+. The corresponding structure at the level of Lie algebras is given as follows.

DEFINITION 7.3.– A Lie algebra $\mathfrak{g}$ is called a *symplectic Lie algebra* if there is a non-degenerate skew-symmetric 2-cocycle ω (the symplectic form) on $\mathfrak{g}$, that is,

$$\omega([x,y],z) + \omega([y,z],x) + \omega([z,x],y) = 0, \quad \forall x,y,z \in \mathfrak{g}. \qquad [7.19]$$

We denote it by $(\mathfrak{g}, \omega)$.

THEOREM 7.1.– Let $(\mathfrak{g}, \omega)$ be a symplectic Lie algebra. Then there exists a compatible pre-Lie algebra structure "$*$" on $\mathfrak{g}$ given by

$$\omega(x * y, z) = -\omega(y, [x,z]), \quad \forall x,y,z \in \mathfrak{g}. \qquad [7.20]$$

COROLLARY 7.1.– Let G be a symplectic Lie group with a left-invariant symplectic form ω^+. Then there is a left-invariant affine structure on G defined by

$$\omega^+(\nabla_{x^+} y^+, z^+) = -\omega^+(y^+, [x^+, z^+])$$

for any left-invariant vector fields x^+, y^+, z^+.

Conversely, there is a symplectic Lie algebra structure on the direct sum $A \oplus A^*$ of underlying space of a pre-Lie algebra A and its dual space A^*.

PROPOSITION 7.5.– Let A be a pre-Lie algebra. Set $T^*\mathfrak{g}(A) = \mathfrak{g}(A) \ltimes_{L^*} A^*$, where L^* is the dual representation of the representation L induced by left multiplication operators. Define the following bilinear form on $A \oplus A^*$

$$\omega_p(x + a^*, y + b^*) = \langle a^*, y \rangle - \langle x, b^* \rangle, \quad \forall x, y \in A, a^*, b^* \in A^*, \tag{7.21}$$

where $\langle , \rangle$ is the ordinary pair between A and A^*. Then $(T^*\mathfrak{g}(A), \omega_p)$ is a symplectic Lie algebra.

REMARK 7.4.– The above construction $(T^*\mathfrak{g}(A), \omega_p)$ is a *phase space* of the Lie algebra $\mathfrak{g}(A)$ in Kupershmidt (1994). Moreover, Kupershmidt pointed out that pre-Lie algebras appear as an underlying structure of those Lie algebras that possess a phase space and thus they form a natural category from the point of view of classical and quantum mechanics (Kupershmidt 1999a).

Kähler structures on Lie algebras are closely related to the study of Kähler Lie groups and Kähler manifolds (Lichnerowicz and Medina 1988).

DEFINITION 7.4.– Let $\mathfrak{g}$ be a real Lie algebra. If there exists a complex structure J and a non-degenerate skew-symmetric bilinear form ω, the following conditions are satisfied:

1) ω is a symplectic form on $\mathfrak{g}$;

2) $\omega(J(x), J(y)) = \omega(x, y)$ for any $x, y \in \mathfrak{g}$;

3) $\omega(x, J(x)) > 0$, for any $x \in \mathfrak{g}$ and $x \neq 0$,

then $\{J, \omega\}$ is called a *Kähler structure* on $\mathfrak{g}$.

PROPOSITION 7.6.– Let $(A, \cdot)$ be a left-symmetric algebra with a symmetric and positive definite bilinear form $B(\ , \ )$. Suppose the bilinear form B satisfies the following condition:

$$B(x \cdot y, z) + B(y, x \cdot z) = 0, \quad \forall x, y, z \in A. \tag{7.22}$$

Then there exists a complex structure J on the phase space $T^*\mathfrak{g}(A) = \mathfrak{g}(A) \ltimes_{L^*} A^*$ given by

$$J(x + y^*) = -y + x^*, \quad \forall x, y \in A, \tag{7.23}$$

where for any $x = \sum_{i=1}^n \lambda_i e_i \in A$, set $x^* = \sum_{i=1}^n \lambda_i e_i^* \in A^*$. Here, $\{e_1, \cdots, e_n\}$ is a basis of A such that $B(e_i, e_j) = \delta_{ij}$ and $\{e_1^*, \cdots, e_n^*\}$ is its dual basis. Furthermore, there exists a Kähler structure $\{-J, \omega_p\}$ on $T^*\mathfrak{g}(A)$, where ω_p is given by equation [7.21].

REMARK 7.5.– In fact, the positive definite bilinear form $\mathfrak{B}$ satisfying equation [7.22] on a pre-Lie algebra induces a left-invariant Hessian metric on the corresponding connected real Lie group G, thus making it a Hessian manifold. Recall that a *Hessian manifold* M is a flat affine manifold provided with a Hessian metric. Note that a *Hessian metric* on a smooth manifold M is a Riemannian metric g such that for each point $p \in M$, there exists a C^∞-function φ defined on a neighborhood of p such that $g_{ij} = \frac{\partial^2 \varphi}{\partial x^i \partial x^j}$ (Shima 1980).

See Chu 1974; Shima 1980; Lichnerowicz and Medina 1988; Kupershmidt 1994, 1999a; Diatta and Medina 2004 and Bai 2006 for more details.

7.2.6. *Vertex algebras*

Vertex algebras are fundamental algebraic structures in conformal field theory.

DEFINITION 7.5.– A *vertex algebra* is a vector space V equipped with a linear map

$$Y : V \to \mathrm{Hom}(V, V((x))), v \to Y(v, x) = \sum_{n\in\mathbb{Z}} v_n x^{-n-1} (\text{where } v_n \in \mathrm{End}V)$$

and equipped with a distinguished vector $\mathbf{1} \in V$ such that

$$Y(\mathbf{1}, x) = 1;$$

$$Y(v, x)\mathbf{1} \in V[[x]] \text{ and } Y(v, x)\mathbf{1}|_{x=0} (= v_{-1}\mathbf{1}) = v, \ \forall v \in V,$$

and for $u, v \in V$, there is Jacobi identity:

$$x_0^{-1}\delta(\frac{x_1 - x_2}{x_0})Y(u, x_1)Y(v, x_2) - x_0^{-1}\delta(\frac{x_2 - x_1}{-x_0})Y(v, x_2)Y(u, x_1) \qquad [7.24]$$
$$= x_2^{-1}\delta(\frac{x_1 - x_0}{x_2})Y(Y(u, x_0)v, x_2),$$

where $\delta(x) = \sum_{n\in\mathbb{Z}} x^n$.

PROPOSITION 7.7.– Let $(V, Y, \mathbf{1})$ be a vertex algebra. Then

$$a * b = a_{-1}b \qquad [7.25]$$

defines a pre-Lie algebra.

In fact, by Borcherd's identities:

$$(a_m(b))_n(c) = \sum_{i\geq 0}(-1)^i C_m^i((a_{m-i}(b_{n+i}(c) - (-1)^m b_{m+n-i}(a_i(c))),$$

let $m = n = -1$, we have

$$(a_{-1}b)_{-1}c - a_{-1}(b_{-1}c) = \sum_{i\geq 0}(a_{-2-i}(b_i c) + b_{-2-i}(a_i c)).$$

Then the conclusion follows.

REMARK 7.6.– Conversely, a vertex algebra is equivalent to a pre-Lie algebra and a Lie conformal algebra with some compatible conditions.

See Bakalov and Kac (2003) for more details.

7.3. Some basic results and constructions of pre-Lie algebras

In section 7.3.1, we introduce some basic properties of pre-Lie algebras including some comments on the studies of structure theory and representation theory and some classification results. In section 7.3.2, we give the constructions of pre-Lie algebras from some known structures like commutative associative algebras, Lie algebras, associative algebras and linear functions.

7.3.1. *Some basic results of pre-Lie algebras*

7.3.1.1. *Some studies on structure theory*

A "good" structure theory for an algebraic system usually means that there is a well-defined "radical" and the quotient by the radical is "semisimple" or "reductive", which is roughly a direct sum of "simple" ones, like associative and Lie algebras. Usually, a classification of these "simple" objectives in a certain sense should also be obtained. Unfortunately, pre-Lie algebras do not belong to this case.

In fact, there are a lot of studies on this subject (see Burde 2006 and the references therein). Roughly speaking, there are several different approaches to define a "radical" of a pre-Lie algebra (Chang *et al.* 1999). However, none of them is satisfactory enough to give a "good" structure theory in the above sense. For example, certain "radical" is an ideal (like the so-called Jacobson radical), but it seems very difficult to provide a further study on the quotient by it. It also leads to the fact that there is not a complete theory of semisimple (simple) pre-Lie algebras, except for some classification results in low dimensions. Even the authors in Chang *et al.* (1999) suggested to give up such efforts since they consider identity [7.1] too weak and some additional identities are necessary for a better structure theory.

Nevertheless, the following definition seems acceptable to a certain extent.

DEFINITION 7.6.– Let A be a pre-Lie algebra and $T(A) = \{x \in A | \mathrm{tr}R(x) = 0\}$. The largest left ideal of A contained in $T(A)$ is called the *radical* of A and is denoted by $\mathrm{rad}(A)$.

REMARK 7.7.– Note that the above $\mathrm{rad}(A)$ is only a left ideal of A and one cannot take the quotient $A/\mathrm{rad}(A)$ if it is not an ideal. However, a pre-Lie algebra is transitive if and only if $A = \mathrm{rad}(A)$.

DEFINITION 7.7.– A pre-Lie algebra A is called *simple* if A has no ideals except zero and itself and $AA \neq 0$. A is called *semisimple* if A is a direct sum of simple ideals.

Moreover, the complexity of this problem can be seen from the following example.

EXAMPLE 7.1.– There exists a transitive simple pre-Lie algebra that combines "simplicity" and certain "nilpotence". For example, let A be a three-dimensional pre-Lie algebra with a basis $\{e_1, e_2, e_3\}$ whose non-zero products are given by

$$e_1e_2 = e_2, e_1e_3 = -e_3, e_2e_3 = e_3e_2 = e_1.$$

However, the structure theory for some subclasses have been constructed.

1) *Novikov algebra*: if A is a Novikov algebra, then $\mathrm{rad}(A) = T(A)$ is an ideal. Over an algebraically closed field of characteristic zero, $A/\mathrm{rad}(A)$ is a direct sum of fields and a finite-dimensional simple Novikov algebra is isomorphic to the field (Zelmanov 1987).

2) *Bi-symmetric algebra*: if A is a bi-symmetric algebra, then $\mathrm{rad}(A) = T(A)$ is an ideal. Over a field of characteristic which is not 2 or 3, $A/\mathrm{rad}(A)$ is a semisimple associative algebra and a simple bisymmetric algebra is isomorphic to a simple associative algebra (Kleinfeld 1957; Bai and Meng 2000).

7.3.1.2. *Some comments on representation theory*

The beauty of a representation theory is to study the algebras in terms of the computation of matrices. However, there has not yet been a "natural" pre-Lie algebra structure on the vector space $\mathrm{End}(V)$. However, we can consider the following definition of "representation":

DEFINITION 7.8.– Let $(A, \circ)$ be a pre-Lie algebra and V be a vector space. Let $l, r : A \to gl(V)$ be two linear maps. (l, r, V) is called a *module of* $(A, \circ)$ if

$$l(x)l(y) - l(x \circ y) = l(y)l(x) - l(y \circ x), \quad [7.26]$$

$$l(x)r(y) - r(y)l(x) = r(x \circ y) - r(y)r(x), \forall x, y \in A. \quad [7.27]$$

However, they are a kind of "bimodule" structure, which are too formal to give a direct and computable study in terms of matrices. So up to now, there has not been a suitable (and computable) representation theory of pre-Lie algebras.

7.3.1.3. *Some classification results*

The classification of algebras in the sense of algebraic isomorphisms is always one of the key problems, and also always difficult. There have been certain progresses for the classification of pre-Lie algebras, however, it is impossible to list every classification result here. We only choose to list some classification results as follows. We would like to emphasize again that **there has not been a complete classification of simple pre-Lie algebras** yet.

1) The classification of pre-Lie algebras in low dimensions:

- The classification of two-dimensional complex pre-Lie algebras was given in Bai and Meng (1996) and Burde (1998). The method is basically the computation of structure constants.

- The classification of three-dimensional complex pre-Lie algebras was given in Bai (2009). It depends on a detailed study of 1-cocycles that divides the corresponding classification problem into solving a series of linear problems. It includes the classification of three-dimensional complex Novikov algebras (Bai and Meng 2001a), bi-symmetric algebras (Bai and Meng 2000) and simple pre-Lie algebras (Burde 1998), which have been obtained independently.

- The classification of three-dimensional real simple pre-Lie algebras was given in Kong *et al.* (2012). It depends on the study of the relationships between real and complex pre-Lie algebras.

- The classification of four-dimensional complex transitive pre-Lie algebras on nilpotent Lie algebras was given in Kim (1986). The method is to use an extension theory of pre-Lie algebras.

- The classification of four-dimensional complex transitive simple pre-Lie algebras was given in Burde (1998).

There are also some related classification results, like the classification of three-dimensional pre-Lie superalgebras (Zhang and Bai 2012) and 2|2-dimensional Balinsky-Novikov superalgebras (Wang *et al.* 2012).

2) Some infinite dimensional pre-Lie algebras:

- The classification of infinite dimensional simple Novikov algebras was studied in Osborn (1994) and Xu (1996, 1997).

- The classification of compatible pre-Lie algebras on the Witt and Virasoro algebras satisfying certain natural gradation conditions was given in Kong *et al.* (2011) through the representation theory of the Virasoro algebra, which includes the results given in Chapoton (2004) and Kupershmidt (1999a). The "super" version of the classification result on the super-Virasoro algebras was given in Kong and Bai (2008). Moreover, a class of non-graded compatible pre-Lie algebras on the Witt algebra was given in Tang and Bai (2012). Note that the compatible pre-Lie algebras on the Witt algebra are *simple*.

3) Free pre-Lie algebras: these algebras are the "biggest" pre-Lie algebras and every pre-Lie algebra is a quotient of a free pre-Lie algebra.

- From a pure algebraic point of view, the basis of a free pre-Lie algebra was given explicitly in Segal (1994).

- A free pre-Lie algebra with one generator interpreted in terms of rooted trees was given in Chapoton and Livernet (2001) and in Dzhumadil'daev and Lofwall (2002).

7.3.1.4. *Summary: main problems and ideas*

In summary, due to the non-associativity, we think that these are the main difficulties in the study of pre-Lie algebras:

1) there is not a suitable (and computable) representation theory;

2) there is not a complete (and good) structure theory.

The *main ideas* are to try to find more examples. This includes two key points:

1) pay attention to the relations with other topics (including application);

2) realized or constructed by some known structures.

7.3.2. *Constructions of pre-Lie algebras from some known structures*

7.3.2.1. *Constructions from commutative associative algebras*

PROPOSITION 7.8.– S. Gel'fand, in Gel'fand and Dorfman (1979). Let $(A, \cdot)$ be a commutative associative algebra, and D be its derivation. Then the new product

$$a * b = a \cdot Db, \quad \forall a, b \in A \qquad [7.28]$$

makes $(A, *)$ become a Novikov algebra.

REMARK 7.8.– There are some generalizations of the above result. Let $(A, \cdot)$ be a commutative associative algebra and D be a derivation. Then the new product

$$x *_a y = x \cdot Dy + a \cdot x \cdot y, \;\; \forall\, x, y \in A \tag{7.29}$$

makes $(A, \; *_a)$ become a Novikov algebra for $a \in \mathbb{F}$ by Filippov (1989) and for a fixed element $a \in A$ by Xu (1996).

DEFINITION 7.9.– A *linear deformation* of a Novikov algebra $(A, *)$ is a binary operation $G_1 : A \times A \to A$ such that a family of algebras $g_q : A \times A \to A$ defined by

$$g_q(a, b) = a * b + qG_1(a, b) \tag{7.30}$$

are still Novikov algebras (for every q). If G_1 is commutative, then G_1 is called *compatible*.

REMARK 7.9.– The two kinds of Novikov algebras given by Filippov and Xu are the special compatible linear deformations of the algebras given by S. Gel'fand.

PROPOSITION 7.9.– (Bai and Meng 2001b,c) Novikov algebras in dimension ≤ 3 can be realized as the algebras defined by S. Gel'fand and their compatible linear deformations.

However, let Ω be any set. Let $NP(\Omega)$ be the commutative associative polynomial ring over a ring R with the set of variables equal to

$$\{a[i] | a \in \Omega, i \geq -1\}. \tag{7.31}$$

Let $D : NP(\Omega) \to NP(\Omega)$ be the R-derivation defined by

$$D(a[i]) = a[i + 1], \tag{7.32}$$

and let $\circ$ be the binary operation on $NP(\Omega)$ defined by

$$a \circ b = aD(b). \tag{7.33}$$

THEOREM 7.2.– (Dzhumadil'daev and Lofwall 2002) Free Novikov algebra generated by any set Ω is isomorphic to $(NP(\Omega)_0, \circ)$, where $NP(\Omega)_0$ is the set of elements in $NP(\Omega)$ of weight -1 (it is a subalgebra of $NP(\Omega)$).

COROLLARY 7.2.– Any Novikov algebra is a quotient of a subalgebra of an (infinite-dimensional) algebra given by equation [7.28].

7.3.2.2. *Constructions from Lie algebras*

PROPOSITION 7.10.– (Golubschik and Sokolov 2000) Let $(\mathfrak{g}, [,])$ be a Lie algebra and $R : \mathfrak{g} \to \mathfrak{g}$ be a linear map satisfying the following equation:

$$[R(x), R(y)] = R([R(x), y] + [x, R(y)]), \quad \forall\, x, y \in \mathfrak{g}. \qquad [7.34]$$

Then

$$x * y = [R(x), y], \quad \forall\, x, y \in \mathfrak{g} \qquad [7.35]$$

defines a pre-Lie algebra.

REMARK 7.10.– The linear operator satisfying equation [7.34] is called the *operator form of the CYBE* or the *Rota–Baxter operator of weight zero* in the context of Lie algebras.

We give a more detailed interpretation of the above constructions in the next section.

7.3.2.3. *Constructions from associative algebras*

There are two approaches. One is a direct consequence of proposition 7.10.

COROLLARY 7.3.– Let $(A, \cdot)$ be an associative algebra and $R : A \to A$ be a linear map satisfying

$$R(x) \cdot R(y) = R(R(x) \cdot y + x \cdot R(y)), \forall x, y \in A. \qquad [7.36]$$

Then

$$x * y = R(x) \cdot y - y \cdot R(x), \quad \forall x, y \in A \qquad [7.37]$$

defines a pre-Lie algebra.

Another approach is given as follows.

PROPOSITION 7.11.– (Golubschik and Sokolov 2000; Ebrahimi-Fard 2002) Let $(A, \cdot)$ be an associative algebra and $R : A \to A$ be a linear map satisfying

$$R(x) \cdot R(y) + R(x \cdot y) = R(R(x) \cdot y + x \cdot R(y)), \forall x, y \in A. \qquad [7.38]$$

Then

$$x * y = R(x) \cdot y - y \cdot R(x) - x \cdot y, \quad \forall x, y \in A \qquad [7.39]$$

defines a pre-Lie algebra.

REMARK 7.11.– The linear operators defined by equations [7.36] and [7.38] are called a *Rota–Baxter operator of weight zero and weight 1*, respectively. The Rota–Baxter operators were introduced to solve analytic and combinatorial problems and attract more attention in many fields in mathematics and mathematical physics (Guo 2009).

REMARK 7.12.– The linear operator defined by equation [7.38] is related to the so-called "modified classical Yang–Baxter equation" (Semonov-Tian-Shansky 1983).

7.3.2.4. *Constructions from linear functions*

PROPOSITION 7.12.– (Svinolupov and Sokolov 1994) Let V be a vector space over the complex field $\mathbb{C}$ with the ordinary scalar product $(\, , \,)$ and a a fixed vector in V. Then

$$u * v = (u, v)a + (u, a)v, \forall u, v \in V, \tag{7.40}$$

defines a pre-Lie algebra on V.

REMARK 7.13.– The above construction gives the integrable (generalized) Burgers equation

$$U_t = U_{xx} + 2U * U_x + (U * (U * U)) - ((U * U) * U). \tag{7.41}$$

REMARK 7.14.– In Bai (2004), the above construction was generalized to get pre-Lie algebras from linear functions.

COROLLARY 7.4.– The pre-Lie algebras given by equation [7.40] are simple.

7.4. Pre-Lie algebras and CYBE

In this section, we give a further detailed interpretation of the construction of pre-Lie algebras given in proposition 7.10, which is a direct consequence of the close relationships between pre-Lie algebras and CYBE. Most of the study in this section can be found in Bai (2007).

7.4.1. *The existence of a compatible pre-Lie algebra on a Lie algebra*

PROPOSITION 7.13.– (Medina 1981) The sub-adjacent Lie algebra of a finite-dimensional pre-Lie algebra A over an algebraically closed field with characteristic 0 satisfies

$$[\mathfrak{g}(A), \mathfrak{g}(A)] \neq \mathfrak{g}(A). \tag{7.42}$$

REMARK 7.15.– Therefore, there exists a Lie algebra without a compatible pre-Lie algebra. In particular, there is not a compatible pre-Lie algebra on a semisimple Lie algebra.

A natural question arises: *what is a necessary and sufficient condition under which there exists a compatible pre-Lie algebra on a Lie algebra?*

DEFINITION 7.10.– Let $\mathfrak{g}$ be a Lie algebra and $\rho : \mathfrak{g} \to gl(V)$ be a representation of $\mathfrak{g}$. A *1-cocycle q associated to ρ* (denoted by (ρ, q)) is defined as a linear map from $\mathfrak{g}$ to V satisfying

$$q[x, y] = \rho(x)q(y) - \rho(y)q(x), \forall x, y \in \mathfrak{g}. \qquad [7.43]$$

PROPOSITION 7.14.– There is a compatible pre-Lie algebra on a Lie algebra $\mathfrak{g}$ if and only if there exists a bijective 1-cocycle of $\mathfrak{g}$.

In fact, let (ρ, q) be a bijective 1-cocycle of $\mathfrak{g}$. Then

$$x * y = q^{-1}\rho(x)q(y), \quad \forall x, y \in A, \qquad [7.44]$$

defines a pre-Lie algebra structure on $\mathfrak{g}$. Conversely, for a pre-Lie algebra A, (L, id) is a bijective 1-cocycle of $\mathfrak{g}(A)$.

REMARK 7.16.– There are several equivalent conditions such as the existence of an étale affine representation (Medina 1981). However, such a conclusion provides a linearization procedure of classifying pre-Lie algebras, that is, it divides the corresponding classification problem into solving a series of linear problems, which leads to the classification of three-dimensional complex pre-Lie algebras (Bai 2009).

7.4.2. CYBE: unification of tensor and operator forms

DEFINITION 7.11.– Let $\mathfrak{g}$ be a Lie algebra and $r = \sum_i a_i \otimes b_i \in \mathfrak{g} \otimes \mathfrak{g}$. r is called a solution of *CYBE* in $\mathfrak{g}$ if

$$[r_{12}, r_{13}] + [r_{12}, r_{23}] + [r_{13}, r_{23}] = 0 \text{ in } U(\mathfrak{g}), \qquad [7.45]$$

where $U(\mathfrak{g})$ is the universal enveloping algebra of $\mathfrak{g}$ and

$$r_{12} = \sum_i a_i \otimes b_i \otimes 1; r_{13} = \sum_i a_i \otimes 1 \otimes b_i; r_{23} = \sum_i 1 \otimes a_i \otimes b_i. \qquad [7.46]$$

r is said to be *skew-symmetric* if

$$r = \sum_i (a_i \otimes b_i - b_i \otimes a_i). \qquad [7.47]$$

We also denote $r^{21} = \sum_i b_i \otimes a_i$.

Let r be a solution of CYBE. Set $r = \sum_{i,j} r_{ij} e_i \otimes e_j$, where $\{e_1, \cdots, e_n\}$ is a basis of the Lie algebra $\mathfrak{g}$. Then the matrix

$$r = (r_{ij}) = \begin{pmatrix} r_{11} & \cdots & r_{1n} \\ \cdots & \cdots & \cdots \\ r_{n1} & \cdots & r_{nn} \end{pmatrix}, \qquad [7.48]$$

is called a *classical* r*-matrix*.

Natural question: *if a linear transformation (or generally, a linear map)* R *is given by the classical* r*-matrix under a basis, what should* R *satisfy?*

The first answer was given by Semonov-Tian-Shansky (1983).

PROPOSITION 7.15.– Let $\mathfrak{g}$ be a Lie algebra, and let $r \in \mathfrak{g} \otimes \mathfrak{g}$. Suppose that the following two conditions are satisfied:

1) there exists a non-degenerate symmetric invariant bilinear form B on $\mathfrak{g}$, that is,

$B([x, y], z) = B(x, [y, z]), \ \forall x, y, z \in \mathfrak{g}$;

2) r is skew-symmetric.

Let $R : \mathfrak{g} \to \mathfrak{g}$ be a linear map corresponding to r under an orthonormal basis associated with B. Then r is a solution of CYBE if and only if R satisfies equation [7.34].

REMARK 7.17.– Therefore, in the above sense, equation [7.34] is called the *operator form of the CYBE*.

Another approach was given by Kupershmidt (1999b) by canceling the above condition (1), but replacing the linear transformation $R : \mathfrak{g} \to \mathfrak{g}$ by a linear map $r : \mathfrak{g}^* \to \mathfrak{g}$, where $\mathfrak{g}^*$ is the dual space of $\mathfrak{g}$. Note that

$$\mathfrak{g} \otimes \mathfrak{g} \cong \mathrm{Hom}(\mathfrak{g}^*, \mathfrak{g}). \qquad [7.49]$$

PROPOSITION 7.16.– Let $\mathfrak{g}$ be a Lie algebra, and let $r \in \mathfrak{g} \otimes \mathfrak{g}$. Suppose that r is skew-symmetric. Under the isomorphism given by equation [7.49], we still denote the corresponding linear map from $\mathfrak{g}^*$ to $\mathfrak{g}$ by r. Then r is a solution of CYBE if and only if r satisfies

$$[r(x), r(y)] = r(\mathrm{ad}^* r(x)(y) - \mathrm{ad}^* r(y)(x)), \forall x, y \in \mathfrak{g}^*, \qquad [7.50]$$

where ad^* is the dual representation of adjoint representation (coadjoint representation).

DEFINITION 7.12.– Let $\mathfrak{g}$ be a Lie algebra and $\rho : \mathfrak{g} \to gl(V)$ be a representation of $\mathfrak{g}$. A linear map $T : V \to \mathfrak{g}$ is called an $\mathcal{O}$-*operator* if T satisfies

$$[T(u), T(v)] = T(\rho(T(u))v - \rho(T(v))u), \forall u, v \in V. \qquad [7.51]$$

REMARK 7.18.– Kupershmidt introduced the notion of an $\mathcal{O}$-operator as a natural generalization of CYBE because equations [7.34] and [7.50] express that R and r are $\mathcal{O}$-operators associated with ad and ad^*, respectively.

There is a unification of the tensor and operator forms of CYBE given as follows.

PROPOSITION 7.17.– (Bai 2007) Let $\mathfrak{g}$ be a Lie algebra, $\rho : \mathfrak{g} \to gl(V)$ be a representation of $\mathfrak{g}$ and $\rho^* : \mathfrak{g} \to gl(V^*)$ be the dual representation and $T : V \to \mathfrak{g}$ be a linear map, which is identified as an element in $\mathfrak{g} \otimes V^* \subset (\mathfrak{g} \ltimes_{\rho^*} V^*) \otimes (\mathfrak{g} \ltimes_{\rho^*} V^*)$. Then $r = T - T^{21}$ is a skew-symmetric solution of CYBE in $\mathfrak{g} \ltimes_{\rho^*} V^*$ if and only if T is an $\mathcal{O}$-operator.

7.4.3. *Pre-Lie algebras, $\mathcal{O}$-operators and CYBE*

PROPOSITION 7.18.– Let $\mathfrak{g}$ be a Lie algebra and $\rho : \mathfrak{g} \to gl(V)$ be a representation. Let $T : V \to \mathfrak{g}$ be an $\mathcal{O}$-operator associated to ρ. Then

$$u * v = \rho(T(u))v, \quad \forall u, v \in V \qquad [7.52]$$

defines a pre-Lie algebra on V.

REMARK 7.19.– When we take the adjoint representation, we get the construction of pre-Lie algebras given in proposition 7.10.

LEMMA 7.1.– Let $\mathfrak{g}$ be a Lie algebra and (ρ, V) be a representation. Suppose $f : \mathfrak{g} \to V$ is invertible. Then f is a 1-cocycle of $\mathfrak{g}$ associated with ρ if and only if f^{-1} is an $\mathcal{O}$-operator.

COROLLARY 7.5.– Let $\mathfrak{g}$ be a Lie algebra. There is a compatible pre-Lie algebra structure on $\mathfrak{g}$ if and only if there exists an invertible $\mathcal{O}$-operator of $\mathfrak{g}$.

By proposition 7.17 and since id is **an $\mathcal{O}$-operator associated with** L, we give the following construction of solutions of CYBE from pre-Lie algebras.

PROPOSITION 7.19.– Let A be a pre-Lie algebra. Then

$$r = \sum_{i=1}^{n}(e_i \otimes e_i^* - e_i^* \otimes e_i) \qquad [7.53]$$

is a solution of the CYBE in the Lie algebra $\mathfrak{g}(A) \ltimes_{L^*} A^*$, where $\{e_1, ..., e_n\}$ is a basis of A and $\{e_1^*, ..., e_n^*\}$ is the dual basis.

7.4.4. An algebraic interpretation of "left-symmetry": construction from Lie algebras revisited

We come back to the construction given in proposition 7.10. In fact, it is a direct consequence of proposition 7.18 or the following result.

LEMMA 7.2.– Let $\mathfrak{g}$ be a Lie algebra and f be a linear transformation on $\mathfrak{g}$. Then on $\mathfrak{g}$ the new product

$$x * y = [f(x), y], \forall x, y \in \mathfrak{g} \qquad [7.54]$$

defines a pre-Lie algebra if and only if

$$[f(x), f(y)] - f([f(x), y] + [x, f(y)]) \in C(\mathfrak{g}), \forall x, y \in \mathfrak{g}, \qquad [7.55]$$

where $C(\mathfrak{g}) = \{x \in \mathfrak{g} | [x, y] = 0, \ \forall y \in \mathfrak{g}\}$ is the center of $\mathfrak{g}$.

Furthermore, there is an algebraic interpretation of "left-symmetry" as follows. Let $\{e_i\}$ be a basis of a Lie algebra $\mathfrak{g}$ and $r : \mathfrak{g} \to \mathfrak{g}$ be an $\mathcal{O}$-operator associated with ad, that is, r satisfies equation [7.34]. Set $r(e_i) = \sum_{j \in I} r_{ij} e_j$. Then the basis-interpretation of equation [7.35] is given as

$$e_i * e_j = \sum_{l \in I} r_{il}[e_l, e_j]. \qquad [7.56]$$

Such a construction of left-symmetric algebras (pre-Lie algebras) can be regarded as a Lie algebra "left-twisted" by a classical r-matrix.

On the other hand, let us consider the right-symmetry. We set

$$e_i \cdot e_j = [e_i, r(e_j)] = \sum_{l \in I} r_{jl}[e_i, e_l]. \qquad [7.57]$$

Then the above product defines a right-symmetric algebra on $\mathfrak{g}$, which can be regarded as a Lie algebra "right-twisted" by a classical r-matrix.

7.5. A larger framework: Lie analogues of Loday algebras

Pre-Lie algebras can be put into a bigger framework as one of the algebraic structures of the Lie analogues of Loday algebras.

7.5.1. *Pre-Lie algebras, dendriform algebras and Loday algebras*

DEFINITION 7.13.– (Loday 2001) A *dendriform algebra* $(A, \prec, \succ)$ is a vector space A with two binary operations denoted by $\prec$ and $\succ$ satisfying (for any $x, y, z \in A$)

$$(x \prec y) \prec z = x \prec (y * z), (x \succ y) \prec z = x \succ (y \prec z), x \succ (y \succ z) = (x * y) \succ z,$$

where $x * y = x \prec y + x \succ y$.

PROPOSITION 7.20.– Let $(A, \prec, \succ)$ be a dendriform algebra.

1) The binary operation $* : A \otimes A \to A$ given by

$$x * y = x \prec y + x \succ y, \forall x, y \in A, \qquad [7.58]$$

defines an associative algebra.

2) The binary operation $\circ : A \otimes A \to A$ given by

$$x \circ y = x \succ y - y \prec x, \forall x, y \in A, \qquad [7.59]$$

defines a pre-Lie algebra.

3) Both $(A, *)$ and $(A, \circ)$ have the same sub-adjacent Lie algebra $\mathfrak{g}(A)$ defined by

$$[x, y] = x \succ y + x \prec y - y \succ x - y \prec x, \forall x, y \in A. \qquad [7.60]$$

Relationship among Lie algebras, associative algebras, pre-Lie algebras and dendriform algebras is given as follows in the sense of a commutative diagram of categories (Chapoton 2001):

$$\begin{array}{ccc} \text{Lie algebra} & \leftarrow & \text{Pre-Lie algebra} \\ \uparrow & & \uparrow \\ \text{Associative algebra} & \leftarrow & \text{Dendriform algebra} \end{array} \qquad [7.61]$$

There are many similar algebra structures that have a common property of "splitting associativity", that is, expressing the multiplication of an associative algebra as the sum of a string of binary operations. Explicitly, let $(X, *)$ be an associative algebra over a field $\mathbb{F}$ of characteristic zero and $(*_i)_{1\leq i\leq N} : X \otimes X \to X$ be a family of binary operations on X. Then the operation $*$ splits into the N operations $*_1, \cdots, *_N$ if

$$x * y = \sum_{i=1}^{N} x *_i y, \quad \forall x, y \in X. \qquad [7.62]$$

EXAMPLE 7.2.– For example,

1) $N = 2$: dendriform (di)algebra;

2) $N = 3$: dendriform trialgebra;

3) $N = 4$: quadri-algebra;

4) $N = 8$: octo-algebra;

5) $N = 9$: ennea-algebra;

All of these algebras are called *Loday algebras*.

REMARK 7.20.– For the case $N = 2^n$, $n = 0, 1, 2, \cdots$, there is the following "rule" of constructing Loday algebras:

1) operation axioms can be summarized to be a set of "associativity" relations;

2) by induction, for the algebra $(A, *_i)_{1\leq i\leq 2^n}$, besides the natural (regular) module of A on the underlying vector space of A itself given by the left and right multiplication operators, one can introduce the 2^{n+1} operations $\{*_{i_1}, *_{i_2}\}_{1\leq i\leq 2^n}$ such that

$$x *_i y = x *_{i_1} y + x *_{i_2} y, \quad \forall x, y \in A, \ 1 \leq i \leq 2^n, \qquad [7.63]$$

and their left and right multiplication operators can give a module of $(A, *_i)_{1\leq i\leq 2^n}$ by acting on the underlying vector space of A itself.

7.5.2. L-dendriform algebras

Most of the study in this section can be found in Bai *et al.* (2010).

DEFINITION 7.14.– Let A be a vector space with two binary operations denoted by $\triangleright$ and $\triangleleft : A \otimes A \to A$. $(A, \triangleright, \triangleleft)$ is called an *L-dendriform algebra* if for any $x, y, z \in A$,

$$x \triangleright (y \triangleright z) = (x \triangleright y) \triangleright z + (x \triangleleft y) \triangleright z + y \triangleright (x \triangleright z) - (y \triangleleft x) \triangleright z - (y \triangleright x) \triangleright z, \qquad [7.64]$$

$$x \triangleright (y \triangleleft z) = (x \triangleright y) \triangleleft z + y \triangleleft (x \triangleright z) + y \triangleleft (x \triangleleft z) - (y \triangleleft x) \triangleleft z. \qquad [7.65]$$

PROPOSITION 7.21.– Let $(A, \triangleright, \triangleleft)$ be an L-dendriform algebra.

1) The binary operation $\bullet : A \otimes A \to A$ given by

$$x \bullet y = x \triangleright y + x \triangleleft y, \forall x, y \in A, \qquad [7.66]$$

defines a (horizontal) pre-Lie algebra.

2) The binary operation $\circ : A \otimes A \to A$ given by

$$x \circ y = x \triangleright y - y \triangleleft x, \forall x, y \in A, \qquad [7.67]$$

defines a (vertical) pre-Lie algebra.

3) Both $(A, \bullet)$ and $(A, \circ)$ have the same sub-adjacent Lie algebra $\mathfrak{g}(A)$ defined by

$$[x, y] = x \triangleright y + x \triangleleft y - y \triangleright x - y \triangleleft x, \forall x, y \in A. \qquad [7.68]$$

REMARK 7.21.– Let $(A, \triangleright, \triangleleft)$ be an L-dendriform algebra. Then equations [7.64] and [7.65] can be rewritten as (for any $x, y, z \in A$)

$$x \triangleright (y \triangleright z) - (x \bullet y) \triangleright z = y \triangleright (x \triangleright z) - (y \bullet x) \triangleright z, \qquad [7.69]$$

$$x \triangleright (y \triangleleft z) - (x \triangleright y) \triangleleft z = y \triangleleft (x \bullet z) - (y \triangleleft x) \triangleleft z. \qquad [7.70]$$

Both sides of the above two equations can be regarded as a kind of "generalized associator". In this sense, equations [7.69] and [7.70] express certain "generalized left-symmetry" of the "generalized associators".

The "rule" of introducing the notion of L-dendriform algebra is given as follows.

PROPOSITION 7.22.– Let A be a vector space with two binary operations denoted by $\triangleright, \triangleleft : A \otimes A \to A$.

1) $(A, \triangleright, \triangleleft)$ is an L-dendriform algebra if and only if $(A, \bullet)$ defined by equation [7.66] is a pre-Lie algebra and $(L_{\triangleright}, R_{\triangleleft}, A)$ is a module.

2) $(A, \triangleright, \triangleleft)$ is an L-dendriform algebra if and only if $(A, \circ)$ defined by equation [7.67] is a pre-Lie algebra and $(L_{\triangleright}, -L_{\triangleleft}, A)$ is a module.

PROPOSITION 7.23.– Any dendriform algebra $(A, \succ, \prec)$ is an L-dendriform algebra by letting $x \triangleright y = x \succ y, x \triangleleft y = x \prec y$.

REMARK 7.22.– In the above sense, associative algebras are the special pre-Lie algebras whose associators are zero, whereas dendriform algebras are the special L-dendriform algebras whose "generalized associators" are zero.

Furthermore, there is the following **commutative diagram**:

$$\begin{array}{ccccccc}
\text{Lie} & \overset{-}{\leftarrow} & \text{Pre-Lie} & \overset{-,+}{\leftarrow} & \text{L-dendriform} & & \\
& \overset{-}{\nwarrow} & \Uparrow\in & \overset{-}{\nwarrow} & \Uparrow\in & \overset{-}{\nwarrow} & \\
& & \text{Associative} & \overset{+}{\leftarrow} & \text{Dendriform} & \overset{+}{\leftarrow} & \text{Quadri}
\end{array} \qquad [7.71]$$

where "$\Uparrow\in$" means the inclusion, "$+$" means the binary operation $x \circ_1 y + x \circ_2 y$ and "$-$" means the binary operation $x \circ_1 y - y \circ_2 x$.

REMARK 7.23.– In fact, except for the above motivation, there are some more motivations to introduce the notion of an L-dendriform algebra. For example, it is the underlying algebraic structure of a pseudo-Hessian structure on a Lie group.

7.5.3. *Lie analogues of Loday algebras*

Generalizing the study on pre-Lie algebras and L-dendriform algebras, we give the following structures as Lie analogues of Loday algebras.

Let $(X, [,])$ be a Lie algebra and $(*_i)_{1\leq i\leq N} : X \otimes X \to X$ be a family of binary operations on X. Then the Lie bracket $[\ ,\ ]$ splits into the commutator of N binary operations $*_1, \cdots, *_N$ if

$$[x, y] = \sum_{i=1}^{N}(x *_i y - y *_i x), \quad \forall x, y \in X. \qquad [7.72]$$

7.5.3.1. *"Rule" of construction*

For the case that $N = 2^n$, $n = 0, 1, 2, \cdots$, there is a "rule" of constructing the binary operations $*_i$ as follows: the 2^{n+1} binary operations give a natural module structure of an algebra with the 2^n binary operations on the underlying vector space of the algebra itself, which is the beauty of such algebra structures. That is, by induction, for the algebra $(A, *_i)_{1\leq i\leq 2^n}$, besides the natural module of A on the underlying vector space of A itself given by the left and right multiplication operators, one can introduce the 2^{n+1} binary operations $\{*_{i_1}, *_{i_2}\}_{1\leq i\leq 2^n}$ such that

$$x *_i y = x *_{i_1} y - y *_{i_2} x, \quad \forall x, y \in A,\ 1 \leq i \leq 2^n, \qquad [7.73]$$

and their left or right multiplication operators give a module of $(A, *_i)_{1\leq i\leq 2^n}$ by acting on the underlying vector space of A itself.

EXAMPLE 7.3.– We have the following results:

1) when $N = 1$, the corresponding algebra $(A, *_i)_{1\leq i\leq N}$ is a pre-Lie algebra;

2) when $N = 2$, the corresponding algebra $(A, *_i)_{1\leq i\leq N}$ is an L-dendriform algebra.

REMARK 7.24.– Note that for $n \geq 1$ ($N \geq 2$), in order to make equation [7.72] be satisfied, there is an alternative (sum) form of equation [7.73]

$$x *_i y = x *_{i_1} y + x *'_{i_2} y, \ \ \forall x, y \in A, \ 1 \leq i \leq 2^n, \tag{7.74}$$

by letting $x *'_{i_2} y = -y *_{i_2} x$ for any $x, y \in A$. In particular, in such a situation, it can be regarded as a binary operation $*$ of a pre-Lie algebra that splits into the $N = 2^n$ ($n = 1, 2 \cdots$) binary operations $*_1, ..., *_N$.

DEFINITION 7.15.– (Liu *et al.* 2011) Let A be a vector space with four bilinear products $\searrow, \nearrow, \nwarrow, \swarrow$: $A \otimes A \rightarrow A$. $(A, \searrow, \nearrow, \nwarrow, \swarrow)$ is called an *L-quadri-algebra* if for any $x, y, z \in A$,

$$x \searrow (y \searrow z) - (x * y) \searrow z = y \searrow (x \searrow z) - (y * x) \searrow z, \tag{7.75}$$

$$x \searrow (y \nearrow z) - (x \vee y) \nearrow z = y \nearrow (x \succ z) - (y \wedge x) \nearrow z, \tag{7.76}$$

$$x \searrow (y \nwarrow z) - (x \searrow y) \nwarrow z = y \nwarrow (x * z) - (y \nwarrow x) \nwarrow z, \tag{7.77}$$

$$x \nearrow (y \prec z) - (x \nearrow y) \nwarrow z = y \swarrow (x \wedge z) - (y \swarrow x) \nwarrow z, \tag{7.78}$$

$$x \searrow (y \swarrow z) - (x \succ y) \swarrow z = y \swarrow (x \vee z) - (y \prec x) \swarrow z, \tag{7.79}$$

where

$$x \succ y = x \searrow y + x \nearrow y, x \prec y = x \nwarrow y + x \swarrow y, \tag{7.80}$$

$$x \vee y = x \searrow y + x \swarrow y, x \wedge y = x \nearrow y + x \nwarrow y, \tag{7.81}$$

$$x * y = x \searrow y + x \nearrow y + x \nwarrow y + x \swarrow y = x \succ y + x \prec y = x \vee y + x \wedge y. \tag{7.82}$$

REMARK 7.25.– If both sides of equations [7.75]–[7.79] are zero, we get the identities of the definition of a *quadri-algebra*, which is the Loday algebra with four binary operations (Aguiar and Loday 2004).

There is the following **commutative diagram:**

$$\begin{array}{ccccccccc} \text{Lie} & \leftarrow & \text{Pre-Lie} & \leftarrow & \text{L-dendriform} & \leftarrow & \text{L-quadri} & & \\ & \nwarrow & & \nwarrow & & \nwarrow & & \nwarrow & \\ & & \text{Associative} & \leftarrow & \text{Dendriform} & \leftarrow & \text{Quadri} & \leftarrow & \text{Octo} \end{array} \qquad [7.83]$$

Moreover, there is an operadic interpretation of these algebraic structures, which is related to Manin black products (Bai *et al.* 2013).

7.6. References

Aguiar, M., Loday, J.-L. (2004). Quadri-algebras. *J. Pure Appl. Algebra*, 191, 205–221.

Andrada, A., Salamon, S. (2005). Complex product structure on Lie algebras. *Forum Math.*, 17, 261–295.

Bai, C. (2004). Left-symmetric algebras from linear functions. *J. Algebra*, 281, 651–665.

Bai, C. (2006). A further study on non-abelian phase spaces: Left-symmetric algebraic approach and related geometry. *Rev. Math. Phys.*, 18, 545–564.

Bai, C. (2007). A unified algebraic approach to the classical Yang-Baxter equation. *J. Phy. A: Math. Theor.*, 40, 11073–11082.

Bai, C. (2009). Bijective 1-cocycles and classification of 3-dimensional left-symmetric algebras. *Comm. Algebra*, 37, 1016–1057.

Bai, C., Meng, D. (1996). The classification of left-symmetric algebra in dimension 2. *Chinese Science Bulletin*, 41, 2207 (in Chinese).

Bai, C., Meng, D. (2000). The structure of bi-symmetric algebras and their sub-adjacent Lie algebras. *Comm. Algebra*, 28, 2717–2734.

Bai, C., Meng, D. (2001a). The classification of Novikov algebras in low dimensions. *J. Phys. A: Math. Gen.*, 34, 1581–1594.

Bai, C., Meng, D. (2001b). On the realization of transitive Novikov algebras. *J. Phys. A: Math. Gen.*, 34, 3363–3372.

Bai, C., Meng, D. (2001c). The realizations of non-transitive Novikov algebras. *J. Phys. A: Math. Gen.*, 34, 6435–6442.

Bai, C., Bellier, O., Guo, L., Ni, X. (2013). Splitting of operations, Manin products and Rota-Baxter operators. *Int. Math. Res. Not.*, 485–524.

Bai, C., Liu, L., Ni, X. (2010). Some results on L-dendriform algebras. *J. Geom. Phys.*, 60, 940–950.

Bakalov, B., Kac, V. (2003). Field algebras. *Int. Math. Res. Not.*, 123–159.

Balinskii, A.A., Novikov, S.P. (1985). Poisson brackets of hydrodynamic type, Frobenius algebras and Lie algebras. *Soviet Math. Dokl.*, 32, 228–231.

Burde, D. (1998). Simple left-symmetric algebras with solvable Lie algebra. *Manuscripta Math.*, 95, 397–411.

Burde, D. (2006). Left-symmetric algebras, or pre-Lie algebras in geometry and physics. *Cent. Eur. J. Math.*, 4, 323–357.

Cayley, A. (1890). On the theory of analytic forms called trees. In *Collected Mathematical Papers of Arthur Cayley*, Cambridge University Press, Cambridge, 3, 242–246.

Chang, K., Kim, H., Lee, H. (1999). On radicals of a left-symmetric algebra. *Comm. Algebra*, 27, 3161–3175.

Chapoton, F. (2001). Un endofoncteur de la catégorie des opérades. In *Dialgebras and Related Operads, Lecture Notes in Math.*, Loday, J.-L., Frabetti, A., Chapoton, F., Goichot, F. (eds). Springer, Berlin.

Chapoton, F. (2004). Classification of some simple graded pre-Lie algebras of growth one. *Comm. Algebra*, 32, 243–251.

Chapoton, F., Livernet, M. (2001). Pre-Lie algebras and the rooted trees operad. *Int. Math. Res. Not.*, 395–408.

Chu, B. (1974). Symplectic homogeneous spaces. *Trans. Amer. Math. Soc.*, 197, 145–159.

Connes, A., Kreimer, D. (1998). Hopf algebras, renormalization and noncommutative geometry. *Comm. Math. Phys.*, 199, 203–242.

Diatta, A., Medina, A. (2004). Classical Yang-Baxter equation and left-invariant affine geometry on Lie groups. *Manuscripta Math.*, 114, 477–486.

Dzhumadil'daev, A., Lofwall, C. (2002). Trees, free right-symmetric algebras, free Novikov algebras and identities. *Homology, Homotopy and Applications*, 4, 165–190.

Ebrahimi-Fard, K. (2002). Loday-type algebras and the Rota-Baxter relation. *Lett. Math. Phys.*, 61, 139–147.

Filippov, V. (1989). A class of simple nonassociative algebras. *Mat. Zametki*, 45, 101–105.

Gel'fand, I.M., Dorfman, I.Y. (1979). Hamiltonian operators and algebraic structures related to them. *Funct. Anal. Appl.*, 13, 248–262.

Gerstenhaber, M. (1963). The cohomology structure of an associative ring. *Ann. Math.*, 78, 267–288.

Golubschik, I.Z., Sokolov, V.V. (2000). Generalized operator Yang-Baxter equations, integrable ODEs and nonassociative algebras. *J. Nonlinear Math. Phys.*, 7, 184–197.

Guo, L. (2009). What is a Rota-Baxter algebra? *Notice of Amer. Math. Soc.*, 56, 1436–1437.

Kim, H. (1986). Complete left-invariant affine structures on nilpotent Lie groups. *J. Diff. Geom.*, 24, 373–394.

Kleinfeld, Z. (1957). Assosymmetric rings. *Proc. Amer. Math. Soc.*, 8, 983–986.

Kong, K., Bai, C. (2008). Left-symmetric superalgebra structures on the super-Virasoro algebras. *Pac. J. Math.*, 235, 43–55.

Kong, K., Chen, H., Bai, C. (2011). Classification of graded pre-Lie algebraic structures on Witt and Virasoro algebras. *Inter. J. Math.*, 22, 201–222.

Kong, K., Bai, C., Meng, D. (2012). On real left-symmetric algebras. *Comm. Algebra*, 40, 1641–1668.

Koszul, J.-L. (1961). Domaines bornés homogènes et orbites de groupes de transformation affinés. *Bull. Soc. Math. France*, 89, 515–533.

Kupershmidt, B.A. (1994). Non-abelian phase spaces. *J. Phys. A: Math. Gen.*, 27, 2801–2810.

Kupershmidt, B.A. (1999a). On the nature of the Virasoro algebra. *J. Nonlinear Math. Phys.*, 6, 222–245.

Kupershmidt, B.A. (1999b). What a classical r-matrix really is? *J. Nonlinear Math. Phys.*, 6, 448–488.

Lichnerowicz, A., Medina, A. (1988). On Lie groups with left invariant symplectic or Kahlerian structures. *Lett. Math. Phys.*, 16, 225–235.

Liu, L., Ni, X., Bai, C. (2011). L-quadri-algebras. *Sci. Sin. Math.*, 42, 105–124 (in Chinese).

Loday, J.-L. (2001). Dialgebras. In *Dialgebras and Related Operads*, Loday, J.-L., Frabetti, A., Chapoton, F., Goichot, F. (eds). Springer, Berlin.

Medina, A. (1981). Flat left-invariant connections adapted to the automorphism structure of a Lie group. *J. Diff. Geom.*, 16, 445–474.

Osborn, J.M. (1994). Infinite dimensional Novikov algebras of characteristic 0. *J. Algebra*, 167, 146–167.

Segal, D. (1992). The structures of complete left-symmetric algebras. *Math. Ann.*, 293, 569–578.

Segal, D. (1994). Free left-symmetrical algebras and an analogue of the Poincaré-Birkhoff-Witt theorem. *J. Algebra*, 164, 750–772.

Semonov-Tian-Shansky, M.A. (1983). What is a classical r-matrix? *Funct. Anal. Appl.*, 17, 259–272.

Shima, H. (1980). Homogeneous Hessian manifolds. *Ann. Inst. Fourier*, 30, 91–128.

Svinolupov, S.I., Sokolov, V.V. (1994). Vector-matrix generalizations of classical integrable equations. *Theoret. and Math. Phys.*, 100, 959–962.

Tang, X., Bai, C. (2012). A class of non-graded pre-Lie algebraic structures on the Witt algebra. *Math. Nach.*, 285, 922–935.

Vinberg, E.B. (1963). Convex homogeneous cones. *Transl. Moscow Math. Soc.*, 12, 340–403.

Wang, Y., Chen, Z., Bai, C. (2012). Classification of Balinsky-Novikov superalgebras with dimension 2|2. *J. Phys. A: Math. Theor.*, 45, 225201.

Xu, X. (1996). On simple Novikov algebras and their irreducible modules. *J. Algebra*, 185, 905–934.

Xu, X. (1997). Novikov-Poisson algebras. *J. Algebra*, 190, 253–279.

Zelmanov, E. (1987). On a class of local translation invariant Lie algebras. *Soviet Math. Dokl.*, 35, 216–218.

Zhang, R., Bai, C. (2012). On some left-symmetric superalgebras. *J. Algebra. Appl.*, 11, 1250097.

8

Symplectic, Product and Complex Structures on 3-Lie Algebras

Yunhe SHENG and Rong TANG
Jilin University, Changchun, China

8.1. Introduction

A symplectic structure on a Lie algebra $\mathfrak{g}$ is a non-degenerate 2-cocycle $\omega \in \wedge^2\mathfrak{g}^*$. The underlying structure of a symplectic Lie algebra is a quadratic pre-Lie algebra (Chu 1974). An almost product structure on a Lie algebra $\mathfrak{g}$ is a linear map E satisfying $E^2 = \mathrm{Id}$. If, in addition, E also satisfies the following integrability condition:

$$[Ex, Ey] = E([Ex, y] + [x, Ey] - E[x, y]), \quad \forall x, y \in \mathfrak{g},$$

then E is called a product structure. The above integrability condition is called the Nijenhuis condition. An equivalent characterization of a product structure is that $\mathfrak{g}$ is the direct sum (as vector spaces) of two subalgebras. An almost complex structure on a Lie algebra $\mathfrak{g}$ is a linear map J satisfying $J^2 = -\mathrm{Id}$. A complex structure on a Lie algebra is an almost complex structure that satisfies the Nijenhuis condition. Adding compatibility conditions between a complex structure and a product structure, between a symplectic structure and a paracomplex structure, between a symplectic structure and a complex structure, one obtains a complex product structure, a para-Kähler structure and a pseudo-Kähler structure, respectively. These structures play important roles in algebra, geometry and mathematical physics, and are widely studied. See (Alekseevsky and Perelomov 1997; Andrada 2008; Andrada *et al.* 2011,

Algebra and Applications 1,
coordinated by Abdenacer MAKHLOUF.

2005; Andrada and Salamon 2005; Bai 2008, 2006; Bajo and Benayadi 2011; Benayadi and Boucetta 2015; Calvaruso 2015a,b; Cleyton *et al.* 2010, 2011; Li and Tomassini 2012; Salamon 2001) for more details.

Generalizations of Lie algebras to higher arities, including 3-Lie algebras and more generally, n-Lie algebras (Filippov 1985; Kasymov 1987; Takhtajan 1995), have attracted attention from several fields of mathematics and physics. It is the algebraic structure corresponding to Nambu mechanics (Gautheron 1996; Nambu 1973; Takhtajan 1994). In (Basu and Harvey 2005), Basu and Harvey suggested replacing the Lie algebra appearing in the Nahm equation with a 3-Lie algebra for the lifted Nahm equations. Furthermore, in the context of the Bagger–Lambert–Gustavsson model of multiple M2-branes, Bagger–Lambert managed to construct, using a ternary bracket, an $N = 2$ supersymmetric version of the worldvolume theory of the M-theory membrane (see Bagger and Lambert (2008)). An extensive literature is related to this pioneering work (see Bagger and Lambert (2009); Gomis *et al.* (2008); Ho *et al.* (2008); Papadopoulos (2008)). See the review article (de Azcarraga and Izquierdo 2010) for more details. In particular, metric 3-algebras were deeply studied in the seminal works (de Medeiros and Figueroa-O'Farrill 2008; de Medeiros *et al.* 2009b,a). In Liu *et al.* (2016), the authors introduced the notion a Nijenhuis operator on an n-Lie algebra, which generates a trivial deformation.

We aim to study symplectic structures, product structure and complex structures on 3-Lie algebras and these combined structures. In the case of Lie algebras, pre-Lie algebras play important roles in these studies. It is believable that 3-pre-Lie algebras will play important roles in the corresponding studies. Thus, first, we introduce the notion of a representation of a 3-pre-Lie algebra and construct the associated semidirect product 3-pre-Lie algebra. Several important properties of representations of 3-pre-Lie algebras are studied.

The notion of a symplectic structure on a 3-Lie algebra was introduced in Bai *et al.* (2019) and it is shown that the underlying structure of a symplectic 3-Lie algebra is a quadratic 3-pre-Lie algebra. We introduce the notion of a phase space of a 3-Lie algebra $\mathfrak{g}$, which is a symplectic 3-Lie algebra $\mathfrak{g} \oplus \mathfrak{g}^*$ satisfying some conditions, and show that a 3-Lie algebra has a phase space if and only if it is subadjacent to a 3-pre-Lie algebra. We also introduce the notion of a Manin triple of 3-pre-Lie algebras and show that there is a one-to-one correspondence between Manin triples of 3-pre-Lie algebras and phase spaces of 3-Lie algebras.

An almost product structure on a 3-Lie algebra $\mathfrak{g}$ is defined to be a linear map $E : \mathfrak{g} \longrightarrow \mathfrak{g}$, satisfying $E^2 = \mathrm{Id}$. It is challenging to add an integrability condition on an almost product structure to obtain a product structure on a 3-Lie algebra. We note that the Nijenhuis condition (see equation [8.4]) given in Liu *et al.* (2016) is the correct integrability condition. Let us explain this issue. Denote by $\mathfrak{g}_{\pm}$ the eigenspaces corresponding to eigenvalues ± 1 of an almost product structure E. Then it is obvious that $\mathfrak{g} = \mathfrak{g}_+ \oplus \mathfrak{g}_-$ as vector spaces. The Nijenhuis condition ensures that both $\mathfrak{g}_+$ and $\mathfrak{g}_-$ are subalgebras. This is what "integrability" means. Moreover, we find that there are four types of special integrability conditions, which are called strict product structure, abelian product structure, strong abelian product structure and perfect product structure. Respectively, each of them gives rise to a special decomposition of the original 3-Lie algebra. It is surprising to see that a strong abelian product structure is also an $\mathcal{O}$-operator on a 3-Lie algebra associated with the adjoint representation. Since an $\mathcal{O}$-operator on a 3-Lie algebra associated with the adjoint representation is also a Rota–Baxter operator (Bai *et al.* 2013; Pei *et al.* 2017), it turns out that involutive Rota–Baxter operator can also serve as an integrability condition. This is totally different from the case of Lie algebras. Furthermore, by the definition of a perfect product structure, an involutive automorphism of a 3-Lie algebra can also serve as an integrability condition. This is also a new phenomenon. Note that the decomposition that a perfect product structure gives is exactly the condition required in the definition of a matched pair of 3-Lie algebras (Bai *et al.* 2019). Thus, this kind of product structure will be frequently used in our studies.

An almost complex structure on a 3-Lie algebra $\mathfrak{g}$ is defined to be a linear map $J : \mathfrak{g} \longrightarrow \mathfrak{g}$, satisfying $J^2 = -\mathrm{Id}$. With the above motivation, we define a complex structure on a 3-Lie algebra $\mathfrak{g}$ to be an almost complex structure satisfying the Nijenhuis condition. Then $\mathfrak{g}_i$ and $\mathfrak{g}_{-i}$, which are eigenspaces of eigenvalues $\pm i$ of a complex linear map $J_{\mathbb{C}}$ (the complexification of J), are subalgebras of the 3-Lie algebra $\mathfrak{g}_{\mathbb{C}}$, the complexification of $\mathfrak{g}$. There are also four types of special integrability conditions, and each of them gives rise to a special decomposition of $\mathfrak{g}_{\mathbb{C}}$.

Then we add a compatibility condition between a complex structure and a product structure on a 3-Lie algebra to define a complex product structure on a 3-Lie algebra. We give an equivalent characterization of a complex product structure on a 3-Lie algebra $\mathfrak{g}$ using the decomposition of $\mathfrak{g}$. We add a compatibility condition between a symplectic structure and a paracomplex structure on a 3-Lie algebra to define a para-Kähler structure on a 3-Lie algebra. An equivalent characterization of a para-Kähler structure on a 3-Lie algebra $\mathfrak{g}$ is also given using the decomposition of $\mathfrak{g}$. There is also a pseudo-Riemannian structure that is associated with a para-Kähler structure on a 3-Lie algebra. We introduce the notion of a Levi-Civita product associated with a pseudo-Riemannian 3-Lie algebra, and give its precise formulas. Finally, we add a compatibility condition between a symplectic structure and a complex structure on a 3-Lie algebra to define a pseudo-Kähler structure on a 3-Lie algebra. The relation between a para-Kähler structure and a pseudo-Kähler structure on a 3-Lie algebra is investigated.

In the following, we work over the real field $\mathbb{R}$ and the complex field $\mathbb{C}$ and all the vector spaces are finite-dimensional.

8.2. Preliminaries

First, we recall the notion of a Nijenhuis operator on a 3-Lie algebra, which will be frequently used as the integrability condition in our later studies. Then, we recall the notion of a 3-pre-Lie algebra, which is the main tool we will use to construct examples of symplectic, product and complex structures on 3-Lie algebras.

DEFINITION 8.1.– *A* 3-***Lie algebra*** *is a vector space* $\mathfrak{g}$ *together with a trilinear skew-symmetric bracket* $[\cdot,\cdot,\cdot]_{\mathfrak{g}} : \wedge^3\mathfrak{g} \longrightarrow \mathfrak{g}$ *such that the following* ***fundamental*** *identity holds for all* $x, y, z, w, v \in \mathfrak{g}$,

$$[x, y, [z, w, v]_{\mathfrak{g}}]_{\mathfrak{g}} = [[x, y, z]_{\mathfrak{g}}, w, v]_{\mathfrak{g}} + [z, [x, y, w]_{\mathfrak{g}}, v]_{\mathfrak{g}} + [z, w, [x, y, v]_{\mathfrak{g}}]_{\mathfrak{g}}. \quad [8.1]$$

Let $(\mathfrak{g}, [\cdot,\cdot,\cdot]_{\mathfrak{g}})$ be a 3-Lie algebra, and $N : \mathfrak{g} \longrightarrow \mathfrak{g}$ a linear map. Define a 3-ary bracket $[\cdot,\cdot,\cdot]^1_N : \wedge^3\mathfrak{g} \longrightarrow \mathfrak{g}$ by

$$[x, y, z]^1_N = [Nx, y, z]_{\mathfrak{g}} + [x, Ny, z]_{\mathfrak{g}} + [x, y, Nz]_{\mathfrak{g}} - N[x, y, z]_{\mathfrak{g}}. \quad [8.2]$$

Then we define 3-ary bracket $[\cdot,\cdot,\cdot]^2_N : \wedge^3\mathfrak{g} \longrightarrow \mathfrak{g}$ by

$$[x, y, z]^2_N = [Nx, Ny, z]_{\mathfrak{g}} + [x, Ny, Nz]_{\mathfrak{g}} + [Nx, y, Nz]_{\mathfrak{g}} - N[x, y, z]^1_N. [8.3]$$

DEFINITION 8.2.– *(Liu* et al. *2016). Let* $(\mathfrak{g}, [\cdot,\cdot,\cdot]_{\mathfrak{g}})$ *be a 3-Lie algebra. A linear map* $N : \mathfrak{g} \longrightarrow \mathfrak{g}$ *is called a* ***Nijenhuis operator*** *if the following* ***Nijenhuis condition*** *is satisfied:*

$$[Nx, Ny, Nz]_{\mathfrak{g}} = N[x, y, z]^2_N, \quad \forall x, y, z \in \mathfrak{g}. \quad [8.4]$$

DEFINITION 8.3.– *(Kasymov 1987). A* ***representation*** *of a 3-Lie algebra* $(\mathfrak{g}, [\cdot,\cdot,\cdot]_{\mathfrak{g}})$ *on a vector space* V *is a linear map* $\rho : \wedge^2\mathfrak{g} \longrightarrow \mathfrak{gl}(V)$*, such that for all* $x_1, x_2, x_3, x_4 \in \mathfrak{g}$*, there holds:*

$$\rho([x_1, x_2, x_3]_{\mathfrak{g}}, x_4) + \rho(x_3, [x_1, x_2, x_4]_{\mathfrak{g}}) = [\rho(x_1, x_2), \rho(x_3, x_4)];$$

$$\rho([x_1, x_2, x_3]_{\mathfrak{g}}, x_4) = \rho(x_1, x_2) \circ \rho(x_3, x_4) + \rho(x_2, x_3) \circ \rho(x_1, x_4) + \rho(x_3, x_1) \circ \rho(x_2, x_4).$$

EXAMPLE 8.1.– Let $(\mathfrak{g}, [\cdot,\cdot,\cdot]_{\mathfrak{g}})$ be a 3-Lie algebra. The linear map $\mathrm{ad} : \wedge^2\mathfrak{g} \longrightarrow \mathfrak{gl}(\mathfrak{g})$ defines a representation of the 3-Lie algebra $\mathfrak{g}$ on itself, which we call the **adjoint representation** of $\mathfrak{g}$.

Let A be a vector space. For a linear map $\phi : A \otimes A \rightarrow \mathfrak{gl}(V)$, we define a linear map $\phi^* : A \otimes A \rightarrow \mathfrak{gl}(V^*)$ by

$$\langle \phi^*(x, y)\alpha, v\rangle = -\langle \alpha, \phi(x, y)v\rangle, \quad \forall \alpha \in V^*, x, y \in \mathfrak{g}, v \in V.$$

LEMMA 8.1.– (Bai *et al.* 2019). Let (V, ρ) be a representation of a 3-Lie algebra $(\mathfrak{g}, [\cdot,\cdot,\cdot]_{\mathfrak{g}})$. Then (V^*, ρ^*) is a representation of the 3-Lie algebra $(\mathfrak{g}, [\cdot,\cdot,\cdot]_{\mathfrak{g}})$, which is called the **dual representation**.

LEMMA 8.2.– Let $\mathfrak{g}$ be a 3-Lie algebra, V a vector space and $\rho : \wedge^2\mathfrak{g} \to \mathfrak{gl}(V)$ a skew-symmetric linear map. Then $(V; \rho)$ is a representation of $\mathfrak{g}$ if and only if there is a 3-Lie algebra structure (called the **semidirect product**) on the direct sum of vector spaces $\mathfrak{g} \oplus V$, defined by

$$[x_1 + v_1, x_2 + v_2, x_3 + v_3]_\rho = [x_1, x_2, x_3]_{\mathfrak{g}} + \rho(x_1, x_2)v_3 + \rho(x_2, x_3)v_1 + \rho(x_3, x_1)v_2, \tag{8.5}$$

for all $x_i \in \mathfrak{g}, v_i \in V, 1 \le i \le 3$. We denote this semidirect product 3-Lie algebra by $\mathfrak{g} \ltimes_\rho V$.

DEFINITION 8.4.– *Let A be a vector space with a linear map $\{\cdot,\cdot,\cdot\} : \otimes^3 A \to A$. The pair $(A, \{\cdot,\cdot,\cdot\})$ is called a 3-**pre-Lie algebra** if the following identities hold:*

$$\{x, y, z\} = -\{y, x, z\} \tag{8.6}$$

$$\{x_1, x_2, \{x_3, x_4, x_5\}\} = \{[x_1, x_2, x_3]_C, x_4, x_5\} + \{x_3, [x_1, x_2, x_4]_C, x_5\} + \{x_3, x_4, \{x_1, x_2, x_5\}\} \tag{8.7}$$

$$\{[x_1, x_2, x_3]_C, x_4, x_5\} = \{x_1, x_2, \{x_3, x_4, x_5\}\} + \{x_2, x_3, \{x_1, x_4, x_5\}\} + \{x_3, x_1, \{x_2, x_4, x_5\}\}, \tag{8.8}$$

where $x, y, z, x_i \in A, 1 \le i \le 5$ and $[\cdot,\cdot,\cdot]_C$ is defined by

$$[x, y, z]_C \triangleq \{x, y, z\} + \{y, z, x\} + \{z, x, y\}, \quad \forall x, y, z \in A. \tag{8.9}$$

PROPOSITION 8.1.– (Bai *et al.* 2019, Proposition 3.21). Let $(A, \{\cdot,\cdot,\cdot\})$ be a 3-pre-Lie algebra. Then $(A, [\cdot,\cdot,\cdot]_C)$ is a 3-Lie algebra, which is called the sub-adjacent 3-Lie algebra of A, and denoted by A^c. $(A, \{\cdot,\cdot,\cdot\})$ is called the compatible 3-pre-Lie algebra structure on the 3-Lie algebra A^c.

We define the left multiplication $L : \wedge^2 A \longrightarrow \mathfrak{gl}(A)$ by $L(x, y)z = \{x, y, z\}$ for all $x, y, z \in A$. Then (A, L) is a representation of the 3-Lie algebra A^c. Moreover, we define the right multiplication $R : \otimes^2 A \to \mathfrak{gl}(A)$ by $R(x, y)z = \{z, x, y\}$. If there is a 3-pre-Lie algebra structure on its dual space A^*, we denote the left multiplication and right multiplication by $\mathcal{L}$ and $\mathcal{R}$, respectively.

DEFINITION 8.5.– *(Bai* et al. *2019, Definition 3.16). Let $(\mathfrak{g}, [\cdot,\cdot,\cdot]_{\mathfrak{g}})$ be a 3-Lie algebra and (V, ρ) a representation. A linear operator $T : V \to \mathfrak{g}$ is called an* $\mathcal{O}$-**operator** *associated with (V, ρ) if T satisfies for all $u, v, w \in V$,*

$$[Tu, Tv, Tw]_{\mathfrak{g}} = T(\rho(Tu, Tv)w + \rho(Tv, Tw)u + \rho(Tw, Tu)v). \tag{8.10}$$

PROPOSITION 8.2.– (Bai *et al.* 2019, Proposition 3.27). Let $(\mathfrak{g}, [\cdot,\cdot,\cdot]_{\mathfrak{g}})$ be a 3-Lie algebra. Then there is a compatible 3-pre-Lie algebra if and only if there exists an invertible $\mathcal{O}$-operator $T : V \to \mathfrak{g}$ associated with a representation (V, ρ). Furthermore, the compatible 3-pre-Lie structure on $\mathfrak{g}$ is given by

$$\{x, y, z\} = T\rho(x, y)T^{-1}(z), \quad \forall x, y, z \in \mathfrak{g}. \qquad [8.11]$$

8.3. Representations of 3-pre-Lie algebras

We introduce the notion of a representation of a 3-pre-Lie algebra, construct the corresponding semidirect product 3-pre-Lie algebra and give the dual representation.

DEFINITION 8.6.– *A* ***representation*** *of a 3-pre-Lie algebra $(A, \{\cdot,\cdot,\cdot\})$ on a vector space V consists of a pair (ρ, μ), where $\rho : \wedge^2 A \to \mathfrak{gl}(V)$ is a representation of the 3-Lie algebra A^c on V and $\mu : \otimes^2 A \to \mathfrak{gl}(V)$ is a linear map such that for all $x_1, x_2, x_3, x_4 \in A$, the following equalities hold:*

$$\begin{aligned}
\rho(x_1, x_2)\mu(x_3, x_4) &= \mu(x_3, x_4)\rho(x_1, x_2) - \mu(x_3, x_4)\mu(x_2, x_1)\\
&\quad + \mu(x_3, x_4)\mu(x_1, x_2) + \mu([x_1, x_2, x_3]_C, x_4) + \mu(x_3, \{x_1, x_2, x_4\}),\\
\mu([x_1, x_2, x_3]_C, x_4) &= \rho(x_1, x_2)\mu(x_3, x_4) + \rho(x_2, x_3)\mu(x_1, x_4)\\
&\quad + \rho(x_3, x_1)\mu(x_2, x_4),\\
\mu(x_1, \{x_2, x_3, x_4\}) &= \mu(x_3, x_4)\mu(x_1, x_2) + \mu(x_3, x_4)\rho(x_1, x_2)\\
&\quad - \mu(x_3, x_4)\mu(x_2, x_1) - \mu(x_2, x_4)\mu(x_1, x_3)\\
&\quad - \mu(x_2, x_4)\rho(x_1, x_3) + \mu(x_2, x_4)\mu(x_3, x_1) + \rho(x_2, x_3)\mu(x_1, x_4),\\
\mu(x_3, x_4)\rho(x_1, x_2) &= \mu(x_3, x_4)\mu(x_2, x_1) - \mu(x_3, x_4)\mu(x_1, x_2)\\
&\quad + \rho(x_1, x_2)\mu(x_3, x_4) - \mu(x_2, \{x_1, x_3, x_4\}) + \mu(x_1, \{x_2, x_3, x_4\}).
\end{aligned}$$

It is obvious that (L, R) is a representation of a 3-pre-Lie algebra on itself, which is called the **regular representation**.

Let (V, ρ, μ) be a representation of a 3-pre-Lie algebra $(A, \{\cdot,\cdot,\cdot\})$. Define a trilinear bracket operation $\{\cdot,\cdot,\cdot\}_{\rho,\mu} : \otimes^3(A \oplus V) \to A \oplus V$ by

$$\{x_1 + v_1, x_2 + v_2, x_3 + v_3\}_{\rho,\mu} \triangleq \{x_1, x_2, x_3\} + \rho(x_1, x_2)v_3 + \mu(x_2, x_3)v_1 - \mu(x_1, x_3)v_2.$$

By straightforward computations, we have the following theorem:

THEOREM 8.1.– With the above notation, $(A \oplus V, \{\cdot,\cdot,\cdot\}_{\rho,\mu})$ is a 3-pre-Lie algebra.

This 3-pre-Lie algebra is called the **semidirect product** of the 3-pre-Lie algebra $(A, \{\cdot,\cdot,\cdot\})$ and (V, ρ, μ), and it is denoted by $A \ltimes_{\rho,\mu} V$.

Let V be a vector space. Define the switching operator $\tau : \otimes^2 V \longrightarrow \otimes^2 V$ by

$$\tau(T) = x_2 \otimes x_1, \quad \forall T = x_1 \otimes x_2 \in \otimes^2 V.$$

PROPOSITION 8.3.– Let (ρ, μ) be a representation of a 3-pre-Lie algebra $(A, \{\cdot,\cdot,\cdot\})$ on a vector space V. Then $\rho - \mu\tau + \mu$ is a representation of the sub-adjacent 3-Lie algebra $(A^c, [\cdot,\cdot,\cdot]_C)$ on the vector space V.

PROOF.– By theorem 8.1, we have the semidirect product 3-pre-Lie algebra $A \ltimes_{\rho,\mu} V$. Considering its sub-adjacent 3-Lie algebra structure $[\cdot,\cdot,\cdot]_C$, we have

$$\begin{aligned}
&[x_1 + v_1, x_2 + v_2, x_3 + v_3]_C = \{x_1 + v_1, x_2 + v_2, x_3 + v_3\}_{\rho,\mu} \\
&+ \{x_2 + v_2, x_3 + v_3, x_1 + v_1\}_{\rho,\mu}\{x_3 + v_3, x_1 + v_1, x_2 + v_2\}_{\rho,\mu} \\
&= \{x_1, x_2, x_3\} + \rho(x_1, x_2)v_3 + \mu(x_2, x_3)v_1 - \mu(x_1, x_3)v_2 \\
&+ \{x_2, x_3, x_1\} + \rho(x_2, x_3)v_1 + \mu(x_3, x_1)v_2 - \mu(x_2, x_1)v_3 \\
&+ \{x_3, x_1, x_2\} + \rho(x_3, x_1)v_2 + \mu(x_1, x_2)v_3 - \mu(x_3, x_2)v_1 \\
&= [x_1, x_2, x_3]_C + ((\rho - \mu\tau + \mu)(x_1, x_2))v_3 \\
&+ ((\rho - \mu\tau + \mu)(x_2, x_3))v_1 + ((\rho - \mu\tau + \mu)(x_3, x_1))v_2.
\end{aligned} \quad [8.12]$$

By lemma 8.2, $\rho - \mu\tau + \mu$ is a representation of the sub-adjacent 3-Lie algebra $(A^c, [\cdot,\cdot,\cdot]_C)$ on the vector space V. The proof is completed. □

PROPOSITION 8.4.– Let (ρ, μ) be a representation of a 3-pre-Lie algebra $(A, \{\cdot,\cdot,\cdot\})$ on a vector space V. Then $(\rho^* - \mu^*\tau + \mu^*, -\mu^*)$ is a representation of the 3-pre-Lie algebra $(A, \{\cdot,\cdot,\cdot\})$ on the vector space V^*, which is called the **dual representation** of the representation (V, ρ, μ).

PROOF.– By proposition 8.3, $\rho - \mu\tau + \mu$ is a representation of the sub-adjacent 3-Lie algebra $(A^c, [\cdot,\cdot,\cdot]_C)$ on the vector space V. By lemma 8.1, $\rho^* - \mu^*\tau + \mu^*$ is a representation of the sub-adjacent 3-Lie algebra $(A^c, [\cdot,\cdot,\cdot]_C)$ on the dual vector space V^*. It is straightforward to deduce that other conditions in definition 8.6 also hold. We leave details to readers. □

If $(\rho, \mu) = (L, R)$ is the regular representation of a 3-pre-Lie algebra $(A, \{\cdot,\cdot,\cdot\})$, then $(\rho^* - \mu^*\tau + \mu^*, -\mu^*) = (\mathrm{ad}^*, -R^*)$ and the corresponding semidirect product 3-Lie algebra is $A^c \ltimes_{L^*} A^*$, which is the key object when we construct phase spaces of 3-Lie algebras.

8.4. Symplectic structures and phase spaces of 3-Lie algebras

We introduce the notion of a phase space of a 3-Lie algebra and show that a 3-Lie algebra has a phase space if and only if it is sub-adjacent to a 3-pre-Lie algebra. Moreover, we introduce the notion of a Manin triple of 3-pre-Lie algebras and show that there is a one-to-one correspondence between Manin triples of 3-pre-Lie algebras and perfect phase spaces of 3-Lie algebras.

DEFINITION 8.7.– *(Bai* et al. *2019). A* ***symplectic structure*** *on a 3-Lie algebra* $(\mathfrak{g}, [\cdot,\cdot,\cdot]_{\mathfrak{g}})$ *is a non-degenerate skew-symmetric bilinear form* $\omega \in \wedge^2\mathfrak{g}^*$ *satisfying the following equality:*

$$\omega([x,y,z]_{\mathfrak{g}}, w) - \omega([y,z,w]_{\mathfrak{g}}, x) + \omega([z,w,x]_{\mathfrak{g}}, y) - \omega([w,x,y]_{\mathfrak{g}}, z) = 0. \quad [8.13]$$

EXAMPLE 8.2.– Consider the four-dimensional Euclidean 3-Lie algebra A_4 given in Bagger and Lambert (2008). The underlying vector space is $\mathbb{R}^4$. Relative to an orthogonal basis $\{e_1, e_2, e_3, e_4\}$, the 3-Lie bracket is given by

$$[e_1,e_2,e_3] = e_4, \quad [e_2,e_3,e_4] = e_1, \quad [e_1,e_3,e_4] = e_2, \quad [e_1,e_2,e_4] = e_3.$$

Then it is straightforward to see that any non-degenerate skew-symmetric bilinear form is a symplectic structure on A_4. In particular,

$$\omega_1 = e_3^* \wedge e_1^* + e_4^* \wedge e_2^*, \quad \omega_2 = e_2^* \wedge e_1^* + e_4^* \wedge e_3^*, \quad \omega_3 = e_2^* \wedge e_1^* + e_3^* \wedge e_4^*,$$
$$\omega_4 = e_1^* \wedge e_2^* + e_4^* \wedge e_3^*, \quad \omega_5 = e_1^* \wedge e_2^* + e_3^* \wedge e_4^*, \quad \omega_6 = e_1^* \wedge e_3^* + e_2^* \wedge e_4^*$$

are symplectic structures on A_4, where $\{e_1^*, e_2^*, e_3^*, e_4^*\}$ are the dual basis.

PROPOSITION 8.5.– (Bai *et al.* 2019). Let $(\mathfrak{g}, [\cdot,\cdot,\cdot]_{\mathfrak{g}}, \omega)$ be a symplectic 3-Lie algebra. Then there exists a compatible 3-pre-Lie algebra structure $\{\cdot,\cdot,\cdot\}$ on $\mathfrak{g}$ given by

$$\omega(\{x,y,z\}, w) = -\omega(z, [x,y,w]_{\mathfrak{g}}), \quad \forall x,y,z,w \in \mathfrak{g}. \quad [8.14]$$

A **quadratic 3-pre-Lie algebra** is a 3-pre-Lie algebra $(A, \{\cdot,\cdot,\cdot\})$ equipped with a non-degenerate skew-symmetric bilinear form $\omega \in \wedge^2 A^*$ such that the following invariant condition holds:

$$\omega(\{x,y,z\}, w) = -\omega(z, [x,y,w]_C), \quad \forall x,y,z,w \in A. \quad [8.15]$$

Proposition 8.5 states that quadratic 3-pre-Lie algebras are the underlying structures of symplectic 3-Lie algebras.

Let V be a vector space and $V^* = \mathrm{Hom}(V, \mathbb{R})$ its dual space. Then there is a natural non-degenerate skew-symmetric bilinear form ω on $T^*V = V \oplus V^*$ given by:

$$\omega(x + \alpha, y + \beta) = \langle \alpha, y \rangle - \langle \beta, x \rangle, \quad \forall x, y \in V, \alpha, \beta \in V^*. \qquad [8.16]$$

DEFINITION 8.8.– *Let* $(\mathfrak{h}, [\cdot,\cdot,\cdot]_{\mathfrak{h}})$ *be a 3-Lie algebra and* $\mathfrak{h}^*$ *its dual space.*

i) If there is a 3-Lie algebra structure $[\cdot,\cdot,\cdot]$ *on the direct sum vector space* $T^*\mathfrak{h} = \mathfrak{h} \oplus \mathfrak{h}^*$ *such that* $(\mathfrak{h} \oplus \mathfrak{h}^*, [\cdot,\cdot,\cdot], \omega)$ *is a symplectic 3-Lie algebra, where* ω *is given by* [8.16], *and* $(\mathfrak{h}, [\cdot,\cdot,\cdot]_{\mathfrak{h}})$ *and* $(\mathfrak{h}^*, [\cdot,\cdot,\cdot]|_{\mathfrak{h}^*})$ *are 3-Lie subalgebras of* $(\mathfrak{h} \oplus \mathfrak{h}^*, [\cdot,\cdot,\cdot])$, *then the symplectic 3-Lie algebra* $(\mathfrak{h} \oplus \mathfrak{h}^*, [\cdot,\cdot,\cdot], \omega)$ *is called a* ***phase space*** *of the 3-Lie algebra* $(\mathfrak{h}, [\cdot,\cdot,\cdot]_{\mathfrak{h}})$.

ii) A phase space $(\mathfrak{h} \oplus \mathfrak{h}^*, [\cdot,\cdot,\cdot], \omega)$ *is called* ***perfect*** *if the following conditions are satisfied:*

$$[x, y, \alpha] \in \mathfrak{h}^*, \quad [\alpha, \beta, x] \in \mathfrak{h}, \quad \forall x, y \in \mathfrak{h}, \alpha, \beta \in \mathfrak{h}^*. \qquad [8.17]$$

3-pre-Lie algebras play an important role in the study of phase spaces of 3-Lie algebras.

THEOREM 8.2.– A 3-Lie algebra has a phase space if and only if it is sub-adjacent to a 3-pre-Lie algebra.

PROOF.– Let $(A, \{\cdot,\cdot,\cdot\})$ be a 3-pre-Lie algebra. By proposition 8.1, the left multiplication L is a representation of the sub-adjacent 3-Lie algebra A^c on A. By lemma 8.1, L^* is a representation of the sub-adjacent 3-Lie algebra A^c on A^*. Thus, we have the semidirect product 3-Lie algebra $A^c \ltimes_{L^*} A^* = (A^c \oplus A^*, [\cdot,\cdot,\cdot]_{L^*})$. Then $(A^c \ltimes_{L^*} A^*, \omega)$ is a symplectic 3-Lie algebra, which is a phase space of the sub-adjacent 3-Lie algebra $(A^c, [\cdot,\cdot,\cdot]_C)$. In fact, for all $x_1, x_2, x_3, x_4 \in A$ and $\alpha_1, \alpha_2, \alpha_3, \alpha_4 \in A^*$, we have

$$\begin{aligned}
&\omega([x_1 + \alpha_1, x_2 + \alpha_2, x_3 + \alpha_3]_{L^*}, x_4 + \alpha_4) \\
&= \omega([x_1, x_2, x_3]_C + L^*(x_1, x_2)\alpha_3 + L^*(x_2, x_3)\alpha_1 \\
&\quad + L^*(x_3, x_1)\alpha_2, x_4 + \alpha_4) \\
&= \langle L^*(x_1, x_2)\alpha_3 + L^*(x_2, x_3)\alpha_1 + L^*(x_3, x_1)\alpha_2, x_4 \rangle \\
&\quad - \langle \alpha_4, [x_1, x_2, x_3]_C \rangle \\
&= -\langle \alpha_3, \{x_1, x_2, x_4\} \rangle - \langle \alpha_1, \{x_2, x_3, x_4\} \rangle - \langle \alpha_2, \{x_3, x_1, x_4\} \rangle \\
&\quad - \langle \alpha_4, \{x_1, x_2, x_3\} \rangle - \langle \alpha_4, \{x_2, x_3, x_1\} \rangle - \langle \alpha_4, \{x_3, x_1, x_2\} \rangle.
\end{aligned}$$

Similarly, we have

$$\begin{aligned}
&\omega([x_2+\alpha_2,x_3+\alpha_3,x_4+\alpha_4]_{L^*},x_1+\alpha_1)\\
&=-\langle\alpha_4,\{x_2,x_3,x_1\}\rangle-\langle\alpha_2,\{x_3,x_4,x_1\}\rangle-\langle\alpha_3,\{x_4,x_2,x_1\}\rangle\\
&-\langle\alpha_1,\{x_2,x_3,x_4\}\rangle-\langle\alpha_1,\{x_3,x_4,x_2\}\rangle-\langle\alpha_1,\{x_4,x_2,x_3\}\rangle,\\
&\omega([x_3+\alpha_3,x_4+\alpha_4,x_1+\alpha_1]_{L^*},x_2+\alpha_2)\\
&=-\langle\alpha_1,\{x_3,x_4,x_2\}\rangle-\langle\alpha_3,\{x_4,x_1,x_2\}\rangle-\langle\alpha_4,\{x_1,x_3,x_2\}\rangle\\
&-\langle\alpha_2,\{x_3,x_4,x_1\}\rangle-\langle\alpha_2,\{x_4,x_1,x_3\}\rangle-\langle\alpha_2,\{x_1,x_3,x_4\}\rangle,\\
&\omega([x_4+\alpha_4,x_1+\alpha_1,x_2+\alpha_2]_{L^*},x_3+\alpha_3)\\
&=-\langle\alpha_2,\{x_4,x_1,x_3\}\rangle-\langle\alpha_4,\{x_1,x_2,x_3\}\rangle-\langle\alpha_1,\{x_2,x_4,x_3\}\rangle\\
&-\langle\alpha_3,\{x_4,x_1,x_2\}\rangle-\langle\alpha_3,\{x_1,x_2,x_4\}\rangle-\langle\alpha_3,\{x_2,x_4,x_1\}\rangle.
\end{aligned}$$

Since $\{x_1,x_2,x_3\}=-\{x_2,x_1,x_3\}$, we deduce that ω is a symplectic structure on the semidirect product 3-Lie algebra $A^c\ltimes_{L^*}A^*$. Moreover, $(A^c,[\cdot,\cdot,\cdot]_C)$ is a subalgebra of $A^c\ltimes_{L^*}A^*$ and A^* is an abelian subalgebra of $A^c\ltimes_{L^*}A^*$. Thus, the symplectic 3-Lie algebra $(A^c\ltimes_{L^*}A^*,\omega)$ is a phase space of the sub-adjacent 3-Lie algebra $(A^c,[\cdot,\cdot,\cdot]_C)$.

Conversely, let $(T^*\mathfrak{h}=\mathfrak{h}\oplus\mathfrak{h}^*,[\cdot,\cdot,\cdot],\omega)$ be a phase space of a 3-Lie algebra $(\mathfrak{h},[\cdot,\cdot,\cdot]_{\mathfrak{h}})$. By proposition 8.5, there exists a compatible 3-pre-Lie algebra structure $\{\cdot,\cdot,\cdot\}$ on $T^*\mathfrak{h}$ given by [8.14]. Since $(\mathfrak{h},[\cdot,\cdot,\cdot]_{\mathfrak{h}})$ is a subalgebra of $(\mathfrak{h}\oplus\mathfrak{h}^*,[\cdot,\cdot,\cdot])$, we have

$$\omega(\{x,y,z\},w)=-\omega(z,[x,y,w])=-\omega(z,[x,y,w]_{\mathfrak{h}})=0,\quad \forall x,y,z,w\in\mathfrak{h}.$$

Thus, $\{x,y,z\}\in\mathfrak{h}$, which implies that $(\mathfrak{h},\{\cdot,\cdot,\cdot\}|_{\mathfrak{h}})$ is a subalgebra of the 3-pre-Lie algebra $(T^*\mathfrak{h},\{\cdot,\cdot,\cdot\})$. Its sub-adjacent 3-Lie algebra $(\mathfrak{h}^c,[\cdot,\cdot,\cdot]_C)$ is exactly the original 3-Lie algebra $(\mathfrak{h},[\cdot,\cdot,\cdot]_{\mathfrak{h}})$. □

COROLLARY 8.1.– Let $(T^*\mathfrak{h}=\mathfrak{h}\oplus\mathfrak{h}^*,[\cdot,\cdot,\cdot],\omega)$ be a phase space of a 3-Lie algebra $(\mathfrak{h},[\cdot,\cdot,\cdot]_{\mathfrak{h}})$ and $(\mathfrak{h}\oplus\mathfrak{h}^*,\{\cdot,\cdot,\cdot\})$ the associated 3-pre-Lie algebra. Then both $(\mathfrak{h},\{\cdot,\cdot,\cdot\}|_{\mathfrak{h}})$ and $(\mathfrak{h}^*,\{\cdot,\cdot,\cdot\}|_{\mathfrak{h}^*})$ are subalgebras of the 3-pre-Lie algebra $(\mathfrak{h}\oplus\mathfrak{h}^*,\{\cdot,\cdot,\cdot\})$.

EXAMPLE 8.3.– Let $(A,\{\cdot,\cdot,\cdot\}_A)$ be a 3-pre-Lie algebra. Since there is a semidirect product 3-pre-Lie algebra structure $(A\ltimes_{L^*,0}A^*,\{\cdot,\cdot,\cdot\}_{L^*,0})$ on the phase space $T^*A^c=A^c\ltimes_{L^*}A^*$, one can construct a new phase space $T^*A^c\ltimes_{L^*}(T^*A^c)^*$. This process can be continued indefinitely. Hence, there exists a series of phase spaces $\{A_{(n)}\}_{n\geq 2}$:

$$A_{(1)} = A^c, \quad A_{(2)} = T^*A_{(1)} = A^c \ltimes_{L^*} A^*, \cdots, \quad A_{(n)} = T^*A_{(n-1)}, \cdots.$$

$A_{(n)}$ $(n \geq 2)$ is called the symplectic double of $A_{(n-1)}$.

We introduce the notion of a Manin triple of 3-pre-Lie algebras.

DEFINITION 8.9.– *A* ***Manin triple of*** **3-pre-Lie algebras** *is a triple* $(\mathcal{A}; A, A')$, *where*

– $(\mathcal{A}, \{\cdot,\cdot,\cdot\}, \omega)$ *is a quadratic* 3-pre-Lie *algebra;*

– *both* A *and* A' *are isotropic subalgebras of* $(\mathcal{A}, \{\cdot,\cdot,\cdot\})$*;*

– $\mathcal{A} = A \oplus A'$ *as vector spaces;*

– *for all* $x, y \in A$ *and* $\alpha, \beta \in A'$, *there holds:*

$$\{x, y, \alpha\} \in A', \quad \{\alpha, x, y\} \in A', \quad \{\alpha, \beta, x\} \in A, \quad \{x, \alpha, \beta\} \in A. \quad [8.18]$$

In a Manin triple of 3-pre-Lie algebras, since the skew-symmetric bilinear form ω is non-degenerate, A' can be identified with A^* via

$$\langle \alpha, x \rangle \triangleq \omega(\alpha, x), \quad \forall x \in A, \alpha \in A'.$$

Thus, $\mathcal{A}$ is isomorphic to $A \oplus A^*$ naturally and the bilinear form ω is exactly given by [8.16]. By the invariant condition [8.15], we can obtain the precise form of the 3-pre-Lie structure $\{\cdot,\cdot,\cdot\}$ on $A \oplus A^*$.

PROPOSITION 8.6.– Let $(A \oplus A^*; A, A^*)$ be a Manin triple of 3-pre-Lie algebras, where the non-degenerate skew-symmetric bilinear form ω on the 3-pre-Lie algebra is given by [8.16]. Then we have

$$\{x, y, \alpha\} = (L^* - R^*\tau + R^*)(x, y)\alpha, \quad [8.19]$$

$$\{\alpha, x, y\} = -R^*(x, y)\alpha, \quad [8.20]$$

$$\{\alpha, \beta, x\} = (\mathcal{L}^* - \mathcal{R}^*\tau + \mathcal{R}^*)(\alpha, \beta)x, \quad [8.21]$$

$$\{x, \alpha, \beta\} = -\mathcal{R}^*(\alpha, \beta)x. \quad [8.22]$$

PROOF.– For all $x, y, z \in A, \alpha \in A^*$, we have

$$\begin{aligned}\langle \{x, y, \alpha\}, z\rangle &= \omega(\{x, y, \alpha\}, z) = -\omega(\alpha, [x, y, z]_C) \\ &= -\omega(\alpha, \{x, y, z\} + \{y, z, x\} + \{z, x, y\}) \\ &= -\omega(\alpha, L(x, y)z - R(y, x)z + R(x, y)z) \\ &= -\langle \alpha, L(x, y)z - R(y, x)z + R(x, y)z\rangle \\ &= \langle (L^* - R^*\tau + R^*)(x, y)\alpha, z\rangle,\end{aligned}$$

which implies that [8.19] holds. We have

$$\begin{aligned}\langle \{\alpha, x, y\}, z\rangle &= \omega(\{\alpha, x, y\}, z) = -\omega(y, [\alpha, x, z]_C) \\ &= \omega(y, [z, x, \alpha]_C) = -\omega(\{z, x, y\}, \alpha) \\ &= \langle \alpha, R(x, y)z\rangle = -\langle R^*(x, y)\alpha, z\rangle,\end{aligned}$$

which implies that [8.20] holds. Similarly, we can deduce that [8.21] and [8.22] hold. □

THEOREM 8.3.– There is a one-to-one correspondence between Manin triples of 3-pre-Lie algebras and perfect phase spaces of 3-Lie algebras. More precisely, if $(A \oplus A^*; A, A^*)$ is a Manin triple of 3-pre-Lie algebras, then $(A \oplus A^*, [\cdot,\cdot,\cdot]_C, \omega)$ is a symplectic 3-Lie algebra, where ω is given by [8.16]. Conversely, if $(\mathfrak{h} \oplus \mathfrak{h}^*, [\cdot,\cdot,\cdot], \omega)$ is a perfect phase space of a 3-Lie algebra $(\mathfrak{h}, [\cdot,\cdot,\cdot]_{\mathfrak{h}})$, then $(\mathfrak{h} \oplus \mathfrak{h}^*; \mathfrak{h}, \mathfrak{h}^*)$ is a Manin triple of 3-pre-Lie algebras, where the 3-pre-Lie algebra structure on $\mathfrak{h} \oplus \mathfrak{h}^*$ is given by [8.14].

PROOF.– Let $(A \oplus A^*; A, A^*)$ be a Manin triple of 3-pre-Lie algebras. Denote by $\{\cdot,\cdot,\cdot\}_A$ and $\{\cdot,\cdot,\cdot\}_{A^*}$ the 3-pre-Lie algebra structure on A and A^*, respectively, and denote by $[\cdot,\cdot,\cdot]_A$ and $[\cdot,\cdot,\cdot]_{A^*}$ the corresponding sub-adjacent 3-Lie algebra structure on A and A^*, respectively. By proposition 8.6, it is straightforward to deduce that the corresponding 3-Lie algebra structure $[\cdot,\cdot,\cdot]_C$ on $A \oplus A^*$ is given by

$$\begin{aligned}&[x+\alpha, y+\beta, z+\gamma]_C \\ = &[x, y, z]_A + \mathfrak{L}^*(\alpha, \beta)z + \mathfrak{L}^*(\beta, \gamma)x + \mathfrak{L}^*(\gamma, \alpha)y \\ &+[\alpha, \beta, \gamma]_{A^*} + L^*(x, y)\gamma + L^*(y, z)\alpha + L^*(z, x)\beta.\end{aligned} \quad [8.23]$$

For all $x_1, x_2, x_3, x_4 \in A$ and $\alpha_1, \alpha_2, \alpha_3, \alpha_4 \in A^*$, we have

$$\begin{aligned}
&\omega([x_1+\alpha_1, x_2+\alpha_2, x_3+\alpha_3]_C, x_4+\alpha_4)\\
&= \omega([x_1,x_2,x_3]_A + \mathcal{L}^*(\alpha_1,\alpha_2)x_3 + \mathcal{L}^*(\alpha_2,\alpha_3)x_1 + \mathcal{L}^*(\alpha_3,\alpha_1)x_2\\
&+ [\alpha_1,\alpha_2,\alpha_3]_{A^*} + L^*(x_1,x_2)\alpha_3 + L^*(x_2,x_3)\alpha_1 + L^*(x_3,x_1)\alpha_2, x_4+\alpha_4)\\
&= \langle [\alpha_1,\alpha_2,\alpha_3]_{A^*} + L^*(x_1,x_2)\alpha_3 + L^*(x_2,x_3)\alpha_1 + L^*(x_3,x_1)\alpha_2, x_4\rangle\\
&- \langle \alpha_4, [x_1,x_2,x_3]_A + \mathcal{L}^*(\alpha_1,\alpha_2)x_3 + \mathcal{L}^*(\alpha_2,\alpha_3)x_1 + \mathcal{L}^*(\alpha_3,\alpha_1)x_2\rangle\\
&= \langle [\alpha_1,\alpha_2,\alpha_3]_{A^*}, x_4\rangle - \langle \alpha_3, \{x_1,x_2,x_4\}_A\rangle - \langle \alpha_1, \{x_2,x_3,x_4\}_A\rangle\\
&- \langle \alpha_2, \{x_3,x_1,x_4\}_A\rangle\\
&- \langle \alpha_4, [x_1,x_2,x_3]_A\rangle + \langle \{\alpha_1,\alpha_2,\alpha_4\}_{A^*}, x_3\rangle + \langle \{\alpha_2,\alpha_3,\alpha_4\}_{A^*}, x_1\rangle\\
&+ \langle \{\alpha_3,\alpha_1,\alpha_4\}_{A^*}, x_2\rangle.
\end{aligned}$$

Similarly, we have

$$\begin{aligned}
&\omega([x_2+\alpha_2, x_3+\alpha_3, x_4+\alpha_4], x_1+\alpha_1)\\
&= \langle [\alpha_2,\alpha_3,\alpha_4]_C, x_1\rangle - \langle \alpha_4, \{x_2,x_3,x_1\}_A\rangle - \langle \alpha_2, \{x_3,x_4,x_1\}_A\rangle\\
&- \langle \alpha_3, \{x_4,x_2,x_1\}_A\rangle - \langle \alpha_1, [x_2,x_3,x_4]_C\rangle + \langle \{\alpha_2,\alpha_3,\alpha_1\}_{A^*}, x_4\rangle\\
&+ \langle \{\alpha_3,\alpha_4,\alpha_1\}_{A^*}, x_2\rangle + \langle \{\alpha_4,\alpha_2,\alpha_1\}_{A^*}, x_3\rangle,\\
&\omega([x_3+\alpha_3, x_4+\alpha_4, x_1+\alpha_1], x_2+\alpha_2)\\
&= \langle [\alpha_3,\alpha_4,\alpha_1]_C, x_2\rangle - \langle \alpha_1, \{x_3,x_4,x_2\}_A\rangle - \langle \alpha_3, \{x_4,x_1,x_2\}_A\rangle\\
&- \langle \alpha_4, \{x_1,x_3,x_2\}_A\rangle - \langle \alpha_2, [x_3,x_4,x_1]_C\rangle + \langle \{\alpha_3,\alpha_4,\alpha_2\}_{A^*}, x_1\rangle\\
&+ \langle \{\alpha_4,\alpha_1,\alpha_2\}_{A^*}, x_3\rangle + \langle \{\alpha_1,\alpha_3,\alpha_2\}_{A^*}, x_4\rangle,\\
&\omega([x_4+\alpha_4, x_1+\alpha_1, x_2+\alpha_2], x_3+\alpha_3)\\
&= \langle [\alpha_4,\alpha_1,\alpha_2]_C, x_3\rangle - \langle \alpha_2, \{x_4,x_1,x_3\}_A\rangle - \langle \alpha_4, \{x_1,x_2,x_3\}_A\rangle\\
&- \langle \alpha_1, \{x_2,x_4,x_3\}_A\rangle - \langle \alpha_3, [x_4,x_1,x_2]_C\rangle + \langle \{\alpha_4,\alpha_1,\alpha_3\}_{A^*}, x_2\rangle\\
&+ \langle \{\alpha_1,\alpha_2,\alpha_3\}_{A^*}, x_4\rangle + \langle \{\alpha_2,\alpha_4,\alpha_3\}_{A^*}, x_1\rangle.
\end{aligned}$$

By $\{x_1,x_2,x_3\}_A = -\{x_2,x_1,x_3\}_A$ and $\{\alpha_1,\alpha_2,\alpha_3\}_{A^*} = -\{\alpha_2,\alpha_1,\alpha_3\}_{A^*}$, we deduce that ω is a symplectic structure on the 3-Lie algebra $(A \oplus A^*, [\cdot,\cdot,\cdot]_C)$. Therefore, it is a phase space.

Conversely, let $(\mathfrak{h}\oplus\mathfrak{h}^*, [\cdot,\cdot,\cdot], \omega)$ be a phase space of the 3-Lie algebra $(\mathfrak{h}, [\cdot,\cdot,\cdot]_{\mathfrak{h}})$. By proposition 8.5, there exists a 3-pre-Lie algebra structure $\{\cdot,\cdot,\cdot\}$ on $\mathfrak{h}\oplus\mathfrak{h}^*$ given by [8.14] such that $(\mathfrak{h}\oplus\mathfrak{h}^*, \{\cdot,\cdot,\cdot\}, \omega)$ is a quadratic 3-pre-Lie algebra. By corollary 8.1, $(\mathfrak{h}, \{\cdot,\cdot,\cdot\}|_{\mathfrak{h}})$ and $(\mathfrak{h}^*, \{\cdot,\cdot,\cdot\}|_{\mathfrak{h}^*})$ are 3-pre-Lie subalgebras of $(\mathfrak{h}\oplus\mathfrak{h}^*, \{\cdot,\cdot,\cdot\})$. It is obvious that both $\mathfrak{h}$ and $\mathfrak{h}^*$ are isotropic. Thus, we only need to show that [8.18] holds. By [8.17], for all $x_1, x_2 \in \mathfrak{h}$ and $\alpha_1, \alpha_2 \in \mathfrak{h}^*$, we have

$$\omega(\{x_1, x_2, \alpha_1\}, \alpha_2) = -\omega(\alpha_1, [x_1, x_2, \alpha_2]_C) = 0,$$

which implies that $\{x_1, x_2, \alpha_1\} \in \mathfrak{h}^*$. Similarly, we can show that the other conditions in [8.18] also hold. The proof is completed. □

REMARK 8.1.– The notions of a matched pair of 3-Lie algebras and a Manin triple of 3-Lie algebras were introduced in (Bai *et al.* 2019). By [8.23], we obtain that $(A^c, A^{*c}; L^*, \mathfrak{L}^*)$ is a matched pair of 3-Lie algebras and the phase space is exactly the double of this matched pair. However, one should note that a Manin triple of 3-pre-Lie algebras does not give rise to a Manin triple of 3-Lie algebras.

8.5. Product structures on 3-Lie algebras

We first introduce the notion of a product structure on a 3-Lie algebra using the Nijenhuis condition as the integrability condition. We find four special integrability conditions, each of them gives a special decomposition of the original 3-Lie algebra. Then, we introduce the notion of a (perfect) paracomplex structure on a 3-Lie algebra and give examples.

DEFINITION 8.10.– *Let* $(\mathfrak{g}, [\cdot,\cdot,\cdot]_\mathfrak{g})$ *be a 3-Lie algebra. An* ***almost product structure*** *on the 3-Lie algebra* $(\mathfrak{g}, [\cdot,\cdot,\cdot]_\mathfrak{g})$ *is a linear endomorphism* $E : \mathfrak{g} \to \mathfrak{g}$*, satisfying* $E^2 = \mathrm{Id}$ $(E \neq \pm\mathrm{Id})$*. An almost product structure is called a* ***product*** *structure if the following integrability condition is satisfied:*

$$\begin{aligned} E[x, y, z]_\mathfrak{g} = {} & [Ex, Ey, Ez]_\mathfrak{g} + [Ex, y, z]_\mathfrak{g} + [x, Ey, z]_\mathfrak{g} + [x, y, Ez]_\mathfrak{g} \\ & -E[Ex, Ey, z]_\mathfrak{g} - E[x, Ey, Ez]_\mathfrak{g} - E[Ex, y, Ez]_\mathfrak{g}. \end{aligned} \qquad [8.24]$$

REMARK 8.2.– One can understand a product structure on a 3-Lie algebra as a Nijenhuis operator E on a 3-Lie algebra, satisfying $E^2 = \mathrm{Id}$.

THEOREM 8.4.– Let $(\mathfrak{g}, [\cdot,\cdot,\cdot]_\mathfrak{g})$ be a 3-Lie algebra. Then $(\mathfrak{g}, [\cdot,\cdot,\cdot]_\mathfrak{g})$ has a product structure if and only if $\mathfrak{g}$ admits a decomposition: $\mathfrak{g} = \mathfrak{g}_+ \oplus \mathfrak{g}_-$, where $\mathfrak{g}_+$ and $\mathfrak{g}_-$ are subalgebras of $\mathfrak{g}$.

PROOF.– Let E be a product structure on $\mathfrak{g}$. By $E^2 = \mathrm{Id}$, we have $\mathfrak{g} = \mathfrak{g}_+ \oplus \mathfrak{g}_-$, where $\mathfrak{g}_+$ and $\mathfrak{g}_-$ are the eigenspaces of $\mathfrak{g}$ associated with the eigenvalues ± 1. For all $x_1, x_2, x_3 \in \mathfrak{g}_+$, we have

$$\begin{aligned} & E[x_1, x_2, x_3]_\mathfrak{g} = [Ex_1, Ex_2, Ex_3]_\mathfrak{g} + [Ex_1, x_2, x_3]_\mathfrak{g} + [x_1, Ex_2, x_3]_\mathfrak{g} \\ & + [x_1, x_2, Ex_3]_\mathfrak{g} - E[Ex_1, Ex_2, x_3]_\mathfrak{g} - E[x_1, Ex_2, Ex_3]_\mathfrak{g} - E[Ex_1, x_2, Ex_3]_\mathfrak{g} \\ & = 4[x_1, x_2, x_3]_\mathfrak{g} - 3E[x_1, x_2, x_3]_\mathfrak{g}. \end{aligned}$$

Thus, we have $[x_1, x_2, x_3]_\mathfrak{g} \in \mathfrak{g}_+$, which implies that $\mathfrak{g}_+$ is a subalgebra. Similarly, we can show that $\mathfrak{g}_-$ is a subalgebra.

Conversely, we define a linear endomorphism $E : \mathfrak{g} \to \mathfrak{g}$ by

$$E(x + \alpha) = x - \alpha, \quad \forall x \in \mathfrak{g}_+, \alpha \in \mathfrak{g}_-. \qquad [8.25]$$

Obviously we have $E^2 = \mathrm{Id}$. Since $\mathfrak{g}_+$ is a subalgebra of $\mathfrak{g}$, for all $x_1, x_2, x_3 \in \mathfrak{g}_+$, we have

$$\begin{aligned} & [Ex_1, Ex_2, Ex_3]_\mathfrak{g} + [Ex_1, x_2, x_3]_\mathfrak{g} + [x_1, Ex_2, x_3]_\mathfrak{g} + [x_1, x_2, Ex_3]_\mathfrak{g} \\ & -E[Ex_1, Ex_2, x_3]_\mathfrak{g} - E[x_1, Ex_2, Ex_3]_\mathfrak{g} - E[Ex_1, x_2, Ex_3]_\mathfrak{g} \\ = {} & 4[x_1, x_2, x_3]_\mathfrak{g} - 3E[x_1, x_2, x_3]_\mathfrak{g} = [x_1, x_2, x_3]_\mathfrak{g} \\ = {} & E[x_1, x_2, x_3]_\mathfrak{g}, \end{aligned}$$

which implies that [8.24] holds for all $x_1, x_2, x_3 \in \mathfrak{g}_+$. Similarly, we can show that [8.24] holds for all $x, y, z \in \mathfrak{g}$. Therefore, E is a product structure on $\mathfrak{g}$. □

LEMMA 8.3.– Let E be an almost product structure on a 3-Lie algebra $(\mathfrak{g}, [\cdot,\cdot,\cdot]_\mathfrak{g})$. If E satisfies the following equation:

$$E[x, y, z]_\mathfrak{g} = [Ex, y, z]_\mathfrak{g}, \qquad [8.26]$$

then E is a product structure on $\mathfrak{g}$ such that $[\mathfrak{g}_+, \mathfrak{g}_+, \mathfrak{g}_-]_\mathfrak{g} = 0$ and $[\mathfrak{g}_-, \mathfrak{g}_-, \mathfrak{g}_+]_\mathfrak{g} = 0$, i.e. $\mathfrak{g}$ is the 3-Lie algebra direct sum of $\mathfrak{g}_+$ and $\mathfrak{g}_-$.

PROOF.– By [8.26] and $E^2 = \mathrm{Id}$, we have

$$\begin{aligned} & [Ex, Ey, Ez]_\mathfrak{g} + [Ex, y, z]_\mathfrak{g} + [x, Ey, z]_\mathfrak{g} + [x, y, Ez]_\mathfrak{g} \\ & -E[Ex, Ey, z]_\mathfrak{g} - E[x, Ey, Ez]_\mathfrak{g} - E[Ex, y, Ez]_\mathfrak{g} \\ = {} & [Ex, Ey, Ez]_\mathfrak{g} + E[x, y, z]_\mathfrak{g} + [x, Ey, z]_\mathfrak{g} + [x, y, Ez]_\mathfrak{g} \\ & -[E^2x, Ey, z]_\mathfrak{g} - [Ex, Ey, Ez]_\mathfrak{g} - [E^2x, y, Ez]_\mathfrak{g} \\ = {} & E[x, y, z]_\mathfrak{g}. \end{aligned}$$

Thus, E is a product structure on $\mathfrak{g}$. For all $x_1, x_2 \in \mathfrak{g}_+, \alpha_1 \in \mathfrak{g}_-$, on one hand we have

$$E[\alpha_1, x_1, x_2]_\mathfrak{g} = [E\alpha_1, x_1, x_2]_\mathfrak{g} = -[\alpha_1, x_1, x_2]_\mathfrak{g}.$$

On the other hand, we have

$$E[\alpha_1, x_1, x_2]_{\mathfrak{g}} = E[x_1, x_2, \alpha_1]_{\mathfrak{g}} = [Ex_1, x_2, \alpha_1]_{\mathfrak{g}} = [x_1, x_2, \alpha_1]_{\mathfrak{g}}.$$

Thus, we obtain $[\mathfrak{g}_+, \mathfrak{g}_+, \mathfrak{g}_-]_{\mathfrak{g}} = 0$. Similarly, we have $[\mathfrak{g}_-, \mathfrak{g}_-, \mathfrak{g}_+]_{\mathfrak{g}} = 0$. The proof is completed. □

DEFINITION 8.11.– ***(Integrability condition I)*** *An almost product structure* E *on a 3-Lie algebra* $(\mathfrak{g}, [\cdot,\cdot,\cdot]_{\mathfrak{g}})$ *is called a* ***strict product structure*** *if* [8.26] *holds.*

COROLLARY 8.2.– Let $(\mathfrak{g}, [\cdot,\cdot,\cdot]_{\mathfrak{g}})$ be a 3-Lie algebra. Then $(\mathfrak{g}, [\cdot,\cdot,\cdot]_{\mathfrak{g}})$ has a strict product structure if and only if $\mathfrak{g}$ admits a decomposition: $\mathfrak{g} = \mathfrak{g}_+ \oplus \mathfrak{g}_-$, where $\mathfrak{g}_+$ and $\mathfrak{g}_-$ are subalgebras of $\mathfrak{g}$ such that $[\mathfrak{g}_+, \mathfrak{g}_+, \mathfrak{g}_-]_{\mathfrak{g}} = 0$ and $[\mathfrak{g}_-, \mathfrak{g}_-, \mathfrak{g}_+]_{\mathfrak{g}} = 0$.

LEMMA 8.4.– Let E be an almost product structure on a 3-Lie algebra $(\mathfrak{g}, [\cdot,\cdot,\cdot]_{\mathfrak{g}})$. If E satisfies the following equation:

$$[x, y, z]_{\mathfrak{g}} = -[x, Ey, Ez]_{\mathfrak{g}} - [Ex, y, Ez]_{\mathfrak{g}} - [Ex, Ey, z]_{\mathfrak{g}}, \quad [8.27]$$

then E is a product structure on $\mathfrak{g}$.

PROOF.– By [8.27] and $E^2 = \mathrm{Id}$, we have

$$\begin{aligned}
&[Ex, Ey, Ez]_{\mathfrak{g}} + [Ex, y, z]_{\mathfrak{g}} + [x, Ey, z]_{\mathfrak{g}} + [x, y, Ez]_{\mathfrak{g}}\\
&-E[Ex, Ey, z]_{\mathfrak{g}} - E[x, Ey, Ez]_{\mathfrak{g}} - E[Ex, y, Ez]_{\mathfrak{g}}\\
= &-[Ex, E^2y, E^2z]_{\mathfrak{g}} - [E^2x, Ey, E^2z]_{\mathfrak{g}} - [E^2x, E^2y, Ez]_{\mathfrak{g}}\\
&+[Ex, y, z]_{\mathfrak{g}} + [x, Ey, z]_{\mathfrak{g}} + [x, y, Ez]_{\mathfrak{g}} + E[x, y, z]_{\mathfrak{g}}\\
= &\ E[x, y, z]_{\mathfrak{g}}.
\end{aligned}$$

Thus, E is a product structure on $\mathfrak{g}$. □

DEFINITION 8.12.– ***(Integrability condition II)*** *An almost product structure* E *on a 3-Lie algebra* $(\mathfrak{g}, [\cdot,\cdot,\cdot]_{\mathfrak{g}})$ *is called an* ***abelian product structure*** *if* [8.27] *holds.*

COROLLARY 8.3.– Let $(\mathfrak{g}, [\cdot,\cdot,\cdot]_{\mathfrak{g}})$ be a 3-Lie algebra. Then $(\mathfrak{g}, [\cdot,\cdot,\cdot]_{\mathfrak{g}})$ has an abelian product structure if and only if $\mathfrak{g}$ admits a decomposition: $\mathfrak{g} = \mathfrak{g}_+ \oplus \mathfrak{g}_-$, where $\mathfrak{g}_+$ and $\mathfrak{g}_-$ are abelian subalgebras of $\mathfrak{g}$.

PROOF.– Let E be an abelian product structure on $\mathfrak{g}$. For all $x_1, x_2, x_3 \in \mathfrak{g}_+$, we have

$$\begin{aligned}
[x_1, x_2, x_3]_{\mathfrak{g}} &= -[Ex_1, Ex_2, x_3]_{\mathfrak{g}} - [x_1, Ex_2, Ex_3]_{\mathfrak{g}} - [Ex_1, x_2, Ex_3]_{\mathfrak{g}}\\
&= -3[x_1, x_2, x_3]_{\mathfrak{g}},
\end{aligned}$$

which implies that $[x_1, x_2, x_3]_{\mathfrak{g}} = 0$. Similarly, for all $\alpha_1, \alpha_2, \alpha_3 \in \mathfrak{g}_-$, we also have $[\alpha_1, \alpha_2, \alpha_3]_{\mathfrak{g}} = 0$. Thus, both $\mathfrak{g}_+$ and $\mathfrak{g}_-$ are abelian subalgebras.

Conversely, define a linear endomorphism $E : \mathfrak{g} \to \mathfrak{g}$ by [8.25]. Then it is straightforward to deduce that E is an abelian product structure on $\mathfrak{g}$. □

LEMMA 8.5.– Let E be an almost product structure on a 3-Lie algebra $(\mathfrak{g}, [\cdot,\cdot,\cdot]_{\mathfrak{g}})$. If E satisfies the following equation:

$$[x, y, z]_{\mathfrak{g}} = E[Ex, y, z]_{\mathfrak{g}} + E[x, Ey, z]_{\mathfrak{g}} + E[x, y, Ez]_{\mathfrak{g}}, \qquad [8.28]$$

then we obtain that E is an abelian product structure on $\mathfrak{g}$ such that $[\mathfrak{g}_+, \mathfrak{g}_+, \mathfrak{g}_-]_{\mathfrak{g}} \subset \mathfrak{g}_+$ and $[\mathfrak{g}_-, \mathfrak{g}_-, \mathfrak{g}_+]_{\mathfrak{g}} \subset \mathfrak{g}_-$.

PROOF.– By [8.28] and $E^2 = \mathrm{Id}$, we have

$$\begin{aligned}
&[Ex, Ey, Ez]_{\mathfrak{g}} + [Ex, y, z]_{\mathfrak{g}} + [x, Ey, z]_{\mathfrak{g}} + [x, y, Ez]_{\mathfrak{g}}\\
&-E[Ex, Ey, z]_{\mathfrak{g}} - E[x, Ey, Ez]_{\mathfrak{g}} - E[Ex, y, Ez]_{\mathfrak{g}}\\
=\ & E[x, Ey, Ez]_{\mathfrak{g}} + E[Ex, y, Ez]_{\mathfrak{g}} + E[Ex, Ey, z]_{\mathfrak{g}} + E[x, y, z]_{\mathfrak{g}}\\
&-E[Ex, Ey, z]_{\mathfrak{g}} - E[x, Ey, Ez]_{\mathfrak{g}} - E[Ex, y, Ez]_{\mathfrak{g}}\\
=\ & E[x, y, z]_{\mathfrak{g}}.
\end{aligned}$$

Thus, we obtain that E is a product structure on $\mathfrak{g}$. For all $x_1, x_2, x_3 \in \mathfrak{g}_+$, by [8.28], we have

$$\begin{aligned}
[x_1, x_2, x_3]_{\mathfrak{g}} &= E[Ex_1, x_2, x_3]_{\mathfrak{g}} + E[x_1, Ex_2, x_3]_{\mathfrak{g}} + E[x_1, x_2, Ex_3]_{\mathfrak{g}}\\
&= 3E[x_1, x_2, x_3]_{\mathfrak{g}} = 3[x_1, x_2, x_3]_{\mathfrak{g}}.
\end{aligned}$$

Thus, we obtain $[\mathfrak{g}_+, \mathfrak{g}_+, \mathfrak{g}_+]_{\mathfrak{g}} = 0$. Similarly, we have $[\mathfrak{g}_-, \mathfrak{g}_-, \mathfrak{g}_-]_{\mathfrak{g}} = 0$. By corollary 8.3, E is an abelian product structure on $\mathfrak{g}$. Moreover, for all $x_1, x_2 \in \mathfrak{g}_+, \alpha_1 \in \mathfrak{g}_-$, we have

$$\begin{aligned}
[x_1, x_2, \alpha_1]_{\mathfrak{g}} &= E[Ex_1, x_2, \alpha_1]_{\mathfrak{g}} + E[x_1, Ex_2, \alpha_1]_{\mathfrak{g}} + E[x_1, x_2, E\alpha_1]_{\mathfrak{g}}\\
&= E[x_1, x_2, \alpha_1]_{\mathfrak{g}},
\end{aligned}$$

which implies that $[\mathfrak{g}_+, \mathfrak{g}_+, \mathfrak{g}_-]_{\mathfrak{g}} \subset \mathfrak{g}_+$. Similarly, we have $[\mathfrak{g}_-, \mathfrak{g}_-, \mathfrak{g}_+]_{\mathfrak{g}} \subset \mathfrak{g}_-$. □

DEFINITION 8.13.– ***(Integrability condition III)*** *An almost product structure E on a 3-Lie algebra $(\mathfrak{g}, [\cdot,\cdot,\cdot]_{\mathfrak{g}})$ is called a* ***strong abelian product structure*** *if* [8.28] *holds.*

COROLLARY 8.4.– Let $(\mathfrak{g}, [\cdot,\cdot,\cdot]_{\mathfrak{g}})$ be a 3-Lie algebra. Then $(\mathfrak{g}, [\cdot,\cdot,\cdot]_{\mathfrak{g}})$ has a strong abelian product structure if and only if $\mathfrak{g}$ admits a decomposition: $\mathfrak{g} = \mathfrak{g}_+ \oplus \mathfrak{g}_-$, where $\mathfrak{g}_+$ and $\mathfrak{g}_-$ are abelian subalgebras of $\mathfrak{g}$ such that $[\mathfrak{g}_+, \mathfrak{g}_+, \mathfrak{g}_-]_{\mathfrak{g}} \subset \mathfrak{g}_+$ and $[\mathfrak{g}_-, \mathfrak{g}_-, \mathfrak{g}_+]_{\mathfrak{g}} \subset \mathfrak{g}_-$.

REMARK 8.3.– Let E be a strong abelian product structure on a 3-Lie algebra $(\mathfrak{g}, [\cdot,\cdot,\cdot]_{\mathfrak{g}})$. Then we can define $\nu_+ : \mathfrak{g}_+ \longrightarrow \mathrm{Hom}(\wedge^2\mathfrak{g}_-, \mathfrak{g}_-)$ and $\nu_- : \mathfrak{g}_- \longrightarrow \mathrm{Hom}(\wedge^2\mathfrak{g}_+, \mathfrak{g}_+)$ by

$$\nu_+(x)(\alpha,\beta) = [\alpha,\beta,x]_{\mathfrak{g}}, \quad \nu_-(\alpha)(x,y) = [x,y,\alpha]_{\mathfrak{g}}, \quad \forall x,y \in \mathfrak{g}_+, \alpha,\beta \in \mathfrak{g}_-.$$

It turns out ν_+ and ν_- are generalized representations of abelian 3-Lie algebras $\mathfrak{g}_+$ and $\mathfrak{g}_-$ on $\mathfrak{g}_-$ and $\mathfrak{g}_+$ respectively. See Liu *et al.* (2017) for more details about generalized representations of 3-Lie algebras.

More surprisingly, a strong abelian product structure is an $\mathcal{O}$-operator as well as a Rota–Baxter operator (Bai *et al.* 2013; Pei *et al.* 2017). Thus, some $\mathcal{O}$-operators and Rota–Baxter operators on 3-Lie algebras can serve as integrability conditions.

PROPOSITION 8.7.– Let E be an almost product structure on a 3-Lie algebra $(\mathfrak{g}, [\cdot,\cdot,\cdot]_{\mathfrak{g}})$. Then E is a strong abelian structure on $\mathfrak{g}$ if and only if E is an $\mathcal{O}$-operator associated with the adjoint representation $(\mathfrak{g}, \mathrm{ad})$. Furthermore, there exists a compatible 3-pre-Lie algebra $(\mathfrak{g}, \{\cdot,\cdot,\cdot\})$ on the 3-Lie algebra $(\mathfrak{g}, [\cdot,\cdot,\cdot]_{\mathfrak{g}})$, here the 3-pre-Lie algebra structure on $\mathfrak{g}$ is given by

$$\{x,y,z\} = E[x,y,Ez]_{\mathfrak{g}}, \quad \forall x,y,z \in \mathfrak{g}. \qquad [8.29]$$

PROOF.– By [8.28], for all $x,y,z \in \mathfrak{g}$ we have

$$\begin{aligned}[Ex,Ey,Ez]_{\mathfrak{g}} &= E[E^2x,Ey,Ez]_{\mathfrak{g}} + E[Ex,E^2y,Ez]_{\mathfrak{g}} + E[Ex,Ey,E^2z]_{\mathfrak{g}} \\ &= E(\mathrm{ad}_{Ex,Ey}z + \mathrm{ad}_{Ey,Ez}x + \mathrm{ad}_{Ez,Ex}y).\end{aligned}$$

Thus, E is an $\mathcal{O}$-operator associated with the adjoint representation $(\mathfrak{g}, \mathrm{ad})$.

Conversely, if for all $x,y,z \in \mathfrak{g}$, we have

$$\begin{aligned}[Ex,Ey,Ez]_{\mathfrak{g}} &= E(\mathrm{ad}_{Ex,Ey}z + \mathrm{ad}_{Ey,Ez}x + \mathrm{ad}_{Ez,Ex}y) \\ &= E([Ex,Ey,z]_{\mathfrak{g}} + [x,Ey,Ez]_{\mathfrak{g}} + [Ex,y,Ez]_{\mathfrak{g}}),\end{aligned}$$

then $[x,y,z]_{\mathfrak{g}} = E[x,y,Ez]_{\mathfrak{g}} + E[Ex,y,z]_{\mathfrak{g}} + E[x,Ey,z]_{\mathfrak{g}}$ by $E^{-1} = E$. Thus, E is a strong abelian structure on $\mathfrak{g}$.

Furthermore, by $E^{-1} = E$ and proposition 8.2, there exists a compatible 3-pre-Lie algebra on $\mathfrak{g}$ given by $\{x, y, z\} = E\mathrm{ad}_{x,y}E^{-1}(z) = E[x, y, Ez]_{\mathfrak{g}}$. The proof is completed. □

There is a new phenomenon that an involutive automorphism of a 3-Lie algebra also serves as an integrability condition.

LEMMA 8.6.– Let E be an almost product structure on a 3-Lie algebra $(\mathfrak{g}, [\cdot,\cdot,\cdot]_{\mathfrak{g}})$. If E satisfies the following equation:

$$E[x, y, z]_{\mathfrak{g}} = [Ex, Ey, Ez]_{\mathfrak{g}}, \tag{8.30}$$

then E is a product structure on $\mathfrak{g}$ such that $[\mathfrak{g}_+, \mathfrak{g}_+, \mathfrak{g}_-]_{\mathfrak{g}} \subset \mathfrak{g}_-, \quad [\mathfrak{g}_-, \mathfrak{g}_-, \mathfrak{g}_+]_{\mathfrak{g}} \subset \mathfrak{g}_+$.

PROOF.– By [8.30] and $E^2 = \mathrm{Id}$, we have

$$\begin{aligned}
&[Ex, Ey, Ez]_{\mathfrak{g}} + [Ex, y, z]_{\mathfrak{g}} + [x, Ey, z]_{\mathfrak{g}} + [x, y, Ez]_{\mathfrak{g}} \\
&-E[Ex, Ey, z]_{\mathfrak{g}} - E[x, Ey, Ez]_{\mathfrak{g}} - E[Ex, y, Ez]_{\mathfrak{g}} \\
=\ & E[x, y, z]_{\mathfrak{g}} + [Ex, y, z]_{\mathfrak{g}} + [x, Ey, z]_{\mathfrak{g}} + [x, y, Ez]_{\mathfrak{g}} \\
&-[E^2x, E^2y, Ez]_{\mathfrak{g}} - [Ex, E^2y, E^2z]_{\mathfrak{g}} - [E^2x, Ey, E^2z]_{\mathfrak{g}} \\
=\ & E[x, y, z]_{\mathfrak{g}}.
\end{aligned}$$

Thus, E is a product structure on $\mathfrak{g}$. Moreover, for all $x_1, x_2 \in \mathfrak{g}_+, \alpha_1 \in \mathfrak{g}_-$, we have

$$E[x_1, x_2, \alpha_1]_{\mathfrak{g}} = [Ex_1, Ex_2, E\alpha_1]_{\mathfrak{g}} = -[x_1, x_2, \alpha_1]_{\mathfrak{g}},$$

which implies that $[\mathfrak{g}_+, \mathfrak{g}_+, \mathfrak{g}_-]_{\mathfrak{g}} \subset \mathfrak{g}_-$. Similarly, we have $[\mathfrak{g}_-, \mathfrak{g}_-, \mathfrak{g}_+]_{\mathfrak{g}} \subset \mathfrak{g}_+$. □

DEFINITION 8.14.– ***(Integrability condition IV)*** *An almost product structure E on a 3-Lie algebra $(\mathfrak{g}, [\cdot,\cdot,\cdot]_{\mathfrak{g}})$ is called a* ***perfect product structure*** *if* [8.30] *holds.*

COROLLARY 8.5.– Let $(\mathfrak{g}, [\cdot,\cdot,\cdot]_{\mathfrak{g}})$ be a 3-Lie algebra. Then $(\mathfrak{g}, [\cdot,\cdot,\cdot]_{\mathfrak{g}})$ has a perfect product structure if and only if $\mathfrak{g}$ admits a decomposition: $\mathfrak{g} = \mathfrak{g}_+ \oplus \mathfrak{g}_-$, where $\mathfrak{g}_+$ and $\mathfrak{g}_-$ are subalgebras of $\mathfrak{g}$ such that $[\mathfrak{g}_+, \mathfrak{g}_+, \mathfrak{g}_-]_{\mathfrak{g}} \subset \mathfrak{g}_-$ and $[\mathfrak{g}_-, \mathfrak{g}_-, \mathfrak{g}_+]_{\mathfrak{g}} \subset \mathfrak{g}_+$.

COROLLARY 8.6.– A strict product structure on a 3-Lie algebra is a perfect product structure.

REMARK 8.4.– Let E be a product structure on a 3-Lie algebra $(\mathfrak{g}, [\cdot,\cdot,\cdot]_{\mathfrak{g}})$. By theorem 8.4, $\mathfrak{g}_+$ and $\mathfrak{g}_-$ are subalgebras. However, the brackets of mixed terms are very complicated. But a perfect product structure E on $(\mathfrak{g}, [\cdot,\cdot,\cdot]_{\mathfrak{g}})$ ensures $[\mathfrak{g}_+,\mathfrak{g}_+,\mathfrak{g}_-]_{\mathfrak{g}} \subset \mathfrak{g}_-$ and $[\mathfrak{g}_-,\mathfrak{g}_-,\mathfrak{g}_+]_{\mathfrak{g}} \subset \mathfrak{g}_+$. Note that this is exactly the condition required in the definition of a matched pair of 3-Lie algebras (Bai *et al.* 2019). Thus, E is a perfect product structure if and only if $(\mathfrak{g}_+, \mathfrak{g}_-)$ is a matched pair of 3-Lie algebras. This type of product structure is very important in our later studies.

DEFINITION 8.15.–

*i) A **paracomplex structure** on a 3-Lie algebra $(\mathfrak{g}, [\cdot,\cdot,\cdot]_{\mathfrak{g}})$ is a product structure E on $\mathfrak{g}$ such that the eigenspaces of $\mathfrak{g}$ associated with the eigenvalues ± 1 have the same dimension, i.e.* $\dim(\mathfrak{g}_+) = \dim(\mathfrak{g}_-)$.

*ii) A **perfect paracomplex structure** on a 3-Lie algebra $(\mathfrak{g}, [\cdot,\cdot,\cdot]_{\mathfrak{g}})$ is a perfect product structure E on $\mathfrak{g}$ such that the eigenspaces of $\mathfrak{g}$ associated with the eigenvalues ± 1 have the same dimension, i.e.* $\dim(\mathfrak{g}_+) = \dim(\mathfrak{g}_-)$.

PROPOSITION 8.8.– Let $(A, \{\cdot,\cdot,\cdot\})$ be a 3-pre-Lie algebra. Then, on the semidirect product 3-Lie algebra $A^c \ltimes_{L^*} A^*$, there is a perfect paracomplex structure $E : A^c \ltimes_{L^*} A^* \to A^c \ltimes_{L^*} A^*$ given by

$$E(x + \alpha) = x - \alpha, \quad \forall x \in A^c, \alpha \in A^*. \tag{8.31}$$

PROOF.– It is obvious that $E^2 = \mathrm{Id}$. Moreover, we have $(A^c \ltimes_{L^*} A^*)_+ = A$, $(A^c \ltimes_{L^*} A^*)_- = A^*$ and they are two subalgebras of the semidirect product 3-Lie algebra $A^c \ltimes_{L^*} A^*$. By theorem 8.4, E is a product structure on $A^c \ltimes_{L^*} A^*$. Since A and A^* have the same dimension, E is a paracomplex structure on $A^c \ltimes_{L^*} A^*$. It is obvious that E is perfect. The proof is completed. □

In the following, we give some examples of product structures.

EXAMPLE 8.4.– There is a unique non-trivial three-dimensional 3-Lie algebra. It has a basis $\{e_1, e_2, e_3\}$ with respect to which the non-zero product is given by

$$[e_1, e_2, e_3] = e_1.$$

Then $E = \begin{pmatrix} 1 & 0 & 0 \\ 0 & 1 & 0 \\ 0 & 0 & -1 \end{pmatrix}$ and $E = \begin{pmatrix} 1 & 0 & 0 \\ 0 & -1 & 0 \\ 0 & 0 & 1 \end{pmatrix}$ are strong abelian product structures and $E = \begin{pmatrix} -1 & 0 & 0 \\ 0 & 1 & 0 \\ 0 & 0 & 1 \end{pmatrix}$ is a perfect product structure.

EXAMPLE 8.5.– Consider the four-dimensional Euclidean 3-Lie algebra A_4 given in example 8.2. Then

$$E_1 = \begin{pmatrix} 1 & 0 & 0 & 0 \\ 0 & 1 & 0 & 0 \\ 0 & 0 & -1 & 0 \\ 0 & 0 & 0 & -1 \end{pmatrix}, \ E_2 = \begin{pmatrix} 1 & 0 & 0 & 0 \\ 0 & -1 & 0 & 0 \\ 0 & 0 & 1 & 0 \\ 0 & 0 & 0 & -1 \end{pmatrix}, \ E_3 = \begin{pmatrix} 1 & 0 & 0 & 0 \\ 0 & -1 & 0 & 0 \\ 0 & 0 & -1 & 0 \\ 0 & 0 & 0 & 1 \end{pmatrix},$$

$$E_4 = \begin{pmatrix} -1 & 0 & 0 & 0 \\ 0 & 1 & 0 & 0 \\ 0 & 0 & 1 & 0 \\ 0 & 0 & 0 & -1 \end{pmatrix}, \ E_5 = \begin{pmatrix} -1 & 0 & 0 & 0 \\ 0 & 1 & 0 & 0 \\ 0 & 0 & -1 & 0 \\ 0 & 0 & 0 & 1 \end{pmatrix}, E_6 = \begin{pmatrix} -1 & 0 & 0 & 0 \\ 0 & -1 & 0 & 0 \\ 0 & 0 & 1 & 0 \\ 0 & 0 & 0 & 1 \end{pmatrix}$$

are perfect and abelian product structures.

8.6. Complex structures on 3-Lie algebras

We introduce the notion of a complex structure on a real 3-Lie algebra using the Nijenhuis condition as the integrability condition. Parallel to the case of product structures, we also find four special integrability conditions.

DEFINITION 8.16.– *Let* $(\mathfrak{g}, [\cdot,\cdot,\cdot]_{\mathfrak{g}})$ *be a real 3-Lie algebra. An* ***almost complex structure*** *on* $\mathfrak{g}$ *is a linear endomorphism* $J : \mathfrak{g} \to \mathfrak{g}$*, satisfying* $J^2 = -\mathrm{Id}$*. An almost complex structure is called a* ***complex*** *structure if the following integrability condition is satisfied:*

$$\begin{aligned} J[x,y,z]_{\mathfrak{g}} &= -[Jx,Jy,Jz]_{\mathfrak{g}} + [Jx,y,z]_{\mathfrak{g}} + [x,Jy,z]_{\mathfrak{g}} + [x,y,Jz]_{\mathfrak{g}} \\ &\quad + J[Jx,Jy,z]_{\mathfrak{g}} + J[x,Jy,Jz]_{\mathfrak{g}} + J[Jx,y,Jz]_{\mathfrak{g}}. \end{aligned} \quad [8.32]$$

REMARK 8.5.– One can understand a complex structure on a 3-Lie algebra as a Nijenhuis operator J on a 3-Lie algebra, satisfying $J^2 = -\mathrm{Id}$.

REMARK 8.6.– One can also use definition 8.16 to define the notion of a complex structure on a complex 3-Lie algebra, considering J to be $\mathbb{C}$-linear. However, this is not very interesting since, for a complex 3-Lie algebra, there is a one-to-one correspondence between such $\mathbb{C}$-linear complex structures and product structures (see proposition 8.12).

Consider $\mathfrak{g}_{\mathbb{C}} = \mathfrak{g} \otimes_{\mathbb{R}} \mathbb{C} \cong \{x + iy | x, y \in \mathfrak{g}\}$, the complexification of the real 3-Lie algebra $\mathfrak{g}$, which turns out to be a complex 3-Lie algebra by extending the 3-Lie bracket on $\mathfrak{g}$ complex trilinearly, and we denote it by $(\mathfrak{g}_{\mathbb{C}}, [\cdot,\cdot,\cdot]_{\mathfrak{g}_{\mathbb{C}}})$. We have an equivalent description of the integrability condition given in definition 8.16. We denote by σ the conjugation in $\mathfrak{g}_{\mathbb{C}}$ with respect to the real form $\mathfrak{g}$, that is, $\sigma(x+iy) = x - iy$, $x, y \in \mathfrak{g}$. Then, σ is a complex antilinear, involutive automorphism of the complex vector space $\mathfrak{g}_{\mathbb{C}}$.

THEOREM 8.5.– Let $(\mathfrak{g}, [\cdot,\cdot,\cdot]_{\mathfrak{g}})$ be a real 3-Lie algebra. Then $\mathfrak{g}$ has a complex structure if and only if $\mathfrak{g}_{\mathbb{C}}$ admits a decomposition:

$$\mathfrak{g}_{\mathbb{C}} = \mathfrak{q} \oplus \mathfrak{p}, \qquad [8.33]$$

where $\mathfrak{q}$ and $\mathfrak{p} = \sigma(\mathfrak{q})$ are complex subalgebras of $\mathfrak{g}_{\mathbb{C}}$.

PROOF.– We extend the complex structure J complex linearly, which is denoted by $J_{\mathbb{C}}$, i.e. $J_{\mathbb{C}} : \mathfrak{g}_{\mathbb{C}} \longrightarrow \mathfrak{g}_{\mathbb{C}}$ is defined by

$$J_{\mathbb{C}}(x+iy) = Jx + iJy, \quad \forall x, y \in \mathfrak{g}. \qquad [8.34]$$

Then $J_{\mathbb{C}}$ is a complex linear endomorphism on $\mathfrak{g}_{\mathbb{C}}$ satisfying $J_{\mathbb{C}}^2 = -\mathrm{Id}$ and the integrability condition [8.32] on $\mathfrak{g}_{\mathbb{C}}$. Denote by $\mathfrak{g}_{\pm i}$ the corresponding eigenspaces of $\mathfrak{g}_{\mathbb{C}}$ associated with the eigenvalues $\pm i$ and holds:

$$\mathfrak{g}_{\mathbb{C}} = \mathfrak{g}_i \oplus \mathfrak{g}_{-i}.$$

It is straightforward to see that $\mathfrak{g}_i = \{x - iJx | x \in \mathfrak{g}\}$ and $\mathfrak{g}_{-i} = \{x + iJx | x \in \mathfrak{g}\}$. Therefore, we have $\mathfrak{g}_{-i} = \sigma(\mathfrak{g}_i)$.

For all $X, Y, Z \in \mathfrak{g}_i$, we have

$$\begin{aligned} J_{\mathbb{C}}[X,Y,Z]_{\mathfrak{g}_{\mathbb{C}}} &= -[J_{\mathbb{C}}X, J_{\mathbb{C}}Y, J_{\mathbb{C}}Z]_{\mathfrak{g}_{\mathbb{C}}} + [J_{\mathbb{C}}X, Y, Z]_{\mathfrak{g}_{\mathbb{C}}} + [X, J_{\mathbb{C}}Y, Z]_{\mathfrak{g}_{\mathbb{C}}} + [X, Y, J_{\mathbb{C}}Z]_{\mathfrak{g}_{\mathbb{C}}} \\ &\quad + J_{\mathbb{C}}[J_{\mathbb{C}}X, J_{\mathbb{C}}Y, Z]_{\mathfrak{g}_{\mathbb{C}}} + J_{\mathbb{C}}[X, J_{\mathbb{C}}Y, J_{\mathbb{C}}Z]_{\mathfrak{g}_{\mathbb{C}}} + J_{\mathbb{C}}[J_{\mathbb{C}}X, Y, J_{\mathbb{C}}Z]_{\mathfrak{g}_{\mathbb{C}}} \\ &= 4i[X,Y,Z]_{\mathfrak{g}_{\mathbb{C}}} - 3J_{\mathbb{C}}[X,Y,Z]_{\mathfrak{g}_{\mathbb{C}}}. \end{aligned}$$

Thus, we have $[X,Y,Z]_{\mathfrak{g}_{\mathbb{C}}} \in \mathfrak{g}_i$, which implies that $\mathfrak{g}_i$ is a subalgebra. Similarly, we can show that $\mathfrak{g}_{-i}$ is also a subalgebra.

Conversely, we define a complex linear endomorphism $J_{\mathbb{C}} : \mathfrak{g}_{\mathbb{C}} \to \mathfrak{g}_{\mathbb{C}}$ by

$$J_{\mathbb{C}}(X + \sigma(Y)) = iX - i\sigma(Y), \quad \forall X, Y \in \mathfrak{q}. \qquad [8.35]$$

Since σ is a complex antilinear, involutive automorphism of $\mathfrak{g}_{\mathbb{C}}$, we have

$$\begin{aligned} J_{\mathbb{C}}^2(X + \sigma(Y)) &= J_{\mathbb{C}}(iX - i\sigma(Y)) = J_{\mathbb{C}}(iX + \sigma(iY)) \\ &= i(iX) - i\sigma(iY) = -X - \sigma(Y), \end{aligned}$$

i.e. $J_{\mathbb{C}}^2 = -\mathrm{Id}$. Since $\mathfrak{q}$ is a subalgebra of $\mathfrak{g}_{\mathbb{C}}$, for all $X, Y, Z \in \mathfrak{q}$, we have

$$\begin{aligned}
&- [J_{\mathbb{C}}X, J_{\mathbb{C}}Y, J_{\mathbb{C}}Z]_{\mathfrak{g}_{\mathbb{C}}} + [J_{\mathbb{C}}X, Y, Z]_{\mathfrak{g}_{\mathbb{C}}} + [X, J_{\mathbb{C}}Y, Z]_{\mathfrak{g}_{\mathbb{C}}} + [X, Y, J_{\mathbb{C}}Z]_{\mathfrak{g}_{\mathbb{C}}}\\
&+ J_{\mathbb{C}}[J_{\mathbb{C}}X, J_{\mathbb{C}}Y, Z]_{\mathfrak{g}_{\mathbb{C}}} + J_{\mathbb{C}}[X, J_{\mathbb{C}}Y, J_{\mathbb{C}}Z]_{\mathfrak{g}_{\mathbb{C}}} + J_{\mathbb{C}}[J_{\mathbb{C}}X, Y, J_{\mathbb{C}}Z]_{\mathfrak{g}_{\mathbb{C}}}\\
&= 4i[X, Y, Z]_{\mathfrak{g}_{\mathbb{C}}} - 3J_{\mathbb{C}}[X, Y, Z]_{\mathfrak{g}_{\mathbb{C}}} = i[X, Y, Z]_{\mathfrak{g}_{\mathbb{C}}}\\
&= J_{\mathbb{C}}[X, Y, Z]_{\mathfrak{g}_{\mathbb{C}}},
\end{aligned}$$

which implies that $J_{\mathbb{C}}$ satisfies [8.32] for all $X, Y, Z \in \mathfrak{q}$. Similarly, we can show that $J_{\mathbb{C}}$ satisfies [8.32] for all $\mathcal{X}, \mathcal{Y}, \mathcal{Z} \in \mathfrak{g}_{\mathbb{C}}$. Since $\mathfrak{g}_{\mathbb{C}} = \mathfrak{q} \oplus \mathfrak{p}$, we can write $\mathcal{X} \in \mathfrak{g}_{\mathbb{C}}$ as $\mathcal{X} = X + \sigma(Y)$, for some $X, Y \in \mathfrak{q}$. Since σ is a complex antilinear, involutive automorphism of $\mathfrak{g}_{\mathbb{C}}$, we have

$$\begin{aligned}
&(J_{\mathbb{C}} \circ \sigma)(X + \sigma(Y)) = J_{\mathbb{C}}(Y + \sigma(X)) = iY - i\sigma(X) = \sigma(iX - i\sigma(Y))\\
&= (\sigma \circ J_{\mathbb{C}})(X + \sigma(Y)),
\end{aligned}$$

which implies that $J_{\mathbb{C}} \circ \sigma = \sigma \circ J_{\mathbb{C}}$. Moreover, since $\sigma(\mathcal{X}) = \mathcal{X}$ is equivalent to $\mathcal{X} \in \mathfrak{g}$, we deduce that the set of fixed points of σ is the real vector space $\mathfrak{g}$. By $J_{\mathbb{C}} \circ \sigma = \sigma \circ J_{\mathbb{C}}$, there is a well-defined $J \in \mathfrak{gl}(\mathfrak{g})$ given by

$$J \triangleq J_{\mathbb{C}}|_{\mathfrak{g}}.$$

By knowing that $J_{\mathbb{C}}$ satisfies [8.32] and $J_{\mathbb{C}}^2 = -\mathrm{Id}$ on $\mathfrak{g}_{\mathbb{C}}$, J is a complex structure on $\mathfrak{g}$. The proof is completed. □

LEMMA 8.7.– Let J be an almost complex structure on a real 3-Lie algebra $(\mathfrak{g}, [\cdot, \cdot, \cdot]_{\mathfrak{g}})$. If J satisfies

$$J[x, y, z]_{\mathfrak{g}} = [Jx, y, z]_{\mathfrak{g}}, \quad \forall x, y, z \in \mathfrak{g}, \qquad [8.36]$$

then J is a complex structure on $(\mathfrak{g}, [\cdot, \cdot, \cdot]_{\mathfrak{g}})$.

PROOF.– By [8.36] and $J^2 = -\mathrm{Id}$, we have

$$\begin{aligned}
&-[Jx, Jy, Jz]_{\mathfrak{g}} + [Jx, y, z]_{\mathfrak{g}} + [x, Jy, z]_{\mathfrak{g}} + [x, y, Jz]_{\mathfrak{g}}\\
&+J[Jx, Jy, z]_{\mathfrak{g}} + J[x, Jy, Jz]_{\mathfrak{g}} + J[Jx, y, Jz]_{\mathfrak{g}}\\
&= -[Jx, Jy, Jz]_{\mathfrak{g}} + J[x, y, z]_{\mathfrak{g}} + [x, Jy, z]_{\mathfrak{g}} + [x, y, Jz]_{\mathfrak{g}}\\
&+[J^2x, Jy, z]_{\mathfrak{g}} + [Jx, Jy, Jz]_{\mathfrak{g}} + [J^2x, y, Jz]_{\mathfrak{g}}\\
&= J[x, y, z]_{\mathfrak{g}}.
\end{aligned}$$

Thus, we obtain that J is a complex structure on $\mathfrak{g}$. □

DEFINITION 8.17.– **(Integrability condition I)** *An almost complex structure J on a real 3-Lie algebra $(\mathfrak{g}, [\cdot,\cdot,\cdot]_{\mathfrak{g}})$ is called a* **strict complex structure** *if* [8.36] *holds.*

COROLLARY 8.7.– Let $(\mathfrak{g}, [\cdot,\cdot,\cdot]_{\mathfrak{g}})$ be a real 3-Lie algebra. Then there is a strict complex structure on $(\mathfrak{g}, [\cdot,\cdot,\cdot]_{\mathfrak{g}})$ if and only if $\mathfrak{g}_{\mathbb{C}}$ admits a decomposition: $\mathfrak{g}_{\mathbb{C}} = \mathfrak{q} \oplus \mathfrak{p}$, where $\mathfrak{q}$ and $\mathfrak{p} = \sigma(\mathfrak{q})$ are complex subalgebras of $\mathfrak{g}_{\mathbb{C}}$ such that $[\mathfrak{q}, \mathfrak{q}, \mathfrak{p}]_{\mathfrak{g}_{\mathbb{C}}} = 0$ and $[\mathfrak{p}, \mathfrak{p}, \mathfrak{q}]_{\mathfrak{g}_{\mathbb{C}}} = 0$, i.e. $\mathfrak{g}_{\mathbb{C}}$ is a 3-Lie algebra direct sum of $\mathfrak{q}$ and $\mathfrak{p}$.

PROOF.– Let J be a strict complex structure on a real 3-Lie algebra $(\mathfrak{g}, [\cdot,\cdot,\cdot]_{\mathfrak{g}})$. Then $J_{\mathbb{C}}$ is a strict complex structure on the complex 3-Lie algebra $(\mathfrak{g}_{\mathbb{C}}, [\cdot,\cdot,\cdot]_{\mathfrak{g}_{\mathbb{C}}})$. For all $X, Y \in \mathfrak{g}_i$ and $\sigma(Z) \in \mathfrak{g}_{-i}$, on one hand we have

$$J_{\mathbb{C}}[X, Y, \sigma(Z)]_{\mathfrak{g}_{\mathbb{C}}} = [J_{\mathbb{C}}X, Y, \sigma(Z)]_{\mathfrak{g}_{\mathbb{C}}} = i[X, Y, \sigma(Z)]_{\mathfrak{g}_{\mathbb{C}}}.$$

On the other hand, we have

$$J_{\mathbb{C}}[X, Y, \sigma(Z)]_{\mathfrak{g}_{\mathbb{C}}} = J_{\mathbb{C}}[\sigma(Z), X, Y]_{\mathfrak{g}_{\mathbb{C}}} = [J_{\mathbb{C}}\sigma(Z), X, Y]_{\mathfrak{g}_{\mathbb{C}}} = -i[\sigma(Z), X, Y]_{\mathfrak{g}_{\mathbb{C}}}.$$

Thus, we obtain $[\mathfrak{g}_i, \mathfrak{g}_i, \mathfrak{g}_{-i}]_{\mathfrak{g}_{\mathbb{C}}} = 0$. Similarly, we can show $[\mathfrak{g}_{-i}, \mathfrak{g}_{-i}, \mathfrak{g}_i]_{\mathfrak{g}_{\mathbb{C}}} = 0$.

Conversely, define a complex linear endomorphism $J_{\mathbb{C}} : \mathfrak{g}_{\mathbb{C}} \to \mathfrak{g}_{\mathbb{C}}$ by [8.35]. Then it is straightforward to deduce that $J_{\mathbb{C}}^2 = -\mathrm{Id}$. Since $\mathfrak{q}$ is a subalgebra of $\mathfrak{g}_{\mathbb{C}}$, for all $X, Y, Z \in \mathfrak{q}$, we have

$$J_{\mathbb{C}}[X, Y, Z]_{\mathfrak{g}_{\mathbb{C}}} = i[X, Y, Z]_{\mathfrak{g}_{\mathbb{C}}} = [J_{\mathbb{C}}X, Y, Z]_{\mathfrak{g}_{\mathbb{C}}},$$

which implies that $J_{\mathbb{C}}$ satisfies [8.36] for all $X, Y, Z \in \mathfrak{q}$. Similarly, we can show that $J_{\mathbb{C}}$ satisfies [8.36] for all $\mathcal{X}, \mathcal{Y}, \mathcal{Z} \in \mathfrak{g}_{\mathbb{C}}$. By the proof of theorem 8.5, we obtain that $J \triangleq J_{\mathbb{C}}|_{\mathfrak{g}}$ is a strict complex structure on the real 3-Lie algebra $(\mathfrak{g}, [\cdot,\cdot,\cdot]_{\mathfrak{g}})$. The proof is completed. □

Let J be an almost complex structure on a real 3-Lie algebra $(\mathfrak{g}, [\cdot,\cdot,\cdot]_{\mathfrak{g}})$. We can define a complex vector space structure on the real vector space $\mathfrak{g}$ by

$$(a + bi)x \triangleq ax + bJx, \quad \forall a, b \in \mathbb{R}, x \in \mathfrak{g}. \qquad [8.37]$$

We define two maps $\varphi : \mathfrak{g} \to \mathfrak{g}_i$ and $\psi : \mathfrak{g} \to \mathfrak{g}_{-i}$ as follows:

$$\varphi(x) = \frac{1}{2}(x - iJx),$$

$$\psi(x) = \frac{1}{2}(x + iJx).$$

It is straightforward to deduce that φ is a complex linear isomorphism and $\psi = \sigma \circ \varphi$ is a complex antilinear isomorphism between complex vector spaces.

Let J be a strict complex structure on a real 3-Lie algebra $(\mathfrak{g}, [\cdot, \cdot, \cdot]_{\mathfrak{g}})$. Then with the complex vector space structure defined above, $(\mathfrak{g}, [\cdot, \cdot, \cdot]_{\mathfrak{g}})$ is a complex 3-Lie algebra. In fact, the fact that the 3-Lie bracket is complex trilinear follows from

$$\begin{aligned}[(a+bi)x, y, z]_{\mathfrak{g}} &= [ax + bJx, y, z]_{\mathfrak{g}} = a[x, y, z]_{\mathfrak{g}} + b[Jx, y, z]_{\mathfrak{g}} \\ &= a[x, y, z]_{\mathfrak{g}} + bJ[x, y, z]_{\mathfrak{g}} = (a+bi)[x, y, z]_{\mathfrak{g}}\end{aligned}$$

using [8.36] and [8.37].

Let J be a complex structure on $\mathfrak{g}$. Define a new bracket $[\cdot, \cdot, \cdot]_J : \wedge^3\mathfrak{g} \to \mathfrak{g}$ by

$$[x, y, z]_J \triangleq \frac{1}{4}([x, y, z]_{\mathfrak{g}} - [x, Jy, Jz]_{\mathfrak{g}} - [Jx, y, Jz]_{\mathfrak{g}} - [Jx, Jy, z]_{\mathfrak{g}}). \quad [8.38]$$

PROPOSITION 8.9.– Let J be a complex structure on a real 3-Lie algebra $(\mathfrak{g}, [\cdot, \cdot, \cdot]_{\mathfrak{g}})$. Then $(\mathfrak{g}, [\cdot, \cdot, \cdot]_J)$ is a real 3-Lie algebra. Moreover, J is a strict complex structure on $(\mathfrak{g}, [\cdot, \cdot, \cdot]_J)$ and the corresponding complex 3-Lie algebra $(\mathfrak{g}, [\cdot, \cdot, \cdot]_J)$ is isomorphic to the complex 3-Lie algebra $\mathfrak{g}_i$.

PROOF.– One can show that $(\mathfrak{g}, [\cdot, \cdot, \cdot]_J)$ is a real 3-Lie algebra. Here, we use a different approach to prove this result. By [8.32], for all $x, y, z \in \mathfrak{g}$, we have

$$\begin{aligned}&[\varphi(x), \varphi(y), \varphi(z)]_{\mathfrak{g}_{\mathbb{C}}} = \frac{1}{8}[x - iJx, y - iJy, z - iJz]_{\mathfrak{g}_{\mathbb{C}}} \\ &= \frac{1}{8}([x, y, z]_{\mathfrak{g}} - [x, Jy, Jz]_{\mathfrak{g}} - [Jx, y, Jz]_{\mathfrak{g}} - [Jx, Jy, z]_{\mathfrak{g}}) \\ &\quad - \frac{1}{8}i([x, y, Jz]_{\mathfrak{g}} + [x, Jy, z]_{\mathfrak{g}} + [Jx, y, z]_{\mathfrak{g}} - [Jx, Jy, Jz]_{\mathfrak{g}}) \\ &= \frac{1}{8}([x, y, z]_{\mathfrak{g}} - [x, Jy, Jz]_{\mathfrak{g}} - [Jx, y, Jz]_{\mathfrak{g}} - [Jx, Jy, z]_{\mathfrak{g}}) \\ &\quad - \frac{1}{8}iJ([x, y, z]_{\mathfrak{g}} - [x, Jy, Jz]_{\mathfrak{g}} - [Jx, y, Jz]_{\mathfrak{g}} - [Jx, Jy, z]_{\mathfrak{g}}) \\ &= \varphi[x, y, z]_J. \qquad [8.39]\end{aligned}$$

Thus, we have $[x, y, z]_J = \varphi^{-1}[\varphi(x), \varphi(y), \varphi(z)]_{\mathfrak{g}_{\mathbb{C}}}$. Since J is a complex structure, $\mathfrak{g}_i$ is a 3-Lie subalgebra. Therefore, $(\mathfrak{g}, [\cdot, \cdot, \cdot]_J)$ is a real 3-Lie algebra.

By [8.32], for all $x, y, z \in \mathfrak{g}$, we have

$$\begin{aligned} J[x,y,z]_J &= \frac{1}{4}J([x,y,z]_\mathfrak{g} - [x,Jy,Jz]_\mathfrak{g} - [Jx,y,Jz]_\mathfrak{g} - [Jx,Jy,z]_\mathfrak{g}) \\ &= \frac{1}{4}(-[Jx,Jy,Jz]_\mathfrak{g} + [Jx,y,z]_\mathfrak{g} + [x,Jy,z]_\mathfrak{g} + [x,y,Jz]_\mathfrak{g}) \\ &= [Jx,y,z]_J, \end{aligned}$$

which implies that J is a strict complex structure on $(\mathfrak{g}, [\cdot,\cdot,\cdot]_J)$. By [8.39], φ is a complex 3-Lie algebra isomorphism. The proof is completed. □

PROPOSITION 8.10.– Let J be a complex structure on a real 3-Lie algebra $(\mathfrak{g}, [\cdot,\cdot,\cdot]_\mathfrak{g})$. Then J is a strict complex structure on $(\mathfrak{g}, [\cdot,\cdot,\cdot]_\mathfrak{g})$ if and only if $[\cdot,\cdot,\cdot]_J = [\cdot,\cdot,\cdot]_\mathfrak{g}$.

PROOF.– If J is a strict complex structure on $(\mathfrak{g}, [\cdot,\cdot,\cdot]_\mathfrak{g})$, by $J[x,y,z]_\mathfrak{g} = [Jx,y,z]_\mathfrak{g}$, we have

$$[x,y,z]_J = \frac{1}{4}([x,y,z]_\mathfrak{g} - [x,Jy,Jz]_\mathfrak{g} - [Jx,y,Jz]_\mathfrak{g} - [Jx,Jy,z]_\mathfrak{g}) = [x,y,z]_\mathfrak{g}.$$

Conversely, if $[\cdot,\cdot,\cdot]_J = [\cdot,\cdot,\cdot]_\mathfrak{g}$, we have

$$-3[x,y,z]_\mathfrak{g} = [x,Jy,Jz]_\mathfrak{g} + [Jx,y,Jz]_\mathfrak{g} + [Jx,Jy,z]_\mathfrak{g}.$$

Then by the integrability condition of J, we obtain

$$\begin{aligned} 4J[x,y,z]_J &= -[Jx,Jy,Jz]_\mathfrak{g} + [Jx,y,z]_\mathfrak{g} + [x,Jy,z]_\mathfrak{g} + [x,y,Jz]_\mathfrak{g} \\ &= 3[Jx,y,z]_\mathfrak{g} + [Jx,y,z]_\mathfrak{g} \\ &= 4[Jx,y,z]_\mathfrak{g}, \end{aligned}$$

which implies that $J[x,y,z]_\mathfrak{g} = [Jx,y,z]_\mathfrak{g}$. The proof is completed. □

LEMMA 8.8.– Let J be an almost complex structure on a real 3-Lie algebra $(\mathfrak{g}, [\cdot,\cdot,\cdot]_\mathfrak{g})$. If J satisfies the following equation:

$$[x,y,z]_\mathfrak{g} = [x,Jy,Jz]_\mathfrak{g} + [Jx,y,Jz]_\mathfrak{g} + [Jx,Jy,z]_\mathfrak{g}, \quad [8.40]$$

then J is a complex structure on $\mathfrak{g}$.

PROOF.– By [8.40] and $J^2 = -\mathrm{Id}$, we have

$$\begin{aligned}&-[Jx, Jy, Jz]_{\mathfrak{g}} + [Jx, y, z]_{\mathfrak{g}} + [x, Jy, z]_{\mathfrak{g}} + [x, y, Jz]_{\mathfrak{g}}\\&+J[Jx, Jy, z]_{\mathfrak{g}} + J[x, Jy, Jz]_{\mathfrak{g}} + J[Jx, y, Jz]_{\mathfrak{g}}\\=&-[Jx, J^2y, J^2z]_{\mathfrak{g}} - [J^2x, Jy, J^2z]_{\mathfrak{g}} - [J^2x, J^2y, Jz]_{\mathfrak{g}}\\&+[Jx, y, z]_{\mathfrak{g}} + [x, Jy, z]_{\mathfrak{g}} + [x, y, Jz]_{\mathfrak{g}} + J[x, y, z]_{\mathfrak{g}}\\=&\ J[x, y, z]_{\mathfrak{g}}.\end{aligned}$$

Thus, we obtain that J is a complex structure on $\mathfrak{g}$. □

DEFINITION 8.18.– ***(Integrability condition II)*** *An almost complex structure J on a real 3-Lie algebra $(\mathfrak{g}, [\cdot,\cdot,\cdot]_{\mathfrak{g}})$ is called an* ***abelian complex structure*** *if* [8.40] *holds.*

REMARK 8.7.– Let J be an abelian complex structure on a real 3-Lie algebra $(\mathfrak{g}, [\cdot,\cdot,\cdot]_{\mathfrak{g}})$. Then $(\mathfrak{g}, [\cdot,\cdot,\cdot]_J)$ is an abelian 3-Lie algebra.

COROLLARY 8.8.– Let $(\mathfrak{g}, [\cdot,\cdot,\cdot]_{\mathfrak{g}})$ be a real 3-Lie algebra. Then $\mathfrak{g}$ has an abelian complex structure if and only if $\mathfrak{g}_{\mathbb{C}}$ admits a decomposition: $\mathfrak{g}_{\mathbb{C}} = \mathfrak{q} \oplus \mathfrak{p}$, where $\mathfrak{q}$ and $\mathfrak{p} = \sigma(\mathfrak{q})$ are complex abelian subalgebras of $\mathfrak{g}_{\mathbb{C}}$.

PROOF.– Let J be an abelian complex structure on $\mathfrak{g}$. By proposition 8.9, we obtain that φ is a complex 3-Lie algebra isomorphism from $(\mathfrak{g}, [\cdot,\cdot,\cdot]_J)$ to $(\mathfrak{g}_i, [\cdot,\cdot,\cdot]_{\mathfrak{g}_{\mathbb{C}}})$. Since J is abelian, $(\mathfrak{g}, [\cdot,\cdot,\cdot]_J)$ is an abelian 3-Lie algebra. Therefore, $\mathfrak{q} = \mathfrak{g}_i$ is an abelian subalgebra of $\mathfrak{g}_{\mathbb{C}}$. Since $\mathfrak{p} = \mathfrak{g}_{-i} = \sigma(\mathfrak{g}_i)$, for all $x_1+iy_1, x_2+iy_2, x_3+iy_3 \in \mathfrak{g}_i$, we have

$$\begin{aligned}&[\sigma(x_1 + iy_1), \sigma(x_2 + iy_2), \sigma(x_3 + iy_3)]_{\mathfrak{g}_{\mathbb{C}}}\\=&[x_1 - iy_1, x_2 - iy_2, x_3 - iy_3]_{\mathfrak{g}_{\mathbb{C}}}\\=&([x_1, x_2, x_3]_{\mathfrak{g}} - [x_1, y_2, y_3]_{\mathfrak{g}} - [y_1, x_2, y_3]_{\mathfrak{g}} - [y_1, y_2, x_3]_{\mathfrak{g}})\\&-i([x_1, x_2, y_3]_{\mathfrak{g}} + [x_1, y_2, x_3]_{\mathfrak{g}} + [y_1, x_2, x_3]_{\mathfrak{g}} - [y_1, y_2, y_3]_{\mathfrak{g}})\\=&\ \sigma[x_1 + iy_1, x_2 + iy_2, x_3 + iy_3]_{\mathfrak{g}_{\mathbb{C}}}\\=&\ 0.\end{aligned}$$

Thus, $\mathfrak{p}$ is an abelian subalgebra of $\mathfrak{g}_{\mathbb{C}}$.

Conversely, by theorem 8.5, there is a complex structure J on $\mathfrak{g}$. Moreover, by proposition 8.9, we have a complex 3-Lie algebra isomorphism φ from $(\mathfrak{g}, [\cdot,\cdot,\cdot]_J)$ to $(\mathfrak{q}, [\cdot,\cdot,\cdot]_{\mathfrak{g}_{\mathbb{C}}})$. Thus, $(\mathfrak{g}, [\cdot,\cdot,\cdot]_J)$ is an abelian 3-Lie algebra. By the definition of $[\cdot,\cdot,\cdot]_J$, we obtain that J is an abelian complex structure on $\mathfrak{g}$. The proof is completed. □

LEMMA 8.9.– Let J be an almost complex structure on a real 3-Lie algebra $(\mathfrak{g}, [\cdot,\cdot,\cdot]_{\mathfrak{g}})$. If J satisfies the following equation:

$$[x,y,z]_{\mathfrak{g}} = -J[Jx,y,z]_{\mathfrak{g}} - J[x,Jy,z]_{\mathfrak{g}} - J[x,y,Jz]_{\mathfrak{g}}, \qquad [8.41]$$

then J is a complex structure on $\mathfrak{g}$.

PROOF.– By [8.41] and $J^2 = -\mathrm{Id}$, we have

$$\begin{aligned}
&-[Jx,Jy,Jz]_{\mathfrak{g}} + [Jx,y,z]_{\mathfrak{g}} + [x,Jy,z]_{\mathfrak{g}} + [x,y,Jz]_{\mathfrak{g}} \\
&+J[Jx,Jy,z]_{\mathfrak{g}} + J[x,Jy,Jz]_{\mathfrak{g}} + J[Jx,y,Jz]_{\mathfrak{g}} \\
=\ & J[J^2x,Jy,Jz]_{\mathfrak{g}} + J[Jx,J^2y,Jz]_{\mathfrak{g}} + J[Jx,Jy,J^2z]_{\mathfrak{g}} + J[x,y,z]_{\mathfrak{g}} \\
&+J[Jx,Jy,z]_{\mathfrak{g}} + J[x,Jy,Jz]_{\mathfrak{g}} + J[Jx,y,Jz]_{\mathfrak{g}} \\
=\ & J[x,y,z]_{\mathfrak{g}}.
\end{aligned}$$

Thus, J is a complex structure on $\mathfrak{g}$. □

DEFINITION 8.19.– ***(Integrability condition III)*** *An almost complex structure J on a real 3-Lie algebra $(\mathfrak{g}, [\cdot,\cdot,\cdot]_{\mathfrak{g}})$ is called a* ***strong abelian complex structure*** *if* [8.41] *holds.*

COROLLARY 8.9.– Let $(\mathfrak{g}, [\cdot,\cdot,\cdot]_{\mathfrak{g}})$ be a real 3-Lie algebra. Then $\mathfrak{g}$ has a strong abelian complex structure if and only if $\mathfrak{g}_{\mathbb{C}}$ admits a decomposition: $\mathfrak{g}_{\mathbb{C}} = \mathfrak{q} \oplus \mathfrak{p}$, where $\mathfrak{q}$ and $\mathfrak{p} = \sigma(\mathfrak{q})$ are abelian complex subalgebras of $\mathfrak{g}_{\mathbb{C}}$ such that $[\mathfrak{q},\mathfrak{q},\mathfrak{p}]_{\mathfrak{g}_{\mathbb{C}}} \subset \mathfrak{q}$ and $[\mathfrak{p},\mathfrak{p},\mathfrak{q}]_{\mathfrak{g}_{\mathbb{C}}} \subset \mathfrak{p}$.

Parallel to the case of strong abelian product structures on a 3-Lie algebra, strong abelian complex structures on a 3-Lie algebra are also $\mathcal{O}$-operators associated with the adjoint representation.

PROPOSITION 8.11.– Let J be an almost complex structure on a real 3-Lie algebra $(\mathfrak{g}, [\cdot,\cdot,\cdot]_{\mathfrak{g}})$. Then J is a strong abelian complex structure on a 3-Lie algebra $(\mathfrak{g}, [\cdot,\cdot,\cdot]_{\mathfrak{g}})$ if and only if $-J$ is an $\mathcal{O}$-operator on $(\mathfrak{g}, [\cdot,\cdot,\cdot]_{\mathfrak{g}})$ associated with the adjoint representation $(\mathfrak{g}, \mathrm{ad})$. Furthermore, there exists a compatible 3-pre-Lie algebra $(\mathfrak{g}, \{\cdot,\cdot,\cdot\})$ on the 3-Lie algebra $(\mathfrak{g}, [\cdot,\cdot,\cdot]_{\mathfrak{g}})$; here the 3-pre-Lie algebra structure on $\mathfrak{g}$ is given by

$$\{x,y,z\} = -J[x,y,Jz]_{\mathfrak{g}}, \quad \forall x,y,z \in \mathfrak{g}. \qquad [8.42]$$

PROOF.– By [8.41], for all $x, y, z \in \mathfrak{g}$ we have

$$\begin{aligned}[-Jx, -Jy, -Jz]_\mathfrak{g} &= J[J^2x, Jy, Jz]_\mathfrak{g} + J[Jx, J^2y, Jz]_\mathfrak{g} + J[Jx, Jy, J^2z]_\mathfrak{g} \\ &= -J(\mathrm{ad}_{-Jx,-Jy}z + \mathrm{ad}_{-Jy,-Jz}x + \mathrm{ad}_{-Jz,-Jx}y).\end{aligned}$$

Thus, $-J$ is an $\mathcal{O}$-operator associated with the adjoint representation $(\mathfrak{g}, \mathrm{ad})$.

Conversely, if for all $x, y, z \in \mathfrak{g}$, we have

$$\begin{aligned}[-Jx, -Jy, -Jz]_\mathfrak{g} &= -J(\mathrm{ad}_{-Jx,-Jy}z + \mathrm{ad}_{-Jy,-Jz}x + \mathrm{ad}_{-Jz,-Jx}y) \\ &= -J([-Jx, -Jy, z]_\mathfrak{g} + [x, -Jy, -Jz]_\mathfrak{g} + [-Jx, y, -Jz]_\mathfrak{g}),\end{aligned}$$

then we obtain $[x, y, z]_\mathfrak{g} = -J[x, y, Jz]_\mathfrak{g} - J[Jx, y, z]_\mathfrak{g} - J[x, Jy, z]_\mathfrak{g}$ by $(-J)^{-1} = J$.

Furthermore, by $(-J)^{-1} = J$ and proposition 8.2, there exists a compatible 3-pre-Lie algebra on $\mathfrak{g}$ given by $\{x, y, z\} = -J\mathrm{ad}_{x,y}(-J^{-1}(z)) = -J[x, y, Jz]_\mathfrak{g}$. The proof is completed. □

LEMMA 8.10.– Let J be an almost complex structure on a real 3-Lie algebra $(\mathfrak{g}, [\cdot,\cdot,\cdot]_\mathfrak{g})$. If J satisfies the following equation:

$$J[x, y, z]_\mathfrak{g} = -[Jx, Jy, Jz]_\mathfrak{g}, \qquad [8.43]$$

then J is a complex structure on $\mathfrak{g}$.

PROOF.– By [8.43] and $J^2 = -\mathrm{Id}$, we have

$$\begin{aligned}& -[Jx, Jy, Jz]_\mathfrak{g} + [Jx, y, z]_\mathfrak{g} + [x, Jy, z]_\mathfrak{g} + [x, y, Jz]_\mathfrak{g} \\ & +J[Jx, Jy, z]_\mathfrak{g} + J[x, Jy, Jz]_\mathfrak{g} + J[Jx, y, Jz]_\mathfrak{g} \\ = {} & J[x, y, z]_\mathfrak{g} + [Jx, y, z]_\mathfrak{g} + [x, Jy, z]_\mathfrak{g} + [x, y, Jz]_\mathfrak{g} \\ & -[J^2x, J^2y, Jz]_\mathfrak{g} - [Jx, J^2y, J^2z]_\mathfrak{g} - [J^2x, Jy, J^2z]_\mathfrak{g} \\ = {} & J[x, y, z]_\mathfrak{g}.\end{aligned}$$

Thus, J is a complex structure on $\mathfrak{g}$. □

DEFINITION 8.20.– ***(Integrability condition IV)*** *An almost complex structure J on a real 3-Lie algebra $(\mathfrak{g}, [\cdot,\cdot,\cdot]_\mathfrak{g})$ is called a* ***perfect complex structure*** *if* [8.43] *holds.*

COROLLARY 8.10.– Let $(\mathfrak{g}, [\cdot,\cdot,\cdot]_{\mathfrak{g}})$ be a real 3-Lie algebra. Then $\mathfrak{g}$ has a perfect complex structure if and only if $\mathfrak{g}_{\mathbb{C}}$ admits a decomposition: $\mathfrak{g}_{\mathbb{C}} = \mathfrak{q} \oplus \mathfrak{p}$, where $\mathfrak{q}$ and $\mathfrak{p} = \sigma(\mathfrak{q})$ are complex subalgebras of $\mathfrak{g}_{\mathbb{C}}$ such that $[\mathfrak{q},\mathfrak{q},\mathfrak{p}]_{\mathfrak{g}_{\mathbb{C}}} \subset \mathfrak{p}$ and $[\mathfrak{p},\mathfrak{p},\mathfrak{q}]_{\mathfrak{g}_{\mathbb{C}}} \subset \mathfrak{q}$.

COROLLARY 8.11.– Let J be a strict complex structure on a real 3-Lie algebra $(\mathfrak{g}, [\cdot,\cdot,\cdot]_{\mathfrak{g}})$. Then J is a perfect complex structure on $\mathfrak{g}$.

EXAMPLE 8.6.– Consider the four-dimensional Euclidean 3-Lie algebra A_4 given in example 8.2. Then

$$J_1 = \begin{pmatrix} 0 & 0 & -1 & 0 \\ 0 & 0 & 0 & -1 \\ 1 & 0 & 0 & 0 \\ 0 & 1 & 0 & 0 \end{pmatrix}, \quad J_2 = \begin{pmatrix} 0 & -1 & 0 & 0 \\ 1 & 0 & 0 & 0 \\ 0 & 0 & 0 & -1 \\ 0 & 0 & 1 & 0 \end{pmatrix}, \quad J_3 = \begin{pmatrix} 0 & -1 & 0 & 0 \\ 1 & 0 & 0 & 0 \\ 0 & 0 & 0 & 1 \\ 0 & 0 & -1 & 0 \end{pmatrix},$$

$$J_4 = \begin{pmatrix} 0 & 1 & 0 & 0 \\ -1 & 0 & 0 & 0 \\ 0 & 0 & 0 & -1 \\ 0 & 0 & 1 & 0 \end{pmatrix}, \quad J_5 = \begin{pmatrix} 0 & 1 & 0 & 0 \\ -1 & 0 & 0 & 0 \\ 0 & 0 & 0 & 1 \\ 0 & 0 & -1 & 0 \end{pmatrix}, \quad J_6 = \begin{pmatrix} 0 & 0 & 1 & 0 \\ 0 & 0 & 0 & 1 \\ -1 & 0 & 0 & 0 \\ 0 & -1 & 0 & 0 \end{pmatrix}$$

are abelian complex structures. Moreover, J_1, J_6 are strong abelian complex structures and J_2, J_3, J_4, J_5 are perfect complex structures.

8.7. Complex product structures on 3-Lie algebras

We add a compatibility condition between a complex structure and a product structure on a 3-Lie algebra to introduce the notion of a complex product structure. We construct complex product structures using 3-pre-Lie algebras. First, we illustrate the relation between a complex structure and a product structure on a complex 3-Lie algebra.

PROPOSITION 8.12.– Let $(\mathfrak{g}, [\cdot,\cdot,\cdot]_{\mathfrak{g}})$ be a complex 3-Lie algebra. Then E is a product structure on $\mathfrak{g}$ if and only if $J = iE$ is a complex structure on $\mathfrak{g}$.

PROOF.– Let E be a product structure on $\mathfrak{g}$. We have $J^2 = i^2E^2 = -\mathrm{Id}$. Thus, J is an almost complex structure on $\mathfrak{g}$. Since E satisfies the integrability condition [8.24], we have

$$\begin{aligned} J[x,y,z]_{\mathfrak{g}} &= iE[x,y,z]_{\mathfrak{g}} \\ &= -[iEx, iEy, iEz]_{\mathfrak{g}} + [iEx, y, z]_{\mathfrak{g}} + [x, iEy, z]_{\mathfrak{g}} + [x, y, iEz]_{\mathfrak{g}} \\ &\quad + iE[iEx, iEy, z]_{\mathfrak{g}} + iE[x, iEy, iEz]_{\mathfrak{g}} + iE[iEx, y, iEz]_{\mathfrak{g}}. \end{aligned}$$

Thus, J is a complex structure on the complex 3-Lie algebra $\mathfrak{g}$.

The converse part can be proved similarly and we omit details. □

COROLLARY 8.12.– Let J be a complex structure on a real 3-Lie algebra $(\mathfrak{g}, [\cdot,\cdot,\cdot]_{\mathfrak{g}})$. Then, $-iJ_{\mathbb{C}}$ is a paracomplex structure on the complex 3-Lie algebra $(\mathfrak{g}_{\mathbb{C}}, [\cdot,\cdot,\cdot]_{\mathfrak{g}_{\mathbb{C}}})$, where $J_{\mathbb{C}}$ is defined by [8.34].

PROOF.– By theorem 8.5, $\mathfrak{g}_{\mathbb{C}} = \mathfrak{g}_i \oplus \mathfrak{g}_{-i}$ and $\mathfrak{g}_{-i} = \sigma(\mathfrak{g}_i)$, where $\mathfrak{g}_i$ and $\mathfrak{g}_{-i}$ are subalgebras of $\mathfrak{g}_{\mathbb{C}}$. It is obvious that $\dim(\mathfrak{g}_i) = \dim(\mathfrak{g}_{-i})$. By proposition 8.4, there is a paracomplex structure on $\mathfrak{g}_{\mathbb{C}}$. On the other hand, it is obvious that $J_{\mathbb{C}}$ is a complex structure on $\mathfrak{g}_{\mathbb{C}}$. By proposition 8.12, $-iJ_{\mathbb{C}}$ a product structure on the complex 3-Lie algebra $(\mathfrak{g}_{\mathbb{C}}, [\cdot,\cdot,\cdot]_{\mathfrak{g}_{\mathbb{C}}})$. It is straightforward to see that $\mathfrak{g}_i$ and $\mathfrak{g}_{-i}$ are eigenspaces of $-iJ_{\mathbb{C}}$ corresponding to $+1$ and -1. Thus, $-iJ_{\mathbb{C}}$ is a paracomplex structure. □

DEFINITION 8.21.– *Let $(\mathfrak{g}, [\cdot,\cdot,\cdot]_{\mathfrak{g}})$ be a real 3-Lie algebra. A* ***complex product*** *structure on the 3-Lie algebra $\mathfrak{g}$ is a pair $\{J, E\}$ of a complex structure J and a product structure E satisfying*

$$J \circ E = -E \circ J. \qquad [8.44]$$

If E is perfect, we call $\{J, E\}$ a ***perfect complex product*** *structure on $\mathfrak{g}$.*

REMARK 8.8.– Let $\{J, E\}$ be a complex product structure on a real 3-Lie algebra $(\mathfrak{g}, [\cdot,\cdot,\cdot]_{\mathfrak{g}})$. For all $x \in \mathfrak{g}_+$, by [8.44], we have $E(Jx) = -Jx$, which implies that $J(\mathfrak{g}_+) \subset \mathfrak{g}_-$. Analogously, we obtain $J(\mathfrak{g}_-) \subset \mathfrak{g}_+$. Thus, we get $J(\mathfrak{g}_-) = \mathfrak{g}_+$ and $J(\mathfrak{g}_+) = \mathfrak{g}_-$. Therefore, $\dim(\mathfrak{g}_+) = \dim(\mathfrak{g}_-)$ and E is a paracomplex structure on $\mathfrak{g}$.

THEOREM 8.6.– Let $(\mathfrak{g}, [\cdot,\cdot,\cdot]_{\mathfrak{g}})$ be a real 3-Lie algebra $(\mathfrak{g}, [\cdot,\cdot,\cdot]_{\mathfrak{g}})$. Then the following statements are equivalent:

i) $\mathfrak{g}$ has a complex product structure;

ii) $\mathfrak{g}$ has a complex structure J and can be decomposed as $\mathfrak{g} = \mathfrak{g}_+ \oplus \mathfrak{g}_-$, where $\mathfrak{g}_+, \mathfrak{g}_-$ are 3-Lie subalgebras of $\mathfrak{g}$ and $\mathfrak{g}_- = J\mathfrak{g}_+$.

PROOF.– Let $\{J, E\}$ be a complex product structure and let $\mathfrak{g}_\pm$ denote the eigenspaces corresponding to the eigenvalues ± 1 of E. By theorem 8.4, both $\mathfrak{g}_+$ and $\mathfrak{g}_-$ are 3-Lie subalgebras of $\mathfrak{g}$ and $J \circ E = -E \circ J$ implies $\mathfrak{g}_- = J\mathfrak{g}_+$.

Conversely, we can define a linear map $E : \mathfrak{g} \to \mathfrak{g}$ by

$$E(x + \alpha) = x - \alpha, \quad \forall x \in \mathfrak{g}_+, \alpha \in \mathfrak{g}_-.$$

By theorem 8.4, E is a product structure on $\mathfrak{g}$. By $\mathfrak{g}_- = J\mathfrak{g}_+$ and $J^2 = -\mathrm{Id}$, we have

$$E(J(x + \alpha)) = E(J(x) + J(\alpha)) = -J(x) + J(\alpha) = -J(E(x + \alpha)).$$

Thus, $\{J, E\}$ is a complex product structure on $\mathfrak{g}$. The proof is completed. □

EXAMPLE 8.7.– Consider the product structures and the complex structures on the four-dimensional Euclidean 3-Lie algebra A_4 given in examples 8.5 and 8.6, respectively. Then $\{J_i, E_i\}$ for $i = 1, 2, 3, 4, 5, 6$ are complex product structures on A_4.

We give a characterization of a perfect complex product structure on a 3-Lie algebra.

PROPOSITION 8.13.– Let E be a perfect paracomplex structure on a real 3-Lie algebra $(\mathfrak{g}, [\cdot,\cdot,\cdot]_{\mathfrak{g}})$. Then there is a perfect complex product structure $\{J, E\}$ on $\mathfrak{g}$ if and only if there exists a linear isomorphism $\phi : \mathfrak{g}_+ \to \mathfrak{g}_-$ satisfying the following equation:

$$\begin{aligned}&\phi[x, y, z]_{\mathfrak{g}}\\&= -[\phi(x), \phi(y), \phi(z)]_{\mathfrak{g}} + [\phi(x), y, z]_{\mathfrak{g}} + [x, \phi(y), z]_{\mathfrak{g}} + [x, y, \phi(z)]_{\mathfrak{g}}\\&\quad +\phi[\phi(x), \phi(y), z]_{\mathfrak{g}} + \phi[x, \phi(y), \phi(z)]_{\mathfrak{g}} + \phi[\phi(x), y, \phi(z)]_{\mathfrak{g}}.\end{aligned} \qquad [8.45]$$

PROOF.– Let $\{J, E\}$ be a perfect complex product structure on $\mathfrak{g}$. Define a linear isomorphism $\phi : \mathfrak{g}_+ \to \mathfrak{g}_-$ by $\phi \triangleq J|_{\mathfrak{g}_+} : \mathfrak{g}_+ \to \mathfrak{g}_-$. By the compatibility condition [8.32] that the complex structure J satisfies and the coherence condition [8.30] that a perfect product structure E satisfies, we deduce that [8.45] holds.

Conversely, we define an endomorphism J of $\mathfrak{g}$ by

$$J(x + \alpha) = -\phi^{-1}(\alpha) + \phi(x), \quad \forall x \in \mathfrak{g}_+, \alpha \in \mathfrak{g}_-. \qquad [8.46]$$

It is obvious that J is an almost complex structure on $\mathfrak{g}$ and $J \circ E = -E \circ J$. For all $\alpha, \beta, \gamma \in \mathfrak{g}_-$, let $x, y, z \in \mathfrak{g}_+$ such that $\phi(x) = \alpha, \phi(y) = \beta$ and $\phi(z) = \gamma$. By [8.45] and [8.30], we have

$$\begin{aligned}&-[J\alpha, J\beta, J\gamma]_{\mathfrak{g}} + [J\alpha, \beta, \gamma]_{\mathfrak{g}} + [\alpha, J\beta, \gamma]_{\mathfrak{g}} + [\alpha, \beta, J\gamma]_{\mathfrak{g}}\\&+J[J\alpha, J\beta, \gamma]_{\mathfrak{g}} + J[\alpha, J\beta, J\gamma]_{\mathfrak{g}} + J[J\alpha, \beta, J\gamma]_{\mathfrak{g}}\\&= [x, y, z]_{\mathfrak{g}} - [x, \phi(y), \phi(z)]_{\mathfrak{g}} - [\phi(x), y, \phi(z)]_{\mathfrak{g}} - [\phi(x), \phi(y), z]_{\mathfrak{g}}\\&\quad -\phi^{-1}[x, y, \phi(z)]_{\mathfrak{g}} - \phi^{-1}[\phi(x), y, z]_{\mathfrak{g}} - \phi^{-1}[x, \phi(y), z]_{\mathfrak{g}}\\&= -\phi^{-1}[\phi(x), \phi(y), \phi(z)]_{\mathfrak{g}}\\&= J[\alpha, \beta, \gamma]_{\mathfrak{g}},\end{aligned}$$

which implies that [8.32] holds for all $\alpha, \beta, \gamma \in \mathfrak{g}_-$. Similarly, we can deduce that [8.32] holds for all the other cases. Thus, J is a complex structure and $\{J, E\}$ is a perfect complex product structure on the 3-Lie algebra $\mathfrak{g}$. □

In the following, we construct a perfect complex product structure using 3-pre-Lie algebras.

A non-degenerate symmetric bilinear form $\mathcal{B} \in A^* \otimes A^*$ on a real 3-pre-Lie algebra $(A, \{\cdot,\cdot,\cdot\})$ is called **invariant** if

$$\mathcal{B}(\{x,y,z\},w) = -\mathcal{B}(z,\{x,y,w\}), \quad \forall x,y,z,w \in A. \qquad [8.47]$$

Then $\mathcal{B}$ induces a linear isomorphism $\mathcal{B}^\sharp : A \to A^*$ by

$$\langle \mathcal{B}^\sharp(x), y\rangle = \mathcal{B}(x,y), \quad \forall x,y \in A. \qquad [8.48]$$

PROPOSITION 8.14.– Let $(A, \{\cdot,\cdot,\cdot\})$ be a real 3-pre-Lie algebra with a non-degenerate symmetric bilinear from $\mathcal{B}$. Then there is a perfect complex product structure $\{J, E\}$ on the semidirect product 3-Lie algebra $A^c \ltimes_{L^*} A^*$, where E is given by [8.31] and the complex structure J is given as follows:

$$J(x+\alpha) = -{\mathcal{B}^\sharp}^{-1}(\alpha) + \mathcal{B}^\sharp(x), \quad \forall x \in A, \alpha \in A^*. \qquad [8.49]$$

PROOF.– By proposition 8.8, E is a perfect product structure on $A^c \ltimes_{L^*} A^*$. For all $x,y,z \in A^c$, we have

$$\begin{aligned}
&- [\mathcal{B}^\sharp(x), \mathcal{B}^\sharp(y), \mathcal{B}^\sharp(z)]_{L^*} + [\mathcal{B}^\sharp(x), y, z]_{L^*} + [x, \mathcal{B}^\sharp(y), z]_{L^*} + [x, y, \mathcal{B}^\sharp(z)]_{L^*} \\
&+ \mathcal{B}^\sharp[\mathcal{B}^\sharp(x), \mathcal{B}^\sharp(y), z]_{L^*} + \mathcal{B}^\sharp[x, \mathcal{B}^\sharp(y), \mathcal{B}^\sharp(z)]_{L^*} + \mathcal{B}^\sharp[\mathcal{B}^\sharp(x), y, \mathcal{B}^\sharp(z)]_{L^*} \\
&= [\mathcal{B}^\sharp(x), y, z]_{L^*} + [x, \mathcal{B}^\sharp(y), z]_{L^*} + [x, y, \mathcal{B}^\sharp(z)]_{L^*} \\
&= L^*(x,y)\mathcal{B}^\sharp(z) + L^*(y,z)\mathcal{B}^\sharp(x) + L^*(z,x)\mathcal{B}^\sharp(y).
\end{aligned}$$

By [8.47], we have

$$\begin{aligned}
\langle \mathcal{B}^\sharp[x,y,z]_C, w\rangle &= \langle \mathcal{B}^\sharp\{x,y,z\}, w\rangle + \langle \mathcal{B}^\sharp\{y,z,x\}, w\rangle + \langle \mathcal{B}^\sharp\{z,x,y\}, w\rangle \\
&= \mathcal{B}(\{x,y,z\},w) + \mathcal{B}(\{y,z,x\},w) + \mathcal{B}(\{z,x,y\},w) \\
&= -\mathcal{B}(z,\{x,y,w\}) - \mathcal{B}(x,\{y,z,w\}) - \mathcal{B}(y,\{z,x,w\}) \\
&= -\langle \mathcal{B}^\sharp(z), \{x,y,w\}\rangle - \langle \mathcal{B}^\sharp(x), \{y,z,w\}\rangle - \langle \mathcal{B}^\sharp(y), \{z,x,w\}\rangle \\
&= \langle L^*(x,y)\mathcal{B}^\sharp(z), w\rangle + \langle L^*(y,z)\mathcal{B}^\sharp(x), w\rangle + \langle L^*(z,x)\mathcal{B}^\sharp(y), w\rangle,
\end{aligned}$$

which implies that

$$\mathcal{B}^\sharp[x,y,z]_C = L^*(x,y)\mathcal{B}^\sharp(z) + L^*(y,z)\mathcal{B}^\sharp(x) + L^*(z,x)\mathcal{B}^\sharp(y).$$

Thus, we have

$$\begin{aligned}\mathcal{B}^{\sharp}[x,y,z]_C = &-[\mathcal{B}^{\sharp}(x),\mathcal{B}^{\sharp}(y),\mathcal{B}^{\sharp}(z)]_{L^*} + [\mathcal{B}^{\sharp}(x),y,z]_{L^*} + [x,\mathcal{B}^{\sharp}(y),z]_{L^*}\\ &+ [x,y,\mathcal{B}^{\sharp}(z)]_{L^*} + \mathcal{B}^{\sharp}[\mathcal{B}^{\sharp}(x),\mathcal{B}^{\sharp}(y),z]_{L^*}\\ &+ \mathcal{B}^{\sharp}[x,\mathcal{B}^{\sharp}(y),\mathcal{B}^{\sharp}(z)]_{L^*} + \mathcal{B}^{\sharp}[\mathcal{B}^{\sharp}(x),y,\mathcal{B}^{\sharp}(z)]_{L^*}.\end{aligned}$$

By proposition 8.13, we obtain that $\{J, E\}$ is a perfect complex product structure on $A^c \ltimes_{L^*} A^*$. The proof is completed. □

Let $(A, \{\cdot,\cdot,\cdot\})$ be a real 3-pre-Lie algebra. On the real 3-Lie algebra $\mathrm{aff}(A) = A^c \ltimes_L A$, we consider two endomorphisms J and E given by

$$J(x,y) = (-y,x), \quad E(x,y) = (x,-y), \quad \forall x,y \in A. \qquad [8.50]$$

PROPOSITION 8.15.– With the above notations, $\{J, E\}$ is a perfect complex product structure on the 3-Lie algebra $\mathrm{aff}(A)$.

PROOF.– It is obvious that E is a perfect product structure on $\mathrm{aff}(A)$. Moreover, we have $J^2 = -\mathrm{Id}$ and $J \circ E = -E \circ J$. Obviously $\mathrm{aff}(A)_+ = \{(x,0)|x \in A\}$, $\mathrm{aff}(A)_- = \{(0,y)|y \in A\}$. Define $\phi : \mathrm{aff}(A)_+ \to \mathrm{aff}(A)_-$ by $\phi \triangleq J|_{\mathrm{aff}(A)_+} : \mathrm{aff}(A)_+ \to \mathrm{aff}(A)_-$. More precisely, $\phi(x,0) = (0,x)$. Then for all $(x,0),(y,0)$, $(z,0) \in \mathrm{aff}(A)_+$, we have

$$\begin{aligned}&-[\phi(x,0),\phi(y,0),\phi(z,0)]_L + [\phi(x,0),(y,0),(z,0)]_L + [(x,0),\phi(y,0),(z,0)]_L\\ &+ [(x,0),(y,0),\phi(z,0)]_L + \phi[\phi(x,0),\phi(y,0),(z,0)]_L + \phi[(x,0),\phi(y,0),\phi(z,0)]_L\\ &+ \phi[\phi(x,0),(y,0),\phi(z,0)]_L\\ &= [\phi(x,0),(y,0),(z,0)]_L + [(x,0),\phi(y,0),(z,0)]_L + [(x,0),(y,0),\phi(z,0)]_L\\ &= (0,\{y,z,x\}) + (0,\{z,x,y\}) + (0,\{x,y,z\})\\ &= \phi[(x,0),(y,0),(z,0)]_L.\end{aligned}$$

By proposition 8.13, $\{J, E\}$ is a perfect complex product structure on the 3-Lie algebra $\mathrm{aff}(A)$. □

8.8. Para-Kähler structures on 3-Lie algebras

We add a compatibility condition between a symplectic structure and a paracomplex structure on a 3-Lie algebra to introduce the notion of a para-Kähler structure on a 3-Lie algebra. A para-Kähler structure gives rise to a pseudo-Riemannian structure. We introduce the notion of a Levi-Civita product associated with a pseudo-Riemannian 3-Lie algebra and give its precise formulas using the decomposition of the original 3-Lie algebra.

DEFINITION 8.22.– *Let ω be a symplectic structure and E a paracomplex structure on a 3-Lie algebra $(\mathfrak{g}, [\cdot,\cdot,\cdot]_{\mathfrak{g}})$. The triple $(\mathfrak{g}, \omega, E)$ is called a* **para-Kähler** *3-Lie algebra if the following equality holds:*

$$\omega(Ex, Ey) = -\omega(x, y), \quad \forall x, y \in \mathfrak{g}. \qquad [8.51]$$

If E is perfect, we call $(\mathfrak{g}, \omega, E)$ a **perfect para-Kähler** *3-Lie algebra.*

PROPOSITION 8.16.– Let $(A, \{\cdot,\cdot,\cdot\})$ be a 3-pre-Lie algebra. Then $(A^c \ltimes_{L^*} A^*, \omega, E)$ is a perfect para-Kähler 3-Lie algebra, where ω is given by [8.16] and E is defined by [8.31].

PROOF.– By theorem 8.2, $(A^c \ltimes_{L^*} A^*, \omega)$ is a symplectic 3-Lie algebra. By proposition 8.8, E is a perfect paracomplex structure on the phase space T^*A^c. For all $x_1, x_2 \in A, \alpha_1, \alpha_2 \in A^*$, we have

$$\begin{aligned}\omega(E(x_1 + \alpha_1), E(x_2 + \alpha_2)) &= \omega(x_1 - \alpha_1, x_2 - \alpha_2) = \langle -\alpha_1, x_2\rangle - \langle -\alpha_2, x_1\rangle \\ &= -\omega(x_1 + \alpha_1, x_2 + \alpha_2).\end{aligned}$$

Therefore, $(T^*A^c = A^c \ltimes_{L^*} A^*, \omega, E)$ is a perfect para-Kähler 3-Lie algebra. □

Similar to the case of para-Kähler Lie algebras, we have the following equivalent description of a para-Kähler 3-Lie algebra.

THEOREM 8.7.– Let $(\mathfrak{g}, \omega)$ be a symplectic 3-Lie algebra. Then there exists a paracomplex structure E on the 3-Lie algebra $(\mathfrak{g}, [\cdot,\cdot,\cdot]_{\mathfrak{g}})$ such that $(\mathfrak{g}, \omega, E)$ is a para-Kähler 3-Lie algebra if and only if there exist two isotropic 3-Lie subalgebras $\mathfrak{g}_+$ and $\mathfrak{g}_-$ such that $\mathfrak{g} = \mathfrak{g}_+ \oplus \mathfrak{g}_-$ as the direct sum of vector spaces.

PROOF.– Let $(\mathfrak{g}, \omega, E)$ be a para-Kähler 3-Lie algebra. Since E is a paracomplex structure on $\mathfrak{g}$, we have $\mathfrak{g} = \mathfrak{g}_+ \oplus \mathfrak{g}_-$, where $\mathfrak{g}_+$ and $\mathfrak{g}_-$ are 3-Lie subalgebras of $\mathfrak{g}$. For all $x_1, x_2 \in \mathfrak{g}_+$, by [8.51], we have

$$\omega(Ex_1, Ex_2) = \omega(x_1, x_2) = -\omega(x_1, x_2),$$

which implies that $\omega(\mathfrak{g}_+, \mathfrak{g}_+) = 0$. Thus, $\mathfrak{g}_+$ is isotropic. Similarly, $\mathfrak{g}_-$ is also isotropic.

Conversely, since $\mathfrak{g}_+$ and $\mathfrak{g}_-$ are subalgebras, $\mathfrak{g} = \mathfrak{g}_+ \oplus \mathfrak{g}_-$ as vector spaces, there is a product structure E on $\mathfrak{g}$ defined by [8.25]. Moreover, since $\mathfrak{g} = \mathfrak{g}_+ \oplus \mathfrak{g}_-$ as vector spaces and both $\mathfrak{g}_+$ and $\mathfrak{g}_-$ are isotropic, we obtain that dim $\mathfrak{g}_+$=dim $\mathfrak{g}_-$. Thus, E is a paracomplex structure on $\mathfrak{g}$. For all $x_1, x_2 \in \mathfrak{g}_+, \alpha_1, \alpha_2 \in \mathfrak{g}_-$, since $\mathfrak{g}_+$ and $\mathfrak{g}_-$ are isotropic, we have

$$\begin{aligned}\omega(E(x_1+\alpha_1), E(x_2+\alpha_2)) &= \omega(x_1-\alpha_1, x_2-\alpha_2) = -\omega(x_1,\alpha_2) - \omega(\alpha_1, x_2)\\ &= -\omega(x_1+\alpha_1, x_2+\alpha_2).\end{aligned}$$

Thus, $(\mathfrak{g}, \omega, E)$ is a para-Kähler 3-Lie algebra. The proof is completed. □

EXAMPLE 8.8.– Consider the symplectic structures and the perfect paracomplex structures on the four-dimensional Euclidean 3-Lie algebra A_4 given in examples 8.2 and 8.5, respectively. Then $\{\omega_i, E_i\}$ for $i = 1,2,3,4,5,6$ are perfect para-Kähler structures on A_4.

EXAMPLE 8.9.– Let $(\mathfrak{h}, [\cdot,\cdot,\cdot]_{\mathfrak{h}}$ be a 3-Lie algebra and $(\mathfrak{h}\oplus\mathfrak{h}^*, \omega)$ its (perfect) phase space, where ω is given by [8.16]. Then $E : \mathfrak{h}\oplus\mathfrak{h}^* \longrightarrow \mathfrak{h}\oplus\mathfrak{h}^*$ defined by

$$E(x+\alpha) = x - \alpha, \quad \forall x \in \mathfrak{h}, \alpha \in \mathfrak{h}^*, \qquad [8.52]$$

is a (perfect) paracomplex structure and $(\mathfrak{h}\oplus\mathfrak{h}^*, \omega, E)$ is a (perfect) para-Kähler 3-Lie algebra.

Let $(\mathfrak{g}, \omega, E)$ be a para-Kähler 3-Lie algebra. Then it is obvious that $\mathfrak{g}_-$ is isomorphic to $\mathfrak{g}_+^*$ via the symplectic structure ω. Moreover, it is straightforward to deduce that

PROPOSITION 8.17.– Any para-Kähler 3-Lie algebra is isomorphic to the para-Kähler 3-Lie algebra associated with a phase space of a 3-Lie algebra.

In the sequel, we study the Levi-Civita product associated with a perfect para-Kähler 3-Lie algebra.

DEFINITION 8.23.– *A* ***pseudo-Riemannian 3-Lie algebra*** *is a 3-Lie algebra* $(\mathfrak{g}, [\cdot,\cdot,\cdot]_{\mathfrak{g}})$ *endowed with a non-degenerate symmetric bilinear form* S*. The associated Levi-Civita product is the product on* $\mathfrak{g}$, $\nabla : \otimes^3\mathfrak{g} \longrightarrow \mathfrak{g}$ *with* $(x,y,z) \longmapsto \nabla_{x,y}z$*, given by the following formula:*

$$\begin{aligned}&3S(\nabla_{x,y}z, w) \qquad [8.53]\\ &= S([x,y,z]_{\mathfrak{g}}, w) - 2S([x,y,w]_{\mathfrak{g}}, z) + S([y,z,w]_{\mathfrak{g}}, x) + S([z,x,w]_{\mathfrak{g}}, y).\end{aligned}$$

PROPOSITION 8.18.– Let $(\mathfrak{g}, S)$ be a pseudo-Riemannian 3-Lie algebra. Then the Levi-Civita product $\{\cdot,\cdot,\cdot\}$ satisfies the following equations:

$$\nabla_{x,y}z = -\nabla_{y,x}z, \qquad [8.54]$$

$$\nabla_{x,y}z + \nabla_{y,z}x + \nabla_{z,x}y = [x,y,z]_{\mathfrak{g}}. \qquad [8.55]$$

PROOF.– For all $w \in \mathfrak{g}$, it is obvious that

$$\begin{aligned} 3S(\nabla_{y,x}z, w) &= S([y,x,z]_{\mathfrak{g}}, w) - 2S([y,x,w]_{\mathfrak{g}}, z) \\ &\quad + S([x,z,w]_{\mathfrak{g}}, y) + S([z,y,w]_{\mathfrak{g}}, x) \\ &= -3S(\nabla_{x,y}z, w). \end{aligned}$$

By the non-degeneracy of S, we obtain $\nabla_{x,y}z = -\nabla_{y,x}z$.

For all $x, y, z, w \in \mathfrak{g}$, we have

$$3S(\nabla_{x,y}z, w) = S([x,y,z]_{\mathfrak{g}}, w) - 2S([x,y,w]_{\mathfrak{g}}, z) + S([y,z,w]_{\mathfrak{g}}, x) + S([z,x,w]_{\mathfrak{g}}, y),$$
$$3S(\nabla_{y,z}x, w) = S([y,z,x]_{\mathfrak{g}}, w) - 2S([y,z,w]_{\mathfrak{g}}, x) + S([z,x,w]_{\mathfrak{g}}, y) + S([x,y,w]_{\mathfrak{g}}, z),$$
$$3S(\nabla_{z,x}y, w) = S([z,x,y]_{\mathfrak{g}}, w) - 2S([z,x,w]_{\mathfrak{g}}, y) + S([x,y,w]_{\mathfrak{g}}, z) + S([y,z,w]_{\mathfrak{g}}, x).$$

By adding up the three equations, we have

$$3S(\nabla_{x,y}z + \nabla_{y,z}x + \nabla_{z,x}y, w) = 3S([x,y,z]_{\mathfrak{g}}, w),$$

which implies that $\nabla_{x,y}z + \nabla_{y,z}x + \nabla_{z,x}y = [x,y,z]_{\mathfrak{g}}$. The proof is completed. □

Let $(\mathfrak{g}, \omega, E)$ be a perfect para-Kähler 3-Lie algebra. Define a bilinear form S on $\mathfrak{g}$ by

$$S(x, y) \triangleq \omega(x, Ey), \quad \forall x, y \in \mathfrak{g}. \qquad [8.56]$$

PROPOSITION 8.19.– With the above notations, $(\mathfrak{g}, S)$ is a pseudo-Riemannian 3-Lie algebra. Moreover, the associated Levi-Civita product ∇ and the perfect paracomplex structure E satisfy the following compatibility condition:

$$E\nabla_{x,y}z = \nabla_{Ex,Ey}Ez. \qquad [8.57]$$

PROOF.– Since ω is skew-symmetric and $\omega(Ex, Ey) = -\omega(x, y)$, we have

$$S(y, x) = \omega(y, Ex) = -\omega(Ey, E^2x) = -\omega(Ey, x) = \omega(x, Ey) = S(x, y),$$

which implies that S is symmetric. Moreover, since ω is non-degenerate and $E^2 = \mathrm{Id}$, it is obvious that S is non-degenerate. Thus, S is a pseudo-Riemannian metric on the

3-Lie algebra $\mathfrak{g}$. Moreover, we have

$$\begin{aligned}
&3S(\nabla_{Ex,Ey}Ez, w)\\
&= S([Ex, Ey, Ez]_{\mathfrak{g}}, w) - 2S([Ex, Ey, w]_{\mathfrak{g}}, Ez) + S([Ey, Ez, w]_{\mathfrak{g}}, Ex)\\
&+ S([Ez, Ex, w]_{\mathfrak{g}}, Ey)\\
&= S(E[x, y, z]_{\mathfrak{g}}, w) - 2S(E[x, y, Ew]_{\mathfrak{g}}, Ez) + S(E[y, z, Ew]_{\mathfrak{g}}, Ex)\\
&+ S(E[z, x, Ew]_{\mathfrak{g}}, Ey)\\
&= -(S([x, y, z]_{\mathfrak{g}}, Ew) - 2S([x, y, Ew]_{\mathfrak{g}}, z) + S([y, z, Ew]_{\mathfrak{g}}, x)\\
&+ S([z, x, Ew]_{\mathfrak{g}}, y))\\
&= -3S(\nabla_{x,y}z, Ew)\\
&= 3S(E\nabla_{x,y}z, w).
\end{aligned}$$

Thus, we have $E\nabla_{x,y}z = \nabla_{Ex,Ey}Ez$. □

The following two propositions clarify the relationship between the Levi-Civita product and the 3-pre-Lie multiplication on a para-Kähler 3-Lie algebra.

PROPOSITION 8.20.– Let $(\mathfrak{g}, \omega, E)$ be a para-Kähler 3-Lie algebra and ∇ the associated Levi-Civita product. Then for all $x_1, x_2, x_3 \in \mathfrak{g}_+$ and $\alpha_1, \alpha_2, \alpha_3 \in \mathfrak{g}_-$, we have

$$\nabla_{x_1,x_2}x_3 = \{x_1, x_2, x_3\}, \quad \nabla_{\alpha_1,\alpha_2}\alpha_3 = \{\alpha_1, \alpha_2, \alpha_3\}.$$

PROOF.– Since $(\mathfrak{g}, \omega, E)$ is a para-Kähler 3-Lie algebra, 3-Lie subalgebras $\mathfrak{g}_+$ and $\mathfrak{g}_-$ are isotropic and $\mathfrak{g} = \mathfrak{g}_+ \oplus \mathfrak{g}_-$ as vector spaces. For all $x_1, x_2, x_3, x_4 \in \mathfrak{g}_+$, we have

$$\begin{aligned}
&3\omega(\nabla_{x_1,x_2}x_3, x_4)\\
&= 3S(\nabla_{x_1,x_2}x_3, Ex_4) = 3S(\nabla_{x_1,x_2}x_3, x_4)\\
&= S([x_1, x_2, x_3]_{\mathfrak{g}}, x_4) - 2S([x_1, x_2, x_4]_{\mathfrak{g}}, x_3) + S([x_2, x_3, x_4]_{\mathfrak{g}}, x_1)\\
&+ S([x_3, x_1, x_4]_{\mathfrak{g}}, x_2)\\
&= \omega([x_1, x_2, x_3]_{\mathfrak{g}}, x_4) - 2\omega([x_1, x_2, x_4]_{\mathfrak{g}}, x_3) + \omega([x_2, x_3, x_4]_{\mathfrak{g}}, x_1)\\
&+ \omega([x_3, x_1, x_4]_{\mathfrak{g}}, x_2)\\
&= 0.
\end{aligned}$$

By $(\mathfrak{g}_+)^\perp = \mathfrak{g}_+$, we obtain $\nabla_{x_1,x_2}x_3 \in \mathfrak{g}_+$. Similarly, for all $\alpha_1, \alpha_2, \alpha_3 \in \mathfrak{g}_-$, $\nabla_{\alpha_1,\alpha_2}\alpha_3 \in \mathfrak{g}_-$. Furthermore, for all $x_1, x_2, x_3 \in \mathfrak{g}_+$, and $\alpha \in \mathfrak{g}_-$, we have

$$\begin{aligned}
&3\omega(\nabla_{x_1,x_2}x_3, \alpha)\\
&= 3S(\nabla_{x_1,x_2}x_3, E\alpha) = -3S(\nabla_{x_1,x_2}x_3, \alpha)\\
&= -S([x_1,x_2,x_3]_\mathfrak{g}, \alpha) + 2S([x_1,x_2,\alpha]_\mathfrak{g}, x_3) - S([x_2,x_3,\alpha]_\mathfrak{g}, x_1)\\
&- S([x_3,x_1,\alpha]_\mathfrak{g}, x_2)\\
&= \omega([x_1,x_2,x_3]_\mathfrak{g}, \alpha) + 2\omega([x_1,x_2,\alpha]_\mathfrak{g}, x_3) - \omega([x_2,x_3,\alpha]_\mathfrak{g}, x_1)\\
&- \omega([x_3,x_1,\alpha]_\mathfrak{g}, x_2)\\
&= \omega([\alpha,x_1,x_2]_\mathfrak{g}, x_3) + 2\omega([x_1,x_2,\alpha]_\mathfrak{g}, x_3)\\
&= -3\omega(x_3, [x_1,x_2,\alpha]_\mathfrak{g})\\
&= 3\omega(\{x_1,x_2,x_3\}, \alpha).
\end{aligned}$$

Thus, $\nabla_{x_1,x_2}x_3 = \{x_1, x_2, x_3\}$. Similarly, we have $\nabla_{\alpha_1,\alpha_2}\alpha_3 = \{\alpha_1, \alpha_2, \alpha_3\}$. The proof is completed. □

PROPOSITION 8.21.– Let $(\mathfrak{g}, \omega, E)$ be a perfect para-Kähler 3-Lie algebra and ∇ the associated Levi-Civita product. Then for all $x_1, x_2 \in \mathfrak{g}_+$ and $\alpha_1, \alpha_2 \in \mathfrak{g}_-$, we have

$$\nabla_{x_1,x_2}\alpha_1 = \{x_1,x_2,\alpha_1\} + \frac{2}{3}(\{x_2,\alpha_1,x_1\} + \{\alpha_1,x_1,x_2\}), \quad [8.58]$$

$$\nabla_{\alpha_1,x_1}x_2 = -\frac{1}{3}\{\alpha_1,x_1,x_2\} + \frac{2}{3}\{x_2,\alpha_1,x_1\}, \quad [8.59]$$

$$\nabla_{\alpha_1,\alpha_2}x_1 = \{\alpha_1,\alpha_2,x_1\} + \frac{2}{3}(\{\alpha_2,x_1,\alpha_1\} + \{x_1,\alpha_1,\alpha_2\}), \quad [8.60]$$

$$\nabla_{x_1,\alpha_1}\alpha_2 = -\frac{1}{3}\{x_1,\alpha_1,\alpha_2\} + \frac{2}{3}\{\alpha_2,x_1,\alpha_1\}. \quad [8.61]$$

PROOF.– Since $(\mathfrak{g}, \omega, E)$ is a perfect para-Kähler 3-Lie algebra, 3-Lie subalgebras $\mathfrak{g}_+$ and $\mathfrak{g}_-$ are isotropic and $\mathfrak{g} = \mathfrak{g}_+ \oplus \mathfrak{g}_-$ as vector spaces. Thus, we have $S(\mathfrak{g}_+, \mathfrak{g}_+) = S(\mathfrak{g}_-, \mathfrak{g}_-) = 0$. For all $x_1, x_2 \in \mathfrak{g}_+$ and $\alpha_1, \alpha_2 \in \mathfrak{g}_-$, we have

$$\begin{aligned}
&3S(\nabla_{x_1,x_2}\alpha_1, \alpha_2)\\
&= S([x_1,x_2,\alpha_1]_\mathfrak{g}, \alpha_2) - 2S([x_1,x_2,\alpha_2]_\mathfrak{g}, \alpha_1) + S([x_2,\alpha_1,\alpha_2]_\mathfrak{g}, x_1)\\
&+ S([\alpha_1,x_1,\alpha_2]_\mathfrak{g}, x_2)\\
&= 0.
\end{aligned}$$

Since S is non-degenerate, we have $\nabla_{x_1,x_2}\alpha_1 \in \mathfrak{g}_-$. Moreover, For all $x_1, x_2, x_3 \in \mathfrak{g}_+$ and $\alpha_1 \in \mathfrak{g}_-$, we have

$$\begin{aligned}
&3\omega(\nabla_{x_1,x_2}\alpha_1, x_3)\\
&= 3S(\nabla_{x_1,x_2}\alpha_1, Ex_3) = 3S(\nabla_{x_1,x_2}\alpha_1, x_3)\\
&= S([x_1,x_2,\alpha_1]_{\mathfrak{g}}, x_3) - 2S([x_1,x_2,x_3]_{\mathfrak{g}}, \alpha_1) + S([x_2,\alpha_1,x_3]_{\mathfrak{g}}, x_1) + S([\alpha_1,x_1,x_3]_{\mathfrak{g}}, x_2)\\
&= \omega([x_1,x_2,\alpha_1]_{\mathfrak{g}}, Ex_3) - 2\omega([x_1,x_2,x_3]_{\mathfrak{g}}, E\alpha_1)\\
&\quad + \omega([x_2,\alpha_1,x_3]_{\mathfrak{g}}, Ex_1) + \omega([\alpha_1,x_1,x_3]_{\mathfrak{g}}, Ex_2)\\
&= \omega([x_1,x_2,\alpha_1]_{\mathfrak{g}}, x_3) + 2\omega([x_1,x_2,x_3]_{\mathfrak{g}}, \alpha_1) + \omega([x_2,\alpha_1,x_3]_{\mathfrak{g}}, x_1) + \omega([\alpha_1,x_1,x_3]_{\mathfrak{g}}, x_2)\\
&= \omega([x_1,x_2,\alpha_1]_{\mathfrak{g}}, x_3) + 2\omega(\{x_1,x_2,\alpha_1\}, x_3) + \omega(\{x_2,\alpha_1,x_1\}, x_3) + \omega(\{\alpha_1,x_1,x_2\}, x_3).
\end{aligned}$$

Thus, we obtain

$$\nabla_{x_1,x_2}\alpha_1 = \{x_1,x_2,\alpha_1\} + \frac{2}{3}(\{x_2,\alpha_1,x_1\} + \{\alpha_1,x_1,x_2\}),$$

which implies that [8.58] holds.

For all $x_1, x_2 \in \mathfrak{g}_+$ and $\alpha_1, \alpha_2 \in \mathfrak{g}_-$, we have

$$\begin{aligned}
&3S(\nabla_{\alpha_1,x_1}x_2, \alpha_2)\\
&= S([\alpha_1,x_1,x_2]_{\mathfrak{g}}, \alpha_2) - 2S([\alpha_1,x_1,\alpha_2]_{\mathfrak{g}}, x_2) + S([x_1,x_2,\alpha_2]_{\mathfrak{g}}, \alpha_1)\\
&+ S([x_2,\alpha_1,\alpha_2]_{\mathfrak{g}}, x_1)\\
&= 0.
\end{aligned}$$

Since S is non-degenerate, we have $\nabla_{\alpha_1,x_1}x_2 \in \mathfrak{g}_-$. Moreover, for all $x_1, x_2, x_3 \in \mathfrak{g}_+$ and $\alpha_1 \in \mathfrak{g}_-$, we have

$$\begin{aligned}
&3\omega(\nabla_{\alpha_1,x_1}x_2, x_3)\\
&= 3S(\nabla_{\alpha_1,x_1}x_2, Ex_3) = 3S(\nabla_{\alpha_1,x_1}x_2, x_3)\\
&= S([\alpha_1,x_1,x_2]_{\mathfrak{g}}, x_3) - 2S([\alpha_1,x_1,x_3]_{\mathfrak{g}}, x_2) + S([x_1,x_2,x_3]_{\mathfrak{g}}, \alpha_1) + S([x_2,\alpha_1,x_3]_{\mathfrak{g}}, x_1)\\
&= \omega([\alpha_1,x_1,x_2]_{\mathfrak{g}}, x_3) - 2\omega([\alpha_1,x_1,x_3]_{\mathfrak{g}}, x_2) - \omega([x_1,x_2,x_3]_{\mathfrak{g}}, \alpha_1) + \omega([x_2,\alpha_1,x_3]_{\mathfrak{g}}, x_1)\\
&= \omega([\alpha_1,x_1,x_2]_{\mathfrak{g}}, x_3) - 2\omega(\{\alpha_1,x_1,x_2\}, x_3) - \omega(\{x_1,x_2,\alpha_1\}, x_3) + \omega(\{x_2,\alpha_1,x_1\}, x_3).
\end{aligned}$$

Thus, we obtain

$$\nabla_{\alpha_1,x_1}x_2 = -\frac{1}{3}\{\alpha_1,x_1,x_2\} + \frac{2}{3}\{x_2,\alpha_1,x_1\},$$

which implies that [8.59] holds. Note that [8.60] and [8.61] can be proved similarly. We omit details. The proof is completed. □

Under the isomorphism given in proposition 8.17 and the correspondence given in theorem 8.3, using the formulas provided in proposition 8.6, we get the following corollary.

COROLLARY 8.13.– For the perfect para-Kähler 3-Lie algebra $(\mathfrak{h}\oplus\mathfrak{h}^*, \omega, E)$ given in example 8.9, for all $x_1, x_2 \in \mathfrak{h}$ and $\alpha_1, \alpha_2 \in \mathfrak{h}^*$, we have

$$\nabla_{x_1,x_2}\alpha_1 = (L^*(x_1,x_2) - \frac{1}{3}R^*(x_2,x_1) + \frac{1}{3}R^*(x_1,x_2))\alpha_1, \qquad [8.62]$$

$$\nabla_{\alpha_1,x_1}x_2 = (\frac{1}{3}R^*(x_1,x_2) + \frac{2}{3}R^*(x_2,x_1))\alpha_1, \qquad [8.63]$$

$$\nabla_{\alpha_1,\alpha_2}x_1 = (\mathcal{L}^*(\alpha_1,\alpha_2) - \frac{1}{3}\mathcal{R}^*(\alpha_2,\alpha_1) + \frac{1}{3}\mathcal{R}^*(\alpha_1,\alpha_2))x_1, \qquad [8.64]$$

$$\nabla_{x_1,\alpha_1}\alpha_2 = (\frac{1}{3}\mathcal{R}^*(\alpha_1,\alpha_2) + \frac{2}{3}\mathcal{R}^*(\alpha_2,\alpha_1))x_1. \qquad [8.65]$$

8.9. Pseudo-Kähler structures on 3-Lie algebras

We add a compatibility condition between a symplectic structure and a complex structure on a 3-Lie algebra to introduce the notion of a pseudo-Kähler structure on a 3-Lie algebra. The relation between para-Kähler structures and pseudo-Kähler structures on a 3-Lie algebra is investigated.

DEFINITION 8.24.– *Let ω be a symplectic structure and J a complex structure on a real 3-Lie algebra $(\mathfrak{g}, [\cdot,\cdot,\cdot]_{\mathfrak{g}})$. The triple $(\mathfrak{g}, \omega, J)$ is called a real* ***pseudo-Kähler*** *3-Lie algebra if*

$$\omega(Jx, Jy) = \omega(x, y), \quad \forall x, y \in \mathfrak{g}. \qquad [8.66]$$

EXAMPLE 8.10.– Consider the symplectic structures and the complex structures on the four-dimensional Euclidean 3-Lie algebra A_4 given in examples 8.2 and 8.6, respectively. Then $\{\omega_i, J_i\}$ for $i = 1, 2, 3, 4, 5, 6$ are pseudo-Kähler structures on A_4.

PROPOSITION 8.22.– Let $(\mathfrak{g}, \omega, J)$ be a real pseudo-Kähler 3-Lie algebra. Define a bilinear form S on $\mathfrak{g}$ by

$$S(x, y) \triangleq \omega(x, Jy), \quad \forall x, y \in \mathfrak{g}. \qquad [8.67]$$

Then $(\mathfrak{g}, S)$ is a pseudo-Riemannian 3-Lie algebra.

PROOF.– By [8.66], we have

$$S(y, x) = \omega(y, Jx) = \omega(Jy, J^2x) = -\omega(Jy, x) = \omega(x, Jy) = S(x, y),$$

which implies that S is symmetric. Moreover, since ω is non-degenerate and $J^2 = -\mathrm{Id}$, it is obvious that S is non-degenerate. Thus, S is a pseudo-Riemannian metric on the 3-Lie algebra $\mathfrak{g}$. □

DEFINITION 8.25.– *Let $(\mathfrak{g}, \omega, J)$ be a real pseudo-Kähler 3-Lie algebra. If the associated pseudo-Riemannian metric is positive definite, we call $(\mathfrak{g}, \omega, J)$ a real **Kähler** 3-Lie algebra.*

THEOREM 8.8.– Let $(\mathfrak{g}, \omega, E)$ be a complex para-Kähler 3-Lie algebra. Then $(\mathfrak{g}_\mathbb{R}, \omega_\mathbb{R}, J)$ is a real pseudo-Kähler 3-Lie algebra, where $\mathfrak{g}_\mathbb{R}$ is the underlying real 3-Lie algebra, $J = iE$ and $\omega_\mathbb{R} = \mathrm{Re}(\omega)$ is the real part of ω.

PROOF.– By proposition 8.12, $J = iE$ is a complex structure on the complex 3-Lie algebra $\mathfrak{g}$. Thus, J is also a complex structure on the real 3-Lie algebra $\mathfrak{g}_\mathbb{R}$. It is obvious that $\omega_\mathbb{R}$ is skew-symmetric. If for all $x \in \mathfrak{g}$, $\omega_\mathbb{R}(x, y) = 0$. Then we have

$$\omega(x, y) = \omega_\mathbb{R}(x, y) + i\omega_\mathbb{R}(-ix, y) = 0.$$

By the non-degeneracy of ω, we obtain $y = 0$. Thus, $\omega_\mathbb{R}$ is non-degenerate. Therefore, $\omega_\mathbb{R}$ is a symplectic structure on the real 3-Lie algebra $\mathfrak{g}_\mathbb{R}$. By $\omega(Ex, Ey) = -\omega(x, y)$, we have

$$\omega_\mathbb{R}(Jx, Jy) = \mathrm{Re}(\omega(iEx, iEy)) = \mathrm{Re}(-\omega(Ex, Ey)) = \mathrm{Re}(\omega(x, y)) = \omega_\mathbb{R}(x, y).$$

Thus, $(\mathfrak{g}_\mathbb{R}, iE, \omega_\mathbb{R})$ is a real pseudo-Kähler 3-Lie algebra. □

Conversely, we have

THEOREM 8.9.– Let $(\mathfrak{g}, \omega, J)$ be a real pseudo-Kähler 3-Lie algebra. Then $(\mathfrak{g}_\mathbb{C}, \omega_\mathbb{C}, E)$ is a complex para-Kähler 3-Lie algebra, where $\mathfrak{g}_\mathbb{C} = \mathfrak{g} \otimes_\mathbb{R} \mathbb{C}$ is the complexification of $\mathfrak{g}$, $E = -iJ_\mathbb{C}$ and $\omega_\mathbb{C}$ is the complexification of ω, more precisely,

$$\omega_\mathbb{C}(x_1 + iy_1, x_2 + iy_2) = \omega(x_1, x_2) - \omega(y_1, y_2) + i\omega(x_1, y_2) + i\omega(y_1, x_2). \quad [8.68]$$

PROOF.– By corollary 8.12, $E = -iJ_\mathbb{C}$ is a paracomplex structure on the complex 3-Lie algebra $\mathfrak{g}_\mathbb{C}$. It is obvious that $\omega_\mathbb{C}$ is skew-symmetric and non-degenerate. Moreover, since ω is a symplectic structure on $\mathfrak{g}$, we deduce that $\omega_\mathbb{C}$ is a symplectic structure on $\mathfrak{g}_\mathbb{C}$. Finally, by $\omega(Jx, Jy) = \omega(x, y)$, we have

$$\begin{aligned}
&\omega_\mathbb{C}(E(x_1 + iy_1), E(x_2 + iy_2)) = \omega_\mathbb{C}(Jy_1 - iJx_1, Jy_2 - iJx_2) \\
&= \omega(Jy_1, Jy_2) - \omega(Jx_1, Jx_2) - i\omega(Jx_1, Jy_2) - i\omega(Jy_1, Jx_2) \\
&= \omega(y_1, y_2) - \omega(x_1, x_2) - i\omega(x_1, y_2) - i\omega(y_1, x_2) \\
&= -\omega_\mathbb{C}(x_1 + iy_1, x_2 + iy_2).
\end{aligned}$$

Therefore, $(\mathfrak{g}_{\mathbb{C}}, \omega_{\mathbb{C}}, -iJ_{\mathbb{C}})$ is a complex para-Kähler 3-Lie algebra. □

In the end, we construct a Kähler 3-Lie algebra using a 3-pre-Lie algebra with a symmetric and positive definite invariant bilinear form.

PROPOSITION 8.23.– Let $(A, \{\cdot,\cdot,\cdot\})$ be a real 3-pre-Lie algebra with a symmetric and positive definite invariant bilinear form $\mathcal{B}$. Then $(A^c \ltimes_{L^*} A^*, \omega, -J)$ is a real Kähler 3-Lie algebra, where J is given by [8.49] and ω is given by [8.16].

PROOF.– By theorem 8.2 and proposition 8.14, ω is a symplectic structure and J is a perfect complex structure on the semidirect product 3-Lie algebra $(A^c \ltimes_{L^*} A^*, [\cdot,\cdot,\cdot]_{L^*})$. Obviously, $-J$ is also a perfect complex structure on $A^c \ltimes_{L^*} A^*$. Let $\{e_1, \cdots, e_n\}$ be a basis of A such that $\mathcal{B}(e_i, e_j) = \delta_{ij}$ and $e_1^*, \cdots, e_n^*$ be the dual basis of A^*. Then for all i, j, k, l, we have

$$\omega(e_i + e_j^*, e_k + e_l^*) = \delta_{jk} - \delta_{li},$$
$$\omega(-J(e_i + e_j^*), -J(e_k + e_l^*)) = \omega(e_j - e_i^*, e_l - e_k^*) = -\delta_{il} + \delta_{kj},$$

which implies that $\omega(-J(x+\alpha), -J(y+\beta)) = \omega(x+\alpha, y+\beta)$ for all $x, y \in A$ and $\alpha, \beta \in A^*$. Therefore, $(A^c \ltimes_{L^*} A^*, \omega, -J)$ is a pseudo-Kähler 3-Lie algebra. Finally, let $x = \sum_{i=1}^n \lambda_i e_i \in A, \alpha = \sum_{i=1}^n \mu_i e_i^* \in A^*$ such that $x + \alpha \neq 0$. We have

$$\begin{aligned} S(x+\alpha, x+\alpha) &= \omega(x+\alpha, -J(x+\alpha)) \\ &= \omega\Big(\sum_{i=1}^n \lambda_i e_i + \sum_{i=1}^n \mu_i e_i^*, \sum_{i=1}^n \mu_i e_i - \sum_{i=1}^n \lambda_i e_i^*\Big) \\ &= \sum_{i=1}^n \mu_i^2 + \sum_{i=1}^n \lambda_i^2 > 0. \end{aligned}$$

Thus, S is positive definite. Therefore, $\{A^c \ltimes_{L^*} A^*, \omega, -J\}$ is a real Kähler 3-Lie algebra. The proof is completed. □

8.10. References

Alekseevsky, D.V., Perelomov, A.M. (1997). Poisson and symplectic structures on Lie algebras. I. *J. Geom. Phys.*, 22(3), 191–211.

Andrada, A. (2008). Complex product structures on 6-dimensional nilpotent Lie algebras. *Forum Math.*, 20(2), 285–315.

Andrada, A., Salamon, S. (2005). Complex product structures on Lie algebras. *Forum Math.*, 17(2), 261–295.

Andrada, A., Barberis, M.L., Dotti, I., Ovando, G.P. (2005). Product structures on four dimensional solvable Lie algebras. *Homology Homotopy Appl.*, 7(1), 9–37.

Andrada, A., Barberis, M.L., Dotti, I. (2011). Classification of abelian complex structures on 6-dimensional Lie algebras. *J. Lond. Math. Soc.*, 83(1), 232–255.

Bagger, J., Lambert, N. (2008). Gauge symmetry and supersymmetry of multiple M2-branes. *Phys. Rev. D*, 77(6), 065008.

Bagger, J., Lambert, N. (2009). Three-algebras and N=6 Chern-Simons gauge theories. *Phys. Rev. D*, 79(2), 025002.

Bai, C. (2006). A further study on non-abelian phase spaces: Left-symmetric algebraic approach and related geometry. *Rev. Math. Phys.*, 18(5), 545–564.

Bai, C. (2008). Left-symmetric bialgebras and an analogue of the classical Yang-Baxter equation. *Commun. Contemp. Math.*, 10(2), 221–260.

Bai, R,, Guo, L., Li, J., Wu, Y. (2013). Rota-Baxter 3-Lie algebras. *J. Math. Phys.*, 54(56), 063504.

Bai, C., Guo, L., Sheng, Y. (2019). Bialgebras, the classical Yang-Baxter equation and Manin triples for 3-Lie algebras. *Adv. Theo. Math. Phys.*, 23(1), 27–74.

Bajo, I., Benayadi, S. (2011). Abelian para-Kähler structures on Lie algebras. *Diff. Geom. Appl.*, 29(2), 160–173.

Basu, A., Harvey, J.A. (2005). The M2-M5 brane system and a generalized Nahm's equation. *Nuclear Phys. B*, 713(1–3), 136–150.

Benayadi, S., Boucetta, M. (2015). On para-Kähler and hyper-para-Kähler Lie algebras. *J. Algebra*, 436, 61–101.

Calvaruso, G. (2015a). A complete classification of four-dimensional paraKähler Lie algebras. *Complex Manifolds*, 2, 1–10.

Calvaruso, G. (2015b). Four-dimensional paraKähler Lie algebras: Classification and geometry. *Houston J. Math.*, 41(3), 733–748.

Chu, B.Y. (1974). Symplectic homogeneous spaces. *Trans. Amer. Math. Soc.*, 197, 145–159.

Cleyton, R., Lauret, J., Poon, Y. (2010). Weak mirror symmetry of Lie algebras. *J. Symplectic Geom.*, 8(1), 37–55.

Cleyton, R., Poon, Y., Ovando, G.P. (2011). Weak mirror symmetry of complex symplectic Lie algebras. *J. Geom. Phys.*, 61(8), 1553–1563.

de Azcarraga, J.A., Izquierdo, J.M. (2010). n-ary algebras: A review with applications. *J. Phys. A: Math. Theor.*, 43, 293001.

de Medeiros, P., Figueroa-O'Farrill, J. (2008). Metric Lie 3-algebras in Bagger-Lambert theory. *J. High Energy Phys.*, (8), 045.

de Medeiros, P., Figueroa-O'Farrill, J., Mendez-Escobar, E., Ritter, P. (2009a). Metric 3-Lie algebras for unitary Bagger-Lambert theories. *J. High Energy Phys.*, (4), 037.

de Medeiros, P., Figueroa-O'Farrill, J., Mendez-Escobar, E., Ritter, P. (2009b). On the Lie-algebraic origin of metric 3-algebras. *Comm. Math. Phys.*, 290(3), 871–902.

Filippov, V.T. (1985). n-Lie algebras. *Sibirsk. Mat. Zh.*, 26(6), 126–140.

Gautheron, P. (1996). Some remarks concerning Nambu mechanics. *Lett. Math. Phys.*, 37(1), 103–116.

Gomis, J., Rodriguez-Gomez, D., Van Raamsdonk, M., Verlinde, H. (2008). Supersymmetric Yang-Mills theory from Lorentzian three-algebras. *J. High Energy Phys.*, (8), 094.

Ho, P., Hou, R., Matsuo, Y. (2008). Lie 3-algebra and multiple M_2-branes. *J. High Energy Phys.*, (6), 020.

Kasymov, S.M. (1987). On a theory of n-Lie algebras. *Algebra i Logika*, 26(3), 277–297.

Li, T., Tomassini, A. (2012). Almost Kähler structures on four dimensional unimodular Lie algebras. *J. Geom. Phys.*, 62(7), 1714–1731.

Liu, J., Sheng, Y., Zhou, Y., Bai, C. (2016). Nijenhuis operators on n-Lie algebras. *Commun. Theor. Phys. (Beijing)*, 65(6), 659–670.

Liu, J., Makhlouf, A., Sheng, Y. (2017). A new approach to representations of 3-Lie algebras and abelian extensions. *Algebr. Represent. Theory*, 20(6), 1415–1431.

Nambu, Y. (1973). Generalized Hamiltonian dynamics. *Phys. Rev. D.*, 7(3), 2405–2412.

Papadopoulos, G. (2008). M2-branes, 3-Lie algebras and Plucker relations. *J. High Energy Phys.*, (5), 054.

Pei, J., Bai, C., Guo, L. (2017). Splitting of operads and Rota-Baxter operators on operads. *Appl. Categ. Structures*, 25(4), 505–538.

Salamon, S.M. (2001). Complex structures on nilpotent Lie algebras. *J. Pure Appl. Algebra*, 157(2–3), 311–333.

Takhtajan, L. (1994). On foundation of the generalized Nambu mechanics. *Comm. Math. Phys.*, 160(2), 295–315.

Takhtajan, L. (1995). A higher order analog of Chevalley-Eilenberg complex and deformation theory of n-algebras. *St. Petersburg Math. J.*, 6(2), 429–438.

9

Derived Categories

Bernhard KELLER
Department of Mathematics, Paris Diderot University, France

9.1. Introduction

Derived categories were conceived as a "formalism for hyperhomology" (Verdier 1996) in the early 1960s. At that time, they were only used by the circle around Grothendieck but by the 1990s, they had become widespread and had found their way into graduate textbooks (Iversen 1986; Kashiwara and Schapira 1994; Gelfand and Manin 1996; Weibel 1994; Positselski 2011; Zimmermann 2014).

According to Illusie (1990), derived categories were invented by Grothendieck in the early 1960s. He needed them to formulate the duality theory for schemes, which he had announced (Grothendieck 1960) at the International Congress in 1958. Grothendieck's student J.–L. Verdier carried out the essential constructions and, in the course of the year 1963, wrote a summary of the principal results (Verdier 1977). Having at his disposal the required foundations, Grothendieck exposed the duality theory he had conceived in a huge manuscript, which served as a basis for the seminar (Hartshorne 1966) that Hartshorne conducted at Harvard in the autumn of the same year.

Derived categories found their first applications in duality theory in the coherent setting (Hartshorne 1966) and then also in the étale (Verdier 1967; Deligne 1973) and the locally compact setting (Verdier 1963, 1966, 1969; Grivel 1985).

At the beginning of the 1970s, Sato (1969) and Kashiwara (1970) adapted Grothendieck–Verdier's methods to study systems of partial differential equations.

Algebra and Applications 1,

Nowadays, derived categories have become the standard language of microlocal analysis (see Kashiwara and Schapira (1994); Mebkhout (1989); Saito (1986); Borel (1987)). Because of Brylinski–Kashiwara's proof (Brylinski and Kashiwara 1981) of the Kazhdan–Lusztig conjecture, they have penetrated the representation theory of Lie groups (Bernstein and Lunts 1994) and finite Chevalley groups (Scott 1987). In this theory, a central role is played by certain abelian subcategories of derived categories that are modeled on the category of perverse sheaves (Beilinson *et al.* 1982), which originated in the sheaf-theoretic interpretation (Deligne 1979) of intersection cohomology (Goresky and MacPherson 1980, 1983).

In two ground-breaking papers (Beilinson 1978; Bernten *et al.* 1978), Beilinson and Bernstein–Gelfand–Gelfand used derived categories to establish a beautiful relation between coherent sheaves on projective space and representations of certain non-commutative finite-dimensional algebras. Their constructions had numerous generalizations (Geigle and Lenzing 1987; Kapranov 1983, 1986, 1988). They also lead D. Happel to a systematic investigation of the derived category of a finite-dimensional algebra (Happel 1987, 1988). He realized that derived categories provide the proper setting for tilting theory (Brenner and Butler 1980; Happel and Ringel 1982; Bongartz 1981; Angeleri Hügel *et al.* 2007). This theory is the origin of J. Rickard's Morita theory for derived categories (Rickard 1989, 1991) (see also Keller (1991, 1994)). Morita theory has widened the range of applications of derived categories even more. Thus, Broué's conjectures in the modular representation theory of finite groups (Broué 1988) are typical to summarize precision with generality that can be achieved by the systematic use of this language.

In this chapter, we will present Grothendieck's quick definition of the derived category followed by Verdier's more elaborate construction. We will then describe the triangulated structure on the derived category and construct derived functors. These will be applied in derived Morita theory. Finally, we will outline the generalization from rings to differential graded (=dg) categories and conclude by discussing invariants under derived equivalences between dg categories.

9.2. Grothendieck's definition

Let $\mathcal{C}$ be a category and S a set of arrows of $\mathcal{C}$. Then there is a category $\mathcal{C}[S^{-1}]$ and a functor

$$Q : \mathcal{C} \longrightarrow \mathcal{C}[S^{-1}]$$

such that Qs is invertible for each $s \in S$ and each functor F such that Fs is invertible for all $s \in S$ factors uniquely through Q (see Gabriel and Zisman (1967)).

The category $\mathcal{C}[S^{-1}]$ is called the *localization of* $\mathcal{C}$ *at* S and Q is called the *localization functor*. A right or left adjoint to Q is automatically fully faithful.

Now let $\mathcal{A}$ be an abelian category (Grothendieck 1957), for example the category Mod$R$ of all right modules over a ring R. A *complex* over $\mathcal{A}$ is a diagram M of the form

$$\ldots \longrightarrow M^p d^p \xrightarrow{-} M^{p+1} \longrightarrow \ldots$$

where $p \in \mathbb{Z}$ and $d^p d^{p-1} = 0$ for all $p \in \mathbb{Z}$. Thus, M is given by a $\mathbb{Z}$-graded object $(M^p)_{p \in \mathbb{Z}}$ together with a homogeneous endomorphism d of degree 1 such that

$$d^2 = 0.$$

The *homology* of a complex M is the $\mathbb{Z}$-graded object H^*M with components

$$H^p M = (\ker d^p)/(\operatorname{im} d^{p-1}).$$

A *morphism of complexes* $f : L \to M$ is a graded morphism homogeneous of degree 0 and which commutes with the differential. Clearly, the class of complexes and their morphisms form a category $\mathbf{C}(\mathcal{A})$. A morphism of complexes $s : M \to M'$ is a *quasi-isomorphism* if $H^p(s)$ is an isomorphism for all $p \in \mathbb{Z}$. Grothendieck defined the *derived category* $\mathbf{D}(\mathcal{A})$ to be the localization of the category of complexes $\mathbf{C}(\mathcal{A})$ at the class of all quasi-isomorphisms. This definition has the advantage of being quick and elegant but it does not give a useable description of the morphisms in the derived category.

9.3. Verdier's definition

As discussed previously, let $\mathcal{A}$ be an abelian category. Recall that a morphism of complexes $f\colon L \to M$ is *null-homotopic* if there is a graded morphism $h\colon L \to M$ homogeneous of degree -1 such that $f = d \circ h + h \circ d$. Clearly, sums of null-homotopic morphisms are null-homotopic. Moreover, if f is null-homotopic, so are $g \circ f$ and $f \circ k$ for arbitrary morphisms g and k composable with f. The *category up to homotopy* $\mathbf{H}(\mathcal{A})$ is defined as the category whose objects are the complexes and whose morphisms $L \to M$ are classes of morphisms of complexes modulo null-homotopic morphisms. Note that the image of a null-homotopic morphism under homology is zero so that homology induces a well-defined functor on the category up to homotopy. A morphism s of the category up to homotopy is a quasi-isomorphism if its image $H^*(s)$ is invertible.

LEMMA.– The following hold in the category up to homotopy $\mathbf{H}(\mathcal{A})$:

a) all identities are quasi-isomorphisms;

b) if two among s, t, st are quasi-isomorphisms, so is the third;

c) if f is a morphism and s a quasi-isomorphism such that $fs = 0$, then there is a quasi-isomorphism t such that $tf = 0$;

d) each diagram

$$\begin{array}{ccc} L & \xrightarrow{s} & L' \\ {\scriptstyle f}\downarrow & & \\ M & & \end{array}$$

where s is a quasi-isomorphism, can be completed to a commutative square

$$\begin{array}{ccc} L & \xrightarrow{s} & L' \\ {\scriptstyle f}\downarrow & & \downarrow{\scriptstyle f'} \\ M & \xrightarrow[s']{} & M' \end{array}$$

where s' is a quasi-isomorphism.

The properties in the lemma are summed up by stating that the class of quasi-isomorphisms in $\mathbf{H}(\mathcal{A})$ admits a *calculus of left fractions*. For two complexes L and M, define a *left fraction* $s^{-1}f$ to be an equivalence class of diagrams

$$L \xrightarrow{f} M' \xleftarrow{s} M$$

where s is a quasi-isomorphism and two diagrams (f, s) and (g, t) are equivalent if there is a quasi-isomorphism u fitting into a commutative diagram

$$\begin{array}{ccccc} & & M' & & \\ & {\scriptstyle f}\nearrow & \downarrow & \nwarrow{\scriptstyle s} & \\ L & \longrightarrow & M''' & \xleftarrow{u} & M. \\ & {\scriptstyle g}\searrow & \uparrow & \swarrow{\scriptstyle t} & \\ & & M'' & & \end{array}$$

Verdier defines the *derived category* $\mathbf{D}(\mathcal{A})$ to have as objects all complexes and as morphisms $L \to M$ all left fractions from L to M. The *composition* of two left fractions $t^{-1}g$ and $s^{-1}f$ is defined as $(s't)^{-1}(g'f)$ using part (d) of the above lemma to complete the following commutative diagram:

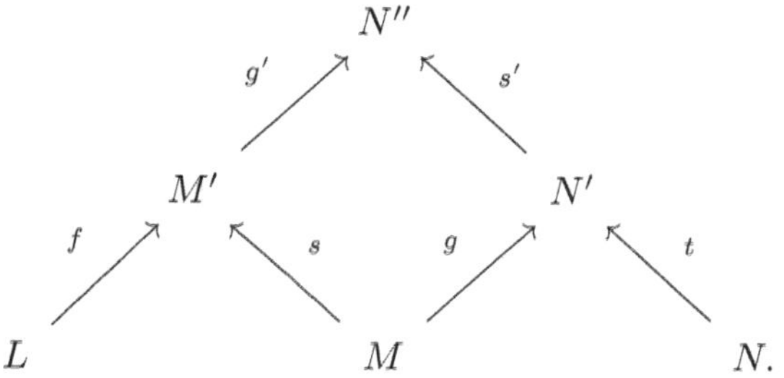

It is not hard to check that Verdier's definition is equivalent to Grothendieck's definition. The following lemma allows us to compute morphisms in the derived category. For a category $\mathcal{C}$, we write $\mathcal{C}(X, Y)$ for the set of morphisms from X to Y.

LEMMA.–

a) If I is a left bounded complex with injective components, the canonical map

$$\mathbf{H}(\mathcal{A})(?, I) \to \mathbf{D}(\mathcal{A})(?, I)$$

is bijective.

b) If P is a right bounded complex with projective components, the canonical map

$$\mathbf{H}(\mathcal{A})(P, ?) \to \mathbf{D}(\mathcal{A})(P, ?)$$

is bijective.

Let $\Sigma\colon \mathbf{C}(\mathcal{A}) \to \mathbf{C}(\mathcal{A})$ be the *suspension functor*, i.e. for a complex X, we have $(\Sigma X)^p = X^{p+1}$ and $d_{\Sigma X} = -d_X$ and for a morphism of complexes f, we have $(\Sigma f)^p = f^{p+1}$. We identify $\mathcal{A}$ with the full subcategory of $\mathbf{C}(\mathcal{A})$ formed by the complexes concentrated in degree 0. Let M be an object of $\mathcal{A}$ and $M \to I$ an injective resolution, i.e. a quasi-isomorphism where I is concentrated in degrees ≥ 0 and has injective components. Then, from the lemma, we find for each complex N and each $n \in \mathbb{Z}$

$$\mathbf{D}(\mathcal{A})(N, \Sigma^n M) \xrightarrow{\sim} \mathbf{D}(\mathcal{A})(N, \Sigma^n I) \xleftarrow{\sim} \mathbf{H}(\mathcal{A})(N, \Sigma^n I)$$

If N is concentrated in degree 0, the last group is easily seen to be isomorphic to the extension group

$$\mathrm{Ext}^n_{\mathcal{A}}(N, M).$$

Here, by convention, the Ext-groups vanish in strictly negative degrees. This result also holds if we do not assume the existence of an injective resolution:

LEMMA 9.1.– For objects N and M in $\mathcal{A}$ and each $n \in \mathbb{Z}$, we have a canonical isomorphism

$$\mathrm{Ext}^n_{\mathcal{A}}(N, M) \xrightarrow{\sim} \mathbf{D}(\mathcal{A})(N, \Sigma^n M).$$

EXAMPLE 9.1.– If $\mathcal{A}$ is the category of vector spaces over a field k, then each object X of $\mathbf{D}(\mathcal{A})$ is canonically isomorphic to $\bigoplus_p \Sigma^{-p} H^p X$ and all extension groups vanish. So $\mathbf{D}(\mathcal{A})$ is equivalent to the category of $\mathbb{Z}$-graded vector spaces.

EXAMPLE 9.2.– If $\mathcal{A}$ is *hereditary*, i.e. we have $\mathrm{Ext}^2_{\mathcal{A}} = 0$, then each object X of $\mathbf{D}(\mathcal{A})$ is non-canonically isomorphic to $\bigoplus_p \Sigma^{-p} H^p X$. The space of morphisms between two objects X and Y is isomorphic to the product over $p \in \mathbb{Z}$ of the groups

$$\mathrm{Hom}_{\mathcal{A}}(H^p X, H^p Y) \oplus \mathrm{Ext}^1_{\mathcal{A}}(H^p X, H^{p-1} Y).$$

9.4. Triangulated structure

As mentioned above, let $\mathcal{A}$ be an abelian category. The categories $\mathbf{H}(\mathcal{A})$ and $\mathbf{D}(\mathcal{A})$ are almost never abelian (they are if and only if all short exact sequences of $\mathcal{A}$ split). However, they do carry a structure induced by the short exact sequences of complexes.

A Σ-*sequence* of $\mathbf{H}(\mathcal{A})$ is a sequence of the form

$$X \longrightarrow Y \longrightarrow Z \longrightarrow \Sigma X.$$

A *morphism of* Σ-*sequences* is a commutative diagram of the form

$$\begin{array}{ccccccc}
X & \longrightarrow & Y & \longrightarrow & Z & \longrightarrow & \Sigma X \\
{\scriptstyle a}\downarrow & & \downarrow & & \downarrow & & \downarrow{\scriptstyle \Sigma a} \\
X' & \longrightarrow & Y' & \longrightarrow & Z' & \longrightarrow & \Sigma X'.
\end{array}$$

For a morphism $f : L \to M$ of $\mathbf{C}(\mathcal{A})$, the *standard triangle* associated with f is the image in $\mathbf{H}(\mathcal{A})$ of the Σ-sequence

$$X \xrightarrow{f} Y \xrightarrow{i} C(f) \xrightarrow{p} \Sigma X ,$$

where $C(f)$ is the *mapping cone* of f, i.e. the graded object $Y \oplus \Sigma X$ endowed with the differential

$$\begin{bmatrix} d_Y & f \\ 0 & d_{\Sigma X} \end{bmatrix} ,$$

where i and p are the canonical injection and projection. A *triangle* of $\mathbf{H}(\mathcal{A})$ is a Σ-sequence isomorphic to a standard triangle.

THEOREM 9.1.– The following hold:

(T0) The triangles are stable under isomorphism of Σ-sequences and for each object X, the following Σ-sequence is a triangle

$$X \xrightarrow{1_X} X \longrightarrow 0 \longrightarrow \Sigma X.$$

(T1) For each morphism $f : X \to Y$, there is a triangle

$$X \xrightarrow{f} Y \longrightarrow Z \longrightarrow \Sigma X.$$

(T2) A Σ-sequence (u, v, w) is a triangle if and only if so is $(v, w, -\Sigma u)$.

(T3) Given two triangles

$$X \longrightarrow Y \longrightarrow Z \longrightarrow \Sigma X \quad \text{and} \quad X' \longrightarrow Y' \longrightarrow Z' \longrightarrow \Sigma X'$$

and a commutative square

$$\begin{array}{ccc} X & \longrightarrow & Y \\ {\scriptstyle a}\downarrow & & \downarrow{\scriptstyle b} \\ X' & \longrightarrow & Y' \end{array}$$

there is a (non-unique) morphism c yielding a morphism of Σ-sequences

$$\begin{array}{ccccccc} X & \longrightarrow & Y & \longrightarrow & Z & \longrightarrow & \Sigma X \\ {\scriptstyle a}\downarrow & & {\scriptstyle b}\downarrow & & {\scriptstyle c}\downarrow & & {\scriptstyle \Sigma a}\downarrow \\ X' & \longrightarrow & Y' & \longrightarrow & Z' & \longrightarrow & \Sigma X' \end{array}$$

(T4) Given two composable morphisms

$$X \xrightarrow{u} Y \xrightarrow{v} Z$$

there is a commutative diagram

$$\begin{array}{ccccccc} X & \xrightarrow{u} & Y & \xrightarrow{x} & Z' & \longrightarrow & \Sigma X \\ {\scriptstyle 1_X}\downarrow & & {\scriptstyle v}\downarrow & & \downarrow & & \downarrow{\scriptstyle 1_{\Sigma X}} \\ X & \longrightarrow & Z & \longrightarrow & Y' & \longrightarrow & \Sigma X \\ & & \downarrow & & \downarrow & & \downarrow{\scriptstyle \Sigma u} \\ & & X' & \xrightarrow{1_{X'}} & X' & \xrightarrow{r} & \Sigma Y \\ & & {\scriptstyle r}\downarrow & & \downarrow & & \\ & & \Sigma Y & \xrightarrow{\Sigma x} & \Sigma Z' & & \end{array}$$

where the first two rows and the two central columns are triangles.

A *triangulated category* is an additive category endowed with an autoequivalence Σ and a class of distinguished Σ-sequences called *triangles* such that the properties T0–T4 of the theorem hold. Thus, the category up to homotopy $\mathbf{H}(\mathcal{A})$ is a triangulated category.

The most important consequence of the axioms T0–T3 is that, for each triangle

$$X \longrightarrow Y \longrightarrow Z \longrightarrow \Sigma X$$

of a triangulated category $\mathcal{T}$, the induced sequences

$$\ldots \longrightarrow \mathcal{T}(?, X) \longrightarrow \mathcal{T}(?, Y) \longrightarrow \mathcal{T}(?, Z) \longrightarrow \ldots$$

and

$$\dots \longrightarrow \mathcal{T}(Z,?) \longrightarrow \mathcal{T}(Y,?) \longrightarrow \mathcal{T}(X,?) \longrightarrow \dots$$

are exact. Via the 5-lemma, this implies that if in a morphism of triangles, two components are invertible, then so is the third. It follows that in a triangle

$$X \xrightarrow{f} Y \longrightarrow Z \longrightarrow \Sigma X \quad ,$$

the third term Z is unique up to (non-unique) isomorphism. One also shows that the direct sum of two Σ-sequences is a triangle if and only if both summands are and that in a triangle (u, v, w), the sequence (u, v) is split exact if and only if $w = 0$. The theory of triangulated categories admitting infinite sums is developed in Neeman (2001). A *triangulated subcategory* of a triangulated category is a full subcategory stable under Σ and Σ^{-1} such that with two terms of a triangle, it also contains the third term. A *thick subcategory* is a triangulated subcategory stable under taking direct summands. An object G is a *generator* of a triangulated category $\mathcal{T}$ if $\mathcal{T}$ coincides with its smallest thick subcategory containing G. Important existence theorems for generators in derived categories appearing in algebraic geometry are given in Bondal and van den Bergh (2003) and Rouquier (2008).

If $\mathcal{S}$ and $\mathcal{T}$ are triangulated categories, a *triangle functor* $\mathcal{S} \to \mathcal{T}$ is a pair (F, ϕ) formed by an additive functor $F : \mathcal{S} \to \mathcal{T}$ and an isomorphism of functors $\phi : F\Sigma \to \Sigma F$ such that for each triangle (u, v, w) of $\mathcal{S}$ the Σ-sequence

$$FX \xrightarrow{Fu} FY \xrightarrow{Fv} FZ(\phi X)(Fw) \xrightarrow{\quad -\quad} \Sigma FX$$

is a triangle of $\mathcal{T}$. Let $Q : \mathbf{H}(\mathcal{A}) \to \mathbf{D}(\mathcal{A})$ be the canonical localization functor. We have a canonical isomorphism $\phi : Q\Sigma \to \Sigma Q$.

LEMMA 9.2.– $\mathbf{D}(\mathcal{A})$ admits a unique structure of triangulated category such that (Q, ϕ) becomes a triangle functor.

The construction of the derived category from the category up to homotopy is a special case of the *localization of triangulated categories*: Let $\mathcal{T}$ be a triangulated category and $\mathcal{N} \subset \mathcal{T}$ a thick subcategory. Define S to be the class of morphisms s such that in a triangle

$$X \xrightarrow{s} Y \longrightarrow N \longrightarrow \Sigma X$$

the cone N belongs to $\mathcal{N}$. Then it is easy to see that a triangle functor $F : \mathcal{T} \to \mathcal{S}$ vanishes on the objects of $\mathcal{N}$ if and only if it makes the morphisms of S invertible. One defines the *Verdier quotient* $\mathcal{T}/\mathcal{N}$ as the localization $\mathcal{T}[S^{-1}]$, which is constructed using a calculus of fractions in complete analogy with Verdier's definition of the derived category. In particular, it inherits a structure of triangulated category from $\mathcal{T}$. By definition, the sequence of triangulated categories

$$0 \longrightarrow \mathcal{N} \longrightarrow \mathcal{T} \longrightarrow \mathcal{T}/\mathcal{N} \longrightarrow 0$$

is *exact*. For example, we obtain the derived category by localizing the category up to homotopy $\mathcal{T} = \mathbf{H}(\mathcal{A})$ at the thick subcategory $\mathcal{N}$ formed by the acyclic complexes, i.e. the complexes with vanishing homology.

If (F, ϕ) and (G, ψ) are triangle functors $\mathcal{S} \to \mathcal{T}$, a *morphism* $(F, \phi) \to (G, \psi)$ is a morphism of functors $\alpha : F \to G$ such that the square

$$\begin{array}{ccc} F\Sigma & \xrightarrow{\phi} & \Sigma F \\ {\scriptstyle \alpha\Sigma}\downarrow & & \downarrow{\scriptstyle \Sigma\alpha} \\ G\Sigma & \xrightarrow[\psi]{} & \Sigma G \end{array}$$

commutes. The *composition* of two triangle functors $(F, \phi) : \mathcal{S} \to \mathcal{T}$ and $(G, \psi) : \mathcal{R} \to \mathcal{S}$ is $(FG, (\phi G)(F\psi))$. Two triangle functors $(F, \phi) : \mathcal{S} \to \mathcal{T}$ and $(G, \psi) : \mathcal{T} \to \mathcal{S}$ are *adjoint*, if there are morphisms $\alpha : (F, \phi)(G, \psi) \to \mathbf{1}_{\mathcal{S}}$ and $\beta : \mathbf{1}_{\mathcal{T}} \to (G, \psi)(F, \phi)$ such that $(G\alpha)(\beta G) = \mathbf{1}_G$ and $(\alpha F)(F\beta) = \mathbf{1}_F$.

LEMMA 9.3.– A triangle functor $(F, \phi) : \mathcal{S} \to \mathcal{T}$ admits a triangle right adjoint if and only if the additive functor $F : \mathcal{S} \to \mathcal{T}$ admits a right adjoint.

THEOREM 9.2.– Let R be a ring and $\mathrm{Mod}R$ the category of right R-modules. The localization functor $\mathbf{H}(\mathrm{Mod}R) \to \mathbf{D}(\mathrm{Mod}R)$ admits a (fully faithful) left adjoint $M \mapsto \mathbf{p}M$ and a (fully faithful) right adjoint $M \mapsto \mathbf{i}M$.

If M is an R-module, then $\mathbf{p}M$ is given by a projective resolution

$$\ldots \longrightarrow P_2 \longrightarrow P_1 \longrightarrow P_0 \longrightarrow 0$$

of M and dually, $\mathbf{i}M$ is given by an injective resolution. We call $\mathbf{p}$ and $\mathbf{i}$ the *resolution functors*.

9.5. Derived functors

We follow Deligne's approach (Deligne 1973) to derived functors. Let $F : \mathcal{A} \to \mathcal{B}$ be an additive functor between abelian categories. It induces functors $\mathbf{C}(\mathcal{A}) \to \mathbf{C}(\mathcal{B})$ and $\mathbf{H}(\mathcal{A}) \to \mathbf{H}(\mathcal{B})$, which we still denote by F. Since we do not assume that F is exact, it does not, in general, induce a functor between the derived categories. Nevertheless, we may look for a functor $\mathbf{R}F : \mathbf{D}(\mathcal{A}) \to \mathbf{D}(\mathcal{B})$, which comes close to making the following square commute:

$$
\begin{array}{ccc}
\mathbf{H}(\mathcal{A}) & \xrightarrow{F} & \mathbf{H}(\mathcal{B}) \\
{\scriptstyle Q}\big\downarrow & & \big\downarrow{\scriptstyle Q} \\
\mathbf{D}(\mathcal{A}) & \xrightarrow[\mathbf{R}F]{} & \mathbf{D}(\mathcal{B}).
\end{array}
$$

For an object Y of $\mathbf{D}(\mathcal{A})$, to define $\mathbf{R}F(Y)$, we define the functor $\mathbf{r}F(Y)$ represented by $\mathbf{R}F(Y)$. Namely, its value at an object $X \in \mathbf{D}(\mathcal{B})$ is formed by the equivalence classes $(f|s)$ of pairs

$$
X \xrightarrow{f} FY' \quad Y' \xleftarrow{s} Y
$$

consisting of a quasi-isomorphism s of $\mathbf{H}(\mathcal{A})$ and a morphism f of $\mathbf{D}(\mathcal{B})$. Two pairs $(f|s)$ and $(f'|s')$ are equivalent if there are commutative diagrams

$$
\begin{array}{ccccccc}
 & & FY' & & Y' & & \\
 & {\scriptstyle f}\nearrow & \big\downarrow{\scriptstyle Fv} & & {\scriptstyle v}\big\downarrow & \nwarrow{\scriptstyle s} & \\
X & \xrightarrow{h} & FY''' & & Y''' & \xleftarrow{u} & Y \\
 & {\scriptstyle f'}\searrow & \big\uparrow{\scriptstyle Fw} & & {\scriptstyle w}\big\uparrow & \swarrow{\scriptstyle s'} & \\
 & & FY'' & & Y'' & &
\end{array}
$$

in $\mathbf{D}(\mathcal{B})$ and $\mathbf{H}(\mathcal{A})$, respectively. The functor $\mathbf{R}F$ is *defined* at Y, if the functor $\mathbf{r}F(Y)$ is representable and in this case, the value $\mathbf{R}F(Y)$ is defined by the isomorphism

$$
\mathrm{Hom}(?, \mathbf{R}F(Y)) = \mathbf{r}F(Y).
$$

The *left derived functor* $\mathbf{L}F$ is defined dually.

LEMMA 9.4.– The domain of definition of $\mathbf{R}F$ is a triangulated subcategory $\mathcal{S}$ of $\mathbf{D}(\mathcal{A})$ and $\mathbf{R}F : \mathcal{S} \to \mathbf{D}(\mathcal{B})$ admits a canonical structure of triangle functor.

LEMMA 9.5.– Suppose that $\mathcal{A} = \mathrm{Mod}R$ for a ring R. Then the left and right derived functors of F are defined on all of $\mathbf{D}(\mathcal{A})$ and we have $\mathbf{R}F(M) = F\mathbf{i}M$ and $\mathbf{L}F(M) = F\mathbf{p}M$ for all $M \in \mathbf{D}(\mathcal{A})$, where $\mathbf{i}$ and $\mathbf{p}$ are the resolution functors defined in the previous section.

9.6. Derived Morita theory

Let B be a ring and T a (right) B-module. Let A be the endomorphism ring of T. Then T becomes an A-B-bimodule and yields the adjoint pair

$$? \otimes_A T : \mathrm{Mod}A \to \mathrm{Mod}B \quad \text{and} \quad \mathrm{Hom}_B(T, ?) : \mathrm{Mod}B \to \mathrm{Mod}A.$$

The following is the main theorem of tilting theory (Angeleri Hügel *et al.* 2007). The module T is called a *tilting module* if it satisfies the properties of (b). We put $\mathbf{D}(A) = \mathbf{D}(\mathrm{Mod}A)$.

THEOREM 9.3.– The following are equivalent:

a) the derived functor $\mathbf{L}(? \otimes_A T) : \mathbf{D}(A) \to \mathbf{D}(B)$ is an equivalence;

b) the module T has the following properties:

i) the module T has a finite resolution by finitely generated projective B-modules,

ii) we have $\mathrm{Ext}^p_B(T,T) = 0$ for all $p > 0$,

iii) there is an exact sequence $0 \to A \to T^0 \to \ldots \to T^N \to 0$ of left A-modules where the T^i are direct summands of finite direct sums of copies of T.

Now let A and B be rings and X a complex of A-B-bimodules. For a complex M of right A-modules, define the complex $M \otimes_A X$ of right B-modules to have the components

$$\bigoplus_{p+q=n} M^p \otimes_A X^q$$

and the differential given by

$$d(m \otimes x) = (dm) \otimes x + (-1)^p m \otimes dx \ ,$$

where $m \in M^p$. For a complex N of right B-modules, define the complex $\mathrm{Hom}_B(X, N)$ of right A-modules to have the components

$$\prod_{-p+q=n} \mathrm{Hom}_B(X^p, N^q)$$

and the differential given by $d(f^p) = (d \circ f^p - (-1)^n f^{p+1} \circ d)$. Then the functors $? \otimes_A X$ and $\mathrm{Hom}_B(X, ?)$ form an adjoint pair between $\mathbf{C}(\mathrm{Mod}A)$ and $\mathbf{C}(\mathrm{Mod}B)$. The following theorem is given by Rickard (1989, 1991). A direct proof is given in Keller (1994, 1998b).

THEOREM 9.4.– Assume that A and B are algebras over a commutative ring k and that A is k-flat. The following are equivalent:

a) there is a triangle equivalence $F : \mathbf{D}(A) \to \mathbf{D}(B)$;

b) there is a complex of B-modules T such that

i) T is quasi-isomorphic to a bounded complex of finitely generated projective B-modules,

ii) we have $\mathrm{Hom}(T, \Sigma^n T) = 0$ for all $n \neq 0$ and $\mathrm{Hom}(T, T) \cong A$,

iii) B belongs to the smallest triangulated subcategory of $\mathbf{D}(B)$ containing T and closed under forming direct summands;

c) there is a complex X of A-B-bimodules such that $\mathbf{L}(? \otimes_A X)$ is an equivalence $\mathbf{D}(A) \to \mathbf{D}(B)$.

The algebras A and B are *derived equivalent* if the conditions of the theorem hold. A complex T as in (b) is called a *(one-sided) tilting complex* and a bimodule complex X as in (c) is called a *two-sided tilting complex*. A direct construction of a two-sided tilting complex from a one-sided one when k is a field is given in Keller (2000).

EXAMPLE 9.3.– Let k be a field of characteristic 0 and V a k-vector space of finite dimension $n+1$. For $p \geq 0$, let S^p be the pth symmetric power of V and Λ^p the pth exterior power of the dual of V. Let A be the algebra of upper triangular $(n+1) \times (n+1)$-matrices whose (i,j)-entry lies in S^{j-i} and B the algebra of lower triangular $(n+1) \times (n+1)$-matrices whose (i,j)-entry lies in Λ^{i-j}. Let S_i be the B-module k, where B acts through the projection onto the ith diagonal entry. Then $T = \Sigma^n S_1 \oplus \Sigma^{n-1} S_2 \oplus \cdots \oplus S_{n+1}$ is a one-sided tilting complex over B with endomorphism algebra A and thus A and B have equivalent derived categories. This is an example of *Koszul duality* (Beilinson *et al.* 1996; Keller 1994). In fact, both derived categories are equivalent to the derived category of quasi-coherent sheaves on the projectivization of V, as shown by Beilinson (1978). Note that for $n \geq 3$, the module categories over A and B are not equivalent.

9.7. Dg categories

Triangulated categories were invented by Grothendieck–Verdier in order to axiomatize the properties of derived categories. While they do capture some key features, they suffer from serious defects. Most importantly, tensor products and functor categories formed from triangulated categories are no longer triangulated. The theory of differential graded (=dg) categories (Keller 2006; Toën 2011) was developed to overcome these limitations.

9.7.1. *Dg categories and functors*

Let k be a commutative ring. A *dg k-module* is a complex of k-modules. Equivalently, it is a $\mathbb{Z}$-graded k-module

$$M = \bigoplus_{n\in\mathbb{Z}} M^n$$

endowed with a differential, i.e. a k-linear endomorphism d homogeneous of degree 1 such that $d^2 = 0$. The *tensor product* $L \otimes M$ of two dg k-modules is the dg k-module with components

$$\bigoplus_{p+q=n} L^p \otimes_k M^q$$

and differential $d_L \otimes \mathbf{1}_M + \mathbf{1}_L \otimes d_M$.

A *dg k-category* is a category $\mathcal{A}$ whose morphism sets $\mathcal{A}(X, Y)$ are dg k-modules and whose compositions are morphisms of dg k-modules

$$\mathcal{A}(Y, Z) \otimes \mathcal{A}(X, Y) \to \mathcal{A}(X, Z).$$

For example, the datum of a dg k-category $\mathcal{A}$ with a single object $*$ is equivalent to that of the dg k-algebra $\mathcal{A}(*, *)$. A typical example with several objects is obtained as follows: let B be a k-algebra. A right dg B-module is a complex of right B-modules. For two dg B-modules, L and M define $\mathrm{Hom}(L, M)^n$ to be the k-module of B-linear maps $f : L \to M$ homogeneous of degree n and make the graded space $\mathrm{Hom}(L, M)$ into a dg k-module by defining

$$d(f) = d_M \circ f - (-1)^n f \circ d_L \,,$$

where f is of degree n. The dg k-category $\mathbf{C}_{dg}(B)$ has as objects all dg B-modules and as morphism spaces the dg k-modules $\mathrm{Hom}(L, M)$ with the natural composition.

Let $\mathcal{A}$ be a dg category. The *opposite dg category* $\mathcal{A}^{op}$ has the same objects as $\mathcal{A}$ and its morphisms are defined by

$$\mathcal{A}(X,Y) = \mathcal{A}(Y,X)\ ;$$

the composition of $f \in \mathcal{A}^{op}(Y,X)^p$ with $g \in \mathcal{A}^{op}(Z,Y)^q$ is given by $(-1)^{pq}gf$. The *category* $Z^0(\mathcal{A})$ has the same objects as $\mathcal{A}$ and its morphisms are defined by

$$(Z^0\mathcal{A})(X,Y) = Z^0(\mathcal{A}(X,Y))\ ,$$

where Z^0 is the kernel of $d : \mathcal{A}(X,Y)^0 \to \mathcal{A}(X,Y)^1$. The *category* $H^0(\mathcal{A})$ has the same objects as $\mathcal{A}$ and its morphisms are defined by

$$(H^0(\mathcal{A}))(X,Y) = H^0(\mathcal{A}(X,Y))\ ,$$

where H^0 denotes the 0th homology of the complex $\mathcal{A}(X,Y)$. For example, if B is a k-algebra, we have isomorphisms of categories

$$Z^0(\mathbf{C}_{dg}(B)) = \mathbf{C}(\mathrm{Mod}B) \quad \text{and} \quad H^0(\mathbf{C}_{dg}(B)) = \mathbf{H}(\mathrm{Mod}B).$$

Let $\mathcal{A}$ and $\mathcal{A}'$ be dg categories. A *dg functor* $F : \mathcal{A} \to \mathcal{A}'$ is given by a map F from the class of objects of $\mathcal{A}$ to the class of objects of $\mathcal{A}'$ and by morphisms of dg k-modules, for all objects X, Y of $\mathcal{A}$,

$$F(X,Y) : \mathcal{A}(X,Y) \to \mathcal{A}'(FX,FY)$$

compatible with the composition and the identities. It is a *quasi-equivalence* if it induces isomorphisms, for X, Y in $\mathcal{A}$,

$$H^*(\mathcal{A}(X,Y)) \to H^*(\mathcal{A}'(FX,FY))\ ,$$

and the induced functor $H^0(\mathcal{A}) \to H^0(\mathcal{A}')$ is an equivalence. The *category of small dg k-categories* dgcat_k has the small dg categories as objects and the dg functors as morphisms. Note that it has an initial object, the empty dg category $\emptyset$, and a final object, the dg category with one object whose endomorphism ring is the zero ring. The *tensor product* $\mathcal{A} \otimes \mathcal{B}$ of two dg categories has as class of objects the product of the class of objects of $\mathcal{A}$ and that of $\mathcal{B}$ and the morphism spaces

$$(\mathcal{A} \otimes \mathcal{B})((X,Y),(X',Y')) = \mathcal{A}(X,X') \otimes \mathcal{B}(Y,Y')$$

with the natural compositions and units.

For two dg functors $F, G : \mathcal{A} \to \mathcal{B}$, the *dg k-module of graded morphisms* $\mathrm{Hom}(F, G)$ has as its nth component the module formed by the families of morphisms

$$\phi_X \in \mathcal{B}(FX, GX)^n$$

such that $(Gf)(\phi_X) = (\phi_Y)(Ff)$ for all $f \in (X, Y)$, $X, Y \in \mathcal{A}$. The differential is induced by that of $\mathcal{B}(FX, GX)$. The set of *morphisms* $F \to G$ is by definition in bijection with $Z^0\mathrm{Hom}(F, G)$. The *dg functor category* $\mathrm{Hom}(\mathcal{A}, \mathcal{B})$ has as objects the dg functors $\mathcal{A} \to \mathcal{B}$ and as morphism complexes the dg k-modules $\mathrm{Hom}(F, G)$.

9.7.2. *The derived category*

Let $\mathcal{A}$ be a dg category. The category of *(right) dg $\mathcal{A}$-modules* is defined as

$$\mathbf{C}(\mathcal{A}) = Z^0\mathrm{Hom}(\mathcal{A}^{op}, \mathbf{C}_{dg}(k)).$$

Thus, a dg $\mathcal{A}$-module is a dg functor $M : \mathcal{A}^{op} \to \mathbf{C}_{dg}(k)$. With each object X of $\mathcal{A}$, it associates a dg k-module $M(X)$ functorial in $X \in \mathcal{A}^{op}$. Its *homology* is the functor $X \mapsto H^*(M(X))$ from $H^0(\mathcal{A})$ to the category of graded k-modules. A morphism of dg modules $s : M \to M'$ is a *quasi-isommorphism* if it induces an isomorphism in homology. The *category up to homotopy of dg modules* is defined as

$$\mathbf{H}(\mathcal{A}) = H^0\mathrm{Hom}(\mathcal{A}^{op}, \mathbf{C}_{dg}(k)).$$

The *derived category* $\mathbf{D}(\mathcal{A})$ is by definition the localization of $\mathbf{H}(\mathcal{A})$ at the class of quasi-isomorphisms. It is not difficult to show that the category up to homotopy and the derived category are canonically triangulated. If $\mathcal{A}$ is the dg category with one object whose endomorphism dg algebra is a k-algebra B (concentrated in degree 0 and endowed with the zero differential), then $\mathbf{C}(\mathcal{A}) = \mathbf{C}(\mathrm{Mod}B)$, $\mathbf{H}(\mathcal{A}) = \mathbf{H}(\mathrm{Mod}B)$ and $\mathbf{D}(\mathcal{A}) = \mathbf{D}(\mathrm{Mod}B)$. For general $\mathcal{A}$, for each object X of $\mathcal{A}$, we have the right module *represented by* X

$$X^\wedge = \mathcal{A}(?, X).$$

For a dg module M and $X \in \mathcal{A}$, we have the Yoneda isomorphism

$$\mathrm{Hom}(X^\wedge, M)_\sim \xrightarrow{-} M(X)$$

which induces an isomorphism

$$\mathbf{D}(\mathcal{A})(X^\wedge, M)_\sim \xrightarrow{\ \sim\ } H^0(M(X)).$$

THEOREM.– Let $\mathcal{A}$ be a dg category. The localization functor $\mathbf{H}(\mathcal{A}) \to \mathbf{D}(\mathcal{A})$ admits a (fully faithful) left adjoint $M \mapsto \mathbf{p}M$ and a (fully faithful) right adjoint $M \mapsto \mathbf{i}M$.

For example, if M is a representable functor $\mathcal{A}(?, X)$, then $\mathbf{p}M = M$. In general, the dg module $\mathbf{p}M$ is constructed via a "resolution" of M by representables (see Keller (1994)).

If $F : \mathcal{A} \to \mathcal{B}$ is a dg functor, the composition with F yields a restriction functor $F^* : \mathbf{D}(\mathcal{B}) \to \mathbf{D}(\mathcal{A})$. The functor F is a *Morita functor* if F^* is an equivalence. It follows from the theorem that all quasi-equivalences are Morita functors.

9.7.3. *Derived functors*

Let $\mathcal{A}$ and $\mathcal{B}$ be small dg categories. Let X be an $\mathcal{A}$-$\mathcal{B}$-*bimodule*, i.e. a dg $\mathcal{A}^{op} \otimes \mathcal{B}$-module. Thus, X is given by complexes $X(B, A)$, for all A in $\mathcal{A}$ and B in $\mathcal{B}$, and morphisms of complexes

$$\mathcal{A}(A, A') \otimes X(B, A) \otimes \mathcal{B}(B', B) \to X(B', A').$$

For each dg $\mathcal{B}$-module M, we obtain a dg $\mathcal{A}$-module

$$GM = \mathrm{Hom}(X, M) : A \mapsto \mathrm{Hom}(X(?, A), M).$$

The functor $G : \mathbf{C}(\mathcal{B}) \to \mathbf{C}(\mathcal{A})$ admits a left adjoint $F : L \mapsto L \otimes_{\mathcal{A}} X$. These functors do not respect quasi-isomorphisms in general, but their derived functors

$$\mathbf{L}F : L \mapsto F(\mathbf{p}L) \text{ and } \mathbf{R}G : M \mapsto G(\mathbf{i}M)$$

form an adjoint pair of functors between $\mathbf{D}(\mathcal{A})$ and $\mathbf{D}(\mathcal{B})$. The following lemma is proved in Keller (1994). A dg $\mathcal{B}$-module is *perfect* if it belongs to the smallest thick subcategory of $\mathbf{D}(\mathcal{B})$ containing the representable $\mathcal{B}$-modules $\mathcal{B}(?, X)$, $X \in \mathcal{B}$. A set of objects $\mathcal{X}$ *generates* $\mathbf{D}(\mathcal{B})$ if $\mathbf{D}(\mathcal{B})$ coincides with its smallest triangulated subcategory stable under forming infinite sums and containing $\mathcal{X}$.

LEMMA 9.6.– The functor $\mathbf{L}F : \mathbf{D}(\mathcal{A}) \to \mathbf{D}(\mathcal{B})$ is an equivalence if and only if

a) the dg $\mathcal{B}$-module $X(?, A)$ is perfect for all A in $\mathcal{A}$;

b) the morphism

$$\mathcal{A}(A, A') \to \mathrm{Hom}(X(?, A), X(?, A'))$$

is a quasi-isomorphism for all A, A' in $\mathcal{A}$;

c) the dg $\mathcal{B}$-modules $X(?, A)$, $A \in \mathcal{A}$, form a generating set for $\mathbf{D}(\mathcal{B})$.

If the conditions of the lemma hold, the dg categories $\mathcal{A}$ and $\mathcal{B}$ are *derived equivalent*. If $\mathcal{A}$ is a dg category, its *perfect derived category* $\mathrm{per}(\mathcal{A})$ is defined as the full subcategory of the derived category formed by the perfect objects. One can show (Neeman 1992) that an object X is perfect in the derived category if and only if it is *compact*, i.e. the functor $\mathrm{Hom}(X, ?)$ commutes with infinite sums. This shows that an equivalence between derived categories induces an equivalence between their perfect subcategories. The *perfect dg category* $\mathrm{per}_{dg}(\mathcal{A})$ is the full dg subcategory of

$$\mathrm{Hom}(\mathcal{A}^{op}, \mathbf{C}_{dg}(k))$$

whose objects are the resolutions $\mathbf{p}P$ of perfect dg modules P.

For two dg categories $\mathcal{A}$ and $\mathcal{B}$, the *category* $\mathrm{rep}(\mathcal{A}, \mathcal{B})$ is defined as the full triangulated subcategory of the derived category $\mathbf{D}(\mathcal{A}^{op} \otimes \mathcal{B})$ formed by the bimodules X such that $X(?, A)$ is perfect in $\mathbf{D}(\mathcal{B})$ for each A in $\mathcal{A}$. These are precisely the bimodules whose associated tensor functor $\mathbf{D}(\mathcal{A}) \to \mathbf{D}(\mathcal{B})$ takes perfect $\mathcal{A}$-modules to perfect $\mathcal{B}$-modules. By the lemma, this always holds when the tensor functor is an equivalence.

9.7.4. *Dg quotients*

Let Hqe denote the category obtained from the category of small dg categories dgcat_k by localizing at the class of all quasi-equivalences. One can show that dgcat_k admits a Quillen model structure whose weak equivalences are the quasi-equivalences (Tabuada 2005). In particular, the morphism spaces of the localized category Hqe are sets and not classes. We need the category Hqe to lift the construction of the Verdier quotient of triangulated categories to the world of dg categories.

Let $\mathcal{A}$ be a small dg category and let $\mathcal{N}$ be a set of objects of $\mathcal{A}$. Let us say that a morphism $Q\colon \mathcal{A} \to \mathcal{B}$ of Hqe *annihilates* $\mathcal{N}$ if the induced functor

$$H^0(\mathcal{A}) \to H^0(\mathcal{B})$$

takes all objects of $\mathcal{N}$ to zero objects (i.e. objects whose identity morphism vanishes in $H^0(\mathcal{B})$). The following theorem is implicit in Keller (1999) and explicit in Drinfeld (2004)).

THEOREM 9.5.– There is a morphism $Q : \mathcal{A} \to \mathcal{A}/\mathcal{N}$ of Hqe that annihilates $\mathcal{N}$ and is universal among the morphisms annihilating $\mathcal{N}$.

We call $\mathcal{A}/\mathcal{N}$ the *dg quotient of* $\mathcal{A}$ *by* $\mathcal{N}$. If $\mathcal{A}$ is k-flat (i.e. $\mathcal{A}(X,Y) \otimes N$ is acyclic for each acyclic dg k-module N), then $\mathcal{A}/\mathcal{N}$ admits a beautiful simple construction (Drinfeld 2004): one adjoins to $\mathcal{A}$ a contracting homotopy for each object of $\mathcal{N}$. The general case can be reduced to this one or treated using orthogonal subcategories (Keller 1999). The following theorem shows the compatibility between dg quotients and Verdier localizations. A sequence of small dg categories in Hqe

$$0 \longrightarrow \mathcal{U} \longrightarrow \mathcal{V} \longrightarrow \mathcal{W} \longrightarrow 0$$

is *exact* if the induced sequence of triangulated categories

$$0 \longrightarrow \mathbf{D}(\mathcal{U}) \longrightarrow \mathbf{D}(\mathcal{V}) \longrightarrow \mathbf{D}(\mathcal{W}) \longrightarrow 0$$

is exact as a sequence of triangulated categories, i.e. the third term identifies with the Verdier quotient of the second term by the first term.

THEOREM 9.6.– Under the hypotheses of the above theorem, the sequence

$$0 \longrightarrow \mathcal{N} \longrightarrow \mathcal{A} \longrightarrow \mathcal{A}/\mathcal{N} \longrightarrow 0$$

is exact.

Using dg quotients, we can construct *dg enhancements* of derived categories. For example, if $\mathcal{E}$ is a small abelian (or, more generally, exact) category, we can take for $\mathcal{A}$ the dg category of bounded complexes $\mathbf{C}^b_{dg}(\mathcal{E})$ over $\mathcal{E}$ and for $\mathcal{N}$ the dg subcategory of acyclic bounded complexes $\mathbf{Ac}^b_{dg}(\mathcal{E})$. Then we obtain the *dg-derived category*

$$\mathbf{D}^b_{dg}(\mathcal{E}) = \mathbf{C}^b_{dg}(\mathcal{E})/\mathbf{Ac}^b_{dg}(\mathcal{E})$$

so that we have

$$\mathcal{D}^b(\mathcal{E}) = H^0(\mathbf{D}^b_{dg}(\mathcal{E})).$$

9.7.5. *Invariants*

K-theory: if $\mathcal{T}$ is a small triangulated category, its *Grothendieck group* $K_0(\mathcal{T})$ is the free abelian group on the set of isomorphism classes of $\mathcal{T}$ modulo the subgroup generated by the elements $[X] - [Y] + [Z]$ associated with the triangles

$$X \longrightarrow Y \longrightarrow Z \longrightarrow \Sigma X$$

of $\mathcal{T}$. If $\mathcal{A}$ is a small dg category, one defines

$$K_0(\mathcal{A}) = K_0(\mathrm{per}(\mathcal{A})).$$

As discussed in section 9.7.3, this is an invariant under derived equivalence. One defines the category Hmo_0 to have as objects all small dg categories and as morphisms the Grothendieck groups

$$K_0(\mathrm{rep}(\mathcal{A}, \mathcal{B}))$$

with the composition induced by the derived tensor product. Then the functor $\mathcal{A} \mapsto K_0(\mathcal{A})$ induces an additive functor defined on Hmo_0 with values in the category of abelian groups. By definition, an *additive invariant* of dg categories is an additive functor defined on Hmo_0. Additive invariants do not distinguish between rather different dg categories. For example, if k is an algebraically closed field, each finite-dimensional algebra of finite global dimension becomes isomorphic in Hmo_0 to a product of copies of k (Keller 1998a) but it is derived equivalent to such a product only if it is semisimple.

One defines the higher K-theory $K(\mathcal{A})$ by applying Waldhausen's construction (Waldhausen 1985) to a suitable category with cofibrations and weak equivalences: here, the category is that of perfect $\mathcal{A}$-modules, the cofibrations are the morphisms $i : L \to M$ of $\mathcal{A}$-modules which admit retractions as morphisms of graded $\mathcal{A}$-modules and the weak equivalences are the quasi-isomorphisms. This construction can be improved so as to yield a functor K from dgcat_k to the homotopy category of spectra. As in Thomason and Trobaugh (1990), from Waldhausen (1985) results, one then obtains the following:

THEOREM 9.7.–

a) The map $\mathcal{A} \mapsto K(\mathcal{A})$ yields a well-defined additive functor on Hmo_0 (Dugger and Shipley 2004).

b) Applied to the bounded dg-derived category $\mathbf{D}^b_{dg}(\mathcal{E})$ of an exact category $\mathcal{E}$, the K-theory defined above agrees with Quillen K-theory.

c) The functor $\mathcal{A} \mapsto K(\mathcal{A})$ is an additive invariant. Moreover, each short exact sequence

$$0 \longrightarrow \mathcal{A} \longrightarrow \mathcal{B} \longrightarrow \mathcal{C} \longrightarrow 0$$

of Hqe yields a long exact sequence

$$\ldots \to K_i(\mathcal{A}) \to K_i(\mathcal{B}) \to K_i(\mathcal{C}) \to \ldots \to K_0(\mathcal{B}) \to K_0(\mathcal{C}).$$

Hochschild homology: let $\mathcal{A}$ be a small k-flat k-category. Following Mitchell (1972), the *Hochschild chain complex* of $\mathcal{A}$ is the complex $C(\mathcal{A})$ concentrated in homological degrees $p \geq 0$ whose pth component is the sum of the

$$\mathcal{A}(X_p, X_0) \otimes \mathcal{A}(X_p, X_{p-1}) \otimes \mathcal{A}(X_{p-1}, X_{p-2}) \otimes \cdots \otimes \mathcal{A}(X_0, X_1) \; ,$$

where $X_0, \ldots, X_p$ range through the objects of $\mathcal{A}$, endowed with the differential

$$d(f_p \otimes \ldots \otimes f_0) = f_{p-1} \otimes \cdots \otimes f_0 f_p + \sum_{i=1}^{p} (-1)^i f_p \otimes \cdots \otimes f_i f_{i-1} \otimes \cdots \otimes f_0 .$$

If $\mathcal{A}$ is a k-flat differential graded category, its Hochschild chain complex $C(\mathcal{A})$ is the sum total complex of the bicomplex obtained as the natural re-interpretation of the above complex. The following theorem is stated for Hochschild homology but analogous theorems hold for all variants of cyclic homology (Keller 1999).

THEOREM 9.8.–

a) The map $\mathcal{A} \mapsto C(\mathcal{A})$ yields an additive functor $\mathrm{Hmo}_0 \to \mathbf{D}(k)$. Moreover, each exact sequence of Hqe yields a canonical triangle of $\mathbf{D}(k)$.

b) If A is a k-algebra, there is a natural isomorphism $C(A) \to C(\mathrm{per}_{dg}(A))$ in $\mathbf{D}(k)$.

The second statement in (a) may be viewed as an excision theorem analogous to Wodzicki (1989).

Hochschild cohomology: let $\mathcal{A}$ be a small dg category over a field k. Its cohomological Hochschild complex $C(\mathcal{A}, \mathcal{A})$ is defined as the product-total complex of the bicomplex whose 0th column is

$$\prod \mathcal{A}(X_0, X_0) \; ,$$

where X_0 ranges over the objects of $\mathcal{A}$, and whose pth column, for $p \geq 1$, is

$$\prod \mathrm{Hom}_k(\mathcal{A}(X_{p-1}, X_p) \otimes \mathcal{A}(X_{p-2}, X_{p-1}) \otimes \cdots \otimes \mathcal{A}(X_0, X_1), \mathcal{A}(X_0, X_p))$$

where $X_0, \ldots, X_p$ range over the objects of $\mathcal{A}$. The horizontal differential is given by the Hochschild differential. This complex carries rich additional structure: as shown in Getzler and Jones (n.d.), it is a B_∞-*algebra*, i.e. its bar construction carries, in addition to its canonical differential and comultiplication, a natural *multiplication* that makes it a dg bialgebra. The B_∞-structure contains in particular the cup product and the Gerstenhaber bracket, which both descend to the Hochschild cohomology

$$\mathrm{HH}^*(\mathcal{A}, \mathcal{A}) = H^*C(\mathcal{A}, \mathcal{A}).$$

Note that $C(\mathcal{A}, \mathcal{A})$ is not functorial with respect to dg functors. However, if $F : \mathcal{A} \to \mathcal{B}$ is a fully faithful dg functor, it clearly induces a restriction map

$$F^* : C(\mathcal{B}, \mathcal{B}) \to C(\mathcal{A}, \mathcal{A})$$

and this map is compatible with the B_∞-structure. This can be used to construct Keller (2003) a morphism

$$\phi_X : C(\mathcal{B}, \mathcal{B}) \to C(\mathcal{A}, \mathcal{A})$$

in the homotopy category of B_∞-algebras associated with each dg $\mathcal{A}$-$\mathcal{B}$-bimodule X such that the functor

$$\mathbf{L}(? \otimes_{\mathcal{A}} X) : \mathrm{per}(\mathcal{A}) \to \mathbf{D}(\mathcal{B})$$

is fully faithful. If moreover the functor $\mathbf{L}(X \otimes_{\mathcal{B}} ?) : \mathrm{per}(\mathcal{B}^{op}) \to \mathbf{D}(\mathcal{A}^{op})$ is fully faithful, then ϕ_X is an isomorphism. We refer to Lowen and van den Bergh (2005) for the closely related study of the Hochschild complex of an abelian category.

9.8. References

Angeleri Hügel, L., Happel, D., Krause, H., (eds). (2007). *Handbook of Tilting Theory*. Cambridge University Press, Cambridge.

Beilinson, A.A. (1978). Coherent sheaves on $\mathbf{P}^n$ and problems in linear algebra. *Funktsional. Anal. i Prilozhen.*, 12(3), 68–69. English translation: *Funct. Anal. Appl.*, 12 (1979), 214–216.

Beilinson, A.A., Bernstein, J., Deligne, P. (1982). *Analyse et topologie sur les espaces singuliers*. Société Mathématique de France, Paris.

Beilinson, A.A., Ginzburg, V., Soergel, W. (1996). Koszul duality patterns in representation theory. *J. Amer. Math. Soc.*, 9(2), 473–527.

Bernstein, J., Lunts, V. (1994). *Equivariant Sheaves and Functors*. Springer, Berlin.

Bernten, I.N., Gelfand, I.M., Gelfand, S.I. (1978). Algebraic vector bundles on $\mathbf{P}^n$ and problems of linear algebra. *Funktsional. Anal. i Prilozhen.*, 12(3), 66–67. English translation: *Funct. Anal. Appl.*, 12(1979), 212–214.

Bondal, A., van den Bergh, M. (2003). Generators and representability of functors in commutative and noncommutative geometry. *Mosc. Math. J.*, 3(1), 1–36.

Bongartz, K. (1981). Tilted algebras. In *Representations of Algebras*, Auslander, M. and Lluis, E. (eds). Springer, Berlin.

Borel, A. (1987). *Algebraic D-Modules*. Academic Press, Boston.

Brenner, S., Butler, M.C.R. (1980). Generalizations of the Bernstein-Gelfand-Ponomarev reflection functors. In *Representation Theory II: Proceedings of the Second International Conference on Representations of Algebras*, Dlab, V. and Gabriel, P (eds). Springer, Berlin.

Broué, M. (1988). Blocs, isométries parfaites, catégories dérivées. *C. R. Acad. Sci. Paris Sér. I Math.*, 307(1), 13–18.

Brylinski, J.-L., Kashiwara, M. (1981). Kazhdan–Lusztig conjecture and holonomic systems. *Inv. Math.*, 64, 387–410.

Deligne, P. (1973). La formule de dualité globale. *Exposé XVIII in SGA 4*. Springer, Berlin, 305, 481–587.

Deligne, P. (1979). Letter to D.Kazhdan and G. Lusztig, April 20.

Drinfeld, V. (2004). DG quotients of DG categories. *J. Algebra*, 272(2), 643–691.

Dugger, D., Shipley, B. (2004). K-theory and derived equivalences. *Duke Math. J.*, 124(3), 587–617.

Gabriel, P., Zisman, M. (1967). *Calculus of Fractions and Homotopy Theory*. Ergebnisse der Mathematik und ihrer Grenzgebiete, Band 35. Springer, New York.

Geigle, W., Lenzing, H. (1987). A class of weighted projective curves arising in representation theory of finite-dimensional algebras. In *Singularities, Representation of Algebras, and Vector Bundles: Proceedings of a Symposium held in Lambrecht*, Greuel, G.-M. and Trautmann, G. (eds). Springer, Berlin.

Gelfand, S.I., Manin, Y.I. (1996). *Methods of Homological Algebra*. Springer-Verlag, Berlin. Translated from the 1988 Russian original.

Getzler, E., Jones, J.D.S. (n.d.). Operads, homotopy algebra, and iterated integrals for double loop spaces. Available at: arXiv: hep-th/9403055.

Goresky, M., MacPherson, R. (1980). Intersection homology theory. *Topology*, 19, 135–162.

Goresky, M., MacPherson, R. (1983). Intersection homology II. *Inv. Math.*, 72, 77–130.

Grivel, P.-P. (1985). Une démonstration du théorème de dualité de Verdier. *Enseign. Math.*, 31(3–4), 227–247.

Grothendieck, A. (1957). Sur quelques points d'algèbre homologique. *Tôhoku Math. J.*, 9, 119–221.

Grothendieck, A. (1960). The cohomology theory of abstract algebraic varieties. In *Proceedings of the International, Congress of Mathematicians 1958*, Todd, J.A. (ed.). Edinburgh, Cambridge University Press, New York.

Happel, D. (1987). On the derived category of a finite-dimensional algebra. *Comment. Math. Helv.*, 62(3), 339–389.

Happel, D. (1988). *Triangulated Categories in the Representation Theory of Finite-dimensional Algebras*. Cambridge University Press, Cambridge.

Happel, D., Ringel, C.M. (1982). Tilted algebras. *Trans. Amer. Math. Soc.*, 274(2), 399–443.

Hartshorne, R. (1966). *Residues and Duality*. Springer, Berlin.

Illusie, L. (1990). Catégories dérivées et dualité : travaux de J.-L. Verdier. *Enseign. Math. (2)*, 36(3–4), 369–391.

Iversen, B. (1986). *Cohomology of Sheaves*. Universitext, Springer-Verlag, Berlin.

Kapranov, M.M. (1983). The derived category of coherent sheaves on Grassmann varieties. *Funktsional. Anal. i Prilozhen.*, 17(2), 78–79. English translation: *Funct. Anal. Appl.*, 17(1983), 145–146.

Kapranov, M.M. (1986). Derived category of coherent bundles on a quadric. *Funktsional. Anal. i Prilozhen.*, 20(2), 67. English Translation: *Funct. Anal. Appl.*, 20, 141–142.

Kapranov, M.M. (1988). On the derived categories of coherent sheaves on some homogeneous spaces. *Invent. Math.*, 92(3), 479–508.

Kashiwara, M. (1970). Algebraic study of systems of partial differential equations. Thesis, University of Tokyo.

Kashiwara, M., Schapira, P. (1994). *Sheaves on Manifolds*. Springer, Berlin. With a chapter in French by Christian Houzel, corrected reprint of the 1990 original.

Keller, B. (1991). Derived categories and universal problems. *Comm. Algebra*, 19, 699–747.

Keller, B. (1994). Deriving DG categories. *Ann. Sci. École Norm. Sup. (4)*, 27(1), 63–102.

Keller, B. (1998a). Invariance and localization for cyclic homology of DG algebras. *J. Pure Appl. Algebra*, 123(1–3), 223–273.

Keller, B. (1998b). On the construction of triangle equivalences. In *Derived Equivalences for Group Rings*, König, S. and Zimmermann, A. (eds). Springer, Berlin.

Keller, B. (1999). On the cyclic homology of exact categories. *J. Pure Appl. Algebra*, 136(1), 1–56.

Keller, B. (2000). Bimodule complexes via strong homotopy actions. *Algebr. Represent. Theory*, 3(4). Special issue dedicated to Klaus Roggenkamp on the occasion of his 60th birthday, 357–376.

Keller, B. (2003). Derived invariance of higher structures on the Hochschild complex. Preprint [Online]. Available at: https://webusers.imj-prg.fr/~bernhard.keller/publ/.

Keller, B. (2006). On differential graded categories. In *International Congress of Mathematicians*, European Mathematical Society, Zürich, II, 151–190.

Lowen, W., van den Bergh, M. (2005). Hochschild cohomology of abelian categories and ringed spaces. *Adv. Math.*, 198(1), 172–221.

Mebkhout, Z. (1989). *Le formalisme des six opérations de Grothendieck pour les* D_X*–modules cohérents*. Travaux en cours, Hermann, Paris.

Mitchell, B. (1972). Rings with several objects. *Adv. Math.*, 8, 1–161.

Neeman, A. (1992). The connection between the K–theory localisation theorem of Thomason, Trobaugh and Yao, and the smashing subcategories of Bousfield and Ravenel. *Ann. Sci. École Normale Supérieure*, 25, 547–566.

Neeman, A. (2001). *Triangulated Categories*. Princeton University Press, Princeton.

Positselski, L. (2011). Two kinds of derived categories, Koszul duality, and comodule-contramodule correspondence. *Mem. Amer. Math. Soc.*, 212(996), vi+133.

Rickard, J. (1989). Morita theory for derived categories. *J. London Math. Soc.*, 39, 436–456.

Rickard, J. (1991). Derived equivalences as derived functors. *J. London Math. Soc. (2)*, 43(1), 37–48.

Rouquier, R. (2008). Dimensions of triangulated categories. *J. K-Theory*, 1(2), 193–256.

Saito, M. (1986). On the derived category of mixed Hodge modules. *Proc. Japan Acad. (A)*, 62, 364–366.

Sato, M. (1969). Hyperfunctions and partial differential equations. In *Proceedings of the International Conference on Functional Analysis and Related Topics*. University of Tokyo Press, Tokyo, 91–94.

Scott, L.L. (1987). Simulating algebraic geometry with algebra I: The algebraic theory of derived categories. In *The Arcata Conference on Representations of Finite Groups (Arcata California 1986), Proc Sympos. Pure Math.* American Mathematical Society, Providence, 47(2), 271–281.

Tabuada, G. (2005). Une structure de catégorie de modèles de Quillen sur la catégorie des dg-catégories. *C. R. Math. Acad. Sci. Paris*, 340(1), 15–19.

Thomason, R.W., Trobaugh, T.F. (1990). Higher algebraic K–theory of schemes and of derived categories. In *The Grothendieck Festschrift*, Cartier, P., Illusie, L., Katz, N.M., Laumon, G., Manin, Y.I., Ribet, K.A. (eds). Birkhäuser, Boston.

Toën, B. (2011). Lectures on dg-categories. In *Topics in Algebraic and Topological K-Theory*, Cortiñas, G. (ed.). Springer, Berlin.

Verdier, J.-L. (1963). Le théorème de dualité de Poincaré. *C.R.A.S. Paris*, 256, 2084–2086.

Verdier, J.-L. (1966). Dualité dans la cohomologie des espaces localement compacts. *Séminaire Bourbaki : années 1964/65 1965/66, exposés 277–312*. Benjamin, 300, 1–13.

Verdier, J.-L. (1967). A duality theorem in the étale cohomology of schemes. In *Conference on Local Fields: NUFFIC Summer School Held at Driebergen (the Netherlands) in 1966*, Springer, T.A. (ed.) Springer, Berlin.

Verdier, J.-L. (1969). Théorème de dualité pour la cohomologie espaces localement compacts. *Dualité de Poincaré, Séminaire Heidelberg–Strasbourg 66/67*, IRMA Strasbourg 3, Strasbourg.

Verdier, J.-L. (1977). Catégories dérivées, état 0. In *SGA*, Springer, Berlin, 262–308.

Verdier, J.-L. (1996). *Des catégories dérivées des catégories abéliennes*. Société Mathématique de France, Paris.

Waldhausen, F. (1985). Algebraic K-theory of spaces. In *Algebraic and Geometric Topology: Proceedings of a Conference held at Rutgers University, New Brunswick*, Ranicki, A., Levitt, N. and Quinn, F. (eds). Springer, Berlin.

Weibel, C.A. (1994). *An Introduction to Homological Algebra*. Cambridge University Press, Cambridge.

Wilkins, D.R. (2020). Letters describing the discovery of quaternions [Online]. Available at: https://www.maths.tcd.ie/pub/HistMath/People/Hamilton/Letters/BroomeBridge.html.

Wodzicki, M. (1989). Excision in cyclic homology and in rational algebraic K-theory. *Ann. Math. (2)*, 129(3), 591–639.

Zimmermann, A. (2014). *Representation Theory, A Homological Algebra Point of View*. Springer, Cham.

List of Authors

Yuri BAHTURIN
Department of Mathematics and Statistics
Memorial University of Newfoundland
St. John's
Canada

Chengming BAI
Chern Institute of Mathematics
Nankai University
Tianjin
China

Miguel CABRERA GARCÍA
Department of Mathematical Analysis
Granada University
Spain

José Antonio CUENCA MIRA
Department of Algebra, Geometry and Topology
Malaga University
Spain

Alberto ELDUQUE
Department of Mathematics
University of Zaragoza
Spain

Bernhard KELLER
Department of Mathematics
Paris Diderot University
France

Mikhail KOCHETOV
Department of Mathematics and Statistics
Memorial University of Newfoundland
St. John's
Canada

Abdenacer MAKHLOUF
IRIMAS-Department of Mathematics
University of Haute Alsace
Mulhouse
France

Consuelo MARTINEZ
Department of Mathematics
University of Oviedo
Spain

Ángel RODRÍGUEZ PALACIOS
Department of Mathematical Analysis
Granada University
Spain

Martin SCHLICHENMAIER
Department of Mathematics
Luxembourg University
Luxembourg

Yunhe SHENG
Jilin University
Changchun
China

Rong TANG
Jilin University
Changchun
China

Mikhail ZAICEV
Faculty of Mechanics and
Mathematics
Lomonosov Moscow State University
Russia

Efim ZELMANOV
Department of Mathematics
University of California San Diego
USA

Index

E, F

G

H

I, J

K, L

M, N

O, P

Q, R

S

T

U, V

W